Two Manuals Combined into One Book

Military Marksmanship, all M16 and M4 Series Weapons

And

Ranger Skills Handbook

Department of Defense

Pentagon
Publishing

***Field Manual**
No. 3-22.9

Headquarters
Department of the Army
Washington, DC, 12 August 2008

Rifle Marksmanship
M16-/M4-Series Weapons

Contents

DISTRIBUTION RESTRICTION: Approved for public release; distribution is unlimited.

***This publication supersedes FM 3-22.9, 24 April 2003.**

Figures

Tables

Preface

This manual provides guidance for planning and executing training on the 5.56-millimeter M16-series rifle (M16A1/A2/A3/A4) and M4 carbine. It is a guide for commanders, leaders, and instructors to develop training programs, plans, and lessons that meet the objectives or intent of the United States Army rifle marksmanship program and FM 7-0.

This manual is organized to lead the trainer through the material needed to conduct training during initial entry training (IET) and unit sustainment training. Preliminary subjects include discussion on the weapon's capabilities, mechanical training, and the fundamentals and principles of rifle marksmanship. Live-fire applications are scheduled after the Soldier has demonstrated preliminary skills.

This manual was revised to include references to new materiel and systems. This revision includes—

- The new Army total marksmanship training strategy, to include specific strategies for the United States Army Reserve (USAR) and the Army National Guard (ARNG).
- Information about the advanced combat optical gunsight (ACOG), the AN/PEQ-15 advanced target pointer/illuminator aiming light (ATPIAL), various thermal sights, and the MK 262 round.
- Information about the alternate qualification record fire courses (known distance [KD] record fire, 25-meter scaled target alternate course, 15-meter scaled target alternate course).
- Information about the rapid magazine change and barricade transition fire for short-range marksmanship (SRM).
- Changes to all of the scorecards.
- Updated terminology.

This publication prescribes DA Form 3595-R (Record Fire Scorecard), DA Form 3601-R (Single Target—Field Fire I Scorecard), DA Form 5239-R (100-, 200-, and 300-Meter Downrange Feedback Scorecard), DA Form 5241-R (Single and Multiple Targets—Field Fire II Scorecard), DA Form 5789-R (Record Fire Scorecard—Known-Distance Course), DA Form 5790-R (Record Fire Scorecard—Scaled Target Alternate Course), DA Form 7489-R (Record Night Fire Scorecard), DA Form 7649-R (Squad Designated Marksman—Record Fire I and II), and DA Form 7650-R (Squad Designated Marksman—Position Evaluation).

This publication applies to the Active Army, the Army National Guard (ARNG)/National Guard of the United States (ARNGUS), and the US Army Reserve (USAR).

Terms that have joint or Army definitions are identified in both the glossary and the text. Terms for which FM 3-22.9 is the proponent FM are indicated with an asterisk in the glossary.

Uniforms depicted in this manual were drawn without camouflage for clarity of the illustration. Unless this publication states otherwise, masculine nouns and pronouns refer to both men and women.

The proponent for this publication is the US Army Training and Doctrine Command. The preparing agency is the United States Army Infantry School (USAIS). You may send comments and recommendations by any means (US mail, e-mail, fax, or telephone) as long as you use DA Form 2028 (Recommended Changes to Publications and Blank Forms) or follow its format. Point of contact information is as follows:

E-mail:	benn.29IN.229-S3-DOC-LIT@conus.army.mil
Phone:	Commercial: 706-545-8623 DSN: 835-8623
Fax:	Commercial: 706-545-8600 DSN: 835-8600
US Mail:	Commandant, USAIS ATTN: ATSH-INB 6650 Wilkin Drive, Bldg 74, Rm 102 Fort Benning, GA 31905-5593

Chapter 1

Marksmanship Training

An effective marksmanship program can be measured by the unit's ability to put effective fire on a target. This chapter provides a proven strategy for establishing and conducting an effective rifle marksmanship training program. The strategy begins with the progressive individual training periods taught during initial entry training (IET) and culminates with advanced rifle marksmanship (ARM) skills. Refresher training is conducted only when necessary.

A Soldier's marksmanship proficiency depends on proper training and application of the basic marksmanship fundamentals. During initial marksmanship training, emphasis is placed on learning the firing fundamentals, which are taught in four phases—preliminary marksmanship instruction (PMI), downrange feedback, field firing, and advanced firing exercises. This prepares Soldiers for advanced optic and laser training for combat-type collective exercises and real-world deployments.

SECTION I. TRAINING STRATEGY

The total Army marksmanship training strategy is the overall concept for integrating resources into a program designed to train, sustain, and improve the individual and collective skills needed to achieve proficiency in individual and collective gunnery tasks. Training strategies for rifle marksmanship are implemented in TRADOC institutions (IET, Noncommissioned Officers Education System [NCOES], Basic Officer Leaders Course [BOLC]) and in units. The overall training strategy is multifaceted and includes supporting strategies that use resources such as publications; ranges; ammunition; and training aids, devices, simulators, and simulations (TADSS). These strategies focus on developing the Soldier and leader skills required for success in combat.

NOTE: See AR 350-1 for specific requirements pertaining to marksmanship training and DA Pam 350-38 for live-fire frequency requirements.

OBJECTIVES

1-1. The procedures and techniques for implementing the total Army rifle marksmanship training strategy are based on the concept that Soldiers must become skilled marksmen. FM 7-0 stresses marksmanship as the paramount Soldier skill. Further, Soldiers should understand common firing principles and be confident in applying their firing skills in combat. Unit leaders accomplish proficiency through practice supervised by qualified instructors/trainers and thorough objective performance assessments.

1-2. The basic firing skills and exercises outlined in this manual must be part of every unit's marksmanship training program. Unit commanders must focus their basic and advanced marksmanship training programs to support their mission-essential task lists (METLs).

MARKSMANSHIP TRAINING STRATEGY

1-3. The following marksmanship training strategy guide contains the tasks that are currently trained in basic rifle marksmanship (BRM) programs, during basic combat training at Army training centers (ATCs), and during infantry one-station unit training (OSUT). It also provides a basis for structuring unit sustainment programs for active Army, Army National Guard, and United States Army Reserve units. Units normally perform diagnostic tests of the tasks and only conduct training on specific periods for Soldiers who must improve their basic firing skills. Unit training is usually conducted in less time than at IET.

1-4. There are two primary components of a marksmanship training strategy: initial training and sustainment training. Both may include individual and collective tasks and skills. Initial training must be taught correctly the first time. A task taught correctly and learned well is retained longer, and skills can be more easily sustained. However, an individual or unit eventually loses skill proficiency if the training is not reinforced. This learning decay depends on many factors, such as the difficulty and complexity of the task. Personnel turnover is a main factor in the decay of collective skills, since the loss of critical team members requires retraining to regain proficiency. If a long period elapses between initial and sustainment sessions or training doctrine is altered, retraining may be required.

Initial Entry Training

1-5. The training strategy for BRM begins in IET and continues in the unit. Figure 1-1 shows the IET training strategy.

1-6. In IET, Soldiers learn how to maintain a rifle, hit a point target, and apply the four marksmanship fundamentals and other skills needed to engage a target. Once Soldiers understand the weapon and have demonstrated skill in zeroing, additional simulations and live-fire training exercises are conducted to prepare Soldiers for qualification. During these exercises, Soldiers master target types and scenarios of increasing difficulty to develop their proficiency.

1-7. IET culminates in the Soldier's proficiency assessment (which is conducted on the standard record fire range or approved alternates), followed by instruction on advanced firing techniques (a night-fire exercise with iron sights [unassisted] or night vision goggles [assisted]). This evaluation allows leaders to determine the effectiveness of the training.

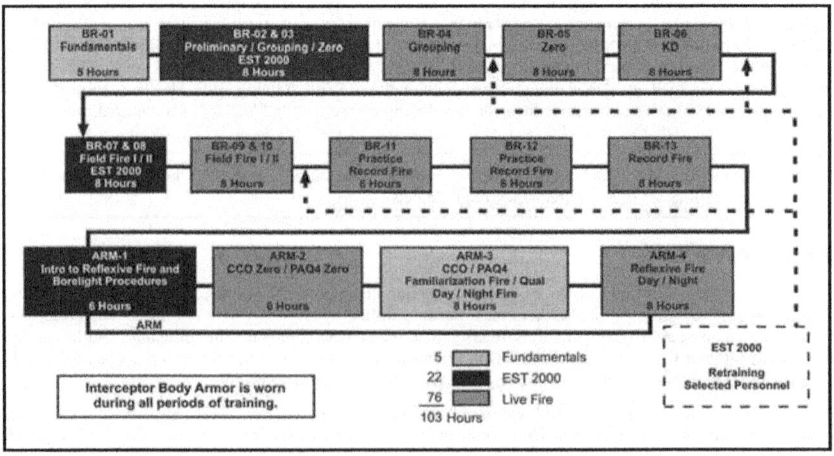

Figure 1-1. Initial entry training marksmanship training strategy.

Sustainment Training

1-8. Training continues in active Army, National Guard, and Army Reserve units using the same basic skills taught in IET. Units must set up a year-round program to sustain skills and have a plan not only for when they are at their home station, but for when they are deployed as well.

1-9. To sustain the basic marksmanship skills taught in IET, periodic PMI is conducted, followed by instructional and qualification range firing. Key elements include—

- The training of trainers.
- Refresher training of nonfiring skills.
- The use of the Engagement Skills Trainer (EST) 2000, Laser Marksmanship Training System (LMTS), or other devices.
- Sustainment training.
- Remedial training.

NOTE: See Appendix A for more information about TADSS.

1-10. Additional skills trained in the unit include—

- Semiautomatic and automatic fires.
- Night fire.
- Mission-oriented protective posture (MOPP) firing.
- Firing using aiming devices.
- Moving target training techniques.
- Squad designated marksman (SDM) training.
- Short-range marksmanship (SRM).

1-11. These skills are trained and integrated into collective training exercises, such as platoon and squad live-fire situation training exercises (STXs). Figure 1-2 shows a year-round training strategy guide.

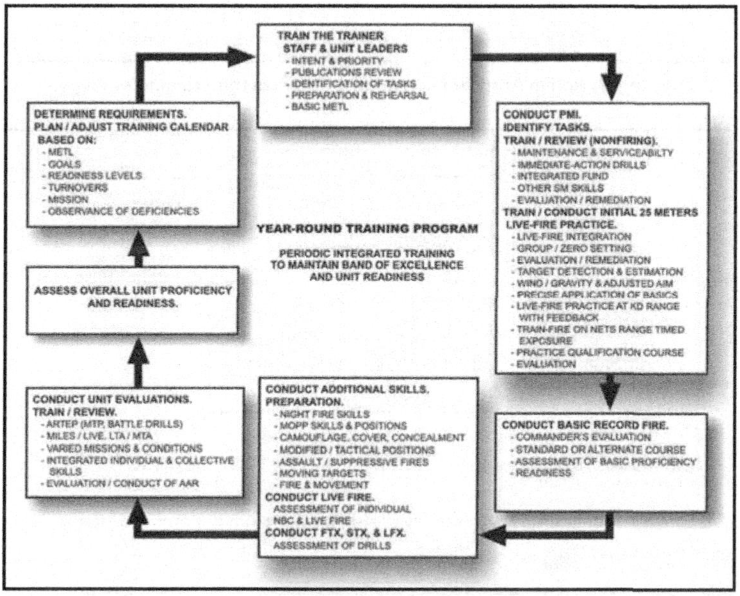

Figure 1-2. Unit marksmanship sustainment strategy.

1-12. General marksmanship knowledge and weapon proficiency are perishable skills. Most units have a readiness requirement that all Soldiers must zero their rifles within a certain time after unit assignment.

Soldiers must confirm the battlesight zeros of their assigned rifles before qualifying. Units should conduct PMI and practice firing throughout the year due to personnel turnover. A year-round marksmanship sustainment program is needed for the unit to maintain the individual and collective firing proficiency requirements to accomplish its mission. The following figures depict marksmanship training programs for active Army home station units (Figure 1-3), National Guard home station units (Figure 1-4), Army Reserve home station units (Figure 1-5), and deployed units (Figure 1-6).

NOTE: While the training strategy depicted in Figures 1-3 through 1-6 is not compulsory, it is strongly recommended for Army-wide application. The strategy has proven success during IET, producing more proficient Soldiers while using less ammunition. Commanders are urged to follow this strategy closely.

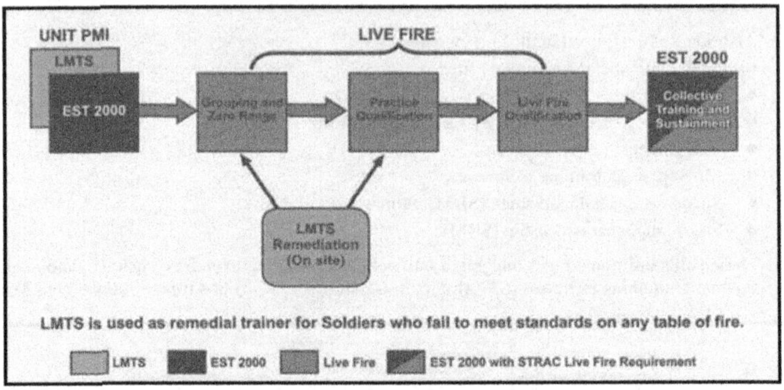

Figure 1-3. Active Army home station marksmanship training strategy.

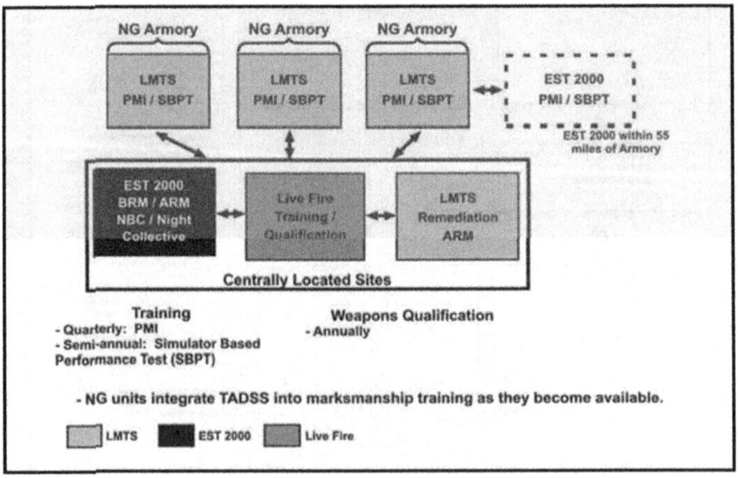

Figure 1-4. National Guard home station marksmanship training strategy.

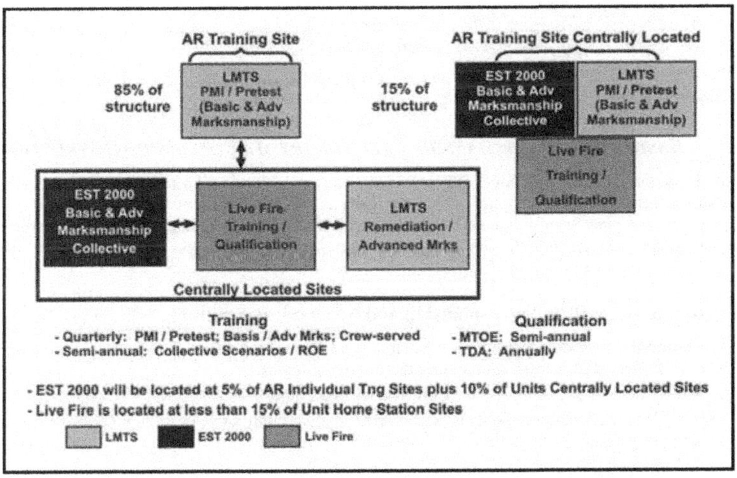

Figure 1-5. Army Reserve home station marksmanship training strategy.

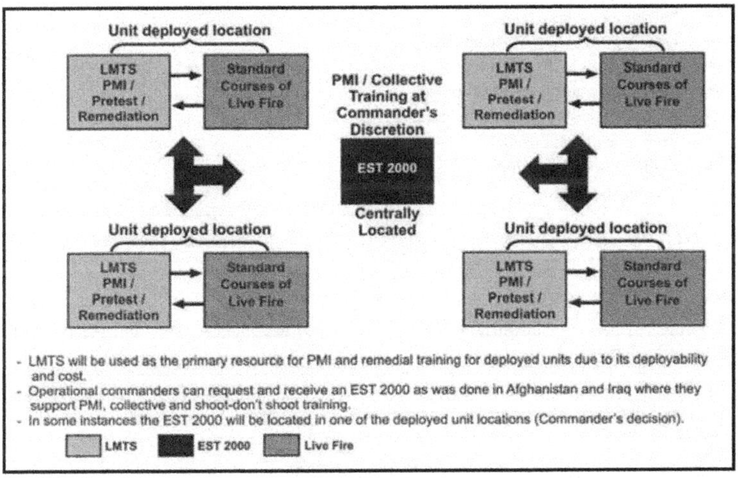

Figure 1-6. Deployed unit marksmanship training strategy.

TRAINING PHASES

1-13. Soldiers progress through five phases of rifle marksmanship training:

- Phase I—Basic Rifle Marksmanship Preliminary Marksmanship Instruction.
- Phase II—Basic Rifle Marksmanship Downrange Feedback Range Firing.
- Phase III—Basic Rifle Marksmanship Field Firing.

- Phase IV—Advanced Rifle Marksmanship.
- Phase V—Advanced Optics, Lasers, and Iron Sights.

1-14. When Soldiers are trained in all phases of rifle marksmanship, a solid sustainment program is key to mission readiness.

Phase I—Basic Rifle Marksmanship Preliminary Marksmanship Instruction

1-15. Understanding the operation and functions of any machine is vital to becoming an expert with that machine. The same theory applies to rifle marksmanship. Soldiers must master weapon maintenance, function checks, and firing fundamentals before progressing to advanced skills and firing exercises under tactical conditions. Armed with this knowledge, a Soldier is able to assess and correct any malfunction to keep the weapon operating properly.

Introduction to Basic Rifle Marksmanship and Mechanical Training

1-16. During this period of instruction, Soldiers are trained to understand the operation and functions of the weapon. This period of instruction includes the following topics:

- General safety rules and weapon clearing procedures.
- Characteristics, components, accessories, and ammunition.
- Disassembly.
- Inspection.
- Lubrication.
- Assembly.
- Function check.
- Loading/unloading the magazine.
- Loading/unloading the weapon.
- Immediate and remedial action.
- Adjusting the front and rear sights.
- Peer coaching.
- Eight cycles of function.
- Troubleshooting.

NOTE: The IET program of instruction (POI) allots four hours for this phase of training.

Marksmanship Fundamentals I

1-17. During this period of instruction, Soldiers are trained to demonstrate the integrated act of shooting using the EST 2000 or LMTS.

NOTE: EST 2000 will not be used in lieu of live-fire qualification except for those outlined in DA Pam 350-38.

1-18. This period of instruction includes the following topics:

- The four fundamentals.
- Basic firing positions.
- Range and safety procedures.
- Dominant eye training.
- Demonstrating the integrated act of shooting during dry-fire exercises.

NOTE: The IET POI allots six hours for this phase of training.

Marksmanship Fundamentals II

1-19. During this period of instruction, Soldiers are trained to demonstrate the integrated act of shooting using the EST 2000 or LMTS.

NOTES: 1. The IET POI allots six hours for this phase of training.

2. See paragraphs 4-76 through 4-80 and Appendix A for more information about EST 2000 and LMTS training.

PHASE II—BASIC RIFLE MARKSMANSHIP DOWNRANGE FEEDBACK RANGE FIRING

1-20. During Phase II, instructors/trainers outline grouping and zeroing procedures and demonstrate how to conduct the three types of known distance (KD) ranges. Downrange feedback provides precise knowledge of what happens to bullets at a given range and allows Soldiers to transition between 25-meter firing and firing on the field fire range. Knowing precisely where bullets hit allows poor firers (with the assistance of instructors/trainers) to improve their performance and good firers to bring their shots to the center of the target. Firers develop the knowledge and skills required to perform with confidence on the field fire range, where only hit-or-miss information is available.

Grouping Procedures

1-21. Grouping is a form of practice firing with two primary objectives: firing tight shot groups and consistently placing those groups in the same location. Frequent use of the EST 2000 or LMTS greatly reduces live-fire grouping time.

Zeroing Procedures

1-22. Zeroing allows firers to use standard issue ammunition to align the sights with the weapon's barrel. When this is accomplished correctly, the aimpoint and the impact point are the same at a given range. For most combat targets, this sight setting provides the greatest probability of a hit with minimum adjustment to the aimpoint. When followed, a properly zeroed rifle for one Soldier is close to zero for another Soldier.

Downrange Feedback

1-23. On KD ranges, Soldiers fire tight shot groups at a known distance and make sight adjustments at a given range while experiencing the effects of wind, gravity, and other environmental factors. The advantage of a KD range is the ability to see precisely where each bullet hits. KD firing is conducted with a single, clearly visible target at a known distance, and the Soldier can establish a position that provides a natural aimpoint on that single target.

NOTES: 1. See Chapter 5 for more information about downrange feedback.

2. See paragraphs 4-76 through 4-80 and Appendix A for more information about EST 2000 and LMTS training.

PHASE III—BASIC RIFLE MARKSMANSHIP FIELD FIRING

1-24. Field fire begins a critical transition from unstressed firing at single targets at known distances to requiring the Soldier to refine techniques for scanning the range for targets, estimating range, and firing quickly and accurately. Phase III includes the following:

- Field Fire I (single timed targets at 75 meters, 175 meters, and 300 meters).
- Field Fire II (single and multiple targets at 75 meters, 175 meters, and 300 meters).
- Practice Record Fire.
- Practice Record Fire II.
- Record Fire.

NOTES: 1. See Chapter 6 for more information about BRM field fire.

2. See paragraphs 4-76 through 4-80 and Appendix A for more information about EST 2000 and LMTS training.

Phase IV—Advanced Rifle Marksmanship

1-25. ARM focuses on the techniques and procedures that the Soldier will need to participate in collective training. This phase addresses the following topics:

- Alternate firing positions.
- Burst fire.
- Quick fire.
- Chemical, biological, radiological, and nuclear (CBRN) fire.
- Moving targets.
- SDM.
- Unassisted night record fire.
- M68 close combat optic (CCO).
- Advanced combat optical gunsight (ACOG).
- AN/PAQ-4C infrared (IR) aiming laser.
- Assisted night record fire IR.
- Assisted night record fire (thermal).
- SRM.

NOTE: See Chapter 7 for more information about ARM.

Phase V—Advanced Optics, Lasers, and Iron Sights

1-26. BRM teaches Soldiers to effectively engage the enemy with the basic rifle using the iron sights (primarily during the day). ARM adds additional situations for the firer. To enhance the lethality of night firing, Soldiers participate in training with optics and lasers. This phase addresses the following topics:

- Backup iron sight (BUIS).
- M68 CCO.
- ACOG.
- AN/PAQ-4C IR aiming laser.
- AN/PEQ-2A/B target pointer illuminator/aiming light (TPIAL).
- Thermal weapon sight (TWS).
- AN/PVS-4 night vision sight.

NOTE: See Chapter 8 for more information about advanced optics, lasers, and iron sights.

SECTION II. UNIT MARKSMANSHIP TRAINING PROGRAM

An effective unit marksmanship program reflects the priority, emphasis, and interest of commanders and trainers. This section outlines a rifle marksmanship training program strategy as guidance in establishing and conducting an effective unit training program. The strategy consists of the individual and leader refresher training for maintaining the basic skills learned during IET. It progresses to training advanced and collective skills under near-combat conditions during live-fire STXs.

MISSION-ESSENTIAL TASKS

1-27. Marksmanship proficiency is critical to soldiering and is required for any unit deployed to a wartime theater. All commanders should develop a METL and organize a training program that devotes adequate time to marksmanship. The unit's combat mission must be considered when establishing training priorities. This not only applies to the tasks selected for the unit's METL, but also to the conditions under which the tasks are to be performed. If a unit may be employed in an urban environment, the effects of range, gravity, and wind may not be as important as automatic or burst fire, quick fire, or assault fire. The reverse may be true of a unit that expects to engage the enemy at long ranges.

TRAINING ASSESSMENT

1-28. To conduct an effective marksmanship program, the unit commander must determine the current marksmanship proficiency of all assigned personnel. Constant evaluation provides commanders understanding of where training emphasis is needed. All results are reviewed to determine any areas that need strengthening, along with any individuals that require special attention. Based on this evaluation, marksmanship training programs are developed and executed. Commanders continually assess the program and modify it as required. To develop a training plan and assess the marksmanship program, commanders should use the following tools:

- Direct observation of training.
- Spot checks.
- Review of past training.

1-29. Based on the commander's evaluation, goals, and missions, quarterly, semiannual, or annual training events are identified. Marksmanship programs must be continuous, and to sustain an effective marksmanship program, resources are required. While the unit may only qualify its Soldiers annually or semiannually, test results show that sustainment training is required at least quarterly to maintain marksmanship skills.

DIRECT OBSERVATION OF TRAINING

1-30. Observing and accurately recording performance reveals the status of weapon maintenance, Soldier zero and qualification results, and each Soldier's ability to hit targets. This also allows the commander to identify Soldiers who need special assistance to reach required standards and those who exceed these standards.

SPOT CHECKS

1-31. Spot checks of individual marksmanship performance, such as interviews and evaluations of Soldiers, provide commanders with valuable information about Soldier proficiency and knowledge of the marksmanship tasks.

REVIEW OF PAST TRAINING

1-32. Commanders review past training to gain valuable information for developing a training plan. The assessment should include—

- The frequency and results of training.
- The basic and advanced record fire results.
- The frequency of unit-conducted collective CBRN or night fire training.

COMMANDER'S EVALUATION GUIDE

1-33. The commander's evaluation guide contains three sections:

- Commander's priorities and intent.
- Soldier assessment.
- Trainer assessment.

1-34. The following is an example of a commander's evaluation guide. Commanders can use this guide not only to assess their unit's marksmanship proficiency, but to assess the unit leaders and their ability to effectively implement a marksmanship program. They can also use it to develop NCOs into subject matter experts.

Commander's Priorities and Intent

1-35. When considering their priorities and intent, commanders answer the following questions:

- Have you clearly stated the priority of rifle (small arms) proficiency in your unit? What is it? Do the staff and subordinates support this priority? Is it based on your METL and an understanding of FM 7-0 and FM 7-1?
- Have you clearly stated the intent of record fire? Are leaders evaluating firing performance based on accurately recorded data and results?
- Have you clearly stated that weapon qualification or record fire is one of the commander's opportunities to assess several skills relating to small arms readiness?
- What qualification course will be used to evaluate your unit's marksmanship readiness (small arms)?
  - Is the standard 300-meter, 300-yard KD, or 25-meter scaled target alternate qualification course used?
  - How will it be conducted? Will the prescribed procedures be followed?
  - Who will collect the data?
- Have you clearly stated the purpose and intent of PMI?
  - What skills will PMI address?
  - Will PMI be performance-oriented? Are tasks integrated?

Soldier Assessment

1-36. During Soldier assessment, commanders answer the following questions:

- Do Soldiers maintain their assigned weapons and magazines IAW the TM? Do they have a manual?
- Do Soldiers conduct serviceability checks of weapons and magazines before training? Were maintenance deficiencies corrected?
- Do Soldiers demonstrate an understanding of the weapon's operation, functioning, and capabilities?
- Can Soldiers correctly apply immediate action procedures to reduce weapon stoppages and then continue to fire? Have they demonstrated this during dry-fire exercises?
- Are Soldiers firing their assigned weapons?
  - How often are weapons reassigned between individuals?
  - What is the value of a recorded zero?
- Can Soldiers precisely and consistently apply the four fundamentals of rifle marksmanship? To what standard have they demonstrated their mastery?
  - During a dry-fire exercise?
  - During a live-fire exercise (LFX)?
  - When firing on the 25-meter course?
  - During KD firing?
- Can Soldiers accurately battlesight zero their assigned rifle to standard?
  - Do they understand sight adjustment procedures?
  - Do they record their rifle's zero? How is it done? Why?
  - Do they record the date the Soldier last zeroed his rifle? What is the sight setting? Are these linked? How do you check this?
- Do Soldiers demonstrate their knowledge of the effects of wind and gravity while firing out to 300 meters? What feedback was provided? How?

- Can Soldiers scan a designated area or sector of fire and detect all targets out to 300 meters? If not, why?
- Can Soldiers quickly engage timed single and multiple targets from both supported and unsupported firing positions out to 300 meters? If not, which targets were not engaged? Which were missed? Why?
- During individual and collective training, do Soldiers demonstrate their ability to manage allocated ammunition and to engage all targets? Do they fire several rounds at one target? Which targets? Why?
- Based on an analysis of individual qualification scores, what is the distribution?
  - Are most Soldiers just meeting the minimum acceptable performance (marksman)?
  - Are most Soldiers distributed in the upper half of the performance spectrum (sharpshooter, expert)?
- What is the hit distribution during collective LFXs?
- Do Soldiers demonstrate proficiency during night-fire, target detection and acquisition, and night fire engagement techniques? When using night vision devices (NVDs)?
- Do Soldiers demonstrate individual marksmanship proficiency during MOPP firing conditions? During collective exercises?
- Do Soldiers demonstrate proficiency during moving target engagements? Do they demonstrate proficiency collectively by hitting moving targets at the multipurpose range complex? If not, is moving target training conducted?
- Are marksmanship skills integrated into tactical exercises and unit LFXs? If so, is suppressive fire, rapid semiautomatic fire, and automatic or burst fire conducted? What tasks in the mission training plan are evaluated?
- Based on onsite observations and analysis of training and firing performance, what skills or tasks show a readiness deficiency?
  - What skills need training emphasis? Individual emphasis? Leader emphasis?
  - What are the performance goals?

Trainer Assessment

1-37. During trainer assessment, commanders answer the following questions:

- Who has trained or will train the trainers?
  - What is the subject matter expertise of the cadre?
  - Are they actually training the critical skills?
  - Have they addressed the nonfiring skills first?
  - What aids and devices are being used? Are EST 2000 and LMTS being properly used?
- What administrative constraints or training distracters can you overcome for the junior officer and NCO? At what level are the resources necessary to train marksmanship controlled (time, training aids, weapons, ammunition, ranges)? Do the sergeants do the job they are charged with?

TRAINERS

1-38. Knowledgeable instructors or cadre are the key to marksmanship performance. All commanders must be aware of maintaining expertise in marksmanship instruction/training.

INSTRUCTOR/TRAINER SELECTION

1-39. Institutional and unit instructors/trainers are selected and assigned from the most highly qualified Soldiers. These Soldiers must demonstrate proficiency in all aspects of rifle marksmanship, be proficient in applying these fundamentals, know the importance of marksmanship training, and have a competent and professional attitude. The commander must ensure that selected instructors/trainers can effectively train other Soldiers. Local instructor/trainer training courses and marksmanship certification programs must be established to ensure that instructor/trainer skills are developed.

CADRE/TRAINER

1-40. Cadre/trainer refers to a marksmanship instructor/trainer that has more experience and expertise than the firer does. He trains Soldiers in the effective use of the rifle by maintaining strict discipline on the firing line, insisting on compliance with range procedures and program objectives, and enforcing safety regulations. A good cadre/trainer must understand the training phases and techniques for developing marksmanship skills. Each cadre/trainer must have the following attributes:

- Knowledge.
- Patience.
- Understanding.
- Consideration.
- Respect.
- Encouragement.

Knowledge

1-41. Effective cadre/trainers must possess a thorough knowledge of the rifle, proficiency in firing, and a complete understanding of this manual and supporting manuals.

Patience

1-42. Effective cadre/trainers relate to the Soldier calmly, persistently, and patiently.

Understanding

1-43. Cadre/trainers enhance success and understanding by emphasizing close observance of rules and instructions.

Consideration

1-44. Most Soldiers, even those who do not fire well, enjoy firing and begin their training with great enthusiasm. A cadre/trainer who is considerate of Soldiers' feelings and who encourages them throughout their training will find training a pleasant and rewarding duty.

Respect

1-45. An experienced cadre is assigned the duties of instructor/trainer, which classifies him as a technical expert and authority. A good cadre/trainer watches for mistakes and patiently makes needed corrections.

Encouragement

1-46. A cadre/trainer can encourage Soldiers by convincing them to achieve good firing performance through practice. He imparts his knowledge and helps Soldiers gain the practical experience necessary.

TRAINING THE TRAINER

1-47. Knowledgeable small-unit leaders are key to marksmanship training. This manual and other training publications provide the unit instructor with the required information for developing a good train-the-trainer program.

1-48. The goal of a progressive train-the-trainer program is to achieve a high state of combat readiness. Through the active and aggressive leadership of the chain of command, a perpetual base of expertise is established and maintained.

1-49. The commander should identify unit personnel who have had assignments as marksmanship instructors. These individuals should be used to train other unit cadre by conducting PMI and LFXs for their Soldiers.

1-50. Assistance and expertise from outside the unit may also be available, such as the Army Marksmanship Unit (AMU) at Fort Benning, Georgia. A suggested train-the-trainer program is outlined below:

- Conduct marksmanship diagnostic test.
- Review operation and function of the rifle and ammunition, immediate action, and safety.
- Conduct PMI; review the four fundamentals.
- Review coaching techniques and device usage.
- Establish grouping and zeroing procedures.
- Review effects of wind and gravity when firing as far away as 300 meters (600 meters for ARM).
- Conduct range operations.
- Conduct qualification/record firing.
- Diagnose firing problems.

DUTIES OF THE INSTRUCTOR/TRAINER

1-51. The instructor/trainer helps the firer master the fundamentals of rifle marksmanship. He ensures that the firer consistently applies what he has learned. When training the beginner, the instructor/trainer could confront problems, such as fear, nervousness, forgetfulness, failure to understand, and a lack of coordination or determination, which may be compounded by arrogance or carelessness. With all types of firers, the instructor/trainer must ensure that firers are aware of their firing errors, understand the causes, and apply remedies. To perform these duties, he—

- Observes the firer.
- Questions the firer.
- Analyzes the shot group.

Observing the Firer

1-52. To pinpoint errors, the instructor/trainer observes the firer during drills and when firing. If there is no indication of probable error, the firer's position, breath control, shot anticipation, and trigger squeeze are closely observed.

Questioning the Firer

1-53. The firer is asked to detect his errors and to explain his firing procedure (position, aiming, breath control, and trigger squeeze).

Analyzing the Shot Group

1-54. Analyzing the shot group is an important step in detecting and correcting errors. When analyzing a target, the instructor/trainer correlates observations of the firer to probable errors in performance, according to the shape and size of shot groups. A poor shot group is usually caused by more than one observable error.

TRAINER CERTIFICATION PROGRAM

1-55. The certification program sustains the trainers' expertise and develops methods of training. The program standardizes procedures for certifying marksmanship trainers. Trainers' technical expertise must be continuously refreshed, updated, and closely managed.

TRAINING BASE

1-56. The training base can expect the same personnel changes as any other organization. Soldiers assigned as marksmanship trainers will have varying degrees of experience and knowledge of training procedures and methods. Therefore, the trainer certification program must be an ongoing process that is tailored to address these variables. At a minimum, formal records should document program progression for each trainer. All marksmanship trainers must complete the four phases of training using the progression steps, and the records should be updated on a quarterly basis.

CERTIFICATION PROGRAM OUTLINE

1-57. Before certification, all trainers must attend all phases of the train-the-trainer program in the following order:

- Phase I—Program Orientation.
- Phase II—Preliminary Marksmanship Training.
- Phase III—Basic Marksmanship Training.
- Phase IV—Advanced Marksmanship Training.

1-58. Then, they conduct all phases to demonstrate their ability to train Soldiers and to diagnose and correct problems. Trainers who fail to attend or do not pass any phase of the diagnostic examination will be assigned to subsequent training.

Phase I—Program Orientation

1-59. During this phase, the trainer must accomplish the following tasks and be certified by the chain of command:

- Be briefed on the concept of the certification program.
- Be briefed on the unit marksmanship training strategy.
- Review the unit marksmanship training outlines.
- Review issued reference material.
- Visit training sites and firing ranges.

Phase II—Preliminary Marksmanship Training

1-60. Phase II should be completed no more than two weeks following the conclusion of Phase I. During Phase II, the trainer demonstrates his ability to master the marksmanship fundamentals, and his performance is reviewed by the chain of command. The results of this review are recorded and maintained on the trainer's progression sheet, which is, in turn, designed IAW the unit SOP.

- Safety, clearing, and unloading procedures.
- Characteristics.
- Capabilities.
- Accessories.
- Ammunition.
- Disassembly.
- Cleaning, inspection, and lubrication procedures.
- Assembly.
- Function check.
- Loading.
- Immediate and remedial actions.
- Four fundamentals.
- Use of EST 2000, LMTS, and other TADSS.
- Sight manipulation.
- Boresighting procedures.
- Zeroing procedures.
- Effect of wind, gravity, and other environmental factors.
- Range determination and estimation.
- Classes of fire.

- Application of fire.
- Range operations, courses of fire, conduct of fire, and fire commands.
- Optics, NVDs, night aiming devices, and thermal devices.
- Scanning techniques.

Phase III—Basic Marksmanship Training

1-61. During this phase, the trainer demonstrates and reinforces what he has learned during Phase II. The trainer sets up and conducts firing on the various ranges. He explains the targets and the zeroing and scoring procedures. The trainer explains the purpose of transition firing, field zeroing procedures, range layout, and the conduct of training on the transition range. This validates that the trainer has gained the knowledge necessary to conduct training. The results of this review are recorded and maintained on the trainer's progression sheet.

Phase IV—Advanced Marksmanship Training

1-62. The final phase of the train-the-trainer program tests the trainer. During this phase, the trainer sets up a range and conducts training for at least one person. If ammunition is available, the trainer conducts a firing exercise. If ammunition is not available, the evaluation is based on the quality of training given.

QUALIFICATION TRAINING

1-63. Although marksmanship is a continuous training requirement, units normally conduct a refresher program before qualification. Soldiers must be well-versed in marksmanship fundamentals and have preparatory marksmanship training before qualification. This applies to qualification for the entire unit or for newly assigned personnel. Trainers must understand that unit rifle marksmanship is not a series of exercises to be trained in a planned sequence as is done during IET, but trainers can use the exercises and POI events covered during IET to identify events that the unit can use for a sustainable and effective unit marksmanship program. The unit must prepare for training by—

- Issuing Soldiers a serviceable weapon.
- Maintaining and replacing bad magazines.
- Issuing and assigning each Soldier his own rifle. Only he will zero and fire the weapon assigned to him.
- Considering available or required resources (targets, ranges, ammunition, training aids, devices, and publications) early in the process.

1-64. Before the Soldier can fire, he must know how to adjust the rifle's sights and should understand ballistics (for example, the effects of wind and gravity on a bullet strike). A refresher training program prevents Soldiers from becoming frustrated and losing confidence, and conserves ammunition and training time. All Soldiers attend this program so they can meet the standards outlined in this manual and supporting manuals.

NOTE: Many individual marksmanship tasks, such as operation and function checks, immediate action, target detection, and dry-fire, do not require live-firing. Live and virtual simulators can be used to reinforce PMI, grouping, zeroing, practice record fire, record fire, CBRN fire, and assisted and unassisted night fire by simulating the LFXs. Building marksmanship confidence by repetition can bring consistency to the unit marksmanship training program.

FEEDBACK

1-65. Feedback must be included in all live-fire training. Soldiers must have precise knowledge of a bullet strike; feedback is not adequate when bullets from previous firings cannot be identified. To provide accurate feedback, trainers ensure that Soldiers triangulate and clearly mark previous shot groups on a zeroing target.

Grouping and Zeroing

1-66. The initial LFX should be a grouping exercise, during which Soldiers apply marksmanship fundamentals to obtain tight, consistent shot groups. Following a successful grouping exercise, Soldiers zero their weapons quickly using only a few rounds.

Downrange Feedback

1-67. After zeroing, downrange feedback should be conducted. If modified field fire or KD ranges are not available, a series of scaled silhouette targets can be used for training on the 25-meter range. A timed-fire scaled silhouette target can add to successful record fire performance, since it represents targets at six different ranges, requires quick response, and allows precise feedback. It also serves as another way to confirm zero and requires the application of the four fundamentals.

Field Fire

1-68. Field fire training is a transitional phase that stresses focusing on a certain area. Soldiers must detect the target as soon as it appears and quickly fire; this is an important combat skill. Soldiers who are exposed to the field fire range before they have refined their basic firing skills cannot benefit from the exercise, and feedback from this exercise is hit-or-miss. If most 175- and 300-meter targets are missed, additional feedback or PMI training should be conducted.

Record Fire

1-69. The intent of record fire is to facilitate the commander's evaluation of several individual tasks and integrated marksmanship skill performances, and to provide unit readiness indicators. The qualification standards are specifically related to a prescribed procedure for the conduct of record fire. Individual performance must be evaluated IAW three components:

- What test was used (standard, known distance, or scaled)?
- How was the test administered?
- How were individual and unit performances distributed (marksman, sharpshooter, and expert) and at what target ranges?

Remedial Training

1-70. For firers who need additional training to meet the requirements, remedial training is conducted using the EST 2000 or LMTS. Soldiers proficient in marksmanship skills can assist in the remedial training effort.

NOTE: Table 1-1 shows the training devices that a commander may use to assist in remedial training and sustain weapon proficiency. See Appendix A for more information about these training devices. The devices replicate, but are not intended to replace, LFXs or qualifications. Active and reserve component units should consult DA Pam 350-38 for regulatory guidance on mandatory live-fire training and qualification events.

Table 1-1. Training simulators, devices, and exercises.

TRAINING DEVICE / EXERCISE	Engagement Skills Trainer (EST 2000)	Laser Marksmanship Training System (LMTS)	Location of Misses and Hits (LOMAH)	Short-Range Training Ammunition (SRTA) and M2 Bolt	Weaponeer
Zero	X	X	X	X	X
Practice Fire	X	X	X	25 m	X
Simulated Record Fire	X	X	X	25 m	X
CBRN Practice	X	X	X	X	X
Simulated CBRN Record	X	X	X	X	X
Unassisted Night Practice	X	X	X	50 m	
Unassisted Simulated Night Record	X	X	X	50 m	
NVD Zero	X	X	X		
NVD Practice	X	X	X	50 m	
NVD Simulated Record	X	X	X	50 m	
Advanced Skills	X	X	X	X	

UNIT LIVE-FIRE EXERCISES

1-71. Unit LFXs are planned, prepared, and performed as outlined in the mission training plan for the infantry platoon and squad. Within the framework of these exercises, Soldiers perform marksmanship tasks under realistic combat conditions.

FUNDAMENTALS

1-72. During training, the fundamentals must apply to combat, as well as to the range. Too often Soldiers disregard the fundamentals while under the pressure of combat, so it is imperative that the Soldier receives feedback regarding his firing results and his use of the fundamentals during collective LFXs. This training should also discuss target acquisition, area fire, quick fire, assuming firing positions, responding to oral fire commands, and safety. To learn SOPs and proper procedures, Soldiers must participate in dry-fire or MILES rehearsals at crawl, walk, and run paces.

EVALUATORS

1-73. During training, enough evaluators must be present to observe each Soldier and provide performance feedback. Evaluators must know the scenario, the location of targets, the friendly plan, and SOPs. They must watch each Soldier to determine if he identifies targets in his sector and successfully engages them. Evaluators must also understand the fundamentals of marksmanship to detect Soldiers' mistakes and review them during the after-action review (AAR).

Chapter 2

Weapon Characteristics, Accessories, and Ammunition

This chapter describes the general components, characteristics, accessories, and ammunition for M16- and M4-series weapons, and includes a brief explanation of how to mount the various accessories.

SECTION I. RIFLES AND CARBINES

All M16-/M4-series weapons are magazine-fed, gas-operated, air-cooled, shoulder-fired 5.56-millimeter weapons. This section describes the general characteristics and components of M16-/M4-series weapons.

CHARACTERISTICS OF M16-/M4-SERIES WEAPONS

2-1. Table 2-1 describes the general characteristics of M16-/M4-series weapons.

Table 2-1. Characteristics of M16-/M4-series weapons.

CHARACTERISTICS	M4-SERIES	M16A2/A3	M16A4	M16A1
WEIGHT (lb)				
Without magazine and sling	6.49	7.78	9.08	6.35
With sling and loaded:				
20-round magazine	7.19	8.48	9.78	6.75
30-round magazine	7.50	8.79	10.09	8.06
Bayonet knife, M9	1.50	1.50	1.50	1.50
Scabbard	0.30	0.30	0.30	0.30
Sling, M1	0.40	0.40	0.40	0.40
LENGTH (in)				
Rifle w/bayonet knife	N/A	44.88	44.88	44.25
Overall rifle length	N/A	39.63	39.63	39.00
Buttstock closed	29.75	N/A	N/A	N/A
Buttstock open	33.0	N/A	N/A	N/A
OPERATIONAL CHARACTERISTICS				
Barrel rifling-righthand 1 twist (in)	7	7	7	12
Muzzle velocity (fps)	2,970	3,100	3,100	3,250
Cyclic rate of fire (rounds per min)	700-900	700-900	800	700-800
MAXIMUM EFFECTIVE RATE OF FIRE (rounds per min)				
Semiautomatic	45	45	45	45-65
3-round burst	90	90 (A2)	90	N/A
Automatic	150-200 A1	150-200 A3	N/A	150-200
Sustained	12-15	12-15	12-15	12-15
RANGE (m)				
Maximum range	3,600	3,600	3,600	2,653
Maximum effective range:				
Point target	500	550	550	460
Area target	600	800	600	N/A

NOTE: For further technical information, refer to TM 9-1005-319-10 and TM 9-1005-249-10.

M4-SERIES CARBINE

2-2. The M4-series carbine consists of the M4, the M4A1, and the M4 modular weapon system (MWS) (Figures 2-1 and 2-2) and features several modifications to equip it for close combat operations:

- The M4-series carbine buttstock has four positions:
  - Closed.
  - ½ open.
  - ¾ open.
  - Fully open.
- The M4 carbine becomes the M4 MWS when the M4 adapter rail system (ARS) is installed (Figure 2-3).
- When operating an M4 or M4 MWS, the firer can move the selector lever (SAFE, SEMI, and BURST) to fire a semiautomatic or a three-round burst.
- The M4A1 is fully automatic.

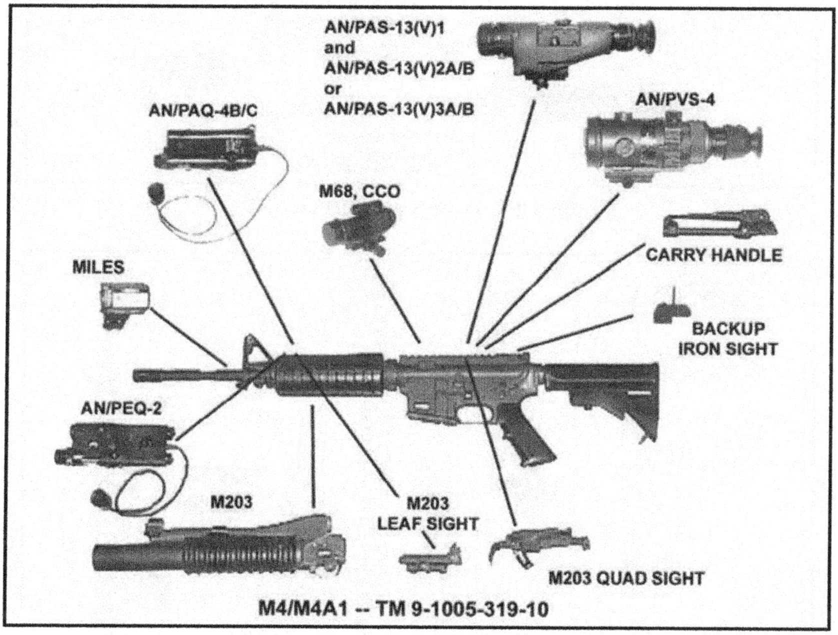

Figure 2-1. M4/M4A1 carbine with accessories.

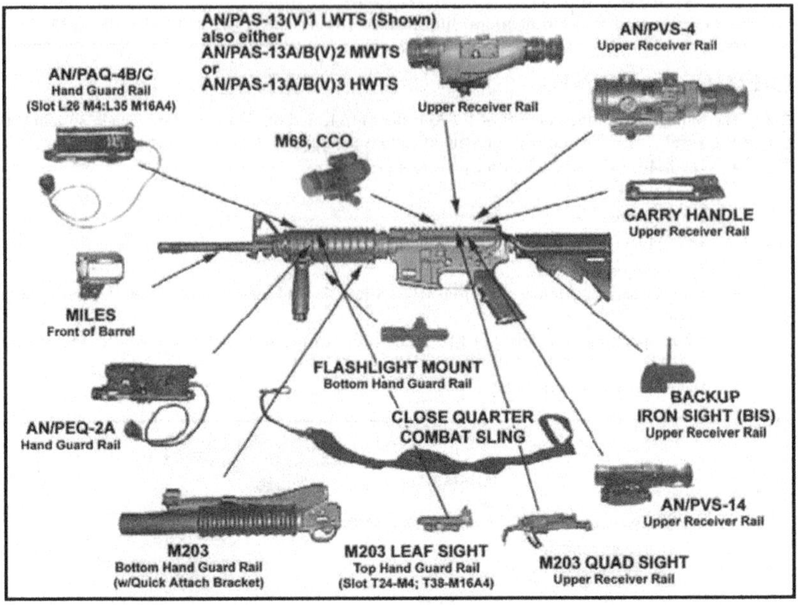

Figure 2-2. M4 MWS with accessories.

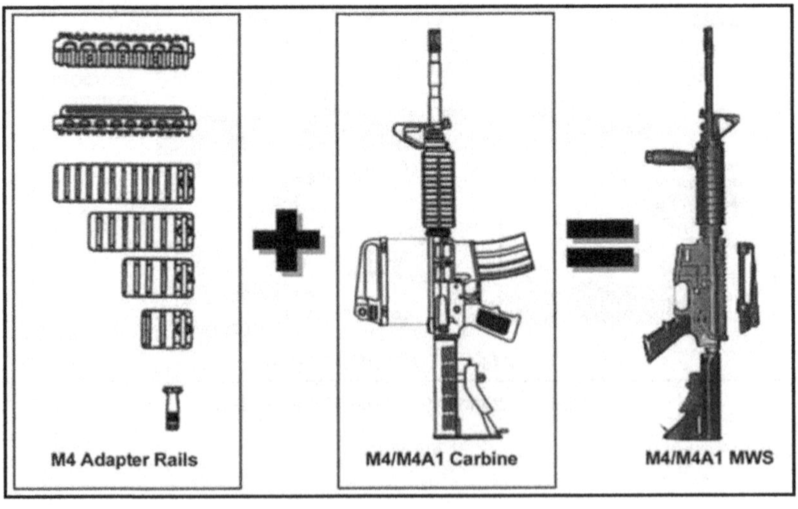

Figure 2-3. M4/M4A1 and M4/M4A1 MWS.

MECHANICALLY ZEROING THE M4/M4A1 CARBINE OR M4 MWS

NOTE: Mechanically zeroing the weapon is only necessary when the weapon zero is questionable, the weapon is newly assigned to the unit, or the weapon sights have been serviced.

2-3. To mechanically zero an M4/M4A1 or M4 MWS—

NOTE: Reference the weapon components using the numbers listed in Figure 2-4.

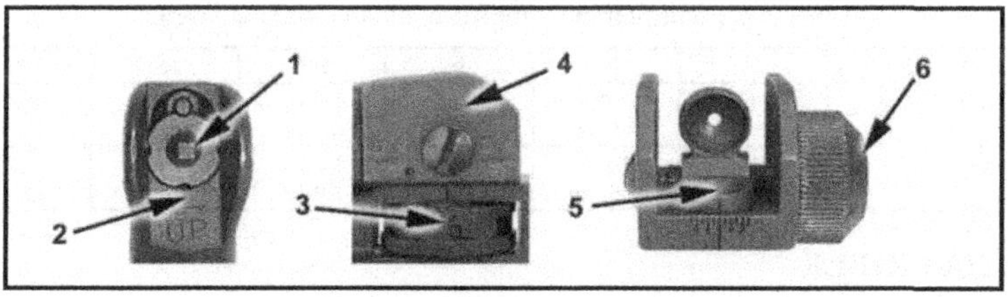

Figure 2-4. M4/M4A1 or M4 MWS mechanical zero.

(1) Adjust the front sightpost (1) until the base of the front sightpost is flush with the front sightpost housing (2).

(2) Turn the elevation knob (3, shown as viewed from above) counterclockwise until the rear sight assembly (4) rests flush with the detachable carrying handle and the 6/3 marking is aligned with the index line (5) on the left side of the carrying handle.

(3) Position the apertures (6) so the unmarked aperture is up and the 0-200 meter aperture is down.

(4) Turn the windage knob (7) to align the index mark (8) on the 0-200 meter aperture with the long center index line on the rear sight assembly.

BATTLESIGHT ZEROING THE M4/M4A1 CARBINE OR M4 MWS

2-4. To battlesight zero an M4/M4A1 or M4 MWS (Figure 2-5)—

NOTE: Reference the weapon components using the numbers listed in Figure 2-5.

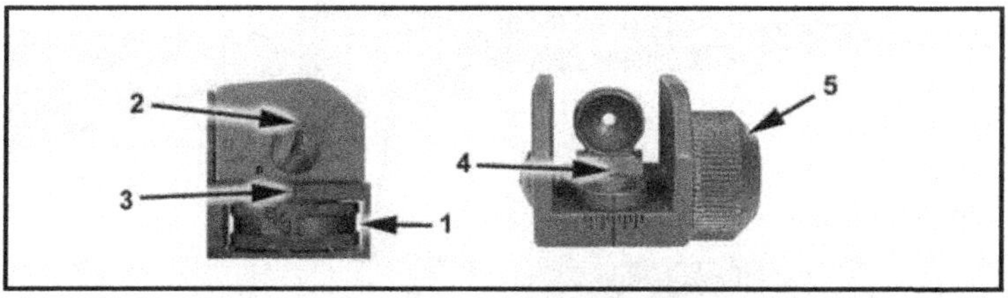

Figure 2-5. M4/M4A1 or M4 MWS battlesight zero.

(1) Turn the elevation knob (1, shown as viewed from above) counterclockwise until the rear sight assembly (2) rests flush with the detachable carrying handle and the 6/3 marking is aligned with the index line (3) on the left side of the detachable carrying handle.

NOTE: The elevation knob remains flush.

(2) Position the apertures (4) so the unmarked aperture is up and the 0-200 meter aperture is down.

(3) Turn the windage knob (5) to align the index mark (6) on the 0-200 meter aperture with the long center index line on the rear sight assembly.

NOTE: The "Z" marking on the elevation knob (employed when using the M4-series weapon's detachable carrying handle) should be ignored. The "Z" marking is only used when the M16A4 is being zeroed.

2-5. Table 2-2 shows how much one click of elevation or windage will move the strike of the round at ranges from 25 to 500 meters.

Table 2-2. Point of impact for the M4/M4A1 and M4 MWS.

RANGE (m)	25	100	200	300	400	500
Elevation	3/4 in 0.9 cm	1 3/8 in 3.5 cm	2 3/4 in 7 cm	4 1/8 in 10.5 cm	5 1/2 in 14 cm	6 7/8 in 17.5 cm
Windage	1/8 in 0.3 cm	1/2 in 1.25 cm	1 in 2.5 cm	1 1/2 in 3.8 cm	2 in 5 cm	2 1/2 in 6.3 cm

M16A2/A3 RIFLE

2-6. Figure 2-6 shows the M16A2/A3 rifle. When operating an M16A2 rifle, the firer can move the selector lever (SAFE, SEMI, and BURST) to fire a semiautomatic or a three-round burst. The M16A3 has the same characteristics as the M16A2, with the exception of the selector lever (SAFE, SEMI, and AUTO) and the addition of the automatic mode.

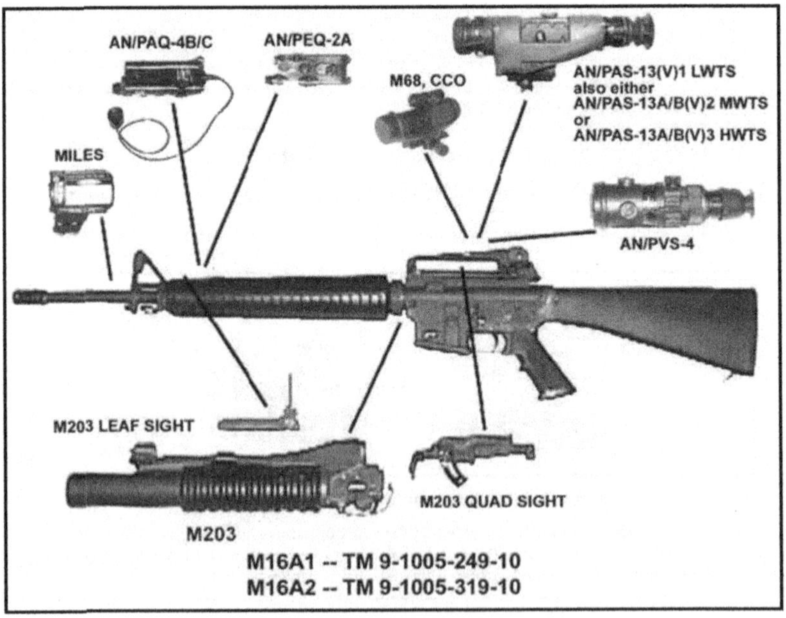

Figure 2-6. M16A2/A3 rifle with accessories.

Mechanically Zeroing the M16A2/A3 Rifle

NOTE: Mechanically zeroing the weapon is only necessary when the weapon zero is questionable, the weapon is newly assigned to the unit, or the weapon sights have been serviced.

2-7. To mechanically zero an M16A2/A3 rifle—

NOTE: Reference the weapon components using the numbers listed in Figure 2-7.

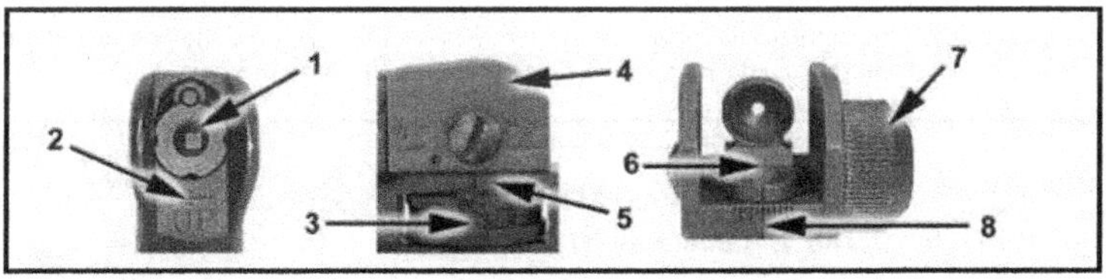

Figure 2-7. M16A2/A3 mechanical zero.

(1) Adjust the front sightpost (1) until the base of the front sightpost is flush with the front sightpost housing (2).

(2) Turn the elevation knob (3, shown as viewed from above) counterclockwise until the rear sight assembly (4) rests flush with the carrying handle and the 8/3 marking is aligned with the index line on the left side of the carrying handle.

(3) Position the apertures (5) so the unmarked aperture is up and the 0-200 meter aperture is down.

(4) Turn the windage knob (6) to align the index mark on the 0-200 meter aperture with the long center index line on the rear sight assembly.

Battlesight Zeroing the M16A2/A3 Rifle

2-8. To battlesight zero an M16A2/A3 rifle—

NOTE: Reference the weapon components using the numbers listed in Figure 2-8.

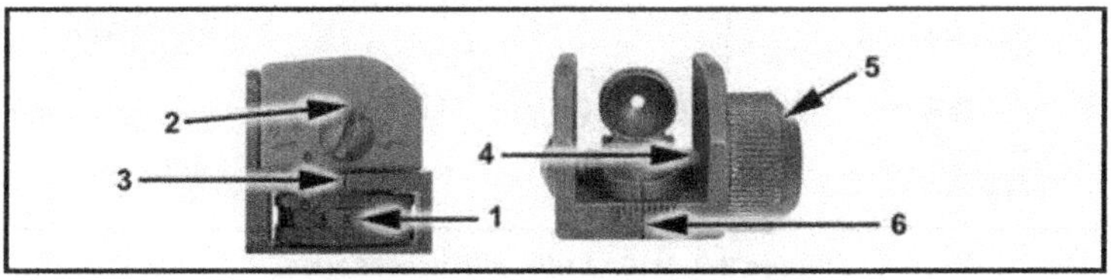

Figure 2-8. M16A2/A3 battlesight zero.

(1) Turn the elevation knob (1, shown as viewed from above) counterclockwise until the rear sight assembly (2) rests flush with the carrying handle and the 8/3 marking is aligned with the index line (3) on the left side of the carrying handle.

(2) Turn the elevation knob one more click clockwise.

(3) Position the apertures (4) so the unmarked aperture is up and the 0-200 meter aperture is down.

(4) Turn the windage knob (5) to align the index mark on the 0-200 meter aperture with the long center index line on the rear sight assembly.

2-9. Table 2-3 shows how much one click of elevation or windage will move the strike of the round at ranges from 25 to 600 meters.

Table 2-3. Point of impact for the M16A2/A3 rifle.

RANGE (m)	25	100	200	300	400	500	600
Elevation	3/4 in 0.9 cm	1 3/8 in 3.5 cm	2 3/4 in 7 cm	4 1/8 in 10.5 cm	5 1/2 in 14 cm	6 7/8 in 17.5 cm	8 1/4 in 20.9 cm
Windage	1/8 in 0.3 cm	1/2 in 1.25 cm	1 in 2.5 cm	1 1/2 in 3.8 cm	2 in 5 cm	2 1/2 in 6.3 cm	3 in 7.6 cm

M16A4 RIFLE

2-10. The M16A4 rifle (Figure 2-9) is the same as the M16A2 rifle, with the addition of a flat top upper receiver (with an integral rail and detachable carrying handle) and the M5 ARS.

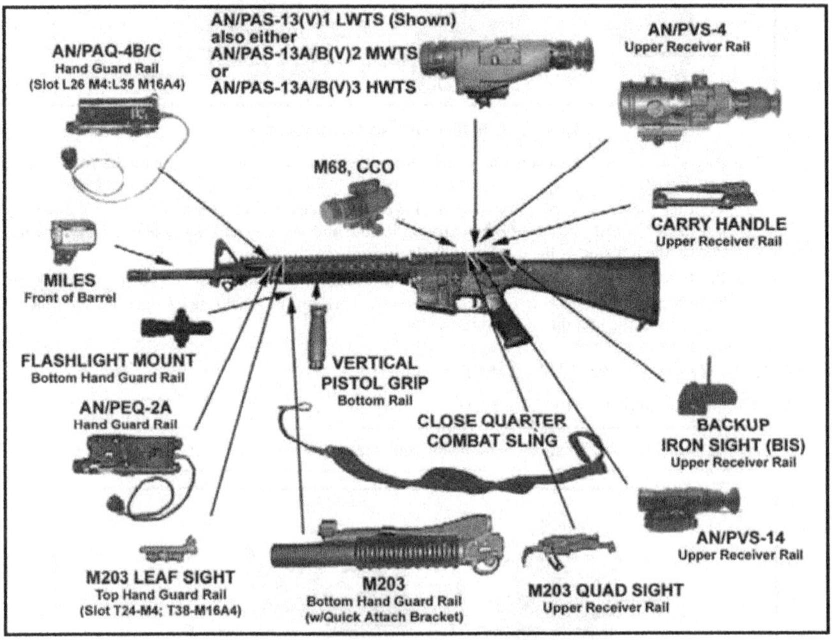

Figure 2-9. M16A4 rifle with accessories.

Mechanically Zeroing the M16A4 Rifle

NOTE: Mechanically zeroing the weapon is only necessary when the weapon zero is questionable, the weapon is newly assigned to the unit, or the weapon sights have been serviced.

2-11. To mechanically zero an M16A4 rifle—

NOTE: Reference the weapon components using the numbers listed in Figure 2-10.

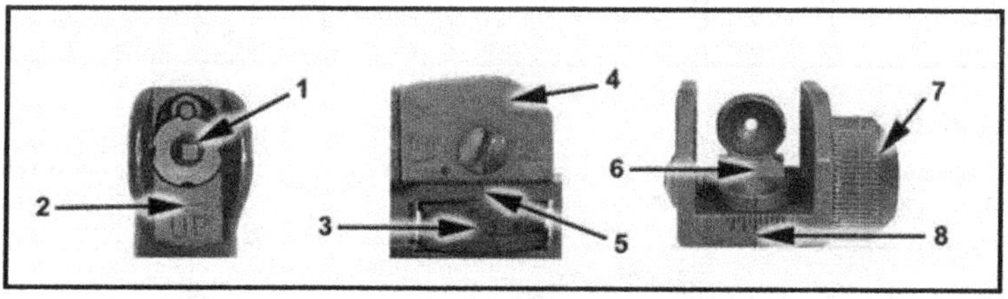

Figure 2-10. M16A4 mechanical zero.

(1) Adjust the front sightpost (1) until the base of the front sightpost is flush with the front sightpost housing (2).

(2) Turn the elevation knob (3, shown as viewed from above) counterclockwise until the rear sight assembly (4) rests flush with the carrying handle and the 6/3 marking is aligned with the index line (5) on the left side of the carrying handle.

(3) Position the apertures (6) so the unmarked aperture is up and the 0-200 meter aperture is down.

(4) Turn the windage knob (7) to align the index mark on the 0-200 meter aperture with the long center index line (8) on the rear sight assembly.

Battlesight Zeroing the M16A4 Rifle

2-12. To battlesight zero an M16A4 rifle—

NOTE: Reference the weapon components using the numbers listed in Figure 2-11.

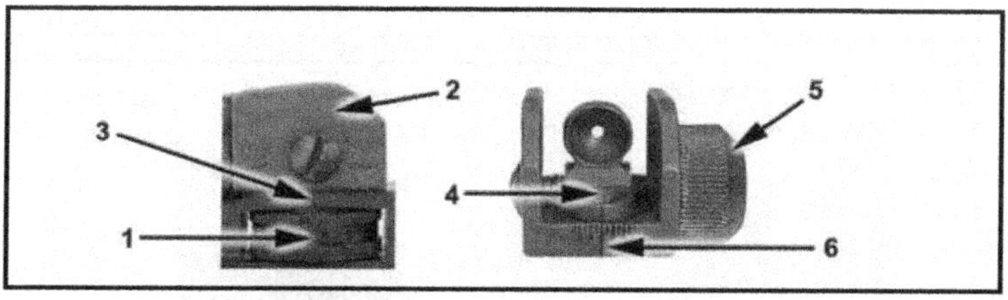

Figure 2-11. M16A4 battlesight zero.

(1) Turn the elevation knob (1, shown as viewed from above) counterclockwise until the rear sight assembly (2) rests flush with the detachable carrying handle and the 6/3 marking is aligned with the index line (3) on the left side of the detachable carrying handle.

(2) Turn the elevation knob two clicks clockwise so the index line on the left side of the detachable carrying handle is aligned with the "Z" on the elevation knob.

(3) Position the apertures (4) so the unmarked aperture is up and the 0-200 meter aperture is down.

(4) Turn the windage knob (5) to align the index mark on the 0-200 meter aperture with the long center index line (6) on the rear sight assembly.

2-13. Table 2-4 shows how much one click of elevation or windage will move the strike of the round at ranges of 25 to 600 meters.

Table 2-4. Point of impact for the M16A4 rifle.

RANGE (m)	25	100	200	300	400	500	600
Elevation	3/4 in 0.9 cm	1 3/8 in 3.5 cm	2 3/4 in 7 cm	4 1/8 in 10.5 cm	5 1/2 in 14 cm	6 7/8 in 17.5 cm	8 1/4 in 20.9 cm
Windage	1/8 in 0.3 cm	1/2 in 1.25 cm	1 in 2.5 cm	1 1/2 in 3.8 cm	2 in 5 cm	2 1/2 in 6.3 cm	3 in 7.6 cm

M16A1 RIFLE

2-14. When operating an M16A1 rifle (Figure 2-12), the firer can move the selector lever (SAFE, SEMI, and AUTO) to fire in the semiautomatic or automatic mode.

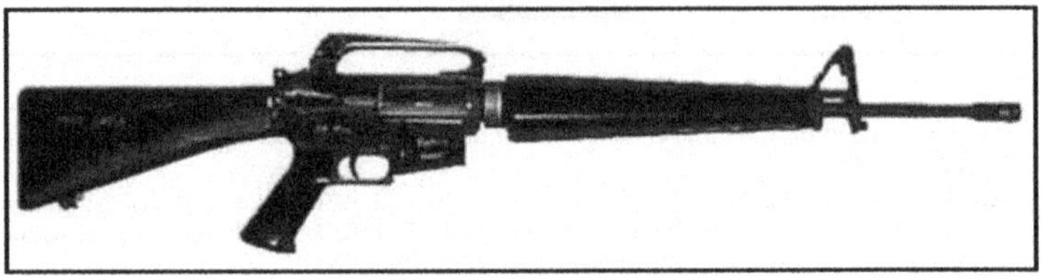

Figure 2-12. M16A1 rifle.

MECHANICALLY ZEROING THE M16A1 RIFLE

NOTE: Mechanically zeroing the weapon is only necessary when the weapon zero is questionable, the weapon is newly assigned to the unit, or the weapon sights have been serviced.

2-15. To mechanically zero an M16A1 rifle—

NOTE: Reference the weapon components using the numbers listed in Figure 2-13.

Figure 2-13. M16A1 mechanical zero.

(1) Adjust the front sightpost (1) until the base of the front sightpost is flush with the front sightpost housing (2).
(2) Adjust the front sightpost 11 clicks in the direction of UP.
(3) Turn the rear sight windage drum (3) left until it stops.
(4) Turn the windage drum right 17 clicks so the rear sight is approximately centered.

BATTLESIGHT ZEROING THE M16A1 RIFLE

2-16. Soldiers should use the aperture marked "L" (Figure 2-14) to battlesight zero their weapons.

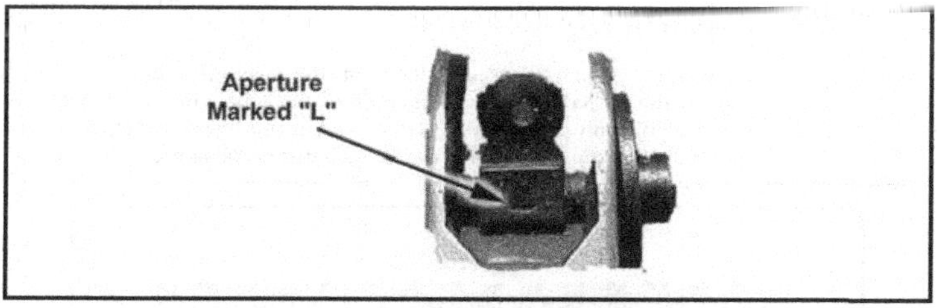

Figure 2-14. M16A1 battlesight zero.

2-17. Table 2-5 shows how much one click of elevation or windage will move the strike of the round at ranges of 25 to 500 meters using standard sights or the low light level sight system.

Table 2-5. Point of impact for the M16A1 rifle.

STANDARD SIGHTS						
Range (m)	25	100	200	300	400	500
Elevation	17/64 in 0.7 cm	1 3/32 in 2.8 cm	2 13/64 in 5.6 cm	3 9/32 in 8.4 cm	4 3/8 in 11.2 cm	5 15/32 in 14 cm
Windage	17/64 in 0.7 cm	1 3/32 in 2.8 cm	2 13/64 in 5.6 cm	3 9/32 in 8.4 cm	4 3/8 in 11.2 cm	5 15/32 in 14 cm
LOW LIGHT LEVEL SIGHT SYSTEM						
Elevation	23/64 in 0.9 cm	1 3/4 in 3.5 cm	2 3/4 in 7 cm	5 1/4 in 10.5 cm	7 in 17.7 cm	8 3/4 in 22.2 cm
Windage	17/64 in 0.7 cm	1 3/32 in 2.8 cm	2 13/64 in 5.6 cm	3 9/32 in 8.4 cm	4 3/8 in 11.2 cm	5 15/32 in 14 cm

SECTION II. ACCESSORY MOUNTING

The M4/M5 ARS and rail grabbers are designed to mount accessories on M16- and M4-series weapons.

M4/M5 ADAPTER RAIL SYSTEM

2-18. The ARS provides a secure mounting point for various accessories that may be mounted on the weapon's top, left, and right. The M4 ARS (Figure 2-15) consists of lightweight sections that replace the standard handguards on the M4 carbine. The M5 ARS replaces the handguards on the M16A4 rifle.

NOTES:

1. Only the armorer is authorized to remove the ARS from a weapon. However, the user may remove only the lower assembly to perform preventive maintenance checks and services (PMCS). If the top section is removed, the accessories mounted on the rail will lose their zero.
2. To prevent the screw from vibrating loose during firing, apply locking compound to the original screw or use a new screw when reinstalling the ARS.
3. For optics and lasers to retain zero, the ARS must be properly installed. Mounted accessories will not retain zero unless proper procedures are followed, as outlined in TM 9-1005-319-23& P.
4. Accessories may be mounted on the right side of the ARS, but are not currently supported with zeroing procedures. Further, the bottom rail of the ARS will not retain zero. Only accessories that do not require zero retention, such as a flashlight or vertical pistol grip, can be mounted on the bottom rail.

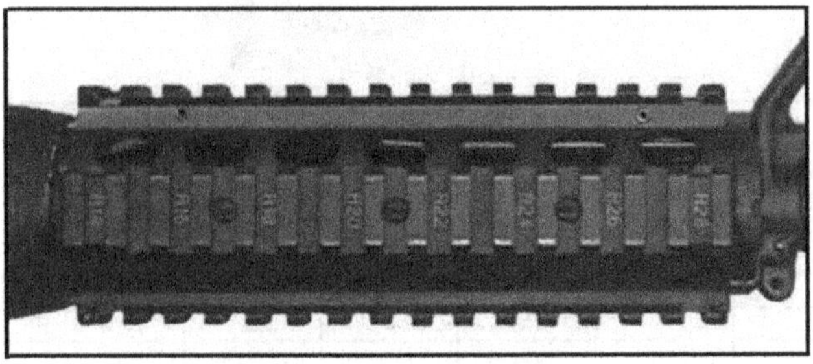

Figure 2-15. Adapter rail system.

Rail Covers

2-19. Rail covers protect the firer's hands from direct contact with the metal parts of the ARS and protect the ARS surfaces from excess wear and damage. The rail covers are available in 11-, 9-, 6-, 5-, and 4-rib sections. For ease of reference, the shorter lengths can be referred to by the number of ribs along their outer surfaces. 11-, 9-, and 5-rib sections are shown in Figure 2-16.

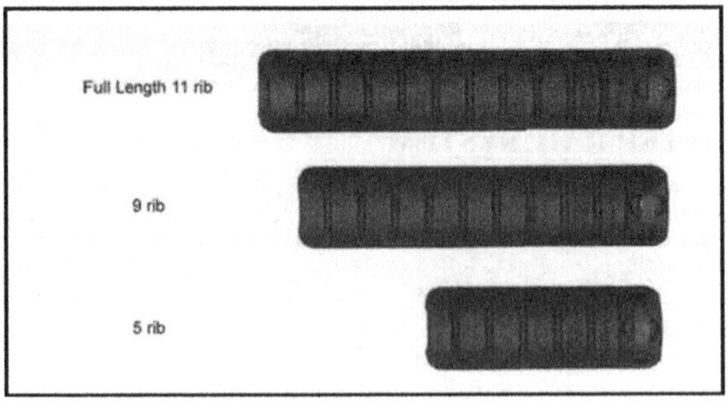

Figure 2-16. M5 rail covers/heat shields.

2-20. ARS rail covers can be quickly attached and detached from the ARS. A spring latch at one end of each rail cover automatically engages cutouts at the end of each rail. To slide the rail cover beyond a cutout or to remove it, apply thumb pressure to the center of the spring latch, and slide it in the desired direction.

NOTES: 1. Keep unused bottom, left, and right rail sections covered with 11-rib rail cover sections.

2. If any accessories are mounted on a rail, cover the remaining rail surface with an appropriately-sized rail cover.

3. If a BUIS is installed, permanently remove the top full length rail cover, and replace it with a shorter rail cover to protect the nonfiring hand when the barrel is hot.

WARNING

When firing at high rates of sustained fire, the barrel and metal components of the ARS can become hot enough to inflict serious burns. Cover exposed metal portions with plastic rail covers. Use the vertical pistol grip (Figure 2-18) during heavy, sustained fire.

NUMBERED RECOIL GROOVES

2-21. The even numbered recoil grooves on each rail of the ARS are sequentially numbered within the recoil grooves themselves (Figure 2-17). Each number is preceded by a letter prefix indicating a specific slot on the ARS. The numbers on the top rail have a "T" prefix, while those on the bottom have a "B" prefix. Additionally, the numbers on the rail to the firer's left have an "L" prefix, while those to the firer's right have an "R" prefix. These addresses—

- Assist the user in remounting an accessory in the same position.
- Allow standardization in the location of mounted accessories.
- Identify reference points for discussions on accessory mounting locations.

Figure 2-17. Address markings on the adapter rail system.

NOTE: Within the notches, each ARS also contains holes that are threaded ¼-inch deep, with 20 threads per inch (Figure 2-17). This is the standard thread size for a camera tripod adapter, which is used to attach standard camera or video accessories. For example, an M4 MWS with an NVD mounted may be attached to a standard camera tripod for hands-free support during long periods of surveillance.

VERTICAL PISTOL GRIP

2-22. Each ARS comes with a vertical pistol grip. To install the pistol grip—

NOTE: Reference the weapon components using the numbers listed in Figure 2-18.

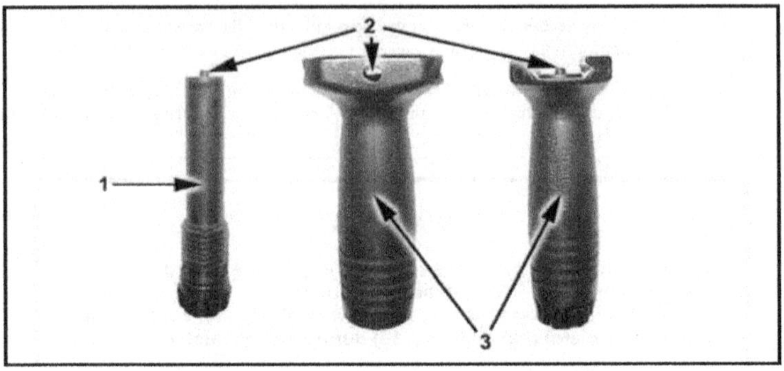

Figure 2-18. Vertical pistol grip.

(1) Remove the rail cover.

(2) Unscrew the pistol grip lock (1) until the tip (2) is no longer visible through the hole in the pistol grip.

(3) Slide the pistol grip (3) onto the ARS.

NOTE: The pistol grip will cover five notches on the ARS.

(4) Align the tip on the top of the pistol grip lock (1) with a notch.

(5) Hand-tighten.

NOTE: For further information about these accessories, refer to TM 9-1005-319-10.

RAIL GRABBERS

2-23. The Insight (Figure 2-19) and Picatinny (Figure 2-21) rail grabbers are designed to mount accessories onto the M16A4 and M4-series weapons. These rail grabbers enable the user to attach accessories on the upper receiver and on all four sides of the ARS, while retaining zero (when installed and tightened properly). Once zeroed, they will retain zero—even when removed from the weapon—as long as the rail grabber is not separated from the accessory and is remounted on the same notch it was zeroed on.

NOTE:
1. Retighten the rail grabber and accessory after firing the first three rounds to fully seat both.
2. The bottom rail of the ARS will not retain zero.

Insight Rail Grabber

2-24. The Insight rail grabber (Figure 2-19) can be mounted where the tightening screw (3, Figure 2-19) is on either the left or right side (when top-mounted) or top or bottom (when mounted on the left side) so it does not interfere with weapon operation.

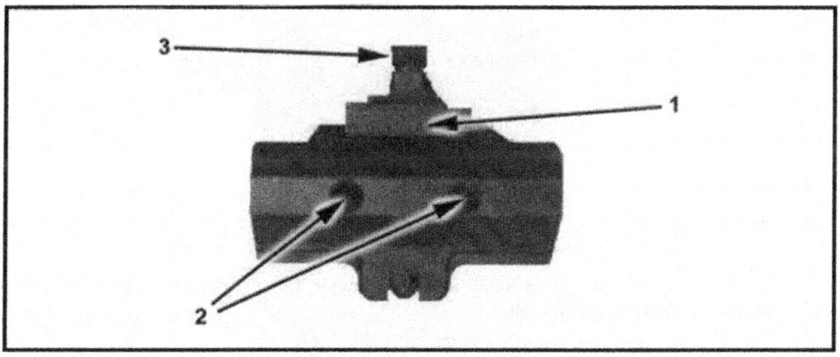

Figure 2-19. Insight rail grabber.

2-25. Unless command-directed, all devices in a unit do not have to be mounted in the same location as long as individual users record or mark (with paint markers or grease pencils) the mounting location on their weapons to avoid unnecessary rezeroing.

CAUTION

Both of the holes (2, Figure 2-19) located in the top of the rail grabber can be used to mount accessories, but the hole closest to the muzzle must be used. This ensures that the majority of the rail grabber is supporting the accessory to prevent damage to the accessory.

NOTES:

1. The Insight rail grabber must fully rest on the ARS in order to retain zero. The locking clamp (1, Figure 2-19) must grasp the ARS, and the screw that tightens the rail grabber must be tightened with a field tool, such as a multipurpose tool.
2. Even if the rail grabber is resting entirely on the ARS, accessories should not make contact with the front sight assembly or the collar of the barrel. The vibrations that occur during firing will interfere with the rail grabber's and accessory's zero retention capabilities.

MILES Training Extender for the Insight Rail Grabber

2-26. The training extender (Figure 2-20) is used to elevate an accessory above the MILES laser during force-on-force training. The training extender is only used when the Insight rail grabber is top-mounted.

2-27. To install the extender and accessories—

NOTE: Reference the weapon components using the numbers listed in Figure 2-20.

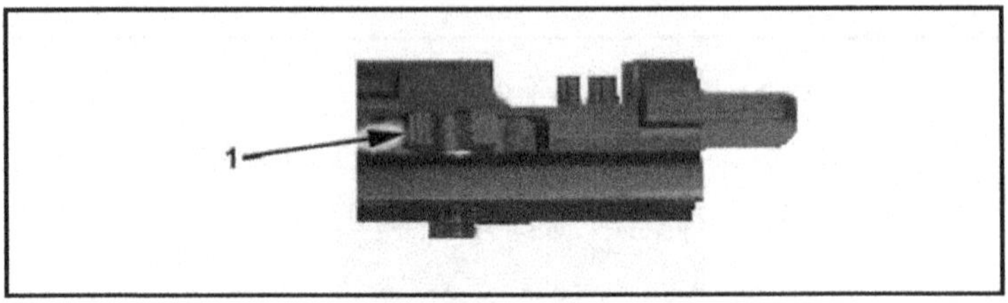

Figure 2-20. Insight rail grabber MILES training extender.

(1) Use the thumbscrew (1) to hand-tighten the extender into the mounting hole closest to the muzzle on the Insight rail grabber.

(2) Install the accessory on top of the extender and tighten.

NOTE: The AN/PEQ-2A/B and AN/PAQ-4B/C must be zeroed before and after using the MILES training extender.

Picatinny Rail Grabber

2-28. The Picatinny rail grabber (Figure 2-21) must fully rest on the ARS in order to retain zero. The locking clamp must grasp the ARS, and the torque-limiting knob (2, Figure 2-21) that tightens the rail grabber must be hand-tightened until it clicks twice.

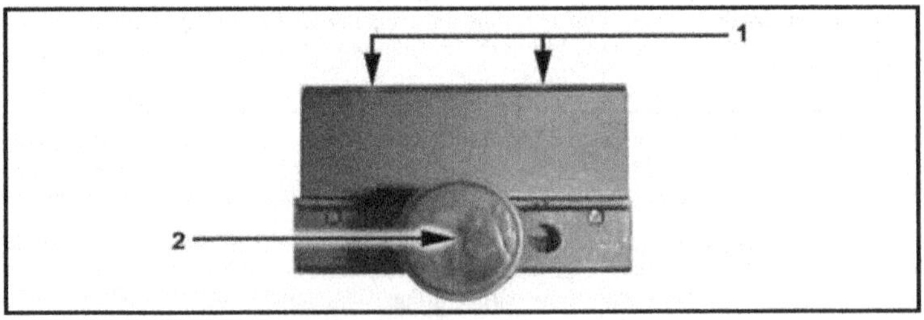

Figure 2-21. Picatinny rail grabber.

CAUTION

Both of the holes located in the top of the rail grabber (1, Figure 2-21) can be used to mount accessories, but the hole closest to the muzzle must be used for the Picatinny rail grabber. This ensures that the majority of the rail grabber is supporting the accessory to prevent damage to the accessory. It also allows the torque-limiting knob to be mounted on either the left or right side (when top-mounted) or top or bottom (when side-mounted) so that it will not interfere with weapon operation.

2-29. Unless command-directed, all devices in a unit do not have to be mounted in the same location as long as individual users record or mark (with paint markers or grease pencils) the mounting location on their weapons to avoid unnecessary rezeroing.

NOTE: Even if the rail grabber is resting entirely on the ARS, accessories should not make contact with the front sight assembly or the collar of the barrel. The vibrations that occur during firing will interfere with the rail grabber's and accessory's zero retention capabilities.

SECTION III. ACCESSORIES

Table 2-6 shows the weapons with which the various accessories are compatible, along with the mounting device and TM number. Table 2-7 shows select characteristics of the various accessories.

Table 2-6. Accessory compatibility and mounting.

ACCESSORY	TECHNICAL MANUAL	M4 MWS , M16A4	M4/M4A1	M16A1/A2/A3
M68, CCO	TM 9-1240-413-13&P	* Upper receiver	* Upper receiver	M16 mount
AN/PVS-14	TM 11-5855-306-10	***Upper receiver	N/A	N/A
AN/PAC-4C	TM 11-5855-301-12&P	** Rail grabber	Bracket assembly	Bracket assembly
AN/PEQ-2A/B	TM 11-5855-308-12&P	** Rail grabber	Bracket assembly	Bracket assembly
AN/PAQ-4	TM 11-5855-261-10	** Rail grabber	Bracket assembly	Bracket assembly
AN/PAS-13B(V1), LWTS	TM 11-5855-312-10	Upper receiver	Upper receiver	M16 mount
AN/PAS-13B(V3), HWTS	TM 11-5855-312-10	Upper receiver	Upper receiver	M16 mount
AN/PAS-13C(V1), LWTS	TM 11-5855-316-10	Upper receiver	Upper receiver	M16 mount
AN/PAS-13C(V3), HWTS	TM 11-5855-316-10	Upper receiver	Upper receiver	M16 mount
AN/PAS-13D(V1), LWTS	TM 11-5855-317-10	Upper receiver	Upper receiver	M16 mount
AN/PAS-13D(V3), HWTS	TM 11-5855-317-10	Upper receiver	Upper receiver	M16 mount
AN/PVS-4(A)	TM 11-5855-213-10	Upper receiver	Upper receiver	M16 mount
BUIS		Upper receiver	Upper receiver	N/A
ACOG		Upper receiver	Upper receiver	M16 mount

* With a half-moon spacer installed.
** Picatinny or Insight rail grabbers may be used.
*** If used in conjunction with the CCO, the CCO will mount on the top rail of the ARS.

Table 2-7. Characteristics of various accessories.

ACCESSORY	WEIGHT	LENGTH	HEIGHT	RANGE
M68, CCO	6.2 oz	4.9 in	2.5 in	300 m
AN/PVS-14	14.8 oz	4.5 in	2.25 in	
AN/PAC-4C	5.78 oz	5.5 in	1.2 in	*600 m +
AN/PEQ-2A/B	7.5 oz	6.4 in	1.2 in	*600 m +
LWTS	4.1 lb	15.5 in	6.25 in	1,600 m +
HWTS	4.5 lb	18 in	6.25 in	1,600 m +
AN/PVS-4	3.8 lb	9.45 in	4.7 in	*600 m
ACOG	15.1 oz	5.8 in	2.17 in	800 m

*Actual range is dependent upon ambient light, NVGs, and background contrast.

M68 CLOSE COMBAT OPTIC

2-30. The M68 CCO (Figure 2-22) is a reflex (nontelescopic) sight that is designed for the "two eyes open" method of sighting, but can be shot with only one eye open. The red dot aiming point follows the horizontal and vertical movement of the firer's eye, while remaining fixed on the target.

NOTES: 1. Retighten the torque-limiting knob after firing the first three rounds to fully seat the M68.

2. No centering or focusing is required beyond 50 meters.

Figure 2-22. M68 close combat optic.

MOUNTING ON THE M16A4 RIFLE OR M4-SERIES CARBINE

2-31. The M68 mounts directly on the integrated rail on top of M16A4 rifles and M4-series carbines (Figure 2-23). The half-moon spacer (1, Figure 23) must be installed to raise the M68 above the front sightpost. The Soldier's preference dictates the notch that the M68 is mounted on. Although any notch is acceptable, testing has shown that the farther away the M68 is from the Soldier's eyes, the better his field of view.

2-32. To mount the M68 on an M16A4 rifle or M4-series carbine (Figure 2-23)—

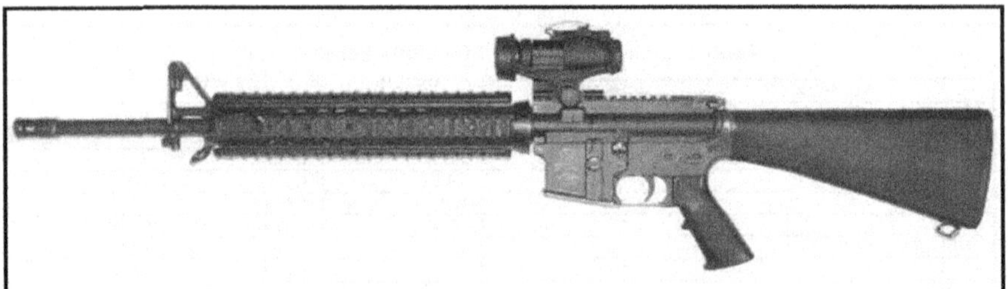

Figure 2-23. Mounting the M68 on an M16A4 rifle or M4-series carbine.

(1) Remove the carrying handle.

(2) Align the locking bar with a notch.

(3) Tighten the torque-limiting knob until it clicks twice.

NOTE: If the M68 is remounted on the same notch, it will retain zero.

MOUNTING ON THE M16A1/A2/A3 RIFLE

2-33. The M68 mounts on the M16A1 mounting bracket (1, Figure 2-24) that attaches to the carrying handle on the M16A1/A2/A3 rifle. The half-moon spacer should not be installed, but will not hinder firing performance.

2-34. To mount the M68 on an M16A1/A2/A3 rifle—

NOTE: Reference the weapon components using the numbers listed in Figure 2-24.

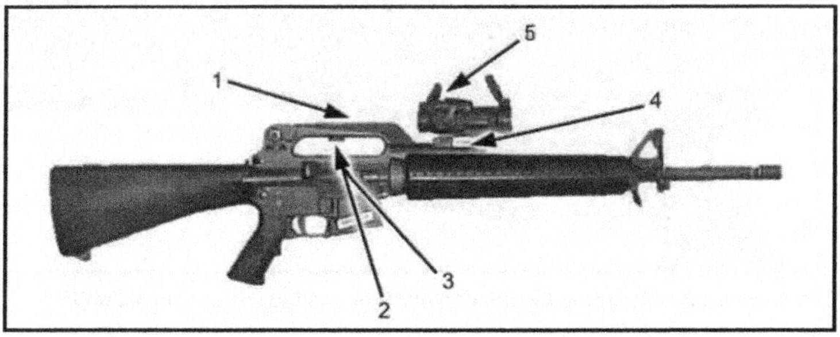

Figure 2-24. Mounting the M68 on an M16A1/A2/A3 rifle.

(1) Firmly hand-tighten the bracket (1), O-ring (2), and machine screw (3).

(2) Align the locking bar (4) under the M68 with the notch in the rail, ensuring that the rotary switch (5) is facing the firer.

(3) Tighten the torque-limiting knob (not shown here) until it clicks twice.

MOUNTING ON THE M16A4 RIFLE OR M4 MWS

2-35. This combination (Figure 2-25) is an effective means of engaging targets during hours of limited visibility. The brightness knob on the M68 should be on the lowest setting (2 or 3) that presents the red dot clearly when viewed through the AN/PVS-14. When employing the AN/PVS-14, Soldiers must consider the following factors:

- The AN/PVS-14 should be mounted where the firer can acquire a good sight picture while performing the integrated act of shooting.
- The M68 can be mounted and zeroed on any slot forward of the AN/PVS-14 as long as the rail grabber fully rests on the ARS and the M68 lens does not rest on the front sightpost.
- The closer the AN/PVS-14 is mounted to the M68, the larger the field of view will be.
- In order to get a clear sight picture with this configuration, fine adjustments must be made to the AN/PVS-14's range focus, gain-control, and diopter.
- The eyecup should be exchanged with the eye guard that is shipped with the AN/PVS-14 to reduce the light signature from the display when not viewing.
- The red dots on settings 2 and 3 project a negligible light signature at night, which can only be seen through an NVD. Settings 4 through 10 project a noticeable signature detectable by opposing forces using NVDs.
- The brighter the dot, the larger the blooming effect becomes in the AN/PVS-14. The blooming effect reduces the Soldier's field of view and prevents him from seeing targets behind the blooming.

2-36. To mount the M68 and AN/PVS-14 on an M16A4 rifle or M4 MWS—

NOTE: Reference the weapon components using the numbers listed in Figure 2-25.

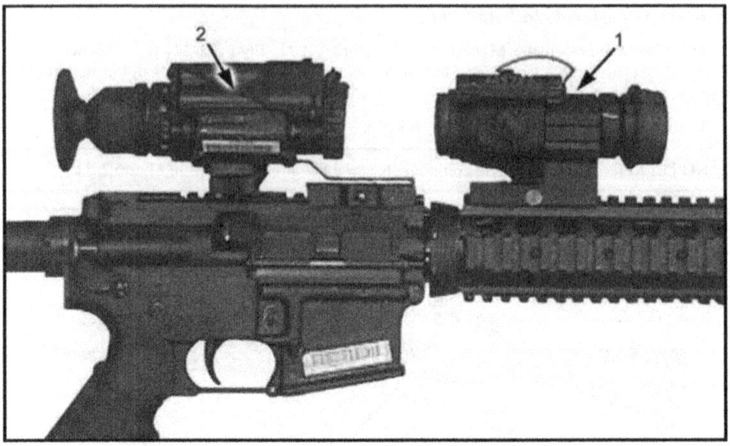

Figure 2-25. Mounting the M68/AN/PVS-14 combination on an M4 MWS.

(1) Remove the carrying handle.

(2) Mount the M68 (1) by tightening the torque-limiting knob.

(3) Mount the AN/PVS-14 (2) where the best field of view is achieved.

(4) Once the preferred location for the M68 is located, zero the M68 to that notch (if different from the notch the M68 was previously zeroed on).

NOTE: For more information about the M68, refer to TM 9-1240-413-13&P. For more information about the AN/PVS-14, refer to TM 11-5855-306-10.

AN/PAQ-4B/C INFRARED AIMING LIGHT

2-37. The AN/PAQ-4B/C IR aiming light (Figure 2-26) projects an IR laser beam that is invisible to the naked eye, but can be seen with NVDs. This aiming light works with the AN/PVS-7-series goggles and the AN/PVS-14. The AN/PAQ-4B/C mounts on various M16-/M4-series weapons with mounting brackets or rail grabbers.

Figure 2-26. AN/PAQ-4B/C infrared aiming light.

MOUNTING ON THE M16A4 RIFLE OR M4 MWS

2-38. The Picatinny (1, Figure 2-27) or Insight (5, Figure 2-27) rail grabber is used to mount the AN/PAQ-4B/C on the ARS.

2-39. To mount the AN/PAQ-4B/C on the ARS—

NOTE: Reference the weapon components using the numbers listed in Figure 2-27.

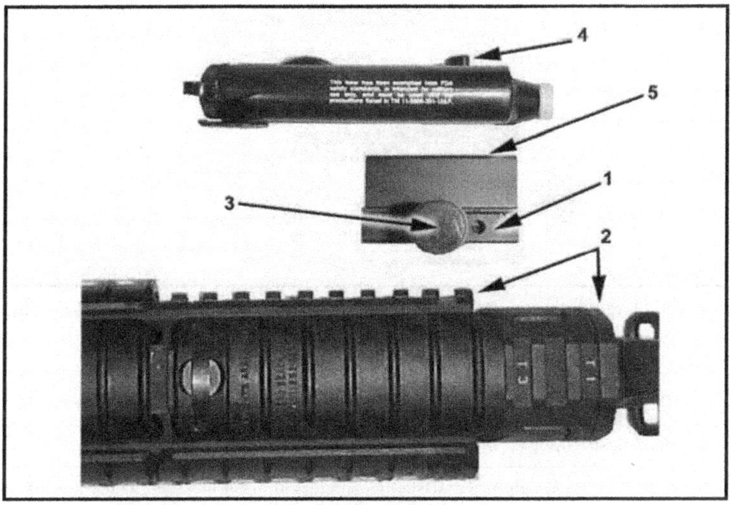

Figure 2-27. Mounting the AN/PAQ-4B/C on the M4 MWS top or left.

(1) Mount the rail grabber all the way forward on the top or either side of the ARS (2), ensuring that it does not extend beyond the end of the ARS.

NOTE: The AN/PAQ-4B/C will not retain zero if the rail grabber extends beyond the end of the integrated rail when mounted.

(2) Tighten the torque-limiting knob (3) until it clicks twice.

(3) Align the thumbscrew (4) on the AN/PAQ-4B/C with the thumbscrew hole in the rail grabber nearest the muzzle.

2-40. The mounting procedures are identical for the M16A4 rifle and M4 MWS. The remote switch should be attached to the weapon where it will not interfere with the functioning of the weapon or hinder the firer's ability to fire the weapon. If the aiming light and rail grabber are removed as a whole unit and mounted onto the same rail, the system will retain zero.

NOTE:
1. If the rail grabber and AN/PAQ-4B/C are separated, the AN/PAQ-4B/C must be rezeroed to the weapon.
2. For further information, refer to TM 11-5855-301-12&P.

MOUNTING ON THE M16A1/A2/A3 RIFLE OR M4 CARBINE

2-41. To mount the AN/PAQ-4B/C on an M16A1/A2/A3 rifle or M4 carbine—

NOTE: Reference the weapon components using the numbers listed in Figure 2-28.

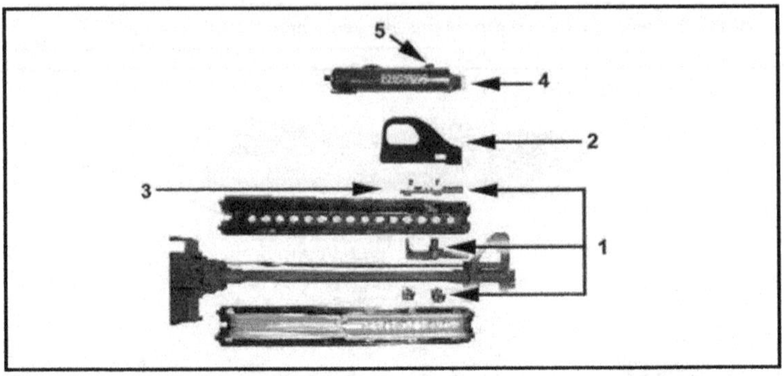

Figure 2-28. Mounting the AN/PAQ-4B/C on the M16A1/A2/A3 rifle and M4 carbine.

(1) Install the bracket assembly (1).
(2) Align the switch lever shroud (2) with the notches on the mounting rail (3).
(3) Lower the ON/OFF switch.
(4) Align the AN/PAQ-4B/C (4) with the notches on the switch lever shroud.
(5) Hand-tighten using the thumbscrew (5).

NOTE:
1. Hand-tighten the plastic thumbscrew to avoid breakage. If the thumbscrew is metal, tool-tightening is recommended to ensure zero retention.
2. To ensure zero retention, retighten the thumbscrew after firing a few rounds.

2-42. The remote switch should be attached to the weapon where it will not interfere with the functioning of the weapon or hinder the firer's ability to fire the weapon.

AN/PEQ-2A/B TARGET POINTER/ILLUMINATOR/AIMING LIGHT

2-43. AN/PEQ-2A and AN/PEQ-2B aiming lights (Figure 2-29) are Class IIIb laser devices that emit a collimated beam of IR light for precise aiming and a separate IR beam for illumination of the target or target area. Both beams can be independently zeroed to the weapon and to each other. The beams can be operated individually or in combination in both high and low power settings.

NOTES:
1. The IR illuminator is equipped with an adjustable bezel to vary the size of the illumination beam based on the size and distance of the target.
2. A safety block is provided for training purposes to limit the operator from selecting high power modes of operation.

2-44. The aiming lights are used with NVDs and can be used as handheld illuminators/ pointers or mounted on the weapon with the included brackets and accessory mounts. In the weapon-mounted mode, the aiming lights can be used to direct fire and to illuminate and designate targets.

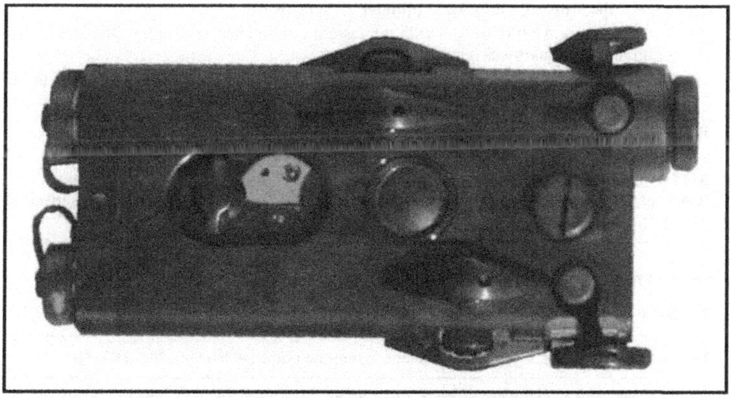

Figure 2-29. AN/PEQ-2A/B target pointer/illuminator/aiming light.

MOUNTING ON THE M16A4 RIFLE OR M4 MWS

2-45. The Picatinny (1, Figure 2-30) or Insight (2, Figure 2-30) rail grabber may be used to mount the AN/PEQ-2A/B on the ARS.

2-46. To mount the AN/PEQ-2A/B on the ARS—

NOTE: Reference the weapon components using the numbers listed in Figure 2-30.

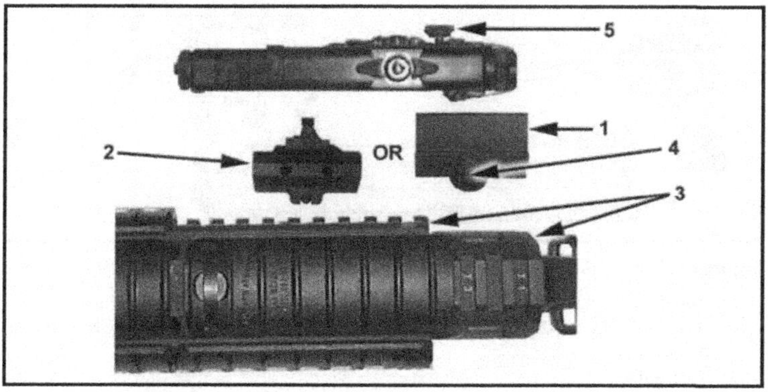

Figure 2-30. Mounting the AN/PEQ-2A/B on the M16A4 rifle and M4 MWS.

(1) Mount the rail grabber all the way forward on the top or either side of the ARS, ensuring that it does not extend beyond the end of the ARS.

NOTE: The AN/PEQ-2A/B will not retain zero if the rail grabber extends beyond the end of the integrated rail when mounted.

(2) Tighten the torque-limiting knob (4) until it clicks twice.

(3) If installing the AN/PEQ-2A/B with the Insight rail grabber, tool-tighten the AN/PEQ-2A/B and rail grabber so that it will not come loose.

(4) Align the thumbscrew (5) on the AN/PEQ-2A/B with the hole that is closest to the front sight assembly located on the top of the rail grabber.

2-47. The mounting procedures are identical for the M16A4 rifle and M4 MWS.

NOTE: If the aiming light and rail grabber are removed as a whole unit and mounted onto the same rail, the system will retain zero. If the rail grabber and AN/PEQ-2A/B are separated, the AN/PEQ-2A/B must be rezeroed to the weapon.

MOUNTING ON THE M16A1/A2/A3 RIFLE OR M4 CARBINE

2-48. To mount the AN/PEQ-2A/B on an M16A1/A2/A3 rifle or M4 carbine—

NOTE: Reference the weapon components using the numbers listed in Figure 2-31.

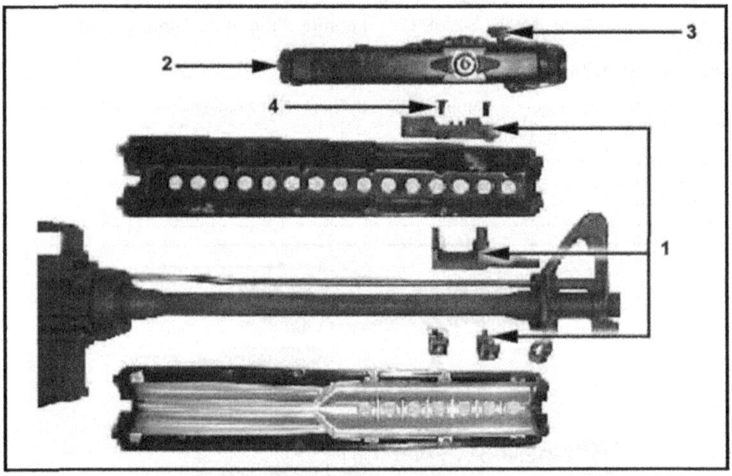

Figure 2-31. Mounting the AN/PEQ-2A/B on M16A1/A2/A3 rifles and M4 carbines.

(1) Install the bracket assembly (1).

(2) Align the AN/PEQ-2A/B (2) thumbscrew (3) with the hole in the mounting rail (4) nearest to the muzzle.

(3) Tool-tighten.

2-49. The remote switch should be attached to the weapon where it will not interfere with the functioning of the weapon or hinder the firer's ability to fire the weapon.

NOTE: To ensure zero retention, retighten the rail grabber and thumbscrew after firing a few rounds.

USE IN CONJUNCTION WITH MILES

2-50. When conducting MILES training with the Insight rail grabber (1, Figure 2-32) or bracket assembly (4, Figure 2-32), the AN/PEQ-2A/B is attached to the M16-/M4-series weapon using the training extender bracket (2, Figure 2-32). The training extender is not required when—

- Mounting the AN/PEQ-2A/B onto the side of the MWS.
- Using the Picatinny rail grabber.

2-51. To attach the training extender, hand-tighten it by turning the thumb wheel (3, Figure 2-32) on the training extender clockwise.

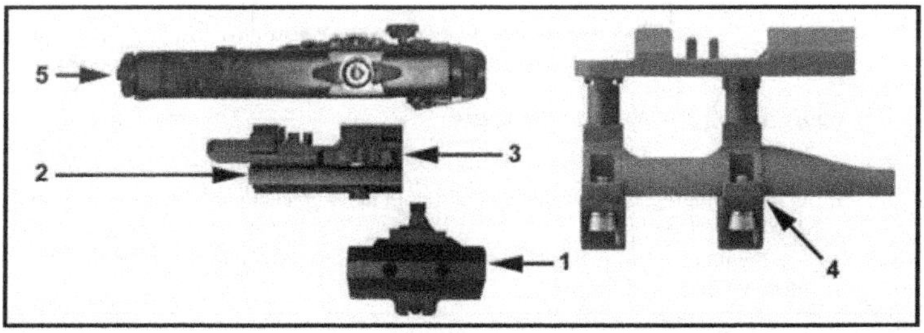

Figure 2-32. MILES training extender bracket installation on M16-/M4-series weapons.

NOTES: 1. The AN/PEQ-2A/B and AN/PAQ-4B/C must be zeroed before and after using the MILES training extender.

2. For further information, refer to TM 11-5855-308-12&P.

AN/PEQ-15 ADVANCED TARGET POINTER/ILLUMINATOR AIMING LIGHT

2-52. The AN/PEQ-15 advanced target pointer/illuminator aiming light (ATPIAL, Figure 2-33) is a multifunctional laser that emits visible or IR light for precise weapon aiming and target/area illumination. This ruggedized system can be used as a handheld illuminator/pointer or can be mounted to weapons equipped with a MIL-STD-1913 rail. Unlike the AN/PAQ-4 and AN/PEQ-2A/2B, the AN/PEQ-15 has an integrated rail grabber molded into the body.

NOTE: The AN/PEQ-15 can be used during force-on-force training in the low power modes only. High power modes can be used on live-fire ranges exceeding 220 meters only. Refer to TM 9-5855-1914-13&P for more information.

2-53. The AN/PEQ-15 ATPIAL's visible aiming laser provides for active target acquisition in low light conditions and close quarters combat situations, and allows users to zero using the borelight without using NVDs. When used in conjunction with NVDs, its IR aiming and illumination lasers provide for active, covert target acquisition in low light or complete darkness.

NOTES: 1. The ATPIAL visible and IR aiming lasers are aligned. A single set of adjusters moves both aiming beams, and the user can boresight/zero using either aiming laser.

2. In addition to momentary and continuous modes of operation, the IR illuminator may be programmed to operate at 1, 2, 4, or 8 pulses per second to eliminate confusion on the battlefield.

Figure 2-33. AN/PEQ-15 advanced target pointer illuminator aiming light.

Mounting on the M16A4 Rifle or M4 MWS

2-54. To mount the AN/PEQ-15 on an M16A4 rifle or M4 MWS (Figure 2-34)—

(1) Loosen the clamping knob on the integral rail grabber bracket until the jaws sufficiently fit over the MIL-STD-1913 rail.

(2) Position the integral rail grabber bracket on the rail, ensuring that the recoil lug is seated in the desired recoil groove of the rail.

NOTE: The ATPIAL may be positioned anywhere on the rail where it is most convenient for the operator; however, ensure that it does not extend beyond the end of the ARS or touch the collar of the barrel. The AN/PEQ-15 will not retain zero if the rail grabber extends beyond the end of the integrated rail when mounted.

---

(3) Turn the clamping knob clockwise to tighten.

(4) Tool-tighten the AN/PEQ-15 onto the MIL-STD-1913 rail to ensure that it does not come loose.

NOTE: Failure to properly secure the AN/PEQ-15 to the rail may lead to zero retention issues.

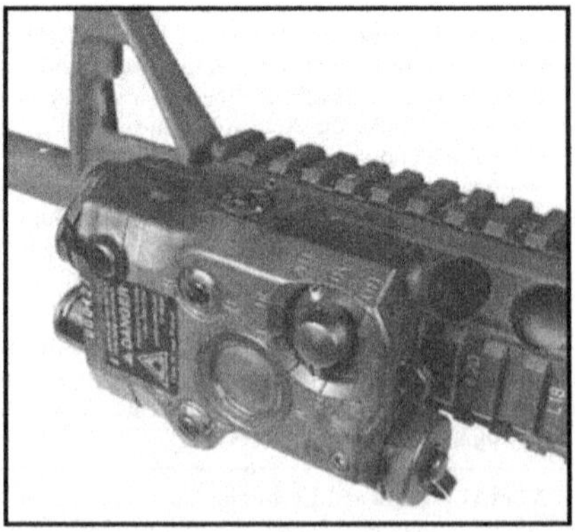

Figure 2-34. AN/PEQ-15 mounted on M4 carbine.

NOTE: If the ATPIAL is removed from the rail, the operator must note the notch that it was zeroed on, and return it to the same position to retain zero.

AN/PAS-13B/C/D (V1) LIGHT WEAPON THERMAL SIGHT AND AN/PAS-13B/C/D (V3) HEAVY WEAPON THERMAL SIGHT

2-55. The AN/PAS-13B/C/D (V1) light weapon thermal sight (LWTS) and the AN/PAS-13B/C/D (V3) heavy weapon thermal sight (HWTS) (Figure 2-35) are silent, lightweight, compact, and durable battery-powered IR imaging sensors that operate with low battery consumption. The AN/PAS-13B/C/D (V1) is used on riflemen's M4s and M16A4s. The AN/PAS-13B/C/D (V3) is used on squad leaders' M4s and M16A4s only.

NOTE: Both the LWTS and the HWTS are referred to henceforth as a singular TWS. For more information, refer to TM 11-5855-312-10, TM 11-5855-316-10, and TM 11-5855-317-10.

2-56. The TWS is capable of target acquisition under conditions of limited visibility, such as darkness, smoke, fog, dust, and haze. The TWS operates effectively both at night and during the day. The TWS is composed of two functional groups: the telescope and the basic sensor.

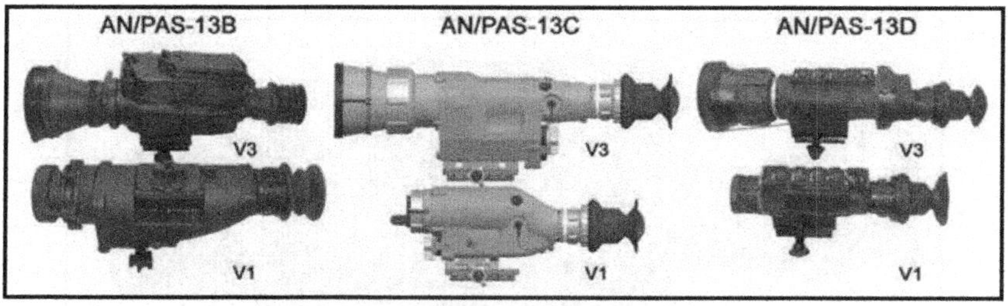

Figure 2-35. AN/PAS-13B/C/D (V1) light weapon thermal sight and AN/PAS-13B/C/D (V3) heavy weapon thermal sight.

Mounting the TWS on an M16A4 Rifle or M4 Carbine

2-57. To mount the TWS on an M16A4 rifle or M4 carbine—

NOTE: Reference the weapon components using the numbers listed in Figure 2-36.

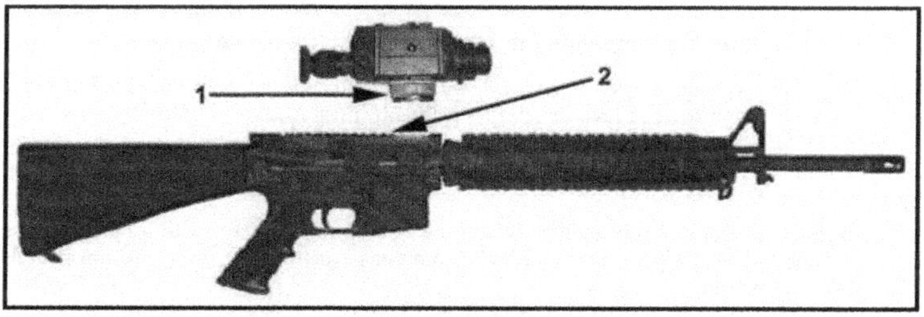

Figure 2-36. Mounting the TWS on M16A4 rifle or M4 carbine.

(1) Align the Picatinny rail grabber with the spacer on the bottom of the TWS with a notch on the integrated rail of the M16A4 rifle or M4 carbine, ensuring that the TWS is positioned to accommodate an effective firing position once the eyecup is depressed.

NOTE: The TWS will not retain zero if the rail grabber extends beyond the end of the integrated rail when mounted.

(2) Tighten the torque-limiting knob clockwise until it clicks twice.

NOTE: Retighten the rail grabber after firing a few rounds to ensure that the sight is fully seated.

2-58. The mounting procedures are identical for the M16A4 rifle and M4 MWS.

Mounting the TWS on an M16A1/A2/A3 Rifle

2-59. To mount the TWS on an M16A1/A2/A3 rifle—

NOTE: Reference the weapon components using the numbers listed in Figure 2-37.

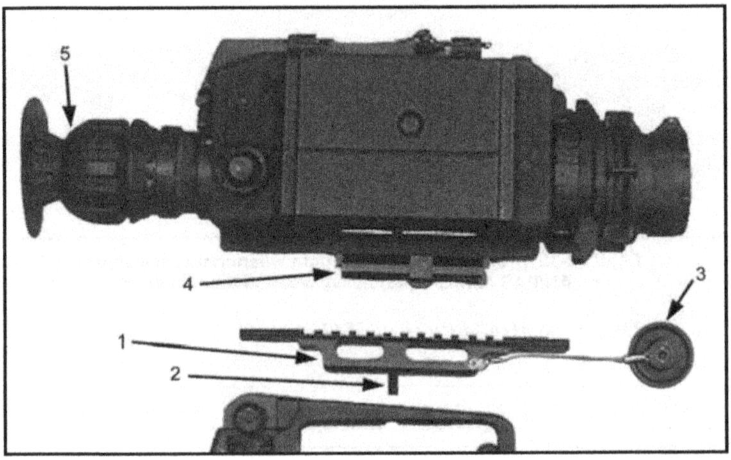

Figure 2-37. Mounting the TWS on an M16A1/A2/A3 rifle.

(1) Insert the weapon bracket's threaded rod (2) through the hole in the carrying handle of the M16A1/A2/A3 rifle, and secure it with the thumb wheel (3).

NOTE: The M16A1/A2/A3 weapon bracket (1) is a standard item in the TWS carrying case.

(2) Align the Picatinny rail grabber (4) on the bottom of the TWS with a notch on the bracket, ensuring that the TWS is positioned to accommodate an effective firing position once the eyecup (5) is depressed.

NOTES: 1. Ensure that the rail grabber fully rests on the bracket when mounting the TWS or the sight will not retain zero.

2. It is not necessary to use a spacer when mounting the HWTS on the M16A1/A2/A3.

AN/PVS-4 NIGHT VISION SIGHT

2-60. The AN/PVS-4 (Figure 2-38) is a passive, battery-operated night sight used for observation and accurate firing of individual-served weapons during nighttime conditions (for example, moonlight, starlight, skyglow) against targets as far away as 400 meters.

NOTE: The sight is less effective when viewing into rain, fog, sleet, snow, smoke, shadows, and other obscurants.

2-61. The sight is portable and mounts to the following weapons:

- M16A1/A2 and M14 rifles.
- M203 and M79 grenade launchers.
- M249 squad automatic weapon (SAW).
- M60 machine gun.
- M67 recoilless rifle.
- M72A1 rocket launcher.

Figure 2-38. AN/PVS-4 night vision sight.

Mounting on the M16A4 Rifle or M4 Carbine

2-62. To mount the AN/PVS-4 night vision sight on an M16A4 rifle or M4 carbine—

NOTE: Reference the weapon components using the numbers listed in Figure 2-39.

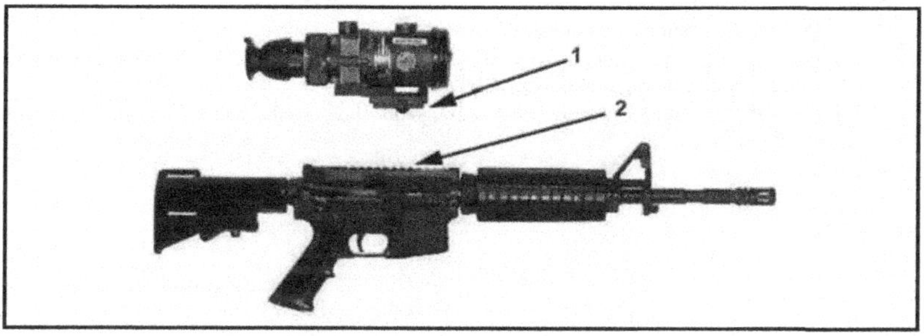

Figure 2-39. Mounting the AN/PVS-4 on an M4 carbine or M4 MWS.

(1) Align the Picatinny rail grabber with a mounting adapter (1) on the bottom of the AN/PVS-4 with a notch on the integrated rail (2) of an M16A4 rifle, M4 carbine, or M4 MWS, ensuring that the AN/PVS-4 is positioned to accommodate an effective firing position once the eyecup is depressed.

NOTE: The AN/PVS-4 will not retain zero if the rail grabber extends beyond the end of the integrated rail when mounted.

(2) Tighten the torque-limiting knob clockwise until it clicks twice.

NOTE: Retighten the rail grabber after firing a few rounds to ensure that the sight is fully seated.

2-63. The mounting procedures are identical for the M4 carbine and M4 MWS.

MOUNTING ON THE M16A1/A2/A3 RIFLE

2-64. The AN/PVS-4 is mounted on the carrying handle of the M16A1/A2/A3 rifle (Figure 2-40). To mount the AN/PVS-4 night vision sight on an M16A1/A2/A3 rifle—

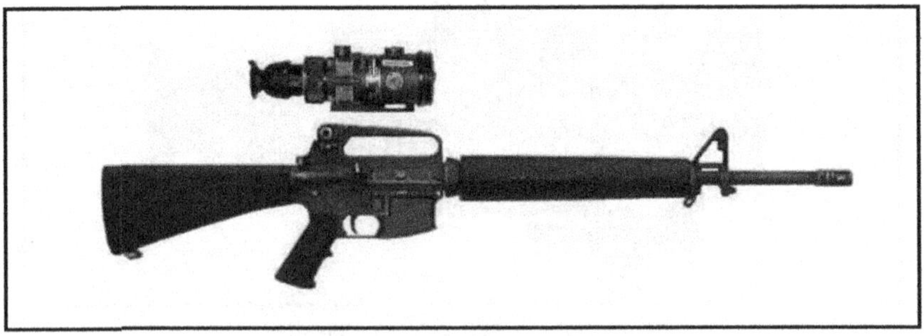

Figure 2-40. Mounting the AN/PVS-4 on an M16A1/A2/A3 rifle.

(1) Position the sight in the groove on the top of the carrying handle.

(2) Align the threaded hole in the base of the sight-mounting adapter over the hole in the handle.

(3) Insert the mounting knob assembly through the hole in the carrying handle, and screw it firmly clockwise into the sight-mounting adapter.

2-65. If difficulty is encountered—

(1) Turn the sight and the rifle upside down.

(2) Place the rifle handle onto the sight-mounting adapter, lining up the hole in the carrying handle with the hole in the sight-mounting adapter.

(3) Place the mounting knob assembly through the hole in the carrying handle, and screw it clockwise.

BACKUP IRON SIGHT

2-66. The BUIS (Figure 2-41) is a semi-permanent flip-up iron sight equipped with a rail-grabbing base. It is intended to remain on the M4 MWS while the M68 CCO is used as the primary means of day fire control. If the M68 fails, the prezeroed BUIS can be flipped up and used to continue the mission. The BUIS provides a backup capability effective when firing 600 meters away and farther and can be installed on the M16A4 rifle and M4 carbine. It provides a sighting capability when all other accessories have been removed, and it can be used to establish approximate zeros for other sighting components without requiring live-fire. Zeros established using this method are only effective to approximately 20 meters and should be refined by a live-fire zero.

2-67. The armorer installs the BUIS on the first notch of the integrated rail, nearest to the charging handle. Before the armorer installs the BUIS, he removes all rail covers/heat shields from the top, except one 4-, 5-, or 6-rib shield. The remaining rail cover/heat shield can be positioned to accommodate accessories and protect the nonfiring hand when the barrel is hot. Once installed and zeroed, the BUIS should be left in the stowed position (collapsing toward the firer, as shown in Figure 2-42) for best durability and minimal interference unless its use is imminent. The BUIS should only be removed by the armorer and remains on the MWS unless the carrying handle/sight is installed.

NOTE: Failure to install the BUIS on the first notch will lead to an improper zero and inaccuracy at longer ranges when the range lever is used.

Figure 2-41. Backup iron sight.

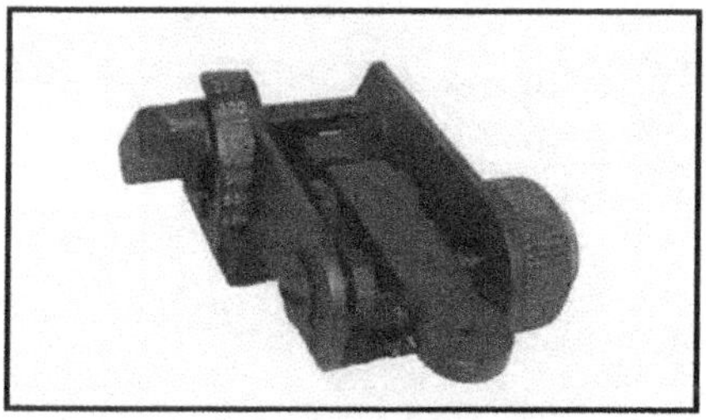

Figure 2-42. Backup iron sight in the stowed position.

ADVANCED COMBAT OPTICAL GUNSIGHT

2-68. The ACOG (Figure 2-43) is designed to provide enhanced target identification and hit probability for the M4A1 or M16-series weapon when firing as far as 800 meters (approximately 870 yards) away. It is designed with dual illuminated technology, using fiber optics for daytime employment and tritium for nighttime and low-light use. The ACOG is a lightweight, rugged, fast, and accurate 3.5 power optic scope. The body is machined from aluminum forgings; both the material and finish are identical to the M4A1. It is internally adjustable to allow the shock of rough handling to be carried by the scope body instead of the adjustment mechanism.

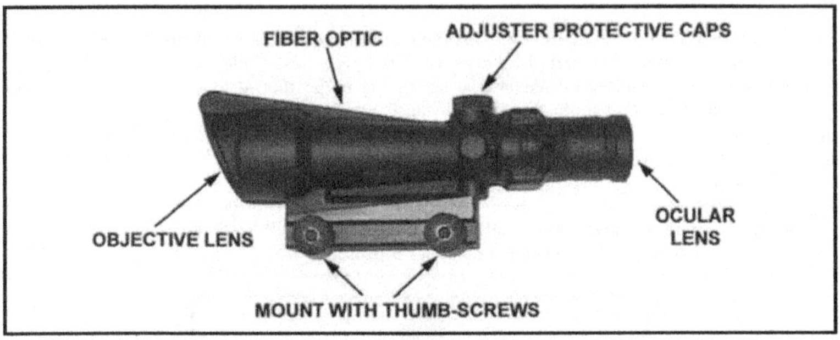

Figure 2-43. Advanced combat optical gunsight.

MOUNTING ON THE M16A4 RIFLE OR M4 CARBINE

2-69. The ACOG can be attached to the flattop easily using the adapter that comes from the factory (Figure 2-44). To attach the ACOG to the flattop with the adapter—

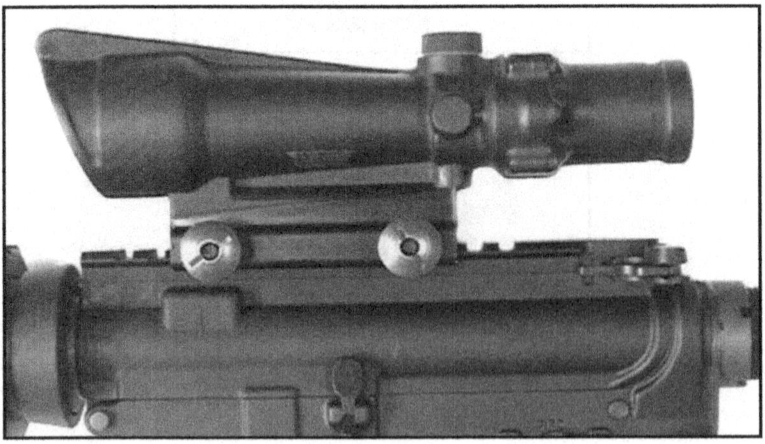

Figure 2-44. Mounting the ACOG on M16A4 rifle and M4 carbine.

(1) Loosen the thumb screws.

(2) Pull the interface clamp bar back against the knobs.

(3) Place the ACOG onto the flattop upper receiver rail surface. Be sure to engage the interface studs on the bottom of the adapter with the grooves on the top mounting surface of the upper receiver.

NOTE: 1. The ACOG can be placed in any of the slots on top of the weapon to allow for eye relief adjustments.

2. To retain zero, place the ACOG into the same slots during each installation.

(4) Using a finger, tighten the knobs firmly.

(5) Add another ¼ of a turn using a coin or a blade screwdriver.

NOTE: The slots on the knobs are only for removing the ACOG from the weapon if the knobs have been attached too tightly for hand-removal. If the slots are used to tighten the knobs, the ACOG will be difficult to remove.

MOUNTING ON THE M16A1/A2/A3 RIFLE

2-70. Although the ACOG is designed for use on the flattop rail of M16A4 rifles and M4 carbines, it can be attached to the carrying handle (Figure 2-45). To attach the ACOG to the carrying handle—

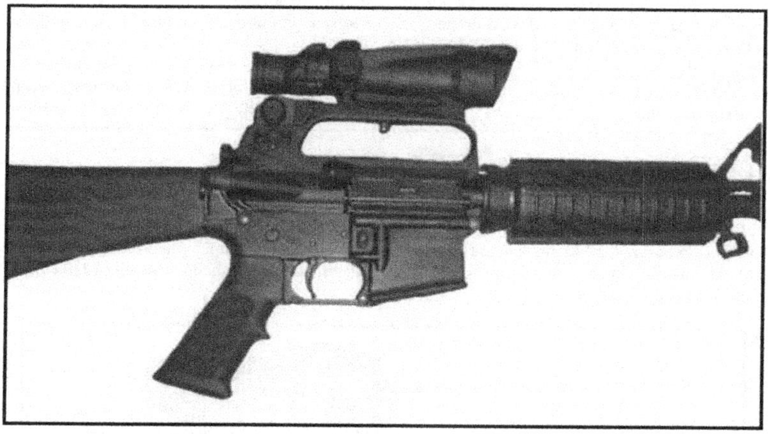

Figure 2-45. Mounting the ACOG on an M16A1/A2/A3 rifle.

(1) Align the forward-most threaded mounting hole on the bottom of the ACOG with the existing hole in the M4/M16 carrying handle.

(2) Press down firmly on the optic until it is seated inside the carrying handle rail.

(3) Insert the locking screw and washer set from the bottom of the carrying handle. Ensure that the locking screw and washer are inserted in the proper order (screw, locking nut, washer), or the screw will eventually shake itself loose. Figure 2-46 shows the proper assembly of the screw set.

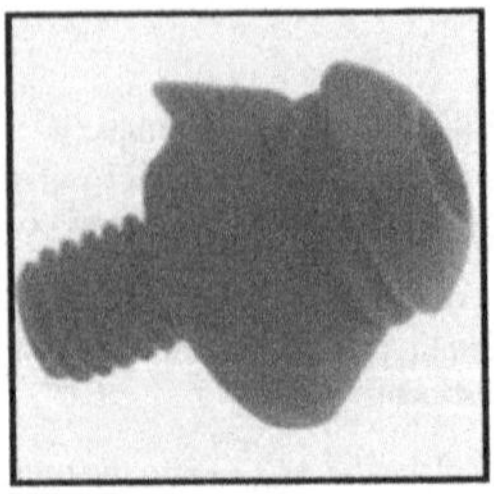

Figure 2-46. ACOG locking screw.

(4) Tighten the locking screw with the factory-supplied ⅛" Allen wrench, and then tighten another ¼ turn.

NOTE: When the ACOG is attached to the carrying handle, the bullet drop compensated (BDC) reticle will be off slightly at extended distances. For example, at 500 meters, the impact will be 5 inches high; at 600 meters, the impact will be 6 inches high. To compensate for this, aim slightly low once the target is ranged beyond 400 meters.

SECTION IV. 10-METER BORESIGHT AND 25-METER ZERO OFFSET

Boresighting is conducted at 10 meters with the borelight, weapon, aiming device, and a 10-meter offset. Each aiming device and weapon combination has a unique 10-meter offset. The 25-meter zero offset target is used when live-firing at 25 meters. When used properly, these offsets will align the aiming device on the selected weapon to engage a target's center of mass at 300 meters.

NOTE: See Appendix F for more information about 10-meter offsets for each weapon combination.

BORELIGHT

2-71. The borelight (Figure 2-47) is an eye-safe laser that is used to zero aiming lasers (such as the AN/PAQ-4 or AN/PEQ-2) without a 25-meter confirmation. The borelight will also boresight optics and iron sights to ensure that the first shot group hits the 25-meter zeroing target when zeroing the weapon. The borelight comes with a 5.56-millimeter, 7.62-millimeter, and .50-caliber mandrel. M203 and MK19 mandrels can be purchased separately.

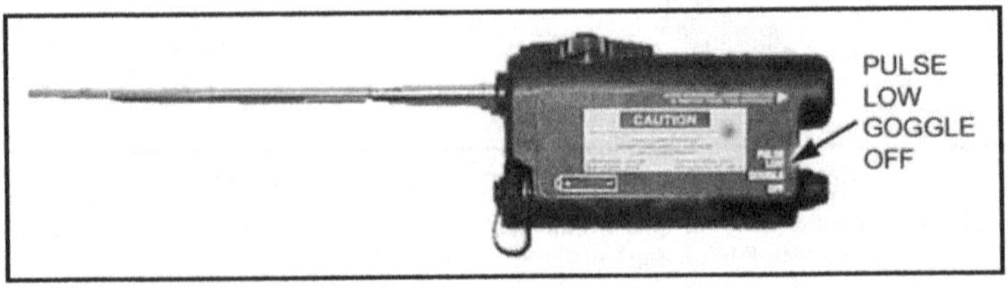

Figure 2-47. Borelight with a 5.56-millimeter mandrel.

2-72. The borelight has four settings:

- OFF.
- GOGGLE.
- LOW.
- PULSE.

OFF SETTING

2-73. This setting is used when the borelight is not in use.

GOGGLE SETTING

2-74. This setting is used when the boresight is being used with NVGs in a tactical environment.

LOW SETTING

2-75. This setting is used during normal operations.

PULSE SETTING

2-76. This setting is used during dry-fire training.

10-METER BORESIGHT

2-77. The 10-meter boresighting target is used in conjunction with the borelight. The 10-meter boresighting target is a 1-centimeter grid system with a crosshair and a circle (Figure 2-48A). The crosshair is the point of aim for the aiming device, and the circle is the point of impact for the borelight.

NOTE: Refer to Chapter 8 for a detailed explanation of boresighting procedures.

25-METER ZERO OFFSET

2-78. The M16A2 25-meter zeroing target is used when live-firing all optics and IR aiming lights at 25 meters, with the appropriate strike zone marked on the target (Figure 2-48B). The M4 zeroing target is only used when zeroing the iron sights on the M4. The M4 25-meter zeroing target will not allow the placement of the correct offset due to the size of the squares, which are 13 millimeters high and wide. The point of aim is always the 300-meter scaled silhouette's center of mass. The designated strike zone is a 4x4 square designating where the rounds should impact when the firer aims at the center of mass.

NOTE: Refer to Chapter 8 for a detailed explanation of the 25-meter offset zeroing procedures.

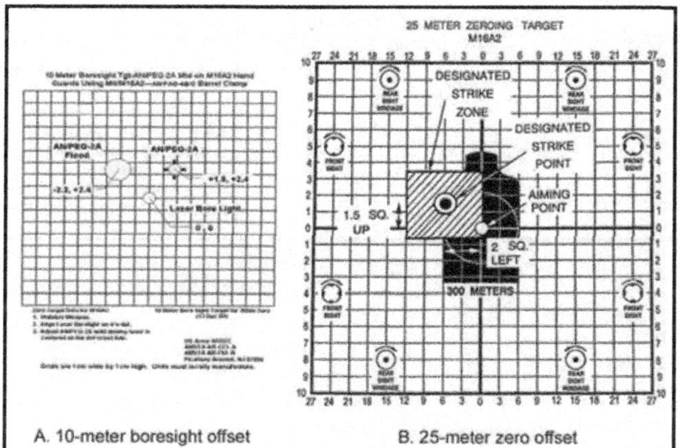

A. 10-meter boresight offset B. 25-meter zero offset

Figure 2-48. 10-meter boresighting target and 25-meter zero offset.

SECTION V. AMMUNITION

This section provides information about different types of standard military ammunition used in M16- and M4-series weapons.

AUTHORIZED AMMUNITION

2-79. Use only authorized ammunition manufactured to U.S. and NATO specifications (Table 2-8).

Table 2-8. Authorized ammunition.

CARTRIDGE/ROUND	IDENTIFICATION	USE	ADDITIONAL INFORMATION
M193 cartridge – 5.56-mm, ball 	Plain tip	The M193 is the standard cartridge for field use with the M16A1 rifle.	The M193 cartridge is a center-fire cartridge with a 55-grain, gilded metal-jacketed, lead alloy core bullet.
M196 cartridge – 5.56-mm, tracer 	Red or orange tip	The M196 cartridge is used only in the M16A1 rifle. Its main uses are for observation of fire, incendiary effect, and signaling.	Soldiers should avoid long-term use of 100 percent tracer rounds, which could cause deposits of incendiary material or chemical compounds that could damage the barrel. When tracer rounds are fired, they are mixed with ball ammunition in a ratio of no greater than one-to-one with a preferred ratio of three or four ball rounds to one tracer round.
M199 cartridge – 5.56-mm, dummy 	Six grooves along the sides of the case beginning about 1/2 inch from its tip	The M199 dummy cartridge is used in all M16-/M4-series weapons during dry-firing and other training.	This cartridge contains no propellant or primer. The primer well is open to prevent damage to the firing pin.
M200 cartridge – 5.56-mm, blank (no projectile) 	Case mouth is closed with a seven-petal rosette crimp, violet tip	The M200 blank cartridge is used in all M16-/M4-series weapons.	N/A
M855 cartridge – 5.56-mm, ball 	Green tip	The M855 cartridge is used in the M16A2/3/4 and in M4-series weapons.	The M855 cartridge has a 62-grain, gilded metal-jacketed, lead alloy core bullet with a steel penetrator. The primer and case are waterproof. This round is also linked and used in the M249. NOTE: This ammunition should not be used in the M16A1 except under emergency conditions, and only at targets less than 90 meters away. The twist of the M16A1 rifling is not sufficient to stabilize the length of the round's projectile.

Table 2-8. Authorized ammunition (continued).

CARTRIDGE/ROUND	IDENTIFICATION	USE	ADDITIONAL INFORMATION
M856 cartridge – 5.56-mm, tracer 	Red tip (orange when linked 4 to 1 for the M249)	The M856 tracer cartridge is used in the M16A2/3/4 and M4-series weapons.	The M856 tracer cartridge has characteristics similar to the M196 tracer, with a slightly longer tracer burnout distance. This cartridge has a 63.7-grain bullet. The M856 does not have a steel penetrator. NOTE: This ammunition should not be used in the M16A1 except under emergency conditions, and only at targets less than 90 meters away. The twist of the M16A1 rifling is not sufficient to stabilize the length of the round's projectile.
M862 cartridge – 5.56-mm, short-range training ammunition (SRTA) 	N/A	The M862 SRTA is used in all rifles and is designed exclusively for training.	The M862 SRTA can be used in lieu of service ammunition on indoor ranges and by units who have a limited range fan that does not allow the firing of service ammunition. If adequate range facilities are not available for sustainment training, SRTA can be used for any firing exercise of 25 meters or less. This includes the 25-meter scaled silhouette, 25-meter alternate qualification course, and quick-fire training. SRTA can also be used for urban operations (UO) training. NOTES: 1. See Appendix A for use of SRTA in training. 2. Although SRTA closely replicates the trajectory and characteristics of service ammunition out to 25 meters, the settings placed on the sights for SRTA could be different for service ammunition. SRTA should not be used to battlesight zero weapons that will fire service ammunition. 3. SRTA ammunition must be used with the M2 training bolt.
M995 cartridge – 5.56-mm, armor-piercing (AP) 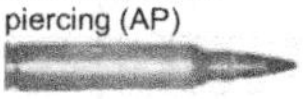	Conventional brass cartridge case Aluminum cup sits at the rear of the projectile (for the purpose of properly locating the penetrator within the projectile)	The M995 cartridge is used by the M249 (SAW), M16/A2/A3/A4, and M4-series weapons. It is intended for use against light armored targets.	The M995 offers the capability to defeat light armored targets at ranges two to three times that of currently available 5.56-mm ammunition. The M995 cartridge consists of a projectile and a propelling charge contained in a brass cartridge case. The projectile is a dense metal penetrator (tungsten carbide) enclosed by a standard gilded metal jacket. The cartridge utilizes a double base propellant. A standard rifle cartridge primer is used in the case to initiate the propelling charge.

TRAJECTORY

2-80. Figures 2-49 through 2-53 show trajectory data for M855 cartridges when fired from M16A2 rifles and M4 carbines.

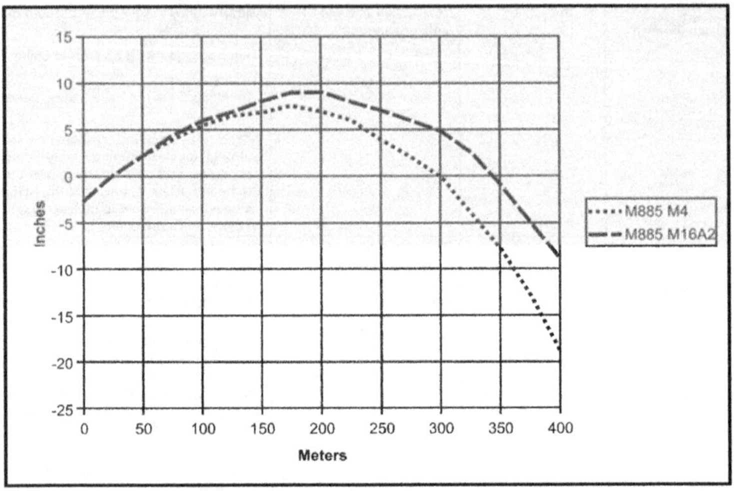

Figure 2-49. M855 drop during 25-meter zeroing (M16A2 at 8/3+1, M4 at 6/3).

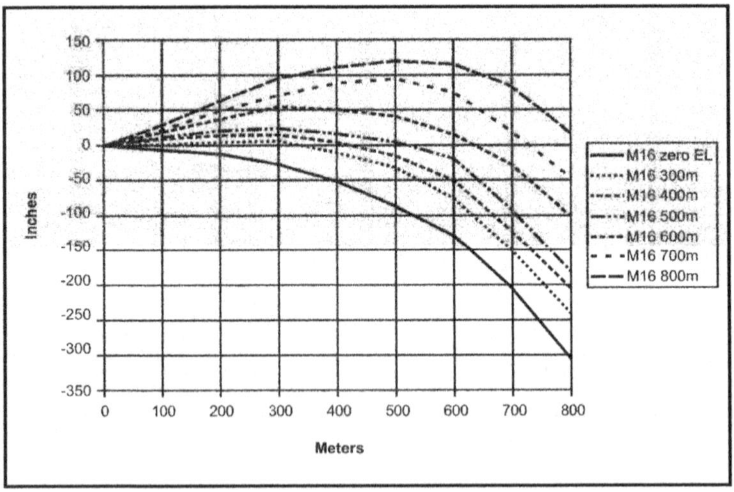

Figure 2-50. Bullet drop of M855 ammunition with M16A2 (8/3).

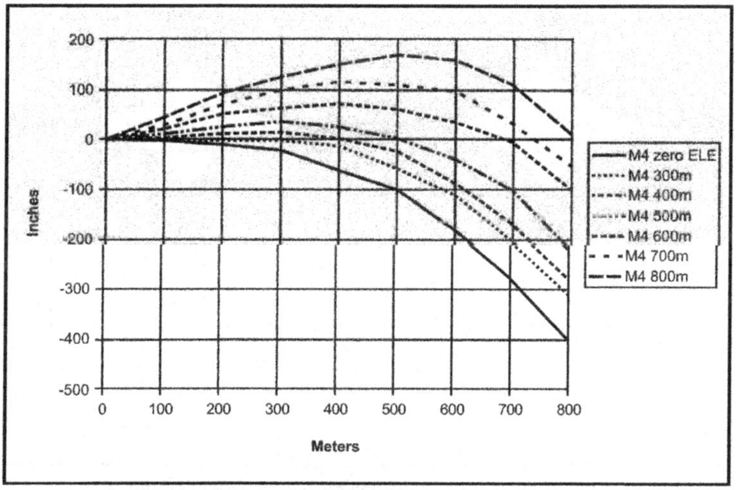

Figure 2-51. Bullet drop of M855 ammunition with M4 (6/3).

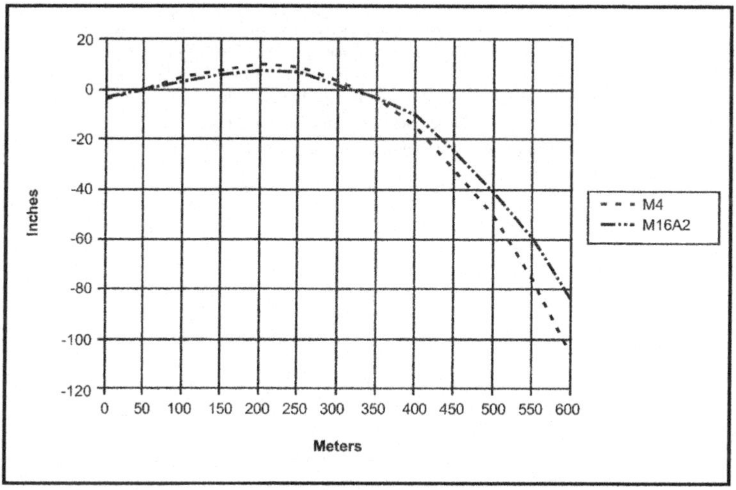

Figure 2-52. M4 carbine and M16A2 rifle bullet trajectory comparison.

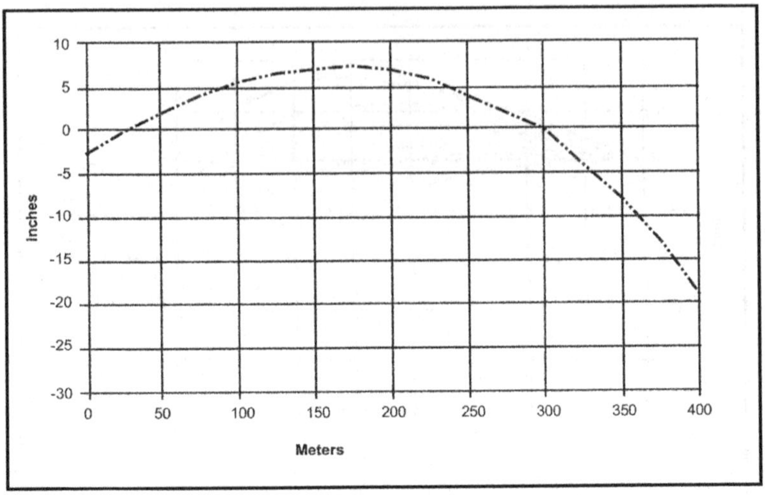

Figure 2-53. Bullet drop of M4/M855 during 25-meter zeroing on 6/3.

STORAGE

2-81. When storing ammunition in the open is necessary, it must be raised on dunnage at least 6 inches from the ground and protected with a cover, leaving enough space for air circulation. Since moisture and high temperatures adversely affect ammunition and explosives, take the following precautions:

- Do not open ammunition boxes until you are ready to use them.
- Protect ammunition from high temperatures and the direct rays of the sun.
- Do not attempt to disassemble ammunition or any of its components.
- Never use lubricants or grease on ammunition.

Chapter 3

Troubleshooting and Destruction

Commanders and unit armorers are responsible for the field level maintenance of weapons and for the destruction of weapons, when necessary. Soldiers are responsible for keeping their weapons clean and operational at all times—in training and in combat—therefore, they should be issued an operator's TM and cleaning equipment for their assigned weapons.

STOPPAGES

3-1. A stoppage is a failure of an automatic or semiautomatic firearm to complete the cycle of operation. The firer can apply immediate or remedial action to clear the stoppage. Some stoppages cannot be cleared by immediate or remedial action and may require weapon repair to correct the problem. A complete understanding of how the weapon functions is an integral part of applying immediate action procedures.

IMMEDIATE ACTION

3-2. Immediate action involves quickly applying a possible correction to reduce a stoppage without performing troubleshooting procedures to determine the actual cause. Apply immediate action only once for a stoppage. If the rifle fails to fire a second time for the same malfunction, inspect the weapon to determine the cause of the stoppage or malfunction, and take the appropriate remedial action. The key word **SPORTS** will help the firer remember the steps for immediate action:

(1) **S**lap gently upward on the magazine to ensure that it is fully seated and that the magazine follower is not jammed.

NOTE: When slapping up on the magazine, be careful not to knock a round out of the magazine into the line of the bolt carrier, causing more problems. Slap only hard enough to ensure that the magazine is fully seated. Ensure that the magazine is locked into place by quickly pulling down on the magazine.

---

(2) **P**ull the charging handle fully to the rear.

(3) **O**bserve the ejection of a live round or expended cartridge.

NOTE: If the weapon fails to eject a cartridge, perform remedial action.

---

(4) **R**elease the charging handle; do not ride it forward.

(5) **T**ap the forward assist assembly to ensure that the bolt is closed.

(6) **S**queeze the trigger and try to fire the rifle.

REMEDIAL ACTION

3-3. Remedial action is the continuing effort to determine the cause of a stoppage or malfunction and attempt to clear the stoppage once it has been identified. To apply the corrective steps for remedial action—

(1) Try to place the weapon on SAFE.

NOTE: A bolt override may not allow the weapon to be placed on SAFE.

---

(2) Remove the magazine.

(3) Lock the bolt to the rear.

(4) Place the weapon on SAFE (if not already done).

MALFUNCTIONS

3-4. Malfunctions are caused by procedural or mechanical failures of the rifle, magazine, or ammunition. Prefiring checks and serviceability inspections identify potential problems before they become malfunctions. This section describes the primary categories of malfunctions.

NOTE: In training, Soldiers must alert other Soldiers and range personnel when experiencing weapon malfunctions.

FAILURE TO FEED, CHAMBER, OR LOCK

3-5. This malfunction can occur when loading the rifle or during the cycle of operation. Once the magazine has been loaded into the rifle, the forward movement of the bolt carrier group could lack enough force (generated by the expansion of the action spring) to feed, chamber, or lock the bolt (Figure 3-1).

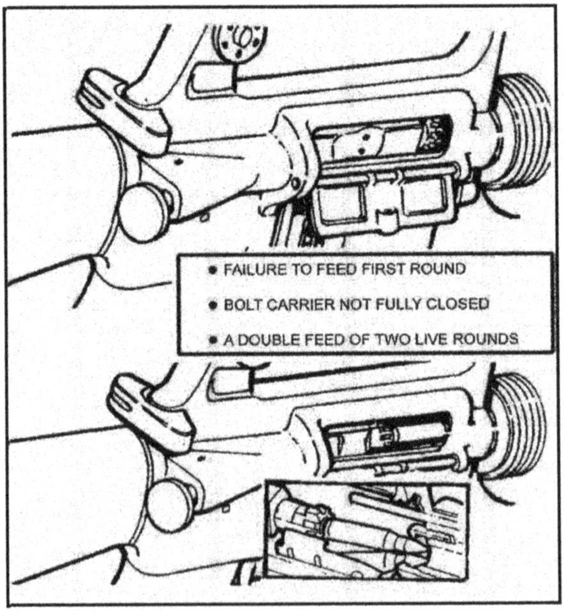

Figure 3-1. Failure to feed, chamber, or lock.

Probable Causes

3-6. The malfunction could be the result of one or more of the following:

- Excess accumulation of dirt or fouling in and around the bolt and bolt carrier.
- Defective magazine (dented, bulged, or a weak magazine spring).
- Improperly loaded magazine.
- Defective round (projectile forced back into the cartridge case, which could result in a stubbed round, or the base of the previous cartridge could be separated, leaving the remainder in the chamber).
- Damaged or broken action spring.
- Exterior accumulation of dirt in the lower receiver extension.
- Fouled gas tube (resulting in short recoil).
- A magazine resting on the ground or pushed forward (causing an improper lock).

Corrective Action

3-7. Applying immediate action usually corrects the malfunction. To avoid the risk of further jamming, watch for ejection of a cartridge and ensure that the upper receiver is free of loose rounds.

3-8. If immediate action fails to clear the malfunction, take remedial action.

NOTE: Do not force the bolt carrier.

3-9. If resistance is encountered (which can occur with an unserviceable round)—

(1) Lock the bolt to the rear.
(2) Remove the magazine.
(3) Clear the malfunction.

3-10. For example, to correct a bolt override (a cartridge has wedged itself between the bolt and charging handle)—

(1) Ensure that the charging handle is pushed forward and locked into place.
(2) Attempt to place the weapon on SAFE.
(3) Secure the rifle, and pull the bolt to the rear until the bolt seats completely into the buffer well.
(4) Turn the rifle upright and allow the overridden cartridge to fall out.

FAILURE TO FIRE CARTRIDGE

3-11. Despite the fact that a round has been chambered, the trigger has been pulled, and the sear has released the hammer, a cartridge may fail to fire. This occurs when the firing pin fails to strike the primer with enough force or when the ammunition is defective. The firer must follow unit safety guidelines until the determining factors of the misfire have been identified and corrected.

Probable Causes

3-12. Excessive carbon buildup on the firing pin (Figure 3-2, A) is often the cause, because the full forward travel of the firing pin is restricted. A defective or worn firing pin can give the same results. Inspection of the ammunition could reveal a shallow indentation or no mark on the primer, indicating a firing pin malfunction (Figure 3-2, B). Cartridges that show a normal indentation on the primer (but did not fire) indicate faulty ammunition or failure of the cartridge to fully seat in the chamber.

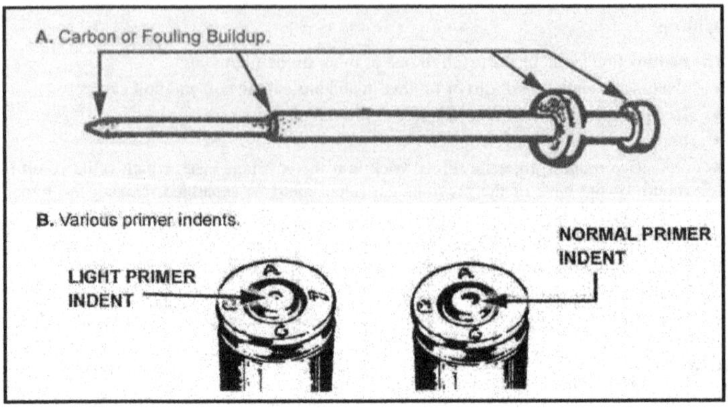

Figure 3-2. Failure to fire.

Corrective Action

3-13. If the malfunction continues—

(1) Inspect the firing pin, bolt, bolt carrier, and locking lug recesses of the barrel extension.

(2) Remove any accumulation of excessive carbon or fouling.

(3) Inspect the firing pin for damage.

3-14. If the round is suspected to be faulty, dispose of it IAW the unit SOP, and ensure that it is reported and returned to the agency responsible for issuing ammunition.

WARNING

If an audible pop or reduced recoil occurs during firing, immediately cease fire. This could be the result of a round being fired without enough force to send the projectile out of the barrel. Do not apply immediate action. Instead, perform the following actions:

1. **Lock the bolt to the rear.**
2. **Place the selector lever in the SAFE position.**
3. **Remove the magazine.**
4. **Visually inspect the bore to ensure that a projectile is not lodged in the barrel.**
5. **If a projectile is lodged in the barrel, do not try to remove it. Turn the rifle in to the unit armorer.**

Failure to Extract

3-15. A failure to extract results when the cartridge case remains in the chamber of the rifle. The bolt and bolt carrier might move rearward only a short distance, but more commonly, the bolt and bolt carrier recoil fully to the rear, leaving the cartridge case in the chamber. A live round is then forced into the base of the cartridge case as the bolt returns in the next feed cycle.

> **WARNING**
>
> **A failure to extract is an extremely serious malfunction, requiring the use of tools to clear. A live round could be left in the chamber and accidentally discharged. If a second live round is fed into the primer of the chambered live round, the rifle could explode and cause personal injury. This malfunction must be properly identified and reported.**
>
> **Ejection failures should not be reported as extraction failures.**

Probable Cause

3-16. Short recoil cycles and fouled or corroded rifle chambers are the most common causes of failures to extract. A damaged extractor or a weak or broken extractor spring can also cause this malfunction.

Corrective Action

3-17. This malfunction is one of the hardest to clear; the severity of the failure determines the corrective action procedures.

NOTE: If the bolt has moved rearward far enough to strip a live round from the magazine in its forward motion, the bolt and bolt carrier must be locked to the rear.

3-18. With the bolt locked to the rear and the weapon on SAFE, perform the following actions:

(1) Remove the magazine and all loose rounds.

(2) Tap the weapon's buttstock on a hard surface to cause the cartridge to fall out of the chamber.

3-19. However, if the cartridge case is ruptured, it can be seized. When this occurs, perform the following actions:

(1) Insert a cleaning rod into the bore from the muzzle end.

(2) Force the cartridge case from the chamber by tapping the cleaning rod against the inside base of the fired cartridge.

3-20. If cleaning and inspecting the mechanism and chamber reveals no defects but failures to extract persist, the extractor and extractor spring should be replaced. If the chamber surface is damaged, the entire barrel must be replaced.

FAILURE TO EJECT

3-21. Ejection of a cartridge is an element in the rifle's cycle of functioning, regardless of the mode of fire. This malfunction occurs when the cartridge is not ejected through the ejection port and either remains partly in the chamber or becomes jammed in the upper receiver as the bolt closes. When the firer initially clears the rifle, the cartridge could strike an inside surface of the receiver and bounce back into the path of the bolt.

Probable Cause

3-22. The cartridge must be extracted before it can eject. Failures to eject can also be caused by a buildup of carbon or fouling on the extractor or by short recoil. Short recoil is usually due to a buildup of fouling in the bolt carrier mechanism or gas tube. Resistance caused by a carbon-coated or corroded chamber can impede the extraction and ejection of a cartridge.

Corrective Action

3-23. While retraction of the charging handle usually frees the cartridge and permits removal, the charging handle must not be released until the position of the next live round is determined.

3-24. If another live round has been sufficiently stripped from the magazine or remains in the chamber, remove the magazine and all live rounds before releasing the charging handle.

3-25. If several malfunctions occur and are not corrected by cleaning and lubricating, replace the ejector spring, extractor spring, and extractor.

OTHER MALFUNCTIONS

3-26. Table 3-1 describes other malfunctions that can occur and the appropriate corrective actions.

Table 3-1. Other malfunctions.

MALFUNCTION	CORRECTIVE ACTION
The bolt fails to lock in the rearward position after the last round in the magazine is fired.	Check for a bad magazine or short recoil.
The bolt fails to lock in the rearward position when the bolt catch has been engaged.	Check the bolt catch; turn in the weapon to the unit armorer.
The weapon fires two or more rounds when the trigger is pulled and the selection lever is in the SEMI position.	This indicates a worn sear, cam, or disconnector. Turn in the weapon to the unit armorer so that he can repair and replace the trigger group parts.
The trigger fails to pull or return after release with the selector set in a firing position.	This indicates that the trigger pin has backed out of the receiver or the hammer spring is broken (A, Figure 3-3). Turn in the weapon to the unit armorer so that he can repair and replace it.
The magazine fails to lock into the magazine well.	Check the magazine and magazine catch for damage (B, Figure 3-3). Turn in to the unit armorer to adjust or replace the catch.
Any part of the bolt carrier group fails to function.	Check for incorrect assembly of components (C, Figure 3-3). Correctly clean and assemble the bolt carrier group, or replace damaged parts.
The ammunition fails to feed from the magazine.	Check for a damaged magazine (D, Figure 3-3). A damaged magazine could cause repeated feeding failures and should be turned in to the unit armorer or exchanged.

NOTE: For more information about troubleshooting malfunctions and replacing components, see organizational and direct support maintenance publications and manuals.

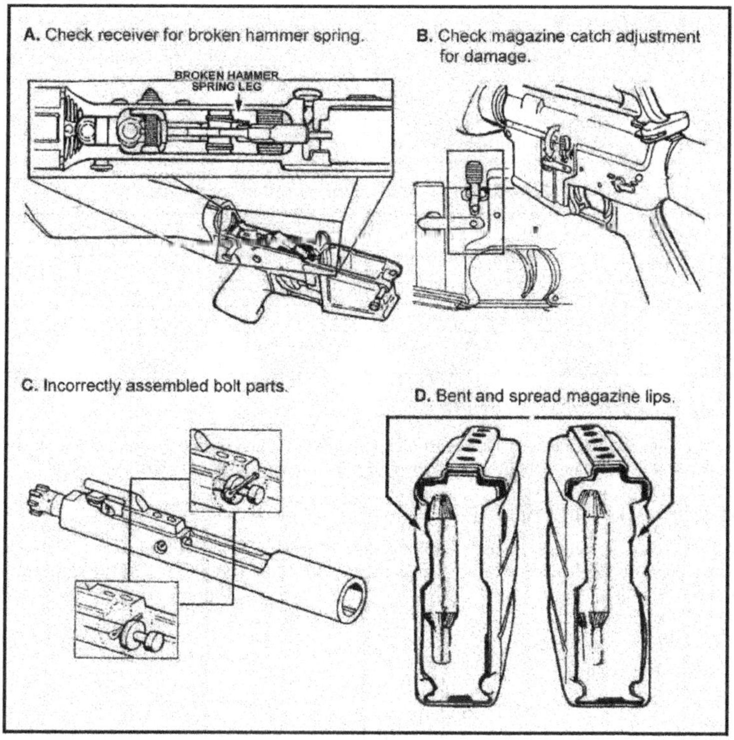

Figure 3-3. Other possible malfunctions.

DESTRUCTION PROCEDURES

3-27. Destruction of any military weapon is authorized only as a last resort to prevent the enemy from capturing or using it. In combat situations, the commander has the authority to destroy weapons, but he must report doing so through the proper channels.

NOTE: Certain procedures outlined require the use of explosives and incendiary grenades. Related principles and the specific conditions under which destruction occurs are command decisions.

METHODS OF DESTRUCTION

3-28. Equipment may be destroyed using several methods. The commander must use his imagination and resourcefulness to select the best method of destruction based on the facilities available. Time is usually critical. Table 3-2 outlines the methods of destruction.

NOTE: If destruction is directed, appropriate safety precautions must be observed.

Table 3-2. Methods of destruction and their applications.

METHOD OF DESTRUCTION	APPLICATION
Mechanical	Use an axe, pick, mattock, sledgehammer, crowbar, or other heavy implement.
Burning	Use gasoline, oil, incendiary grenades, other flammables, or a welding/cutting torch.
Demolition	Use suitable explosives, ammunition or, as a last resort, hand grenades.
Disposal	Bury essential parts, dump them in streams or marshes, or scatter them so widely that recovering them would be impossible.

DEGREE OF DAMAGE

3-29. The method of destruction used must damage equipment and essential spare parts to the extent that they cannot be restored to usable condition (by repair or by cannibalization) in the combat zone.

PRIORITIES OF DESTRUCTION

3-30. When lack of time prevents completely destroying equipment, Soldiers must destroy the same essential parts on all like equipment. The order in which the parts should be destroyed (priority of destruction) is as follows:

(1) Bolt carrier group.
(2) Upper receiver group.
(3) Lower receiver group.

Chapter 4

Preliminary Marksmanship Instruction

An Infantryman's basic battlefield tool is his weapon. To effectively employ his weapon, he must master marksmanship—from the basics of rifle marksmanship to the advanced stages of target engagement.

Understanding the operation and functions of any machine is vital to becoming an expert with that machine. The same theory applies to rifle marksmanship. Commanders must keep this in mind when setting up a training program.

SECTION I. INTRODUCTION TO BASIC RIFLE MARKSMANSHIP AND MECHANICAL TRAINING

This section covers the mechanical training of M16- and M4-series weapons. With this knowledge, a Soldier is able to assess and correct any malfunction to keep the weapon operating properly. This training program (Table 4-1) introduces Soldiers to BRM and teaches them how to maintain, operate, and correct malfunctions on M16- and M4-series weapons. It also teaches peer coaching responsibilities and sight manipulation, while emphasizing safety.

Table 4-1. Introduction to basic rifle marksmanship and mechanical training.

INTRODUCTION TO BASIC RIFLE MARKSMANSHIP AND MECHANICAL TRAINING
Period 1 (4 hours)
Instructional Intent
• Introduce the Soldiers to BRM and teach them how to maintain, operate, and correct malfunctions on a M16-/M4-series weapon. • Teach peer coaching responsibilities and sight manipulation, while emphasizing safety.
Observables
Soldiers— • Emphasize safety throughout training IAW TM 9-1005-319-10 (refer to this TM for more information). • Clear their weapons IAW this manual and TM 9-1005-319-10. • Identify all components of their weapon IAW TM 9-1005-319-10 (refer to this TM for more information). • Handle and identify 5.56-mm ammunition IAW TM 9-1005-319-10 (refer to this TM for more information). • Understand the eight cycles of function IAW this manual. • Understand the modes of fire IAW this manual. • Disassemble and assemble their weapon IAW TM 9-1005-319-10 (refer to this TM for more information). • Perform a function check on their weapon IAW TM 9-1005-319-10 (refer to this TM for more information). • Maintain, load, and unload their magazines IAW TM 9-1005-319-10 (refer to this TM for more information). • Maintain, load, unload, and clear their weapons IAW TM 9-1005-319-10 (refer to this TM for more information). • Perform SPORTS on their weapon within five seconds IAW TM 9-1005-319-10 (refer to this TM for more information). • Correctly manipulate their sights without assistance IAW TM 9-1005-319-10 (refer to this TM for more information). • Are taught peer-coaching techniques and responsibilities IAW this manual.

NOTE: Soldiers who do not meet the standard will receive remedial training before continuing with subsequent instruction.

CLEARING

> **WARNING**
>
> **To maintain safety, the weapon must be cleared before disassembling, cleaning, inspecting, transporting, or storing.**

NOTE: Additional mechanical training, including disassembly, maintenance, assembly, loading, and sight manipulation, is available in TM 9-1005-319-10.

4-1. To clear an M16-/M4-series weapon (Figure 4-1), perform the following actions:

(1) Point the muzzle in a designated safe direction or in a clearing barrel.

(2) Attempt to place the selector lever on SAFE. If the selector lever cannot be placed on SAFE because the weapon is not cocked, proceed to Step 3.

(3) Remove the magazine. To perform this procedure—

- Depress the magazine catch button.
- Pull the magazine down.

(4) Lock the bolt open. To perform this procedure—

- Pull the charging handle rearward.
- Press the bottom of the bolt catch.
- Allow the bolt to move forward until it engages the bolt catch.
- Return the charging handle to the full forward position.
- If you have not done so already, place the selector lever on SAFE.

(5) Visually inspect the receiver and chamber to ensure that these areas contain no ammunition.

(6) Allow the bolt to move forward by pressing the upper portion of the bolt catch.

(7) Place the selector lever on SEMI.

(8) Squeeze the trigger.

(9) Pull the charging handle fully rearward and release it, allowing the bolt to return to the full forward position.

(10) Place the selector lever on SAFE.

NOTE: The weapon is now clear.

(11) Close the ejection port cover.

1. Point the muzzle in a designated safe direction or in a clearing barrel.

2. Attempt to place the selector lever on **SAFE**. If the selector lever cannot be placed on SAFE because the weapon is not cocked, proceed to Step 3.

3. Remove the magazine. To perform this procedure—

a. Depress the magazine catch button.
b. Pull the magazine down.

4. Lock the bolt open. To perform this procedure—

a. Pull the charging handle rearward.
b. Press the bottom of the bolt catch.
c. Allow the bolt to move forward until it engages the bolt catch.
d. Return the charging handle to the full forward position.
e. If you have not done so already, place the selector lever on **SAFE**.

5. Visually inspect the receiver and chamber to ensure that these areas contain no ammo.

6. Allow the bolt to move forward by pressing the upper portion of the bolt catch.

7. Place the selector lever on **SEMI**. Squeeze the trigger.

8. Pull the charging handle fully rearward and release it, allowing the bolt to return to the full forward position.

9. Place the selector lever on **SAFE**. The weapon is now clear.

10. Close the ejection port cover.

Figure 4-1. Clearing.

CYCLES OF FUNCTIONING

4-2. Soldiers must understand the rifle components and the mechanical sequence of events during the firing cycle. After a loaded magazine has been inserted into the weapon, the eight cycles of functioning begin. They include—

- Feeding.
- Chambering.
- Locking.
- Firing.
- Unlocking.
- Extracting.
- Ejecting.
- Cocking.

4-3. The following paragraphs describe the actions that occur during each cycle of functioning.

FEEDING

4-4. During feeding, the following actions occur. This process is shown in Figure 4-2.

(1) As the bolt carrier group moves rearward, it engages the buffer assembly and compresses the action spring into the lower receiver extension.

(2) When the bolt carrier group clears the top of the magazine, the expansion of the magazine spring forces the follower and a new round into the path of the forward movement of the bolt.

(3) The expansion of the action spring sends the buffer assembly and bolt carrier group forward with enough force to strip a new round from the magazine.

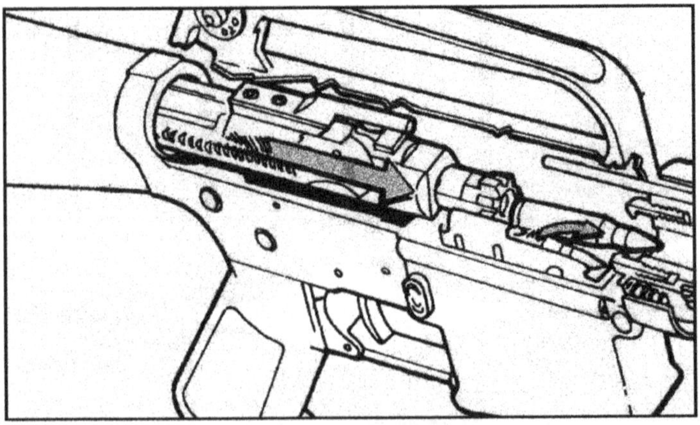

Figure 4-2. Feeding.

CHAMBERING

4-5. During chambering, the following actions occur. This process is shown in Figure 4-3.

(1) As the bolt carrier group continues to move forward, the face of the bolt thrusts the new round into the chamber.

(2) At the same time, thc extractor claw grips the rim of the cartridge and the ejector is compressed.

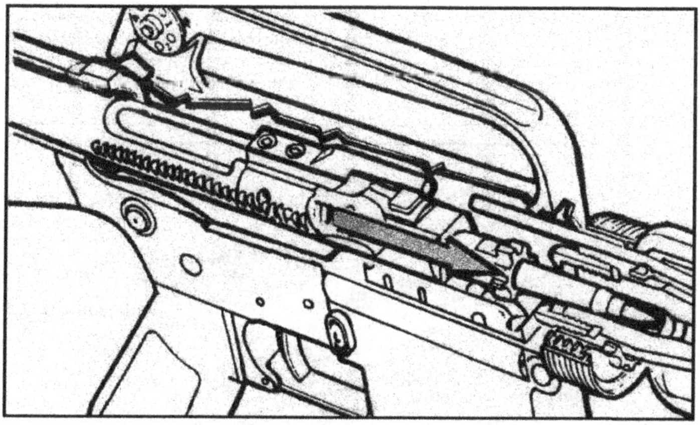

Figure 4-3. Chambering.

LOCKING

4-6. During locking, the following actions occur. This process is shown in Figure 4-4.

(1) As the bolt carrier group moves forward, the bolt cam pin riding in the guide channel in the upper receiver keeps the bolt in its most forward position.

(2) Just before the bolt locking lugs make contact with the barrel extension, the bolt cam pin emerges from the guide channel.

(3) The pressure exerted by the contact of the bolt locking lugs and barrel extension causes the bolt cam pin to move along the cam track (located in the bolt carrier) in a counterclockwise direction, rotating the bolt locking lugs in line behind the barrel extension locking lugs.

NOTE: The rifle is ready to fire.

Figure 4-4. Locking.

FIRING

4-7. During firing, the following actions occur. This process is shown in Figure 4-5.

(1) With a round in the chamber, the hammer cocked, and the selector on SEMI, the firer squeezes the trigger.

(2) The trigger rotates on the trigger pin, depressing the nose of the trigger and disengaging the notch on the bottom of the hammer.

(3) The hammer spring drives the hammer forward.

(4) The hammer strikes the head of the firing pin, driving the firing pin through the bolt and into the primer of the round.

(5) The primer ignites, causing the powder in the cartridge to ignite.

(6) The gas generated by the rapid burning of the powder forces the projectile from the cartridge and propels it through the barrel.

(7) After the projectile has passed the gas port (located on the upper surface of the barrel under the front sight, as shown in Figure 4-5) and before it leaves the barrel, gas enters the gas port and moves into the gas tube.

(8) The gas tube directs the gas into the bolt carrier. It passes down through the key and into a space between the rear of the carrier's bolt cavity and the rear of the bolt itself.

(9) Then, the gas expands.

(10) The bolt is locked into the barrel extension, unable to move forward; the carrier is forced to the rear by the expanding gas.

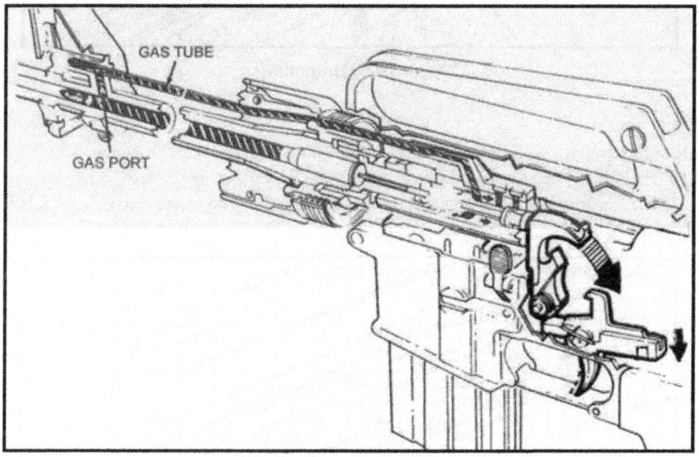

Figure 4-5. Firing.

UNLOCKING

4-8. During unlocking, the following actions occur. This process is shown in Figure 4-6.

(1) As the bolt carrier moves to the rear, the bolt cam pin follows the path of the cam track (located in the bolt carrier).

(2) The cam pin and bolt assembly rotate simultaneously until the locking lugs of the bolt are no longer in line behind the locking lugs of the barrel extension.

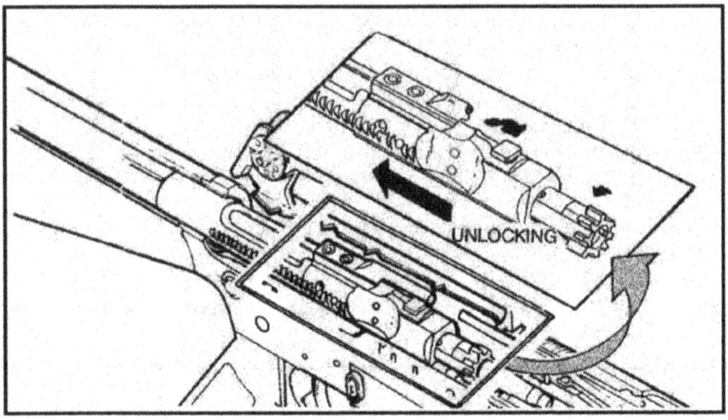

Figure 4-6. Unlocking.

EXTRACTING

4-9. During extracting, the following actions occur. This process is shown in Figure 4-7.

(1) The bolt carrier group continues to move to the rear.

(2) The extractor (which is attached to the bolt) grips the rim of the cartridge case, holds it firmly against the face of the bolt, and withdraws the cartridge case from the chamber.

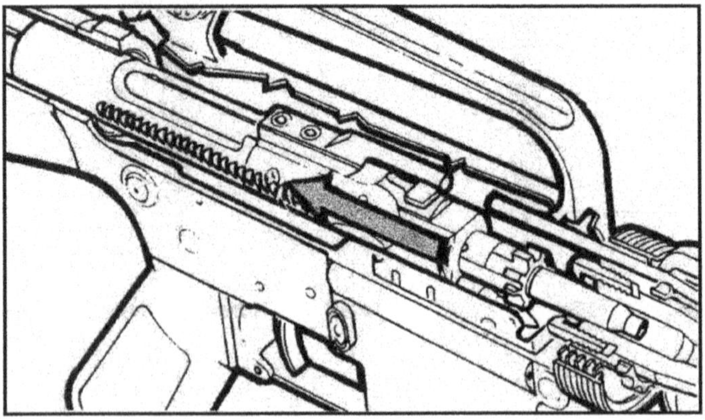

Figure 4-7. Extracting.

EJECTING

4-10. During ejecting, the following actions occur. This process is shown in Figure 4-8.

(1) With the base of a cartridge case firmly against the face of the bolt, the ejector and ejector spring are compressed into the bolt body.

(2) As the rearward movement of the bolt carrier group allows the nose of the cartridge case to clear the front of the ejection port, the cartridge is pushed out by the action of the ejector and ejector spring.

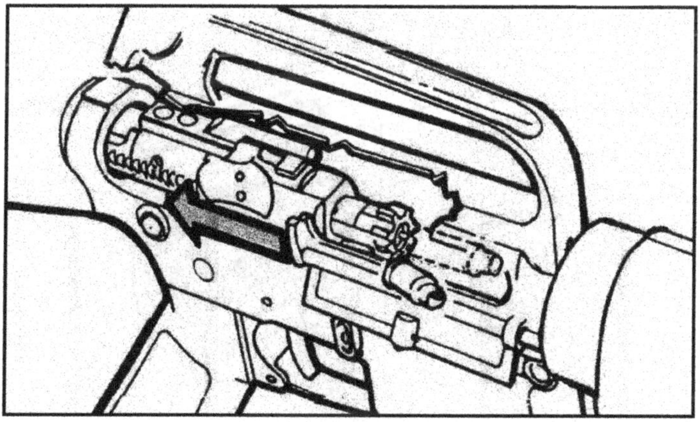

Figure 4-8. Ejecting.

Cocking

4-11. During cocking, the following actions occur. This process is shown in Figure 4-9.

(1) The rearward movement of the bolt carrier overrides the hammer.

(2) The hammer is forced down into the receiver, and the hammer spring is compressed.

NOTE: This action cocks the hammer in the firing position.

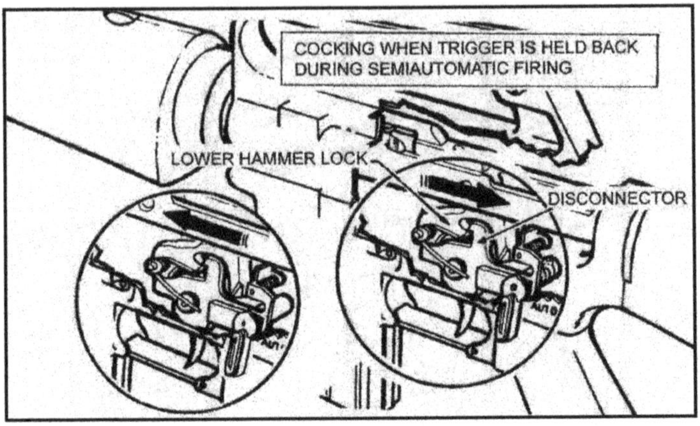

Figure 4-9. Cocking.

MODES OF FIRE

4-12. Weapons may fire using the three modes of fire:

- Semiautomatic (Figure 4-10).
- Automatic (Figure 4-10).
- Burst (Figure 4-11).

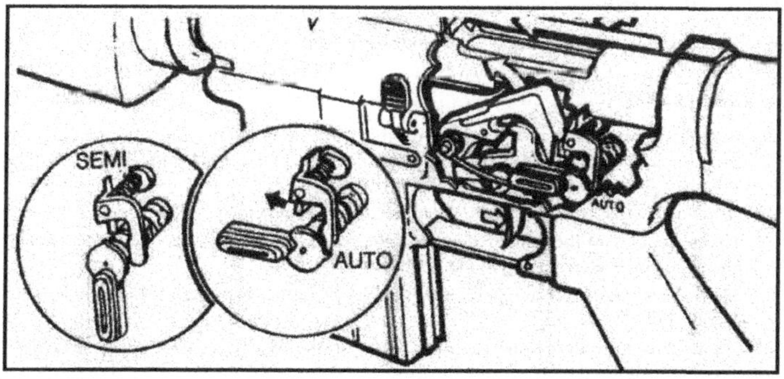

Figure 4-10. Semiautomatic and automatic fire mode selector positions.

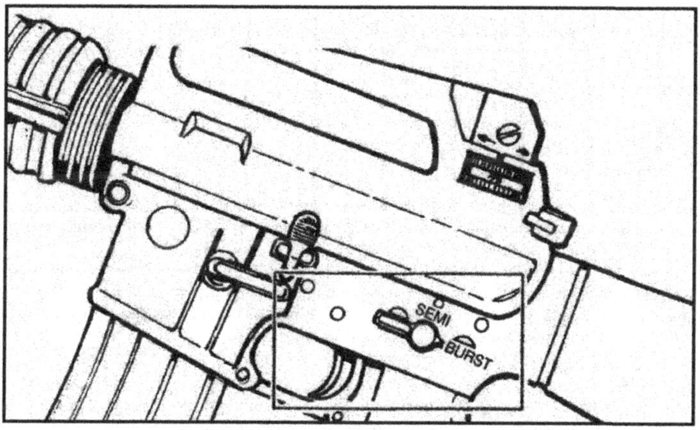

Figure 4-11. Burst fire mode selector position.

SEMIAUTOMATIC MODE OF FIRE

4-13. Weapons that function in the semiautomatic mode of fire are—

- M16A1/A2/A3/A4 rifles.
- M4/M4A1carbines.

4-14. Before a Soldier fires a weapon using the semiautomatic mode of fire, a disconnector (mechanism installed so the firer can fire single rounds) is attached to the trigger and rotated forward by the action of the disconnector spring. When the recoil of the bolt carrier cocks the hammer—

(1) The disconnector engages the lower hook of the hammer and holds it until the trigger is released.

(2) The disconnector rotates to the rear and down, disengaging the hammer and allowing it to rotate forward until caught by the nose of the trigger (this prevents the hammer from following the bolt carrier forward and causing multiple firings).

(3) The trigger must be squeezed again before the next round will fire.

AUTOMATIC MODE OF FIRE

4-15. Weapons that function in the automatic mode of fire are—

- M16A1/A3 rifles.
- M4A1 carbines.

4-16. When the selector lever is in the AUTO position, the rifle continues to fire as long as the trigger is held back and ammunition is in the magazine. The functioning of certain parts of the rifle changes when firing automatically. Once the trigger is squeezed and the round is fired—

(1) The bolt carrier group moves to the rear, and the hammer is cocked.

(2) The center cam of the selector depresses the rear of the disconnector and prevents the nose of the disconnector from engaging the lower hammer hook.

(3) The bottom part of the automatic sear catches the upper hammer hook and holds it until the bolt carrier group moves forward.

(4) The bottom part strikes the top of the sear and releases the hammer, causing the rifle to fire automatically.

(5) If the trigger is released, the hammer moves forward and is caught by the nose of the trigger.

NOTE: This ends the automatic cycle of fire until the trigger is squeezed again.

BURST MODE OF FIRE

4-17. Weapons that function in the burst mode of fire are—

- M16A2/A4 rifles.
- M4 carbines.

4-18. When the selector lever is in the BURST position, the rifle fires a three-round burst if the trigger is held to the rear during the complete cycle. The weapon continues to fire three-round bursts with each separate trigger pull as long as ammunition is in the magazine.

NOTE: Releasing the trigger or exhausting ammunition at any point in the three-round cycle interrupts fire, producing one or two shots. Reapplying the trigger only completes the interrupted cycle; it does not begin a new one. This is not a malfunction.

4-19. The M16A2/4 and M4 disconnectors have a three-cam mechanism that continuously rotates with each firing cycle. Based on the position of the disconnector cam, the first trigger pull (after initial selection of the BURST position) can produce one, two, or three firing cycles before the trigger must be pulled again. The burst cam rotates until it reaches the stop notch.

PEER COACHING

4-20. Peer coaching involves using two Soldiers of equal firing proficiency and experience to assist each other during marksmanship training.

ADVANTAGES AND DISADVANTAGES

4-21. Some problems exist with peer coaching. If the new Soldier does not have adequate guidance, a "blind leading the blind" situation may result, leading to negative training and safety violations. However, when adequate instruction is provided, peer coaching can be helpful even in the IET environment. Since all Soldiers in units have completed BRM, peer coaching within units should yield better results.

Benefits

4-22. The pairing of Soldiers can enhance learning for both participants. The coach learns what to look for as he provides guidance to the firer, and the firer has a chance to ask simple questions and to discuss areas that are not understood. Pairing Soldiers who have demonstrated good firing proficiency with those who have firing problems can improve the performance of problem firers.

Duties

4-23. The peer coach—

- Constantly checks factors that the firer is unable to observe for himself.
- Prevents the firer from repeating errors.
- Assists the firer in applying marksmanship fundamentals during firing.
- Assists the firer in obtaining a good position and in adjusting sandbags.
- Adds to range safety procedures by helping safety personnel with preliminary rifle checks.
- Watches the firer—not the target—to ensure that the firer maintains proper position, breath control, trigger pressure, and trigger squeeze.

NOTE: There are times when the peer coach may be required to observe the target area and not the firer for example, when field fire targets are being engaged and the firer cannot see where he is hitting or missing targets).

Checklist for the Coach

4-24. Coaches follow these procedures to determine and eliminate rifle and firer deficiencies:

- The coach checks to see that the—
  - Rifle is cleared, and defective parts have been replaced.
  - Ammunition is clean, and the magazine is properly placed in the pouch.
  - Sights are blackened and set correctly for long- or short-range firing.
- The coach observes the firer to see if he—
  - Uses the correct position and properly applies the steady position elements.
  - Properly loads the rifle.
  - Obtains the correct sight alignment (with the aid of an M16 sighting device).
  - Holds his breath correctly (by watching his back).
  - Applies proper trigger squeeze (determines whether he flinches or jerks by watching his head, shoulders, trigger finger, and firing hand and arm).
  - Is tense and nervous. If the firer is nervous, the coach has the firer breathe deeply several times to relax.
- Supervisory personnel and peer coaches correct errors as they are detected. If many common errors are observed, it is appropriate to call the group together for discussion, demonstration of proper procedures, and feedback.

Position of the Coach

4-25. During an exercise, the coach should be positioned where he can best observe the firer as he assumes position. Then, he moves to various points around the firer (sides and rear) to check the correctness of the firer's position. The coach requires the firer to make adjustments until the firer obtains a correct position.

4-26. When the coach is satisfied with the firing position, he assumes a coaching position alongside the firer. The coach usually assumes a position like that of the firer on the firing side of the firer (Figure 4-12).

Figure 4-12. Prone position of coach (right-handed firer).

SECTION II. MARKSMANSHIP FUNDAMENTALS I

This training program (Table 4-2) reinforces BRM and trains the four fundamentals to standard through dry-firing and simulation circuit training. It also teaches and reinforces range and safety procedures.

Table 4-2. Marksmanship Fundamentals I training program.

MARKSMANSHIP FUNDAMENTALS I
Period 2 (8 hours)
Instructional Intent
• Reinforce BRM I, and train the four fundamentals to standard with hands-on training, simulation, and dry-firing during circuit training with an M16-/M4-series weapon. • Teach and reinforce range and safety procedures.
Observables
Ensure that— • All equipment (helmet, IBA) is fitted properly to maximize training IAW the local SOP. • Live-fire range procedures are replicated and enforced IAW the local SOP. • The four fundamentals are being integrated into all exercises IAW this manual. • All dry-firing is well-aimed using 25-meter zeroing targets, EST, and LMTS. • Peer coaching is being emphasized IAW this manual.
Tasks
• The four fundamentals (IAW with this manual). • Dominant eye training (IAW with this manual). • Basic firing positions (IAW with this manual). • Range and safety procedures (IAW with the local SOP). • Demonstrate the integrated act of shooting during dry-fire exercises utilizing simulators and training devices (IAW this manual).

NOTES: 1. Simulators and training devices are listed in Appendix A of this manual.

2. Soldiers who do not meet the standard will receive remedial training before continuing with subsequent instruction.

INTERCEPTOR BODY ARMOR

4-27. BRM strategy includes the wearing of interceptor body armor (IBA, shown in Figure 4-13), if it is available—minus the throat, collar, and groin attachments—during all BRM periods and concurrent training. Whether or not IBA is worn, marksmanship fundamentals remain the same.

4-28. Prior to BRM training, use an IBA immersion approach so the Soldier can adapt to weight and movement restrictions. For an easier weight transition, incrementally introduce the outer tactical vest (OTV) and front and back small arms protective insert (SAPI) plates.

Figure 4-13. Interceptor body armor.

PROPER WEAR AND FIT

4-29. When using IBA, adhere to the following guidelines:

- Have adequate IBA quantities on hand for all Soldiers.
- Properly size IBA to the Soldier by conducting deliberate fit procedures to reduce or eliminate fit and size problems.

NOTE: Improper wear and fit of IBA impedes a Soldier's marksmanship ability.

- Ensure that the SAPI plate size corresponds to the OTV.
- Make sure that nothing else is in the OTV/SAPI compartment.

WEAR OF HELMETS WITH INTERCEPTOR BODY ARMOR

4-30. When using helmets with IBA, adhere to the following guidelines:

- When in the prone position, the IBA's back plate tends to shove the personnel armor system for ground troops (PASGT) helmet over Soldiers' eyes. To minimize the PASGT helmet positioning problem, make sure that the helmet is properly sized and fitted. Female shooters with long hair will find that wearing their hair in a bun adds material between the IBA and helmet, further forcing the helmet down over their eyes. Encourage female Soldiers to wear a short (chin length) haircut or cornrow hairstyle. If the female Soldier chooses not to wear short hair, allow her to wear her hair down when firing. Tightening the suspension harness and sweat band (raising the helmet higher on the head) can lessen interference with the IBA, hair, and helmet.
- The Army combat helmet (ACH) is lighter than the PASGT helmet, has better weight distribution, and contains less material that can impede a Soldier's firing vision when in the prone position. The ACH does not interfere with the IBA or block a Soldier's vision while in the prone position.

ADJUSTMENTS TO FIRING POSITIONS

4-31. When using IBA, adhere to the following guidelines:

- To increase comfort and stability while wearing IBA in the prone position, scoop sand or dirt underneath the chest while preparing to fire.
- To alleviate the pain and pressure on elbows and knees that the added weight of IBA causes, use elbow and knee pads. If used in the kneeling position, do not rest the elbow pad on the knee pad; hard plastic on hard plastic is not conducive to a steady position. To help with stability while firing in the kneeling position, squeeze the rifle buttstock between the SAPI plate and bicep. Loosen the firing-side straps and tighten the nonfiring-side straps to shift the SAPI plate away from the firing side.
- Instead of using load-bearing equipment (LBE), attach canteens, ammunition pouches, or first aid pouches directly to the IBA to minimize interference with LBE shoulder straps, IBA, and helmet.
- Reserve IBA firing with throat, collar, or groin protectors for ARM.
- To position themselves more comfortably and be able to reach the handguards, Soldiers of shorter stature may have to increase their body-line-to-rifle axis angle to more of an "L" shape.
- Soldiers should be in a comfortable firing position to leverage the natural point of aim. The more the target and rifle are naturally in line (as when in a relaxed position), the less movement is needed to acquire a proper sight picture.

FOUR FNDAMENTALS

4-32. Before the Soldier approaches the firing line, he must understand and apply the four fundamentals:

- Steady position.
- Aiming.
- Breath control.
- Trigger squeeze.

4-33. Soldiers apply these four fundamentals rapidly and consistently to perform the integrated act of firing. These fundamentals should be practiced while the Soldier is wearing all of his equipment, including his helmet and IBA (if available).

STEADY POSITION

4-34. When the Soldier approaches the firing line, he assumes a comfortable, steady firing position (Figure 4-14). The firer is the best judge of the quality of his position. If he can hold the front sightpost steady through the fall of the hammer, he has a good position. Steady position incorporates the following elements:

- Nonfiring hand grip.
- Rifle's buttstock position.
- Firing hand grip.
- Firing elbow placement.
- Nonfiring-side elbow placement.
- Cheek-to-stock weld.
- Support and muscle relaxation.
- Natural point of aim.

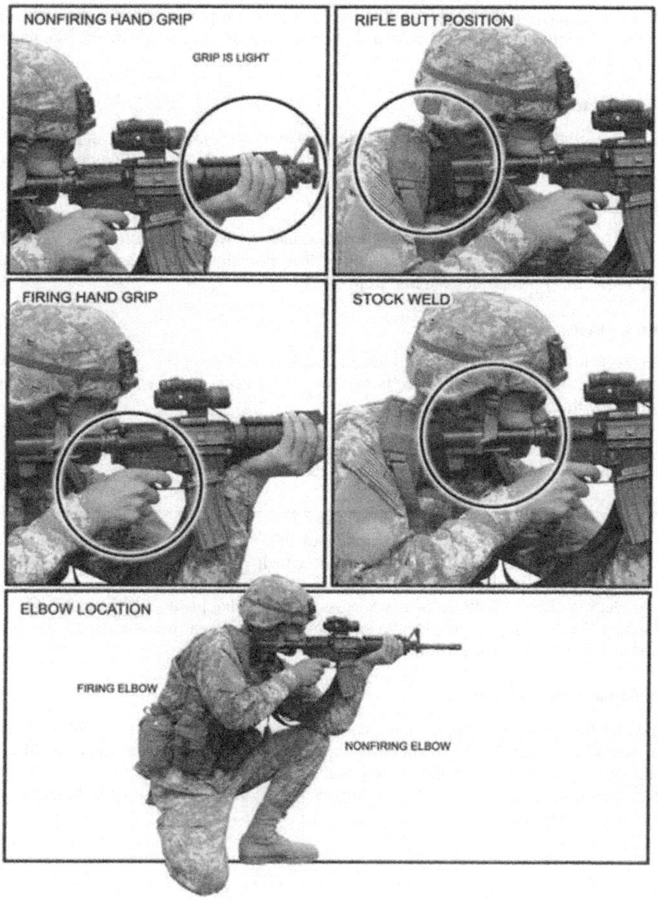

Figure 4-14. Steady position.

Nonfiring Hand Grip

4-35. The weapon's handguard rests lightly on the heel of the nonfiring hand, in the "V" formed by the thumb and fingers.

Rifle's Buttstock Position

4-36. Place the weapon's buttstock into the pocket of the firing shoulder. When wearing IBA, place the weapon's buttstock where the pocket should be; this reduces the effect of recoil and ensures a steady position.

Firing Hand Grip

4-37. The firing hand grasps the pistol grip so that it fits in the "V" formed by the thumb and forefinger. The forefinger is placed on the trigger so that the lay of the weapon is not disturbed when the trigger is squeezed. The remaining three fingers exert a slight rearward pressure to ensure that the buttstock remains in the pocket of the shoulder.

Firing Elbow Placement

4-38. The firing elbow is important in providing balance. Its exact location depends on the firing or fighting position used. Placement of the firing elbow should allow the firer's shoulders to remain level.

Nonfiring-Side Elbow Placement

4-39. The nonfiring-side elbow is positioned firmly under the weapon to allow a comfortable and stable position. When the Soldier engages a wide sector of fire, moving targets, and targets at various elevations, his nonfiring-side elbow should remain free from support.

Cheek-to-Stock Weld

4-40. The cheek-to-stock weld should provide a natural line of sight through the center of the rear sight aperture to the front sightpost and onto the target. The firer's neck should be relaxed, allowing his cheek to fall naturally onto the stock.

NOTE: Proper eye relief is obtained when a Soldier establishes a good cheek-to-stock weld. A small change in eye relief normally occurs each time that the firer assumes a different firing position.

4-41. Through dry-fire training, the Soldier practices this position until he assumes the same cheek-to-stock weld each time he assumes a given position, which provides consistency in aiming. To learn to maintain the same cheek-to-stock weld each time the weapon is aimed, the Soldier should begin by trying to touch the charging handle with his nose when assuming a firing position. The Soldier should be mindful of how the nose touches the charging handle and should be consistent when doing so. This position should be critiqued and reinforced during dry-fire training.

Support and Muscle Relaxation

4-42. When artificial support (for example, sandbags, logs, or stumps) is available, it should be used to steady the position and support the weapon. If support is used properly, the Soldier should be able to relax most of his muscles. If artificial support is not available, the bones—not the muscles—in the firer's upper body must support the weapon. Using muscles to support the rifle can cause muscle fatigue, which in turn, causes the weapon to move.

Natural Point of Aim

4-43. When the Soldier first assumes his firing position, he orients his weapon in the general direction of his target. Then, he adjusts his body to align the weapon and sights with the desired point of aim. When using proper support and consistent cheek-to-stock weld, the Soldier should have his weapon and sights naturally aligned on the target.

4-44. If correct body-rifle-target alignment cannot be achieved, the front sightpost must be held on the target using muscular support and effort. As the weapon fires, muscles tend to relax, causing the front sight to move away from the target, toward the natural point of aim. Adjusting this natural point of aim to the target eliminates this movement. When multiple target exposures are expected or a sector of fire must be covered, the Soldier adjusts his natural point of aim to the center of the expected target exposure area or sector.

AIMING

4-45. Having mastered the task of holding the rifle steady, the Soldier must align the rifle with the target in exactly the same way for each firing. The firer is the final judge as to where his eye is focused. The instructor or trainer emphasizes this point by having the firer focus on the target and then on the front sightpost. He checks the position of the firing eye to ensure that it is in line with the rear sight aperture. The elements of aiming training are as follows:

- Sight alignment.
- Focus of the eye.
- Sight picture.
- Front sightpost.
- Aiming practice.

Sight Alignment

4-46. The weapon must be aligned with the target; to do so, Soldiers place the tip of the front sightpost in the center of the rear sight aperture (Figure 4-15). Any alignment error between the front and rear sights repeats itself for every ½ meter the bullet travels. For example, at the 25-meter line, any error in rifle alignment is multiplied 50 times. If the bullet is misaligned by 1/10 of an inch, it causes a target 300 meters away to be missed by 5 feet.

Figure 4-15. Correct sight alignment.

Focus of the Eye

4-47. A proper firing position aligns the eye with the center of the rear sight aperture. When the eye is focused on the front sightpost, the eye's natural ability to center objects in a circle and to seek the point of greatest light (center of the aperture) aid in providing correct sight alignment. For the average Soldier firing at combat-type targets, the eye's natural ability can accurately align the sights. Therefore, the firer can place the tip of the front sightpost on the point of aim, but the eye must be focused on the tip of the front sightpost. This causes the target to appear blurry, while the front sightpost is seen clearly. Two reasons for focusing on the front sightpost are:

(1) Only a minor aiming error should occur, since the error reflects only as much as the Soldier fails to determine the target's center. A greater aiming error can result if the front sightpost is blurry due to focusing on the target or other objects.

(2) Focusing on the tip of the front sightpost aids the firer in maintaining proper sight alignment.

Sight Picture

4-48. Once the Soldier can correctly align his sights, he can obtain a correct sight picture. A correct sight picture has the target, front sightpost, and rear sightpost aligned. The sight picture includes two basic elements: sight alignment and placement of the point of aim. Placement of the point of aim varies, depending on the engagement range. For example, Figure 4-16 shows a silhouette at 300 meters where the point of aim is the center of mass and the sights are aligned for a correct sight picture.

Figure 4-16. Focus of the eye and correct sight picture.

4-49. The side aiming technique can be used to obtain a correct sight picture (Figure 4-17). It involves positioning the front sightpost to the side of the target in line with the vertical center of mass, keeping the sights aligned. The front sightpost is moved horizontally until the target is directly centered on the front sightpost.

Figure 4-17. Side aiming technique.

Front Sightpost

4-50. The front sightpost is vital to proper firing and should be replaced when damaged. The post should be blackened; when it is shiny, the firer cannot focus precisely on the tip of the front sightpost.

Aiming Practice

4-51. Aiming practice is conducted before firing live rounds. During dry-firing, the Soldier should practice sight alignment and placement of the point of aim. Training aids, such as the M15A1 aiming card, can be used to do this.

BREATH CONTROL

4-52. While sighted on a target, the firer must be aware of the rifle's movement as a result of breathing. Two breath control techniques are practiced during dry-fire:

- Breath control for engaging single targets.
- Breath control for engaging short-exposure targets.

4-53. As the firer's skills improve and as timed or multiple targets are presented, he must learn to control his breath at any part of the breathing cycle. The coach/trainer ensures that the firer uses both breathing techniques and understands them by instructing him to exaggerate his breathing.

Breath Control for Engaging Single Targets

4-54. When zeroing or when time is available to fire a shot, Soldiers fire when there is a natural respiratory pause, when most of the air has been exhaled from the lungs and before inhaling (Figure 4-18). The shot must be fired before the Soldier feels any discomfort.

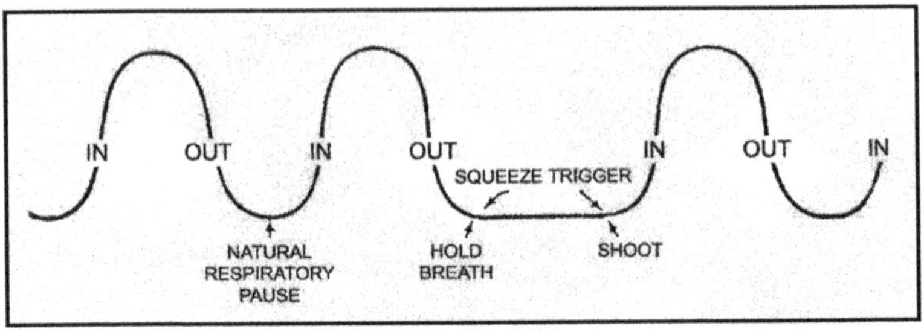

Figure 4-18. Breath control for engaging single targets.

Breath Control for Engaging Short-Exposure Targets

4-55. When employing rapid fire (engaging short-exposure targets), Soldiers stop their breath when they are about to squeeze the trigger (Figure 4-19).

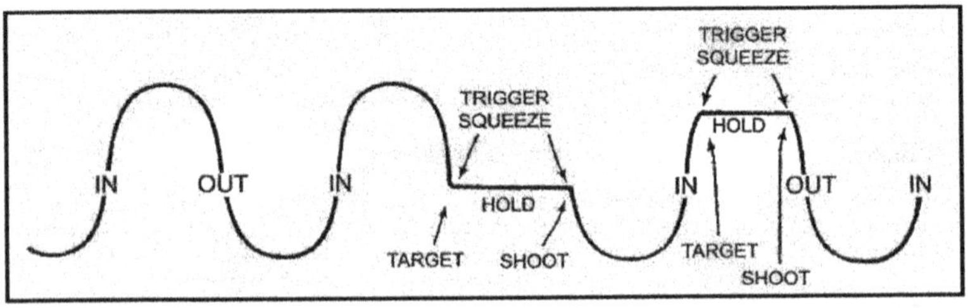

Figure 4-19. Breath control for engaging short-exposure targets.

Trigger Squeeze

4-56. A steady position reduces disturbance of the rifle during trigger squeeze. If the trigger is not properly squeezed, the rifle will be misaligned with the target at the moment of firing. The elements of trigger squeeze training are as follows:

- Rifle movement.
- Trigger finger.
- Trigger squeeze time.
- Coaching trigger squeeze.
- Wobble area.

Rifle Movement

4-57. Trigger squeeze is important for two reasons:

- Any sudden movement of the finger on the trigger can disturb the lay of the rifle and cause the shot to miss the target.
- The precise instant of firing should be a surprise to the Soldier. If a Soldier knows the exact instant that the rifle will fire, the Soldier will naturally compensate for the weapon's noise and recoil, causing him to miss the target. Soldiers usually tense their shoulders when expecting the rifle to fire; it is difficult to detect since the Soldier does not realize that he is flinching.

Trigger Finger

4-58. The Soldier places his trigger finger (index finger on the firing hand) on the trigger between the first joint and the tip of the finger—not the very end of the finger—and adjusts depending on his hand size and grip. The trigger finger must squeeze the trigger to the rear so the hammer falls without disturbing the lay of the rifle.

4-59. When a live round is fired, it is difficult to see the effect that the trigger pull had on the lay of the rifle. It is important to experiment with many finger positions during dry-fire training to ensure that the hammer is falling with little disturbance to the aiming process.

Trigger Squeeze Time

4-60. The proper trigger squeeze should start with slight pressure on the trigger during the initial aiming process. The firer applies more pressure after the front sightpost is steady on the target and he is holding his breath.

4-61. As the firer's skills increase with practice, he needs less time spent on trigger squeeze. A novice firer can take five seconds to perform an adequate trigger squeeze, but as skills improve, he can squeeze the trigger in a second or less.

Coaching Trigger Squeeze

4-62. The coach/trainer—

- Observes the trigger squeeze, emphasizes the correct procedure, and checks the firer's applied pressure.
- Places his finger on the trigger and has the firer squeeze the trigger by applying pressure to his finger.
- Ensures that the firer squeezes straight to the rear on the trigger, avoiding a left or right twisting movement.
- Observes that the firer follows through and holds the trigger to the rear for approximately one second after the round has been fired.

Wobble Area

4-63. Wobble area is the movement of the front sight around the point of aim when the rifle is in the steadiest position.

4-64. The position must provide for the smallest possible wobble area.

- From a supported position, there should be minimal wobble area and little reason to detect movement. If movement of the rifle causes the front sight to leave the target, more practice is needed.
- From an unsupported position, the firer experiences a greater wobble area than from a supported position. If the front sight strays from the target during the firing process, the firer should hold constant pressure on the trigger and resume as soon as he corrects the sighting.

NOTE: The firer should never try to quickly squeeze the trigger while the sight is on the target. The best firing performance results when the trigger is squeezed continuously and the rifle is fired without disturbing its lay.

DOMINANT EYE TRAINING

4-65. This exercise assists the coach and the firer in determining which eye the firer should use when engaging targets. The firer's dominant eye should be identified early in the training process to prevent unnecessary problems, such as a blurred sight picture or the inability to acquire a tight shot group during the grouping exercise. To perform dominant eye training—

(1) The trainer cuts a 1-inch circular hole in the center of an 8- by 10-inch piece of material (can be anything from paper to plywood).

(2) The trainer positions himself approximately 5 feet in front of the Soldier. The trainer closes his nondominant eye and holds his finger in front of and just below his dominant eye to provide the Soldier with a point of aim.

(3) The Soldier holds the training aid with both hands at waist level and looks with both eyes open at the trainer's open eye. With both eyes focused on the trainer's open eye and arms fully extended, the Soldier raises the training aid between himself and the trainer while continuing to look at the trainer's eye through the hole in the training aid. The Soldier's eye that the trainer sees through the hole in the training aid is the Soldier's dominant eye.

BASIC FIRING POSITIONS

4-66. Basic firing positions are taught during PMI. Other advanced firing positions are added later in training to support tactical conditions. The firing positions used during initial training are—

- Individual foxhole supported.

NOTE: Once the individual foxhole supported fighting position has been mastered, the firer should practice various unsupported positions to obtain the smallest possible wobble area during final aiming and hammer fall. The coach/trainer can check the steadiness of the position by observing movement at the forward part of the rifle, by looking through the Ml6 sighting device, or by checking to see if support is being used. The objective is to establish a steady position under various conditions.

- Prone unsupported firing.
- Prone supported firing.
- Kneeling unsupported.

4-67. These firing positions offer a stable platform for firing the rifle. These positions are used during basic record fire.

INDIVIDUAL FOXHOLE SUPPORTED FIRING POSITION

4-68. This position provides the most stable platform for engaging targets. To assume the individual foxhole supported firing position (Figure 4-20)—

(1) Add or remove dirt, sandbags, or other supports to adjust for your height.
(2) Face the target.
(3) Execute a half-face to the firing side.
(4) Lean forward until the chest is against the firing hand corner of the position.
(5) Place the rifle handguard in the "V" formed by the thumb and fingers of the nonfiring hand.
(6) Rest the nonfiring hand on the material (sandbags or berm) to the front of the position.
(7) Place the weapon's buttstock into the pocket of the firing shoulder.
(8) Rest the firing elbow on the ground outside of the position.

NOTES: 1. When prepared positions are not available, the prone supported position can be substituted.

2. The objective is to establish a steady position under various conditions. The ultimate performance of this task is combat. Although the firer must be positioned high enough to observe all targets, he must remain as low as possible to provide added protection from enemy fire.

Figure 4-20. Individual foxhole supported firing position.

PRONE UNSUPPORTED FIRING POSITION

4-69. This firing position offers another stable firing platform for engaging targets. To assume the prone unsupported firing position—

(1) Face the target.
(2) Spread the feet a comfortable distance apart.

(3) Drop to the knees, breaking the fall with the weapon's buttstock.
(4) Using the rifle's buttstock as a pivot, roll onto the nonfiring side, placing the nonfiring-side elbow close to the side of the magazine.
(5) For the basic prone unsupported position (Figure 4-21), spread the legs apart, with the inside of the feet flat on the ground. For the alternate prone unsupported firing position (Figure 4-22), bend the firing leg to relieve pressure on the lower back.
(6) Place the weapon's buttstock between the SAPI plate and bicep to stabilize the weapon and absorb recoil.
(7) Grasp the pistol grip with the firing hand.
(8) Lower the firing elbow to the ground.
(9) Place both elbows on the ground to support the upper body.
(10) Rest the rifle in the "V" formed by the thumb and fingers of the nonfiring hand.
(11) Adjust the position of the firing elbow until the shoulders are approximately level.
(12) Pull back firmly on the rifle with both hands.
(13) Obtain a stock weld and relax, keeping the heels close to the ground.

NOTE: To increase comfort and stability while wearing IBA in the prone position, sand or dirt should be scooped underneath the chest while preparing to fire. Elbow and knee pads can be worn to relieve IBA-induced pain and pressure.

Figure 4-21. Basic prone unsupported firing position.

Figure 4-22. Alternate prone unsupported firing position.

PRONE SUPPORTED FIRING POSITION

4-70. To assume the prone supported firing position—

(1) Face the target and drop to the ground, breaking the fall with the weapon's buttstock.

(2) For the basic prone supported firing position (Figure 4-23), spread the legs apart, with the inside of the feet flat on the ground. For the alternate prone supported firing position (Figure 4-24), bend the firing leg to relieve pressure on the lower back.

(3) Use sandbags or any other suitable object to support the handguard. Keep the nonfiring hand free for use on any part of the rifle.

(4) Place both elbows on the ground to support the upper body.

(5) Place the firing hand on the pistol grip.

(6) Place the nonfiring hand on the upper handguard.

(7) Place the weapon's buttstock between the SAPI plate and bicep to stabilize the weapon and absorb recoil.

NOTE: Elbow and knee pads can be worn to relieve IBA-induced pain and pressure.

Figure 4-23. Basic prone supported firing position.

Figure 4-24. Alternate prone supported firing position.

Kneeling Unsupported Firing Position

4-71. To assume the kneeling unsupported firing position (Figure 4-25)—

(1) Keep the left foot in place.

(2) Step back with the right foot.

(3) Drop to the right knee.

(4) Place the left nonfiring hand on the upper handguard with the upper arm (triceps) on the left knee for support.

(5) Place the right firing hand on the pistol grip, with the weapon's buttstock between the SAPI plate and bicep to stabilize the weapon and absorb recoil.

(6) Rest the ball of the right foot firmly on the ground.

(7) Rest the buttock on the heel.

(8) Relax and lean forward into the position to help absorb recoil.

Figure 4-25. Kneeling unsupported firing position.

TRAINING DEVICES AND EXERCISES

4-72. When used alone or in combinations with the appropriate training strategies, training devices and aids can be used to help individuals or squads sustain or practice basic marksmanship skills. They are beneficial when ammunition is limited for training or practice exercises. Some training devices are complex, costly, and in limited supply, while others are relatively simple, cheap, and in large supply.

M15A1 Aiming Card

4-73. This exercise measures the firer's ability to acquire the same sight picture each time he places his iron sights on a target.

NOTE: Refer to Appendix A for a detailed explanation of training aids and devices.

TARGET BOX AND PADDLE EXERCISE

4-74. This exercise checks the consistency of aiming and placement of three-round shot groups in a dry-fire environment. It incorporates the Soldier's position, breathing, and sight picture to simulate a live-fire 25-meter engagement. Further, it reinforces the basic fundamentals, while refining the Soldier's muscle memory during the integrated act of dry-firing.

NOTE: Refer to Appendix A for a detailed explanation of the target box and paddle exercise and training standards.

DIME/WASHER EXERCISE

4-75. The primary purpose of this exercise is to practice trigger control; it is also used to reinforce good body position and breath control. The Soldier must successfully dry-fire his weapon six consecutive times without the dime or washer falling to the ground.

NOTE: Refer to Appendix A for a detailed explanation of the dime/washer exercise and training standards.

SECTION III. MARKSMANSHIP FUNDAMENTALS II

This training program (Table 4-3) reinforces BRM and the four fundamentals, while demonstrating the integrated act of shooting on the EST 2000 or LMTS.

Table 4-3. Marksmanship Fundamentals II training program.

MARKSMANSHIP FUNDAMENTALS II
Period 3 (8 hours)
Instructional Intent
• Reinforce BRM I and II and the four fundamentals, while demonstrating the integrated act of shooting on the EST or LMTS.
Observables
Ensure that— • All fundamentals are emphasized and applied on the EST or LMTS. • Weapon safety is reinforced on the EST or LMTS. • Peer coaching is emphasized during EST or LMTS firing. • All Soldiers who fail to hit six out of nine shots at the 300-meter EST target receive remedial training.
Tasks
• Demonstrate the integrated act of firing while using the EST.

NOTE: Soldiers who do not meet the standard will receive remedial training before continuing with subsequent instruction.

ENGAGEMENT SKILLS TRAINER 2000

4-76. The EST 2000 is a home station, indoor, multipurpose, multilane, small arms, crew-served, shoulder-launched munitions gunnery simulator with superior accuracy and state-of-the-art graphics.

4-77. The EST 2000 is used to provide small arms weapon training on—

- Marksmanship.
- Squad tactical procedures.
- Close-range shoot/don't shoot techniques and skills.

4-78. Commanders should review DA Pam 350-38 for live-fire events that can be executed using the EST 2000.

NOTE: Refer to Appendix A for a detailed explanation of the EST 2000 training simulator.

LASER MARKSMANSHIP TRAINING SYSTEM

4-79. The LMTS helps bridge the existing gap in individual marksmanship training by providing a more widely available, flexible, transportable, and lightweight means of maintaining marksmanship skills. The LMTS large-company suite with the warrior kit is a home station or deployed, multipurpose, modular, scaleable, individual and unit small arms marksmanship sustainment simulator that accommodates the M16-series rifle and M4 carbine.

4-80. The LMTS supports both IET and unit sustainment training for static mounted and dismounted small units. When incorporated into BRM and Marksmanship Fundamentals I and II, it greatly improves Soldier feedback.

NOTES:

1. The LMTS is not intended to replace live-firing, qualification, or the EST 2000.
2. Refer to Appendix A for a detailed explanation of the LMTS.
3. Soldiers failing to maintain a three-centimeter shot group on the LMTS 25-meter target should receive remedial feedback.

Chapter 5

Downrange Feedback

Downrange feedback provides precise knowledge of what happens to bullets at range. It provides for an effective transition between 25-meter firing and firing on the field fire range. Knowing precisely where all bullets are hitting or missing the target, a poor firer (with instructor/trainer assistance) can improve his performance, and a good firer can bring his shots to the target's center.

This chapter contains guidelines for the instructor/trainer to conduct training on ranges that provide detailed feedback from the targets downrange. This chapter contains grouping and zeroing procedures and procedures for conducting the three types of KD ranges.

SECTION I. GROUPING PROCEDURES

This section provides guidelines for conducting a grouping range (Table 5-1). It includes concept, organization, shot group marking, shot group analysis, multiple shot group analysis, and troubleshooting of the fundamentals.

Table 5-1. Grouping procedures.

GROUPING PROCEDURES
Instructional Intent
• Reinforce PMI by performing the integrated act of shooting. • Shoot two consecutive 3-round shot groups within a 4-centimeter circle at 25 meters.
Special Instructions
Ensure that— • The rear sight is on the proper setting (zero; M16A2/3=8/3+1; M16A4=6/3+2; M4=6/3). • The rear sight aperture is set on 300+1, not 800+1. • The M16A1's rear sight is set on the aperture marked L. • The small aperture is being used. • Proper and accurate shot group marking is enforced. • The firer's name is clearly marked on the target.
Observables
• Coaches analyze the firer's fundamentals. • Majority of the round must be inside the circle to be counted. • Two consecutive 3-round groups are shot with 6 of 6 rounds impacting inside of the 4- centimeter circle.

CONCEPT

5-1. Shot grouping is a form of practice firing with two primary objectives: firing tight shot groups and consistently placing those groups in the same location. Grouping exercises can be conducted anywhere that provides precise location of bullet hits and misses, such as a 25-meter live-fire zeroing range, KD range, EST 2000, LMTS, or location of misses and hits (LOMAH) system.

NOTES: 1. Shot grouping should be conducted between dry-fire training and zeroing.

2. The initial live-fire training should be a grouping exercise with the purpose of practicing and refining marksmanship fundamentals.

ORGANIZATION OF A 25-METER GROUPING RANGE

5-2. The organization and conduct of a grouping range are based on the availability of ammunition, number of personnel, and the firing ability of personnel in training.

5-3. To properly conduct a 25-meter grouping range, perform the following actions:

- Divide the unit into firing orders. The first order fires, while the second order coaches.
- Reserve ten firing points to conduct corrective instruction.
- Provide sandbags at each firing point to accommodate supported firing positions.
- Set up the 25-meter grouping range as depicted in Figure 5-1.

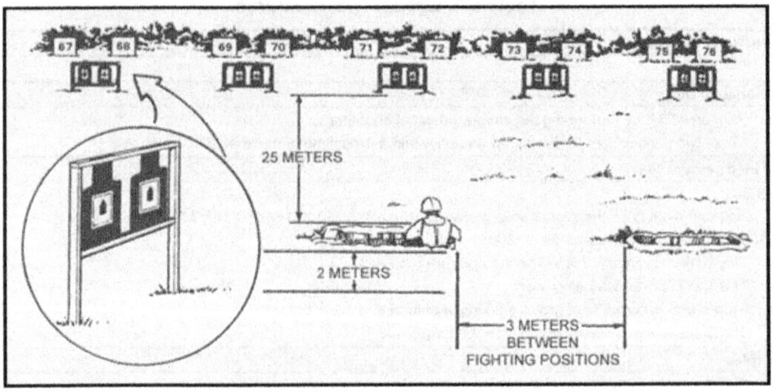

Figure 5-1. 25-meter range.

CONDUCT OF A 25-METER GROUPING FIRING

5-4. Each shot is fired from a supported firing position using the same point of aim (25-meter zeroing target's center of mass). The objective is to fire tight shot groups and to place those shot groups inside a 4-centimeter circle (the actual location of groups on the target is not important).

NOTES: 1. Since this is not a zeroing exercise, few sight adjustments are made unless the shot group is off of or barely on the 25-meter zeroing target.

2. No sight adjustments should be made until the firer can shoot six consecutive shots (two shot groups) inside a 4-centimeter circle. Once this is accomplished, the Soldier is ready to conduct zeroing procedures.

5-5. To conduct a 25-meter grouping firing—

NOTE: Before beginning the 25-meter grouping firing, each Soldier ensures that his sights are set for 25-meter firing.

(1) The Soldier fires a three-round shot group at the 25-meter zeroing target.
(2) The firing line is cleared, and the Soldier and coach move downrange to examine the shot group for fundamental errors, triangulate the shot group, and put the number 1 in the center of the shot group (Figures 5-2 and 5-3).

NOTE: If the shot group is off of the 25-meter zeroing target, the Soldier should mechanically zero the weapon. If the shot group is barely on the target, the Soldier should make a bold adjustment.

(3) The Soldier returns to the firing line and fires a second three-round shot group.
(4) The firing line is cleared, and the Soldier moves downrange to examine the second shot group, triangulate, and mark the center of the shot group with the number 2.
(5) The Soldier groups the two shot groups and marks the center.

5-6. The Soldier repeats Steps 1 through 5 until he places six out of six consecutive rounds inside a 4-centimeter circle. If the Soldier has not grouped in 18 rounds, he should be removed from the firing line and given remedial training before attempting to group again.

NOTE: To be counted, the majority of the round must be inside of the circle.

5-7. Once the Soldier has demonstrated firing proficiency from the supported firing position, grouping exercises can be conducted from the unsupported firing position. For example, 27 rounds are allocated for the grouping exercise (IET only). If the Soldier groups in 18 rounds, he can fire the remaining 9 rounds from the unsupported firing position.

SHOT GROUP MARKING

5-8. If the Soldier is to benefit from this exercise and if the instructor/trainer (or coach) is to provide useful guidance, the Soldier must mark each shot group for a clear record of his firing practice. The instructor/trainer must understand how to analyze shot groups correctly.

5-9. To properly mark the shot groups (Figure 5-2)—

(1) Connect the three bullet holes on the target with a straight line.
(2) Place a number inside of the shot group.

NOTES: 1. The number represents the center of the three shots.

2. When two shots are near one end of the group and the third shot is toward the other end, the number is placed closer to the two near shots (Figure 5-3).

3. This is not a precise marking that requires a measurement, but a procedure to help with shot group analysis.

5-10. The three-round shot group allows the firer's performance to be evaluated.

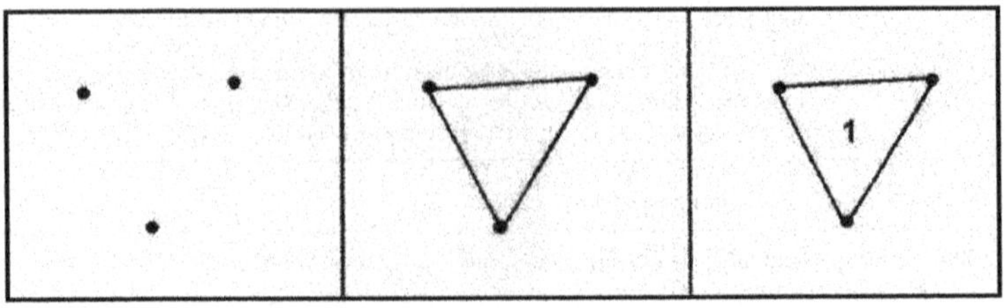

Figure 5-2. Shot group marking.

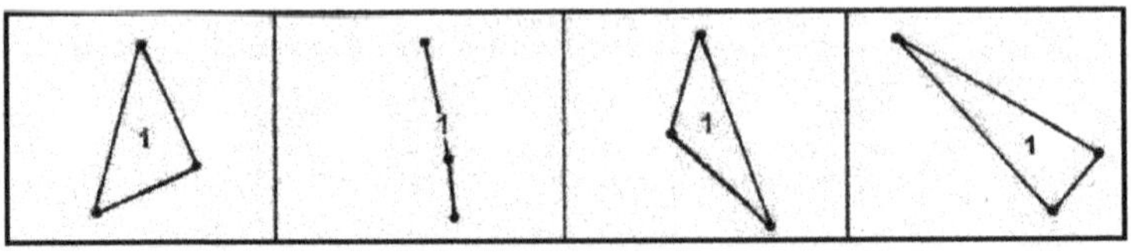

Figure 5-3. Central point of an odd-shaped group.

SINGLE SHOT GROUP ANALYSIS

5-11. The purpose of single shot group analysis is to identify firer errors on the single shots of a shot group so the Soldier can correct these errors while firing the next shot group.

5-12. Shot group analysis begins with the instructor/trainer observing the Soldier while he fires, looking for proper position, aim, trigger squeeze, and breathing. Then, the instructor/trainer analyzes the shot group to confirm problem areas.

NOTE: Coaches should not use shot group analysis without observing the firer.

5-13. The ideal shot group will have all three rounds within a 2-centimeter circle. Three rounds within a 4-centimeter circle is the minimum standard.

NOTE: M16A2 zeroing target squares are .96 centimeter in size, while M4 zeroing target squares are 1.3 centimeters in size.

MATCH-GRADE PERFORMANCE

5-14. The target shown in Figure 5-4 illustrates a match-grade quality weapon/ammunition combination. This combination places all bullets in almost the same hole and helps detect the firer's slightest errors.

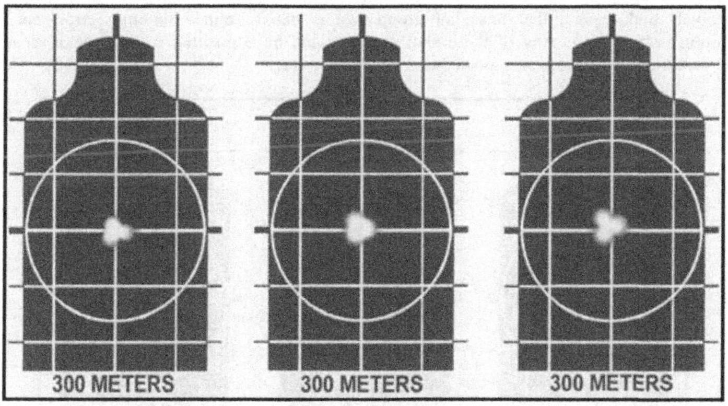

Figure 5-4. 25-meter match grade performance.

2-CENTIMETER SHOT GROUPS

5-15. The variances of standard weapons and ammunition must be considered during shot group analysis. When firing a standard service weapon/ammunition combination, the dispersion pattern may be up to 2 centimeters apart without human error. The instructor/trainer must ensure the Soldier understands that his weapon or ammunition may not be capable of placing three rounds within a 1-centimeter square.

NOTE: The dispersion pattern is not considered a firer error.

5-16. The targets shown in Figure 5-5 reflect proper 25-meter shot group performances using standard weapon/ammunition combinations.

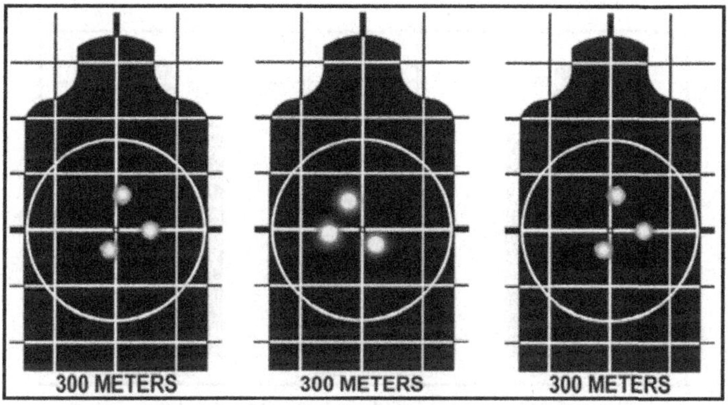

Figure 5-5. Shot groups with no firer error.

3- TO 4-CENTIMETER SHOT GROUPS

5-17. The targets shown in Figure 5-6 represent minimum acceptable firing performances; a better firing performance should be expected.

5-18. The instructor/trainer should ensure that the Soldier is properly applying the four marksmanship fundamentals and explain that this shot group size is due to minor shooting error, not weapon or ammunition performance. Any of these shot groups could have resulted from a minor change in sight picture, breathing, trigger squeeze, position, or an erratic round.

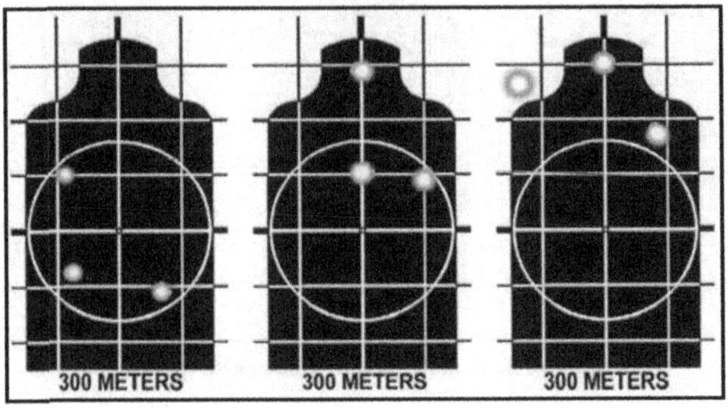

Figure 5-6. Shot groups with minor shooting error.

OVER 4- TO 5-CENTIMETER SHOT GROUPS

5-19. The targets shown in Figure 5-7 represent unacceptable firing performance.

5-20. The instructor/trainer should ensure that the Soldier is properly applying the four marksmanship fundamentals and explain that this shot group size is due to considerable shooting error, not weapon or ammunition performance. Any of these shot groups could have resulted from a change in position, sight picture, breathing, trigger squeeze, position, or an erratic round.

5-21. Soldiers who fire these shot groups should receive dry-fire training or remedial training on the EST 2000 or LMTS to help correct firing problems.

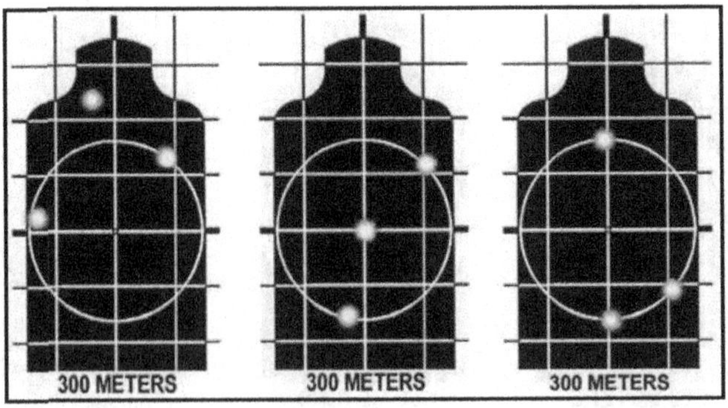

Figure 5-7. Shot groups with considerable shooting error.

Shot Groups Larger than 5 Centimeters

5-22. The targets shown in Figure 5-8 represent unacceptable firing performance; a better firing performance should be expected.

5-23. The instructor/trainer should ensure that the Soldier is properly applying the four marksmanship fundamentals and explain that this shot group size is due to major shooting error, not weapon or ammunition performance. Any of these shot groups could have resulted from a change in position, sight picture, breathing, or trigger squeeze, or the firer may have anticipated the shot.

5-24. Soldiers who fire these shot groups should receive extensive dry-fire training or remedial training on the EST 2000 or LMTS to help correct firing problems.

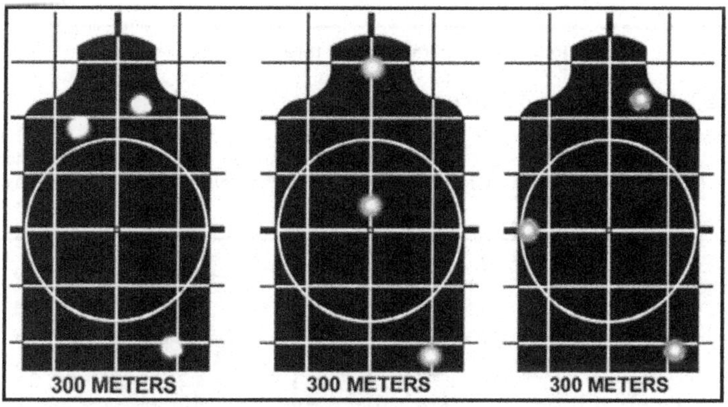

Figure 5-8. Shot groups with major shooting error.

MULTIPLE SHOT GROUP ANALYSIS

5-25. Multiple shot group analysis involves performing ongoing analysis of individual shot groups, while comparing them to each other for consistent aiming.

NOTE: If the Soldier is to benefit from this exercise, and if the instructor/trainer (or coach) is to provide useful guidance, the Soldier must mark each shot group individually and locate the center of more than one shot group.

5-26. To properly mark the shot groups (Figure 5-9)—

(1) Connect each individual shot group on the target with a straight line.

(2) Place a number inside the shot group.

NOTE: The number represents the center of the three shots.

(3) Connect the numbers, and place an X in the center.

NOTE: The X represents the center of the shot groups.

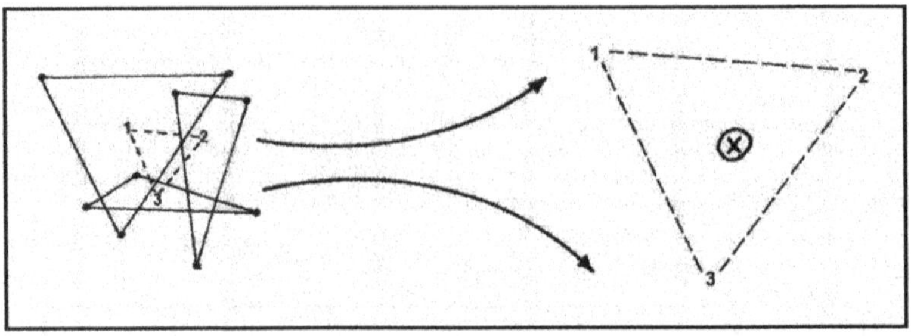

Figure 5-9. Central point of three shot groups.

ACCEPTABLE SHOT GROUPING PERFORMANCE

5-27. The shot groups in Figure 5-10 represent acceptable shot groups (4 centimeters or less) in the same location. The Soldier firing this shot grouping should make a sight change of left 10 and down 4. Any change should be clearly marked on the target and saved for reference. The Soldier is then ready to zero his weapon.

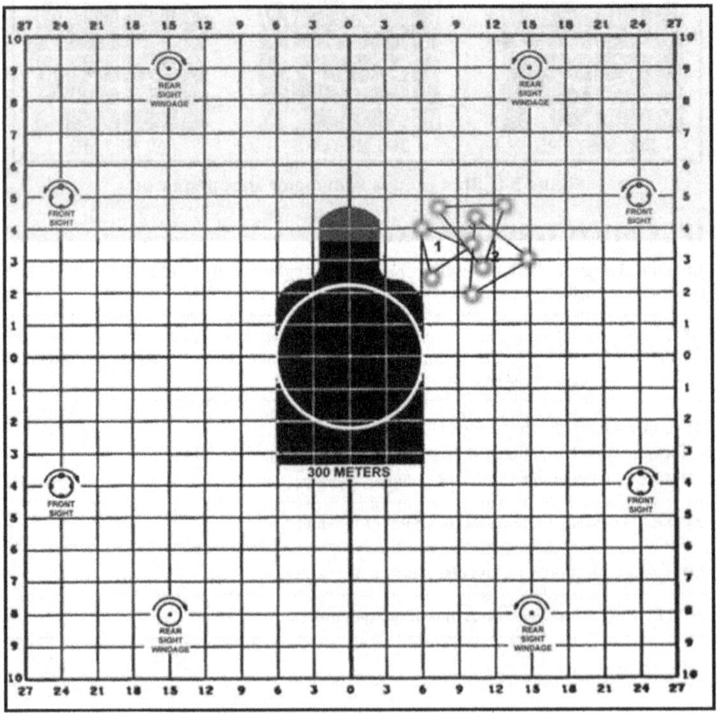

Figure 5-10. Acceptable shot grouping performance.

NOTES: 1. Location of the shot group on the 25-meter target is not important when conducting a grouping exercise. The size and dispersion of the shot groups are the main focus of this exercise.

2. Before the Soldier should be allowed to make any adjustments or start zeroing procedures, two consecutive shot groups must fall within a 4-centimeter circle when fired at 25 meters.

SHOT GROUPS WITH INCONSISTENT AIMING

5-28. The groups in Figure 5-11 indicate that the Soldier firing the shot groups is applying proper firing fundamentals, but is using a different point of aim each time a shot group is fired.

5-29. The instructor/trainer should question the Soldier's understanding of the aiming process and check his position for consistency. The instructor/trainer cannot determine which shot group best represents the firer's zero.

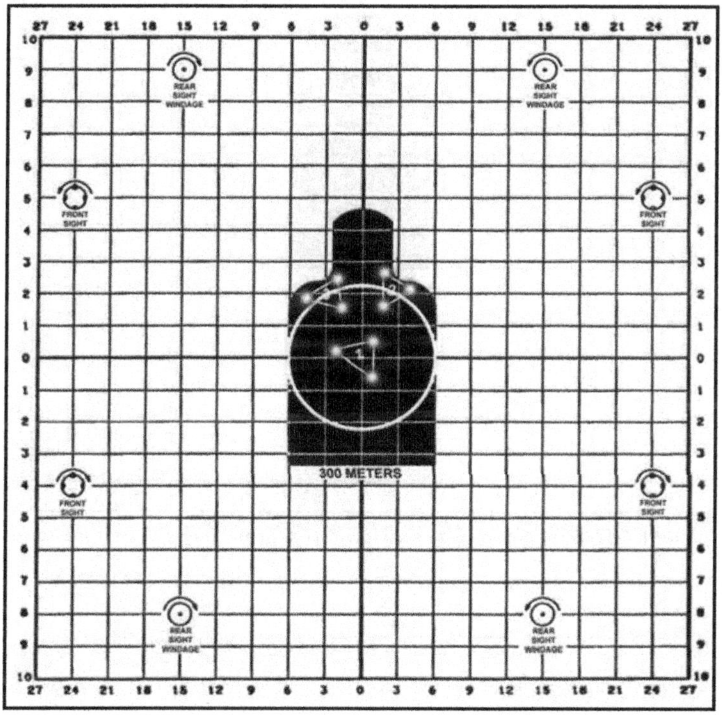

Figure 5-11. Shot groups with inconsistent aiming.

Shot Groups with Consistent Aiming and Major Shooting Error

5-30. The groups in Figure 5-12 indicate consistent aiming, but the Soldier is not applying the four fundamentals properly while firing each shot group.

5-31. The firer should be assigned a coach to troubleshoot his application of the four fundamentals in an attempt to isolate his firing errors.

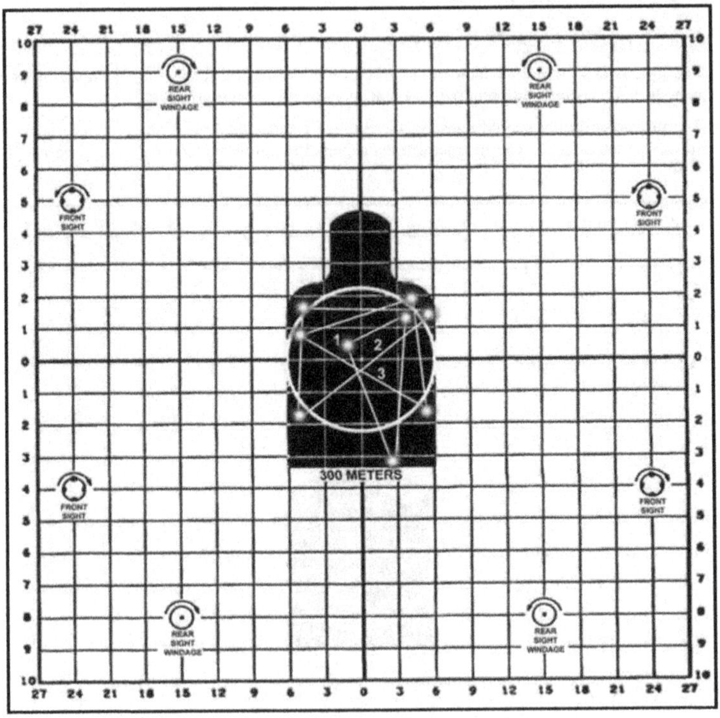

Figure 5-12. Shot groups with consistent aiming and major shooting error.

SHOT GROUPS WITH INCONSISTENT AIMING AND MAJOR SHOOTING ERROR

5-32. The groups shown in Figure 5-13 indicate inconsistent aiming and major shooting error.

5-33. The firer should be assigned a coach to troubleshoot his application of the four fundamentals in an attempt to isolate his firing errors.

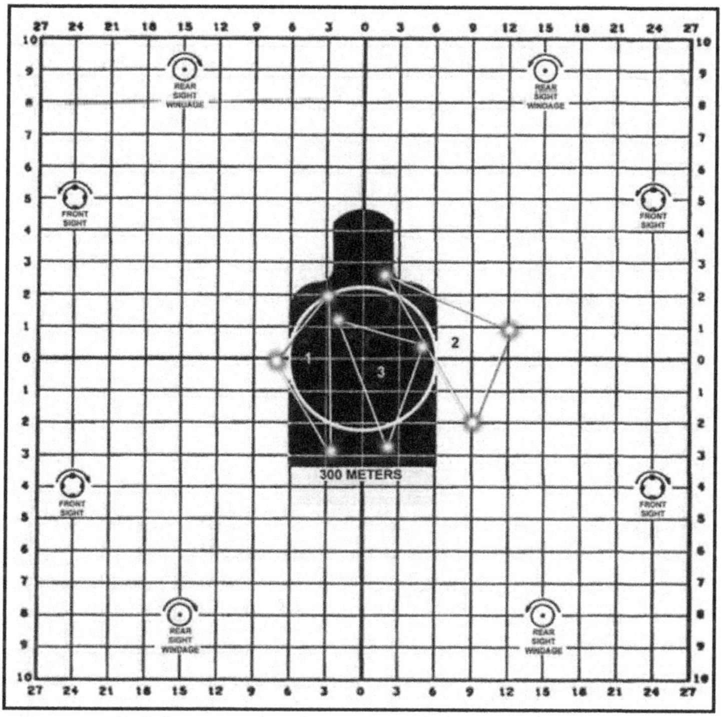

Figure 5-13. Shot groups with inconsistent aiming and major shooting error.

SHOT GROUPS WITH IMPROPER VERTICAL PLACEMENT

5-34. When viewed as nine shots, the shot groups shown in Figure 5-14 reflect proper horizontal placement of shots, but unsatisfactory vertical dispersion. This indicates a failure to vertically aim at the target's center of mass for each shot.

5-35. The instructor/trainer should check the Soldier's aiming procedure and adherence to marksmanship fundamentals.

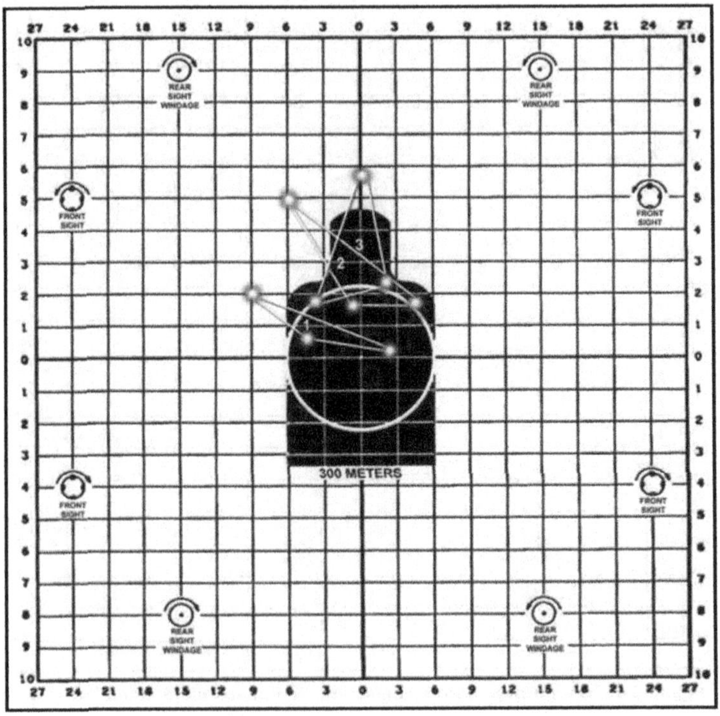

Figure 5-14. Shot groups with improper vertical placement.

IMPROPER SHOT GROUPS ON THE EDGE OF THE TARGET

5-36. The shot groups shown in Figure 5-15 are improper groups.

5-37. The Soldier should make a bold sight change to bring the groups closer to the target's center and ensure that the shot groups remain on the 25-meter zeroing target.

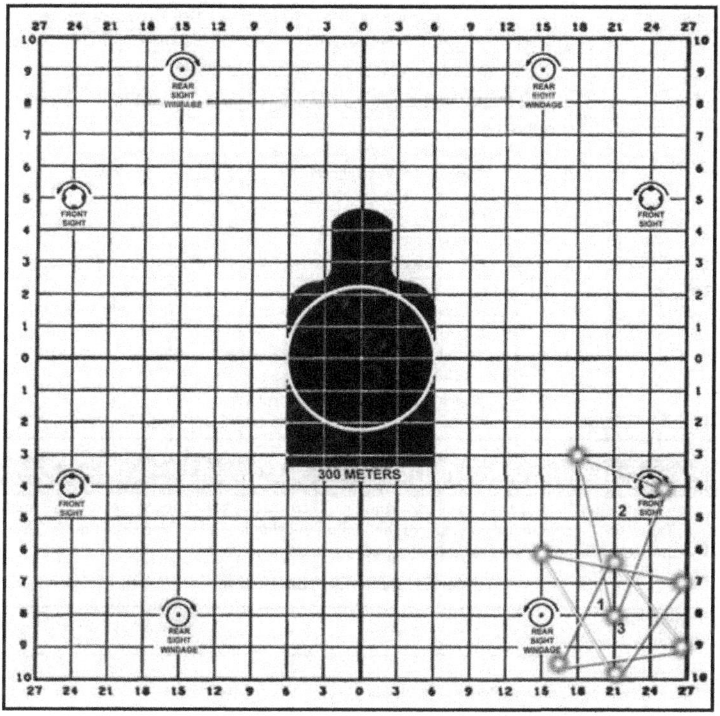

Figure 5-15. Improper shot groups on the edge of the target.

TROUBLESHOOTING THE FUNDAMENTALS

5-38. When troubleshooting the fundamentals, the coach's imagination is the only limiting factor. Table 5-2 outlines the techniques that can be used to identify errors in Soldiers' application of the fundamentals.

Table 5-2. Techniques used to identify errors in Soldiers' application of the fundamentals.

PROBLEM AREA	TECHNIQUE USED TO IDENTIFY THE PROBLEM AREA
Aiming	Attach the M16 sighting device, and observe the Soldier while he fires.
Breathing	Watch the rise and fall of the firer's chest for consistency.
Trigger squeeze	Place your finger over the firer's finger while he fires, feeling for jerking and smooth follow-through. Watch for jerking of the trigger and smooth follow-through.
Position	Observe the following areas for consistency: • Placement of the tip of the nose. • Placement of the trigger finger. • Placement of the nonfiring hand. • Placement of the legs. • Cheek-to-stock weld. • Positioning of equipment.
Other potential problem areas	Ensure that the— • Nonfiring-side eye is not shuttering. • Equipment is fitted properly. • Soldier is not flinching when the trigger is pulled. • Soldier is firing with the dominant eye. • Soldier is wearing glasses (if applicable). • Soldier is maximizing use of the supported position.

SECTION II. ZEROING PROCEDURES

This section provides guidelines for the instructor/trainer to zero M16-/M4-series weapons at 25 meters and at actual range (Table 5-3). It includes concept, organization, mechanical zero, zero recording, 25-meter sight settings, field fire sight settings, and troubleshooting of the fundamentals.

Table 5-3. Zeroing procedures.

ZEROING PROCEDURES
Instructional Intent
• Reinforce PMI while adjusting the confirmed shot groups (center of mass of the 4-centimeter circle with 5 out of 6 consecutive rounds at 25 meters).
Special Instructions
Ensure that— • The rear sight is on the proper setting (zero; M16A2/3=8/3+1; M16A4=6/3+2; M4=6/3). • The rear sight aperture is set on 300+1, not 800+1. • The M16A1's rear sight is set on the aperture marked L. • The small aperture is being used. • Proper and accurate shot group marking is enforced. • The firer's name is clearly marked on the target. • M16A2/A3/A4s zero on M16A2 zeroing targets. • M4s zero on M4 zeroing targets. • M16A1s zero on M16A1 zeroing targets.
Observables
• Coaches analyze the firer's application of the fundamentals. • The majority of the round must be inside of the circle to be counted. • Two consecutive 3-round shot groups are shot with 5 of 6 rounds in the 4- centimeter circle.

PURPOSE

5-39. The purpose of battlesight zeroing is to align the sights with the weapon's barrel given standard issue ammunition. When this is accomplished correctly, the point of aim and point of impact are the same at a given range (250 meters for the M16A1, 300 meters for the M16A2/A3/A4 and M4-series weapons). This sight setting provides the highest hit probability for most combat targets with minimum adjustment to the point of aim.

5-40. When standard zeroing procedures are followed, a properly zeroed weapon for one Soldier is close to the zero for another Soldier. When a straight line is drawn from the target's center to the tip of the front sightpost and through the center of the rear aperture, it makes little difference whose eye is looking along this line. There are many subtle factors that result in differences among individual zeros. Instructors/trainers should emphasize the similarity of individual zeros instead of the differences.

5-41. Most firers can fire with the same zeroed weapon if they properly apply marksmanship fundamentals. This information can be useful in three ways:

(1) If a Soldier has difficulty zeroing and the problem cannot be diagnosed, a good firer could zero the weapon to find the problem and eliminate the weapon as part of the problem.

(2) When a Soldier must fire another Soldier's weapon without opportunity to verify the zero by firing for example, picking up another man's weapon on the battlefield), the weapon will be closer to actual zero if the sights are left unchanged. This information is useful in deciding initial sight settings and recording zeros.

(3) All weapons in the arms room, even those not assigned, should have been previously zeroed by the last Soldier they were assigned to. Zeroing this newly assigned weapon should start with the sights left where they are.

SIGHT VARIANCE

5-42. There is no relationship between the specific sight settings a Soldier uses on his weapon to the sight settings he would use to zero another weapon, which makes it essential that each Soldier zeros the weapon that he is assigned. For example, a Soldier could zero his assigned weapon 10 clicks left of center; when zeroing another weapon, his adjustments could be 10 clicks right of center. This is due to the manufacturing difference between the weapons. Therefore, all newly assigned personnel should be required to zero their weapon as soon as possible after assignment to the unit. The same rule applies anytime a Soldier is assigned a weapon that is returned from field level or sustainment level maintenance, or when the zero is in question.

ORGANIZATION OF A 25-METER ZERO RANGE

NOTES: 1. All Soldiers should successfully group prior to zeroing.

2. If the Soldier is proficient at grouping, two shot groups should be fired to confirm proficiency prior to making any sight adjustments during zeroing procedures.

5-43. To properly conduct a 25-meter zero range, perform the following actions:

- Divide the unit into firing orders. The first order fires, while the second order coaches.
- Reserve firing points to conduct corrective instruction.
- Provide sandbags at each firing point to accommodate supported firing positions.

CONDUCT OF A 25-METER ZERO FIRING

5-44. On the 25-meter zero range, the Soldier applies the fundamentals while consistently aiming at the target's center of mass (A, Figure 5-16). The Soldier fires two separate three-round shot groups (B, Figure 5-16) and groups them. Based on the location of these groups, the Soldier makes the appropriate sight adjustments. Then, the Soldier fires two additional three-round shot groups to confirm that the adjustments have aligned the sights with the center of the target and that the bullets are in the 4-centimeter circle (Figure 5-17).

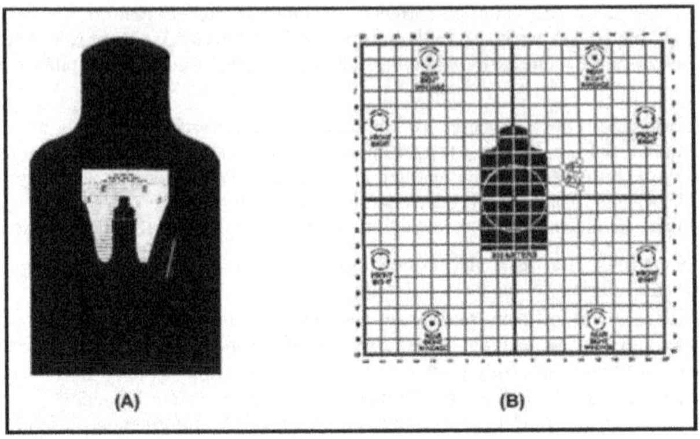

Figure 5-16. Correct aiming (A), initial shot group results (B).

Figure 5-17. Final shot group results.

5-45. To conduct a 25-meter zero range—

NOTES: 1. Each Soldier ensures that his sights are set for 25-meter zeroing.

2. Soldiers fire each shot from a supported firing position using the same point of aim (target's center of mass).

3. Ensure that the correct 25-meter zero target is being used. For M16A1s, use NSN 6920-01-167-1392 (Figure 5-18); for M16A2s, M16A3s, M16A4s, M4s, and M4As, use NSN 6920-01-395-2949 (Figure 5-19; M16A2/A3 is printed on one side, and M16A4/M4/M4A is printed on the other).

---

(1) The Soldier fires a three-round shot group at the 25-meter zeroing target.

(2) The firing line is cleared, and he moves downrange to examine the shot group for fundamental errors, triangulates the shot group, and puts the number 1 in the center of the shot group.

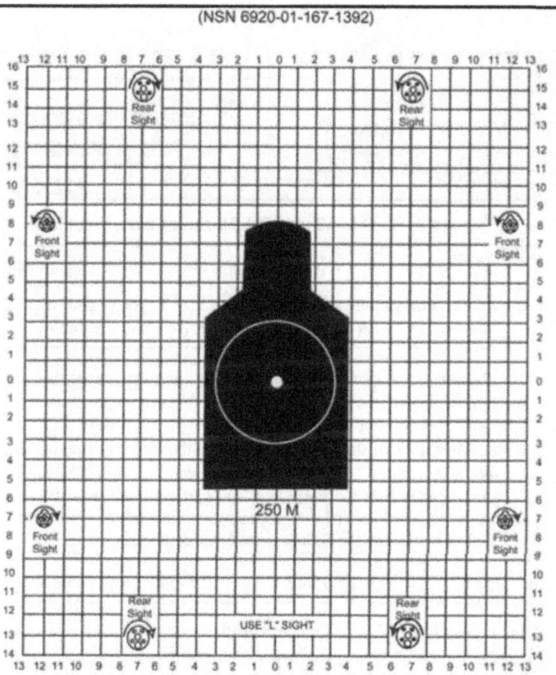

1. **Aim at target center. Adjust sights to move shot group center as close as possible to white dot.**

2. **At completion of zero, rotate rear sight to unmarked aperture and weapon will be battlesight zero for 250 meters.**

Figure 5-18. M16A1 25-meter zero target.

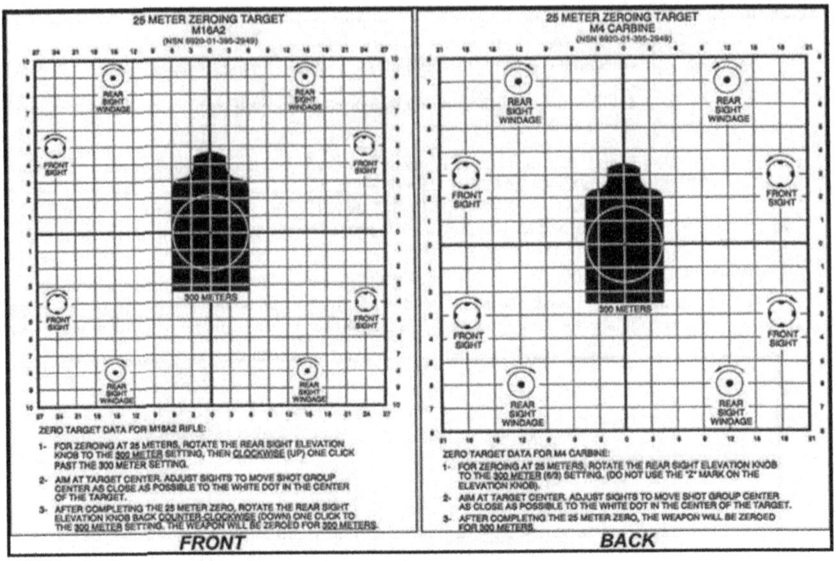

Figure 5-19. M16A2 and M4 25-meter zero target.

NOTES: 1. The Soldier fires two individual shot groups before a sight change is considered.

2. If the initial shot group is not on the target paper, the weapon should be mechanically zeroed before the Soldier fires this weapon again.

---

(3) The Soldier returns to the firing line and fires a second three-round shot group.

(4) The firing line is cleared, and the Soldier moves downrange to examine the second shot group, triangulate, and mark the center of the shot group with the number 2.

(5) The Soldier groups the two shot groups and marks the center of the two shot groups with an X.

(6) If the two shot groups fall within a 4-centimeter circle, the firer determines the sight adjustments he needs to make, identifies the horizontal and vertical lines closest to the X, and reads the 25-meter zeroing target to determine the proper sight adjustments. If the two shot groups do not fall within a 4-centimeter circle, the Soldier continues grouping.

NOTE: The majority of the round must be inside of the circle to be counted.

---

(7) The Soldier annotates any sight adjustments that need to be made to the weapon on the 25-meter zeroing target and ensures that his name is on the target.

(8) If five out of six rounds fell within the 4-centimeter circle, the Soldier is zeroed and can be removed from the firing line. If not, the Soldier returns to the firing line and makes sight adjustments.

5-46. Steps 1 through 8 are repeated until the Soldier places five out of six consecutive rounds inside the 4-centimeter circle. If the Soldier is not zeroed in 18 rounds, he should be removed from the firing line and given remedial training before attempting to zero again.

5-47. Once firing proficiency has been demonstrated from the supported firing position, zeroing exercises can be conducted from the unsupported firing position. For example, 18 rounds are allocated for the zeroing exercise. If the Soldier zeroes in 9 rounds, the Soldier can fire the remaining 9 rounds from the unsupported firing position.

Conduct of a 25-Meter Zero Firing Using the Location of Misses and Hits System

5-48. When using the LOMAH system on a KD range, zero confirmation is part of the program and will be shot as the first scenario. To conduct a 25-meter zero range using the LOMAH system, the Soldier shoots six rounds at the 175-meter/200-yard target while aiming at the target's center of mass. The outcome is evaluated using the following guidelines:

- If the shot group falls within the 11-inch circle on the LOMAH monitor, the Soldier continues the programmed scenario, which is identical to the downrange feedback scenario without LOMAH.
- If the Soldier shoots a shot group that is 11 inches or smaller but is clearly not zeroed, the instructor/trainer assists the Soldier in making sight adjustments based upon the data provided on the LOMAH monitor.
- If the shot group is not tight (greater than 11 inches), the Soldier should be removed from the firing line and given remedial training on the four fundamentals of marksmanship.

SECTION III. KNOWN DISTANCE RANGE

This section provides guidelines for the instructor/trainer to conduct a KD range and apply the effects of wind and gravity. This section also addresses three types of KD ranges: the standard KD range, the KD record fire range, and the modified field fire range.

NOTE: See Table 5-4 for the current training program.

Table 5-4. Downrange feedback.

DOWNRANGE FEEDBACK
Instructional Intent
• Reinforce PMI while shooting from the prone supported and unsupported firing positions. • Build the Soldier's confidence in his ability to hit where he aims while applying the effects of wind and gravity at range.
Special Instructions
Ensure that— • The effects of wind and gravity are thoroughly explained. • The rear sight is on the proper setting (M16A2/3=8/3; M16A4 and M4=6/3 flush; M16A1=the unmarked aperture, short-range). • The rear sight aperture is set on 300, not 800.
Observables
• Spotters provide correct feedback to firers. • Soldiers hit 8 of 10 targets at 100 meters. • Soldiers hit 14 of 20 targets at 200 meters. • Soldiers hit 5 of 10 targets at 300 meters.

CONCEPT

5-49. A KD range has three primary objectives:

(1) Fire tight shot groups at a known distance.

(2) Make sight adjustments at range (not an objective for IET Soldiers) while experiencing the effects of wind and gravity.

(3) Participate in marksmanship testing.

5-50. KD firing brings the Soldier one step closer to being able to fire during combat. The Soldier is provided information concerning the precise hit-or-miss location of every bullet fired. KD firing is conducted with a single, clearly visible target at a known distance, and the Soldier can establish a position that provides a natural point of aim on that single target. Consider the following:

- On the standard KD range, Soldiers fire at 100-, 200-, and 300-meter targets without any time constraints.
- On the KD record fire range, Soldiers fire at 100-, 200-, and 300-meter targets with time constraints.
- On the modified field fire range, Soldiers fire at 100-, 200-, and 300-meter targets on a standard 50- to 300-meter field fire qualification range.

NOTES: 1. If a qualification range is not available, this exercise may be shot on a standard 75- to 300-meter field fire range. Targets and target frames must be set up to accommodate this training.

2. On ranges that are built in yards instead of meters, the same KD targets will be used. The difference is so small that it does not need to be considered.

- The KD range does not require Soldiers to detect targets, estimate ranges to targets, scan sectors of fire, respond to surprise targets, respond to short-exposure targets, or engage multiple targets.
- An advantage of a KD range is the ability to see precisely where each bullet hits. To benefit from this training, Soldiers must clearly see the results of each firing, whether a group, single shot, or 10-round exercise.

KNOWN DISTANCE TARGET DESCRIPTION

5-51. Downrange feedback training should include detailed explanations of the targets. Consider the following:

- KD targets are large enough to capture all bullets fired. Standard E-type and F-type silhouettes can be used if standard KD targets are not available.
- The 16-centimeter circle on 100-meter targets, the 32-centimeter circle on 200-meter targets, and the 48-centimeter circle on 300-meter targets equate to the 4-centimeter zeroing target at 25 meters. If the Soldier's shot group falls within the 4-centimeter circle at 25 meters, it will fall within the circle on the target being shot. If the round falls outside of the circle, the round will clearly miss the 300-meter target (Figure 5-20).
- An X is located in the bottom portion of the circle to show the firer the target's center of mass, where he must aim when his weapon is zeroed.
- The grid system on the targets in Figure 5-20 equates to that of the 25-meter zeroing target. For example, one click on the front sightpost equals one square on the 25-meter zeroing target and also equals one square on the target being shot. Information similar to that on the zeroing target has been overprinted to help Soldiers apply sight adjustments.

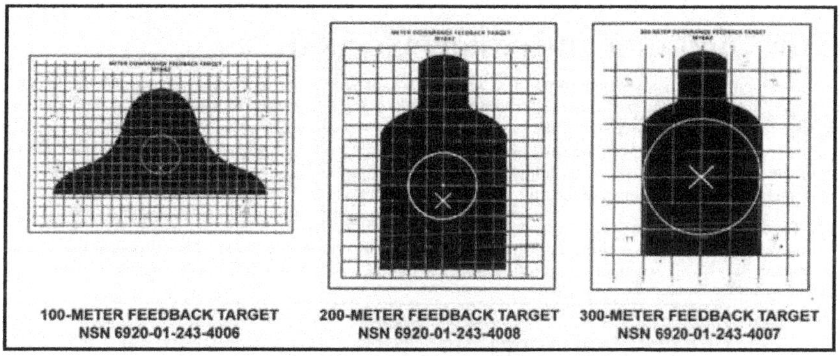

Figure 5-20. Downrange feedback targets.

MARKING KNOWN DISTANCE RANGE TARGETS

5-52. When the initial shot group is fired, target spotters/markers (Figure 5-21) should be placed in each bullet hole, placing the white side on the silhouette and the black side off of the silhouette. This procedure ensures that the firer can see where the rounds impacted and has two benefits:

- Instructors/trainers can observe the firer's performance and focus their attention on the Soldiers having the greatest problems.
- Soldiers are motivated to fire better since their peers can observe their performance.

5-53. On the second and subsequent shot groups, the target spotters/markers should be moved and placed in the holes of the new shot group. The old holes must be pasted using black pastors on black and white pastors on white. Failure to paste all bullet holes makes it difficult to determine one shot group from another.

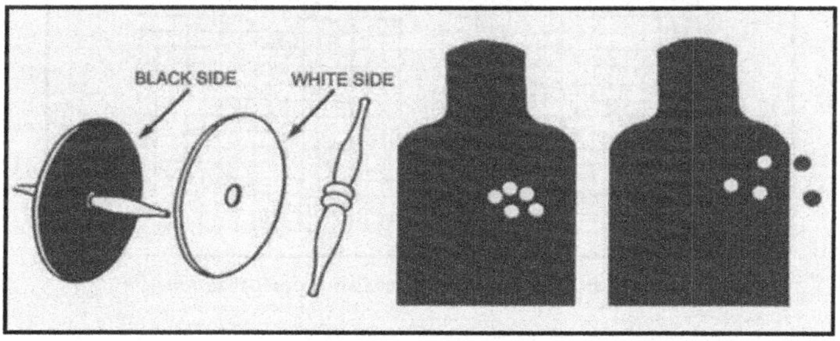

Figure 5-21. Target marking with spotters (markers).

KNOWN DISTANCE SHOT GROUPING ANALYSIS

5-54. Figure 5-22 shows two targets that were both shot with three individual rounds (A). On a pop-up target, these two firing performances would provide the same information to the firing line; each target was hit once and missed twice. Once the targets are properly marked with spotters, it becomes clear why only one round hit either target: The firer on the left is failing to properly apply the four fundamentals; the firer on the right needs to make an adjustment to his iron sights (assuming that wind was not a factor), triangulate the shot group, and read the appropriate adjustments from the target.

5-55. Figure 5-22 shows another two targets that were both shot with three individual rounds (B). On a pop-up target, these two firing performances would appear to be the same. Once properly marked with spotters, it is obvious that the firer on the left needs more training on the four fundamentals.

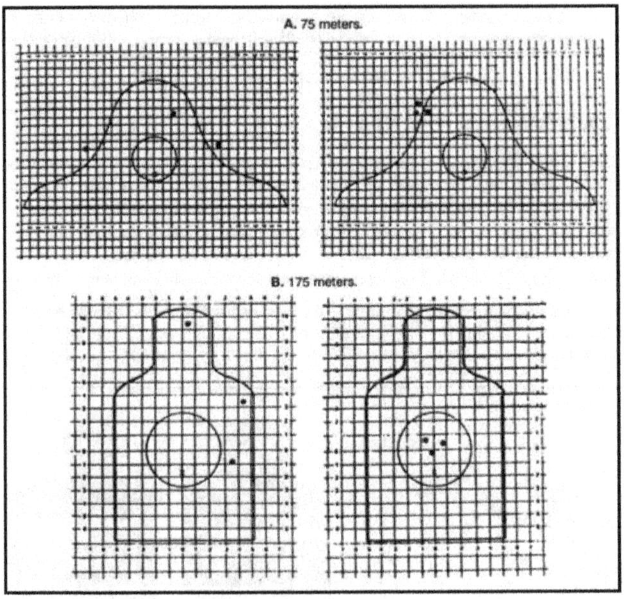

Figure 5-22. Comparison of firing performance.

KNOWN DISTANCE ZEROING

5-56. The 300-meter target can be used to confirm weapon zero or to refine the zero obtained on the 25-meter range. When Soldiers properly compensate for the wind, the zero on this target is more valid than the zero obtained on the 25-meter range. Soldiers should fire two five-round shot groups to confirm zero or three-round shot groups to refine their zero. The pit crews should spot targets after each shot group is fired. If the crosswind exceeds five miles per hour, KD zeroing should not be attempted.

NOTES: 1. IET Soldiers only zero on the 25-meter range.

2. For M16A2/3/4, M4, and M4A1 weapons only. Soldiers should use the unmarked aperture for zeroing and target engagement at all distances on the KD range. When engaging targets beyond 300 meters, the windage knob should be adjusted to the range of the target. 400-meter targets will be set on 4 flush, and 450-meter targets will be set on 4 plus two clicks.

3. For M16A1 rifles only: Soldiers should use the unmarked aperture (short-range) for refining zero at 300 meters. For target engagements beyond the 300-meter line, Soldiers should use the long-range aperture (L).

ELEVATION AND WINDAGE ADJUSTMENTS AT A DISTANCE

5-57. The elevation and windage rule states that one click of elevation or windage moves the strike of the bullet a specific distance at a specific range. At a range of 25 meters, one click of windage moves the strike of the bullet .33 centimeters, and one click of elevation on the front sight moves the strike of the bullet .83 centimeters. To compute the distance (D) one click of elevation (front sight) or windage moves the strike of a bullet at a given range (R)—

(1) Divide the range (expressed in meters) by 25.

(2) Multiply by either .33 centimeters for windage or .83 centimeters for elevation.

Windage: D = R ÷ 25 m x .33 (D = distance in cm)

Elevation: D = R ÷ 25 m x .83 (R = distance in cm)

Windage

5-58. To compute the distance one click of windage moves the strike of the bullet at a range of 300 meters—

(1) Divide 300 meters by 25 meters.

(2) Multiply by .33 centimeters.

D = 300 m ÷ 25 m = 12

12 x .33 = 3.96 cm

Round 3.96 cm up to 4.0 cm.

5-59. One click of windage moves the strike of the bullet 4 centimeters at 300 meters.

NOTE: Tables 5-5 and 5-6 show the amount of change in windage of the strike of the bullet at various ranges.

Elevation—Front Sight

5-60. To compute the distance one click of elevation (front sight) moves the strike of the bullet at a range of 300 meters—

(1) Divide 300 meters by 25 meters.

(2) Multiply by .83 centimeters.

D = 300 m ÷ 25 m = 12

12 x .83 = 9.96 cm

Round 9.96 cm up to 10 cm.

5-61. One click of elevation on the front sight moves the strike of the bullet 10 centimeters at a range of 300 meters.

NOTE: Tables 5-5 and 5-6 show the amount of change in elevation of the strike of the bullet at various ranges.

Elevation—Rear Sight

5-62. The elevation knob adjusts elevation 1.1 inch for each click at 100 meters, with each dot and each number representing one click of elevation.

Table 5-5. M16A2/3 and front sightpost of an M16A4.

DISTANCE (m)	DISTANCE 1 CLICK WILL ADJUST THE POINT OF IMPACT		
	Front Sightpost	Windage Knob	Elevation Wheel
25	.83 cm (3/8 in)	.33 cm (1/8 in)	.5 cm (1/4 in)
50	1.5 cm (5/8 in)	.5 cm (1/4 in)	1.5 cm (1/2 in)
75	2.5 cm (1 in)	1.0 cm (3/8 in)	2.0 cm (3/4 in)
100	3.5 cm (1 3/8 in)	1.5 cm (1/2 in)	2.75 cm (1 in)
150	5.0 cm (2 in)	2.0 cm (3/4 in)	4.0 cm (1 1/2 in)
175	6.0 cm (2 3/8 in)	2.25 cm (7/8 in)	5.0 cm (2 in)
200	6.5 cm (2 5/8 in)	2.5 cm (1 in)	5.5 cm (2 1/4 in)
250	8.5 cm (3 3/8 in)	3.5 cm (1 1/4 in)	7.0 cm (2 3/4 in)
300	10.0 cm (4 in)	4.0 cm (1 1/2 in)	8.5 cm (3 1/4 in)
400	13.5 cm (5 3/8 in)	5.5 cm (2 1/4 in)	11.0 cm (4 1/2 in)
500	17.0 cm (6.69 in)	6.5 cm (2 1/2 in)	14.0 cm (5 1/2 in)
600	20.5 cm (8.05 in)	8.0 cm (3 1/8 in)	16.75 cm (6 1/2 in)
700	24.0 cm (9.45 in)	9.0 cm (3 5/8 in)	19.5 cm (7 1/2 in)
800	27.5 cm (10.83 in)	10.5 cm (4 1/8 in)	22.5 cm (8 3/4 in)
NOTE: All values are rounded off.			

Table 5-6. M4/M4A1 and windage of an M16A4.

DISTANCE (m)	DISTANCE 1 CLICK WILL ADJUST THE POINT OF IMPACT		
	Front Sightpost	**Windage Knob**	**Elevation Wheel**
25	1.2 cm (1/2 in)	.5 cm (1/4 in)	.5 cm (1/4 in)
50	2.4 cm (1 in)	1.5 cm (1/2 in)	1.5 cm (1/2 in)
75	3.6 cm (1 1/2 in)	2.0 cm (3/4 in)	2.0 cm (3/4 in)
100	4.8 cm (1 7/8 in)	2.75 cm (1 in)	2.75 cm (1 in)
150	7.2 cm (2 7/8 in)	4.0 cm (1 1/2 in)	4.0 cm (1 1/2 in)
175	8.4 cm (3 3/8 in)	5.0 cm (2 in)	4.0 cm (2 in)
200	9.6 cm (3 3/4 in)	5.5 cm (2 1/4 in)	5.5 cm (2 1/4 in)
250	12.0 cm (4 3/4 in)	7.0 cm (2 3/4 in)	7.0 cm (2 3/4 in)
300	14.4 cm (5 3/4 in)	8.5 cm (3 1/4 in)	8.5 cm (3 1/4 in)
400	19.2 cm (7 1/2 in)	11.0 cm (4 1/2 in)	11.0 cm (4 1/2 in)
500	24.0 cm (9 1/2 in)	14.0 cm (5 1/2 in)	14.0 cm (5 1/2 in)
600	28.8 cm (11 1/4 in)	16.75 cm (6 1/2 in)	16.75 cm (6 1/2 in)
NOTE: All values are rounded off.			

CONDUCT OF A STANDARD KNOWN DISTANCE RANGE

NOTE: If the range is equipped with the LOMAH system, a firing order will be used to operate the LOMAH throughout the period of instruction and will be fired last.

5-63. Standard KD ranges are conducted using the following considerations:

- The standard KD range is conducted with paper targets at 100, 200, and 300 meters to obtain downrange feedback at 75, 175, and 300 meters (Figure 5-23). Soldiers fire shot groups progressively at the targets.
- Half of the bullets are fired from the supported firing position, and the other half are fired from the unsupported firing position.
- The wind speed and direction must be determined before firing, and the firer must know the distance to the target.
- Soldiers mark the targets after firing each shot group. Based on this feedback, Soldiers receive a critique from their instructor/trainer or coach and apply the adjusted point of aim (as necessary).
- The downrange feedback exercise must be conducted within the constraints of time, ammunition, and available ranges.
- If 30 rounds of ammunition are available for training, firing three-round shot groups 10 times is preferable to firing five-round shot groups 6 times.
- Once the Soldier understands the concept for adjusting the point of aim to compensate for the effects of wind and gravity, he is ready to apply his knowledge on the field fire range.

75-Meter Targets

5-64. Instructors/trainers can provide feedback after each round, each three-round shot group, or each five-round shot group on the 100-meter feedback targets. No time limit is placed on the firer. Soldiers fire from the supported firing position and from the unsupported firing position. Then, the targets are marked and evaluated. Feedback consists of a critique of performance, adjustments to the point of aim, effects of wind and gravity, and shot placement. Target spotters mark the bullet holes so hits can be viewed from the firing line.

NOTE: IET Soldiers fire one five-round shot group from the supported firing position and one five-round shot group from the unsupported firing position. They must hit 8 out of 10 targets.

175-METER TARGETS

5-65. Firers engage the 200-meter target using the same downrange procedures as when engaging the 100-meter target.

NOTE: IET Soldiers fire 10 rounds from the supported firing position and 10 rounds from the unsupported firing position. They must hit 14 out of 20 targets.

300-METER TARGETS

5-66. Firers engage the 300-meter target using the same downrange procedures as when engaging the 100-meter target.

NOTE: IET Soldiers fire one five-round shot group from the supported firing position and one five-round shot group from the unsupported firing position. They must hit 5 out of 10 targets.

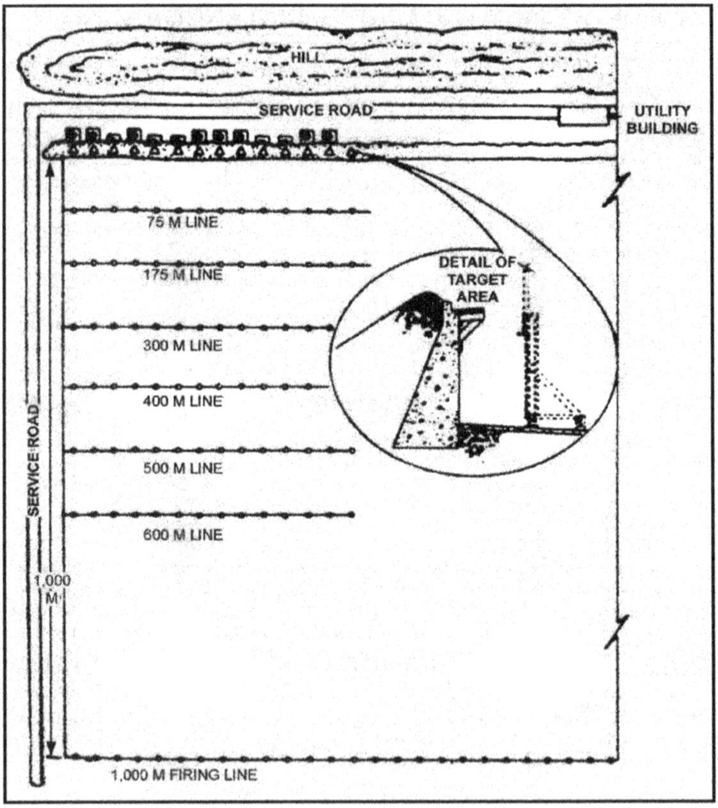

Figure 5-23. Known distance range.

KNOWN DISTANCE RECORD FIRE RANGE

NOTE: See paragraphs 6-79 through 6-82 of Chapter 6 for information about the alternate course KD record fire range.

MODIFIED FIELD FIRE RANGE

5-67. A modified field fire range can be used for downrange feedback. To conduct downrange feedback, minor changes must be made to a standard field fire range. Target frames, like those used on the 25-meter range, are placed on a standard qualification range at 100, 200, and 300 meters. The standard KD range or the KD record fire range can be conducted on the modified field fire range.

NOTE: The firing line must be cleared, moved to the targets for marking, and returned each time a firing order fires.

RECORD OF PERFORMANCE

5-68. During the conduct of downrange feedback, a record of performance should be kept on DA Form 5239-R (100-, 200-, and 300-Meter Downrange Feedback Scorecard) for the AAR.

NOTE: See Appendix B for a sample completed form and the end of this publication for a blank, reproducible copy.

5-69. As Soldiers complete each phase and achieve the performance standard for that range, they should receive a critique. Instructors/trainers must ensure that Soldiers do not progress to a greater range until they become proficient at closer ranges.

25-METER ZERO STANDARD

5-70. A standard E-type silhouette is 48.26 centimeters wide; a circle (angle) that is 48.26 centimeters at 300 meters is 4 centimeters at 25 meters. A Soldier who can fire all bullets in a 4-centimeter circle at 25 meters and adjust the sights for zero will hit the target at ranges as far away as 300 meters (Figure 5-24).

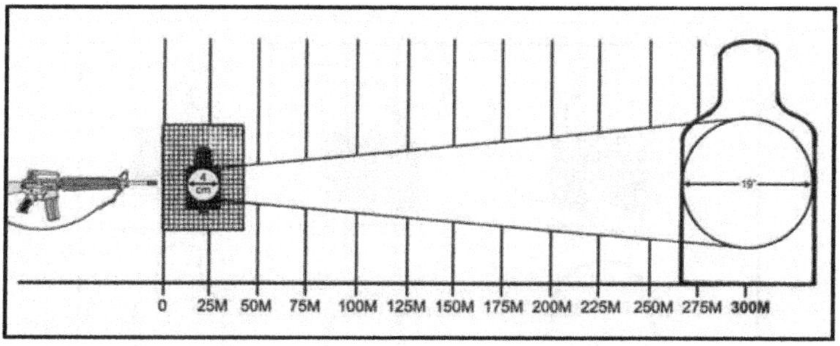

Figure 5-24. 25-meter zero standard.

SECTION IV. EFFECTS OF WIND AND GRAVITY

Marksmanship instructors/trainers should know how the effects of wind and gravity influence the flight of the bullet, and Soldiers should know how to compensate for such bullet displacement. This instruction is appropriate for all marksmanship training and concurrent training.

EFFECTS OF GRAVITY

5-71. Gases created by gunpowder push each round out of the end of the barrel. The barrel must be elevated slightly to allow the round to travel farther, creating an arc. The round will travel straight until it slows down and is gradually pulled to the ground by gravity. Each round fired will be pushed approximately the same distance and will roughly follow the same path.

NOTE: The farther the round travels, the faster it begins to fall.

5-72. When the firer zeroes his weapon, he aligns his line of sight to cross the path of the round at the distance at which he wants to zero his weapon. For example, a 300-meter zero means that the line of sight crosses the path of the round at 300 meters. If the firer engages a target at a distance other than 300 meters (excluding 25 meters), the path of the round hits the target either before or after it crosses the line of sight. If the firer wants his rounds to impact the center of mass, he must adjust his point of aim up or down to account for gravity.

ADJUSTED POINT OF AIM BASED ON GRAVITY

5-73. An adjusted point of aim (Figure 5-25) is intended to increase hit probability when properly presented. However, Soldiers can become confused, which could result in degraded performance. All Soldiers should be taught to aim at the target's center of mass unless they are confident that they know the range to the target. If adjusting the point of aim confuses the Soldier, he should aim at the target's center of mass. These points of aim place the center of each shot group in the target's center of mass (assuming a perfect zero and no firer error).

NOTES: 1. These adjustments are small and should only be applied by competent firers who wish to improve their firing performance.

2. Because the difference between M16- and M4-series weapons is so small and to avoid confusion, the same adjusted points of aim should be used regardless of the weapon being fired.

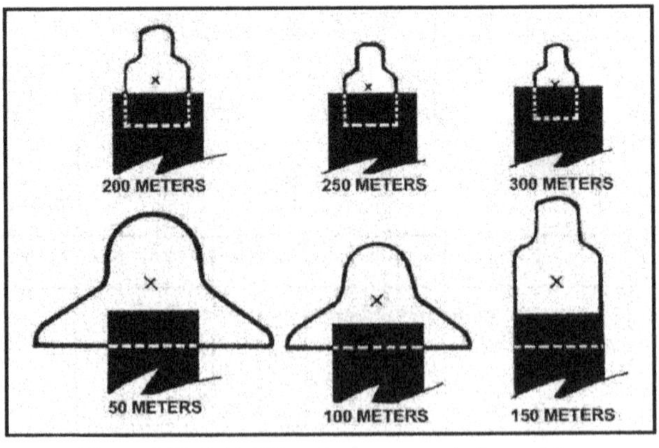

Figure 5-25. M16-/M4-series weapon aiming points.

EFFECTS OF WIND

5-74. Wind affects the bullet similar to the way gravity does: the farther the round travels, the farther the wind will push the round in the direction the wind is blowing. The faster the wind is blowing, the farther the wind will push the bullet.

WIND DIRECTION

5-75. The effects of wind vary depending on changes in wind speed and direction. Wind is classified by the direction it is blowing in relationship to the firer/target line. The clock system is used to indicate wind direction and value (Figure 5-26). This system works as follows:

- Winds that blow from the left (9 o'clock) or right (3 o'clock) are called full-value winds because they have the most effect on the bullet.
- Winds that blow at an angle from the front or rear are called half-value winds because they have about half the effect on the bullet as full-value winds.
- Winds that blow straight into the firer's face or winds that blow straight into the target are termed no-value winds because they have minimal effect on the bullet.

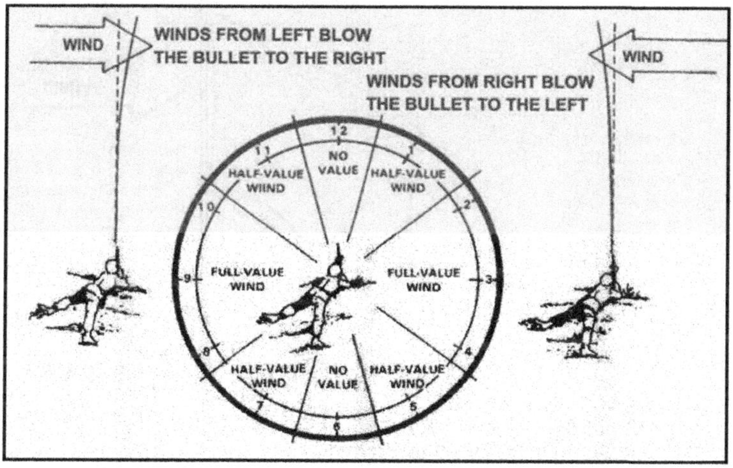

Figure 5-26. Determine wind value using the clock method.

WIND SPEED

5-76. Wind is variable and sometimes quite different at the firing position than at the target's position. Consider the following:

- When wind is blowing hard at the firing line, trees, brush, or terrain could protect the bullet's path.
- Wind can vary by several miles per hour between the time a measurement is taken and when the bullet is fired.

5-77. Therefore, training time should not be wasted trying to teach Soldiers an exact way to measure wind speed. Soldiers should understand that wind can blow a bullet off course, but they should not overcompensate and miss targets by applying too much hold-off.

5-78. A wind gauge can be used for precise measurement of wind velocity. When a gauge is not available, velocity is estimated using one of the following methods:

- Flag method.
- Pointing method.
- Observation method.

Flag Method

5-79. To perform the flag method (Figure 5-27)—

(1) Observe a flag or any cloth-like material hanging from a pole.
(2) Estimate the angle formed at the juncture of the flag and pole.
(3) Divide this angle by the number 4.

NOTE: The answer is the wind velocity expressed in miles per hour.

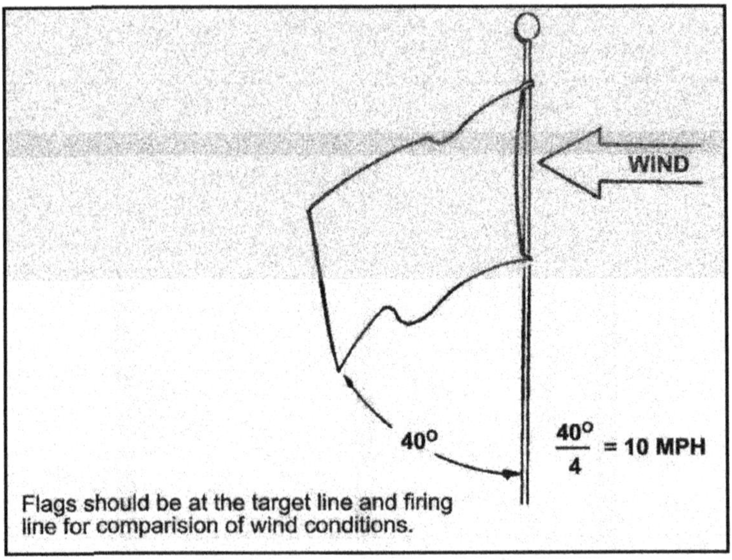

Figure 5-27. Determine wind speed using the flag method.

Pointing Method

5-80. If a flag is not visible, the firer can use the pointing method. To perform the pointing method (Figure 5-28)—

(1) Drop a piece of paper, leaf, or other light material from the shoulder.
(2) Point directly at the place where it lands.
(3) Estimate the angle created by the pointing arm.
(4) Divide this angle by the number 4.

NOTE: The answer is the approximate wind speed at the firing position expressed in miles per hour.

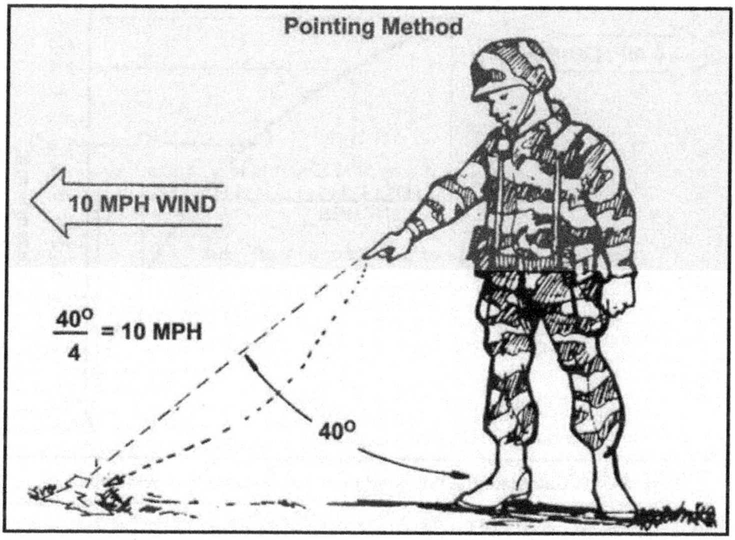

Figure 5-28. Determine wind speed using the pointing method.

Observation Method

5-81. If the flag or pointing methods cannot be used, the following information can assist in determining wind velocities:

- Winds less than 3 miles per hour can barely be felt by the firer, but the presence of slight wind can be determined by drifting smoke.
- Winds of 3 to 5 miles per hour can be felt lightly over the firer's face.
- Winds of 5 to 8 miles per hour constantly move the leaves of trees.
- Winds of 8 to 12 miles per hour raise dust and loose paper.
- Winds of 12 to 15 miles per hour cause small trees to sway.

ADJUSTED POINT OF AIM BASED ON WIND SPEED

5-82. Figure 5-29 illustrates how the effects of wind on the bullet are similar to the effects of gravity—as range increases, the effect of wind increases. For example, a 10 mile-per-hour full-value wind moves an M16A1 (M193) bullet from about ½ of an inch at 25 meters to about 15 inches at 300 meters.

5-83. Table 5-7 displays the wind effects for all conditions—a wind of greater speed increases bullet movement by a uniform amount. For example, a 15 mile-per-hour wind moves the bullet ¾ of an inch at 25 meters and about 22 ½ inches at 300 meters. A half-value wind moves the strike of the round in a 10 mile-per-hour wind ¼ of an inch at 25 meters and 7 ½ inches at 300 meters.

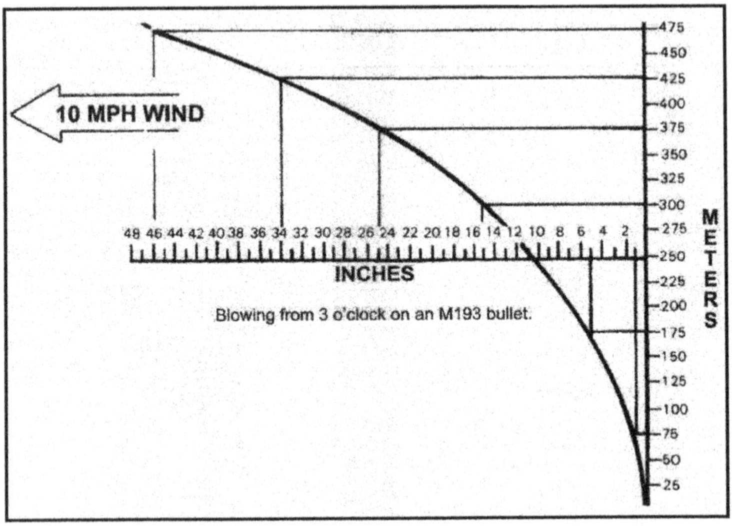

Figure 5-29. Calculate the adjusted point of aim based on wind speed.

NOTE: Table 5-7 can be used to calculate the M193 adjusted point of aim based on wind speed.

Table 5-7. M193 calculated adjusted point of aim based on wind speed (full value).

WIND SPEED (mph)	RANGE (m)								
	25	50	75	100	150	175	200	250	300
	DISTANCE MOVED (in)								
5	1/4	3/8	1/2	1	2	2.5	3.5	5	7.5
10	1/2	3/4	1	2	4	5	7	10	15
15	3/4	1-1/8	1.5	3	6	7.5	10.5	15	22.5

Drift for a 10 Mile-per-Hour Wind Using 5.56-Millimeter M855 Ammunition

5-84. Table 5-8 illustrates the drift using M855 5.56-millimeter ball ammunition fired in an M16A2 rifle with a 300-meter battlesight zero.

Table 5-8. Drift for 10 mile-per-hour wind using M855 ammunition.

DRIFT FOR 10 mph WIND USING M855 5.56-mm BALL AMMUNITION FIRED IN M16A2 RIFLE WITH A 300-METER BATTLESIGHT ZERO				
Range (m)	**Velocity (fps)**	**Trajectory (in)**	**Drop (in)**	**Drift (in)**
0	3,100	-2.5	0.0	0.0
100	2,751	+4.4	-2.3	1.1
200	2,420	+5.8	-10.2	4.9
300	2,115	0.0	-25.3	11.8
400	1,833	-15.0	-49.5	22.4
500	1,569	-42.9	-86.7	38.0
600	1,323	-88.2	-141.3	59.5
700	1,106	-156.1	-220.9	88.4
800	1,010	-267.7	-339.2	124.9

Adjusted Point of Aim Based on Gravity and Wind Speed

5-85. Wind has a minor effect on the M16 bullet (relative to the size of the target) at ranges out to 100 meters. When engaging targets in excess of 150 meters in heavy winds, Soldiers adjust the point of aim for the wind to increase the probability of a hit. Wind effects are uniform in relation to speed—that is, a 5 mile-per-hour wind has half the effect of a 10 mile-per-hour wind, and a 20 mile-per-hour wind has twice the effect of a 10 mile-per-hour wind.

5-86. Firers must adjust their points of aim into the wind to compensate for its effects. If they miss a distant target and wind is blowing from the right, they should aim to the right for the next shot. A guide for the initial adjustment is to split the front sightpost on the edge of the target facing the wind (Figure 5-30).

5-87. Newly assigned Soldiers should aim at the target's center of mass for the first shot, and then adjust for wind when they are confident that wind caused the miss. Experienced firers should apply the appropriate hold-off for the first shot, but should follow the basic rule—when in doubt, aim at the center of mass.

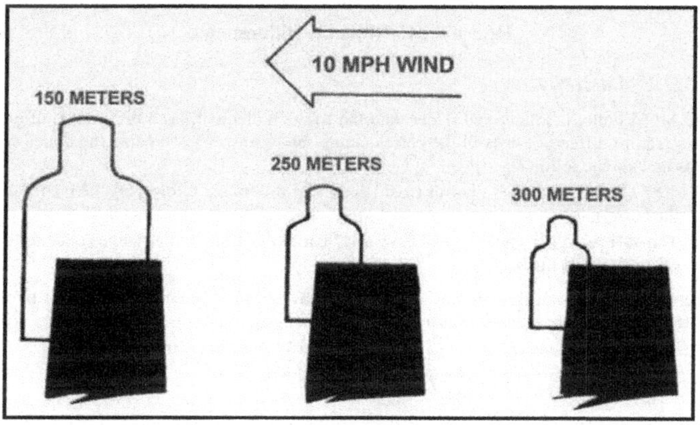

Figure 5-30. M16-/M4-series weapons adjusted point of aim based on wind speed.

SECTION V. BALLISTICS

Commanders and marksmanship trainers must understand some aspects of ballistics to teach the principles of zeroing and engagement of long-range targets. Ballistics is a science dealing with the motion and flight characteristics of projectiles. The study of ballistics in rifles and carbines is divided into three categories:

- Internal ballistics.
- External ballistics.
- Terminal ballistics.

INTERNAL BALLISTICS

5-88. Internal ballistics deals with what happens to the bullet before it leaves the weapon's muzzle.

5-89. The overall dimensions of the combat service 5.56-millimeter cartridges are the same, which allows cartridges to be fired safely in M16-series rifles and M4 carbines, but there are internal differences that affect firing accuracy (Figure 5-31).

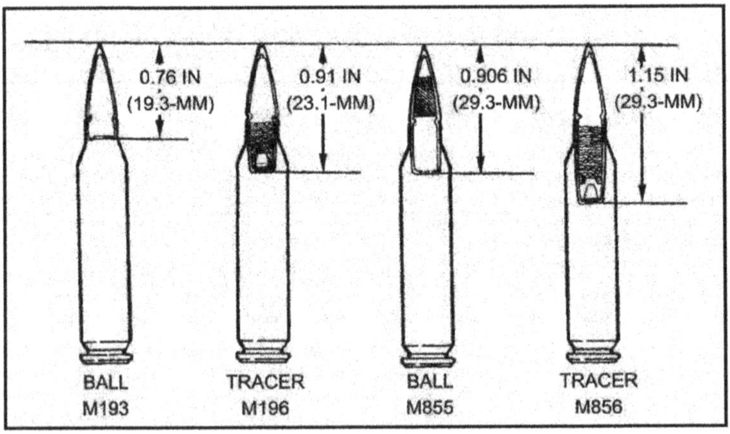

Figure 5-31. Projectile differences.

M855 AND M193 AMMUNITION

5-90. The M855 bullet is longer and wider than the M193 bullet and has a different configuration. These differences require different twists in the barrels, lands, and grooves to stabilize the bullet in flight. These differences include the following:

- The M16A1 has a 1:12 barrel twist (the bullet rotates once for every 12 inches of travel down the barrel).
- The M16A2/A3/A4 and the M4 have a 1:7 barrel twist (the bullet rotates once for every 7 inches of travel down the barrel).

5-91. The M16A1 does not put enough spin on the heavier M855 bullet to stabilize it in flight, causing erratic performance and inaccuracy (Figure 5-32). The shot groups are—

- 30.48 to 35.56 centimeters (12 to 14 inches) at 91.4 meters (100 yards).
- 182.88 centimeters (72 inches) at 274.2 meters (300 yards).

NOTE: Although firing the M855 cartridge in the M16A1 rifle is safe, it should only be used in a combat emergency, and then only for close ranges of 91.4 meters (100 yards) or less.

5-92. The M16A2/A3/A4 rifle and M4/M4A1 carbine fire both M193 and M855 ball ammunition with little difference in accuracy to a range of 500 meters. The M16A2/A3/A4 and M4/M4A1 and their ammunition are more effective than the M16A1 at ranges out to and beyond 500 meters due to better stabilization of the round.

5-93. The three 10-round shot groups in Figure 5-32 (A) were fired by a skilled marksman at a distance of 274.2 meters (300 yards) and 91.4 meters (100 yards) using the same M16A1 rifle.

- At 300 yards, the 25.4-centimeter shot group (shown on the left) was fired (and zeroed) with M193 ammunition.
- The 6-foot shot group (shown on the right) was fired with M855 ammunition.
- At 100 yards, the 35.56-centimeter (14-inch) shot group (shown in the center) was fired with M855 ammunition.

5-94. Figure 5-32 (B) shows two 25.4-centimeter (12-inch) shot groups fired by the same skilled marksman at a distance of 274.2 meters (300 yards) using an M16A2 rifle.

- The shot group on the left was fired and zeroed with M855 ammunition.
- The shot group on the right was fired using M193 ammunition.

NOTE: Both M193 and M855 ball ammunition can be used in training and accurately function in M16A2/3/4 rifles and M4/M4A1 carbines. Due to the different characteristics of each round, zero with the type of ammunition used for training. Do not switch between the types during firing. Do not zero with one type, and then fire the other for any type of training.

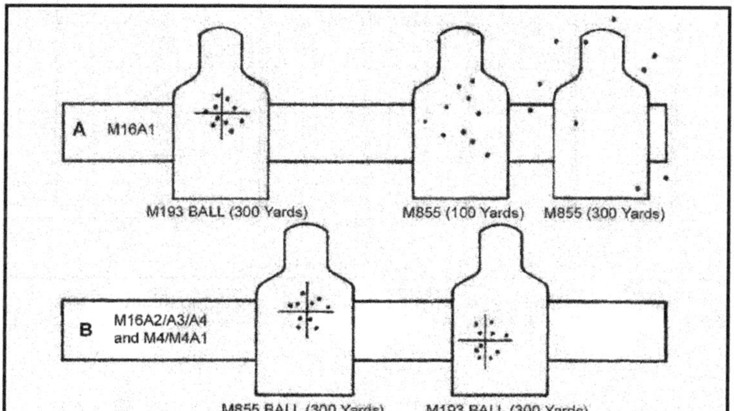

Figure 5-32. Ammunition impact comparison.

5-95. Table 5-9 and Figure 5-33 show the difference between a M16A2/A3/A4 and a M4/M4A1 weapon zeroed with M855 ammunition and then re-zeroed with M193 ammunition at 300 meters. There is practically no difference between the trajectory of the rounds and the impact of the rounds on the target.

Table 5-9. M855 zeroed at 300 meters; M193 re-zeroed at 300 meters using an M16A2/A3/A4 rifle and an M4 carbine.

RANGE (m)	FRONT SIGHT HEIGHT	THETA		BULLET DROP		BULLET LOCATION	
		M193	M855	M193	M855	M193	M855
0	2.568	0.0021987	0.00235575	0	0	-2.57	-2.57
25	2.568	0.0021987	0.00235575	-0.125	-0.14	-0.53	-0.39
50	2.568	0.0021987	0.00235575	-0.52	-0.58	1.24	1.49
75	2.568	0.0021987	0.00235575	-1.21	-1.33	2.71	3.06
100	2.568	0.0021987	0.00235575	-2.1	-2.4	3.99	4.31
125	2.568	0.0021987	0.00235575	-3.18	-3.82	5.07	5.21
150	2.568	0.0021987	0.00235575	-4.91	-5.61	5.51	5.73
175	2.568	0.0021987	0.00235575	-6.89	-7.78	5.69	5.88
200	2.568	0.0021987	0.00235575	-9.16	-10.4	5.58	5.58
225	2.568	0.0021987	0.00235575	-11.8	-13.4	5.11	4.90
250	2.568	0.0021987	0.00235575	-15.4	-16.9	3.67	3.72
275	2.568	0.0021987	0.00235575	-19.2	-20.8	2.04	2.14
300	2.568	0.0021987	0.00235575	-23.44	-25.3	-0.04	-0.04
325	2.568	0.0021987	0.00235575	-28.3	-30.4	-2.74	-2.83
350	2.568	0.0021987	0.00235575	-33.8	-36.0	-6.07	-6.11
375	2.568	0.0021987	0.00235575	-40.13	-42.3	-10.24	-10.09
400	2.568	0.0021987	0.00235575	-46.78	-49.3	-14.72	-14.77

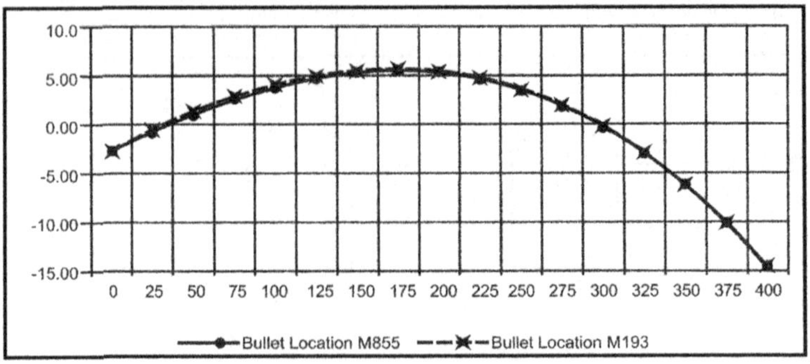

Figure 5-33. M855 zeroed at 300 meters; M193 re-zeroed at 300 meters.

5-96. When zeroing using M855 and M193 ammunition at 25 meters, the differences in the ammunition becomes apparent (Table 5-10 and Figure 5-34). If firing M193 ammunition from an M16A2/A3/A4 rifle or an M4/M4A1 carbine, the weapon should be zeroed with M193 ammunition.

Table 5-10. M855 zeroed at 25 meters; M193 re-zeroed at 25 meters.

RANGE (m)	FRONT SIGHT HEIGHT	THETA		BULLET DROP		BULLET LOCATION	
		M193	M855	M193	M855	M193	M855
0	2.568	0.0023383	0.00235575	0	0	-2.57	-2.57
25	2.568	0.0023383	0.00235575	-0.125	-0.14	-0.39	-0.39
50	2.568	0.0023383	0.00235575	-0.52	-0.58	1.51	1.49
75	2.568	0.0023383	0.00235575	1.21	-1.33	3.13	3.06
100	2.568	0.0023383	0.00235575	-2.1	-2.4	4.54	4.31
125	2.568	0.0023383	0.00235575	-3.18	-3.82	5.76	5.21
150	2.568	0.0023383	0.00235575	-4.91	-5.61	6.33	5.73
175	2.568	0.0023383	0.00235575	-6.89	-7.78	6.65	5.88
200	2.568	0.0023383	0.00235575	-9.16	-10.4	6.68	5.58
225	2.568	0.0023383	0.00235575	-11.8	-13.4	6.35	4.90
250	2.568	0.0023383	0.00235575	-15.4	-16.9	5.05	3.72
275	2.568	0.0023383	0.00235575	-19.2	-20.8	3.55	2.14
300	2.568	0.0023383	0.00235575	-23.44	-25.3	1.61	-0.04
325	2.568	0.0023383	0.00235575	-28.3	-30.4	-0.95	-2.83
350	2.568	0.0023383	0.00235575	-33.8	-36.0	-4.15	-6.11
375	2.568	0.0023383	0.00235575	-40.13	-42.3	-8.18	-10.09
400	2.568	0.0023383	0.00235575	-46.78	-49.3	-12.52	-14.77

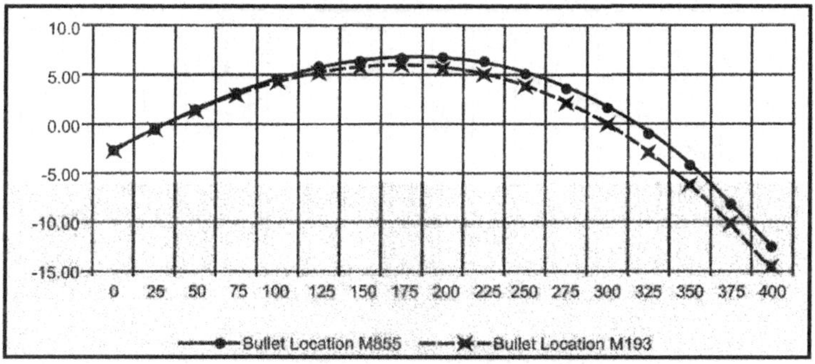

Figure 5-34. M855 zeroed at 25 meters; M193 re-zeroed at 25 meters.

5-97. If an M16A2/A3/A4 rifle or an M4/M4A1 carbine has been zeroed at 25 meters using M855 ammunition and M193 ammunition is fired without re-zeroing, the difference in the impact of the rounds on the target varies between ammunition (Table 5-11 and Figure 5-35).

Table 5-11. M855 zeroed at 25 meters; M193 fired using M855 zero.

RANGE (m)	FRONT SIGHT HEIGHT	THETA		BULLET DROP		BULLET LOCATION	
		M193	M855	M193	M855	M193	M855
0	2.568	0.00235575	0.00235575	0	0	-2.57	-2.57
25	2.568	0.00235575	0.00235575	-0.125	-0.14	-0.37	-0.39
50	2.568	0.00235575	0.00235575	-0.52	-0.58	1.55	1.49
75	2.568	0.00235575	0.00235575	-1.21	-1.33	3.18	3.06
100	2.568	0.00235575	0.00235575	-2.1	-2.4	4.16	4.31
125	2.568	0.00235575	0.00235575	-3.18	-3.82	5.85	5.21
150	2.568	0.00235575	0.00235575	-4.91	-5.61	6.43	5.73
175	2.568	0.00235575	0.00235575	-6.89	-7.78	6.77	5.88
200	2.568	0.00235575	0.00235575	-9.16	-10.4	6.82	5.58
225	2.568	0.00235575	0.00235575	-11.8	-13.4	6.50	4.90
250	2.568	0.00235575	0.00235575	-15.4	-16.9	5.22	3.72
275	2.568	0.00235575	0.00235575	-19.2	-20.8	3.74	2.14
300	2.568	0.00235575	0.00235575	-23.44	-25.3	1.82	-0.04
325	2.568	0.00235575	0.00235575	-28.3	-30.4	-0.73	-2.83
350	2.568	0.00235575	0.00235575	-33.8	-36.0	-3.91	-6.11
375	2.568	0.00235575	0.00235575	-40.13	-42.3	-7.92	-10.09
400	2.568	0.00235575	0.00235575	-46.78	-49.3	-12.25	-14.77

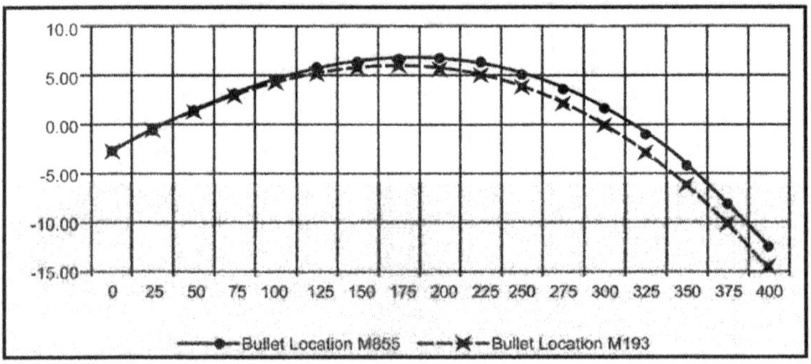

Figure 5-35. M855 zeroed at 25 meters; M193 fired using M855 zero.

5-98. A simple rule of thumb that will preclude any problem is to use only the ammunition specifically designed for each weapon (M193 ball ammunition for M16A1 rifles; M855 ball ammunition for M16A2/3/4 rifles and M4 carbines).

EXTERNAL BALLISTICS

5-99. External ballistics deals with factors affecting the flight path of the bullet between the weapon's muzzle and the target.

5-100. Soldiers must understand the basics of external ballistics so they can make necessary scope adjustments or hold compensations to allow them to hit the target. The external ballistic factors that affect bullet trajectory are:

- Gravity.
- Muzzle velocity.
- Air resistance (drag).
- Altitude or air density.
- Temperature.
- Trajectory.
- Wind.
- Angles.

GRAVITY

5-101. The force of gravity on a bullet is constant regardless of its weight, shape, or velocity. The longer a bullet is in the air or the greater its angle from the vertical, the more effect gravity will have on its trajectory.

NOTE: See paragraphs 5-71 through 5-72 for more information about the effects of gravity.

MUZZLE VELOCITY

5-102. Muzzle velocity is the speed of a bullet as it leaves the barrel, measured in feet per second. Muzzle velocity diminishes as the bullet gets farther away. The bullet reaches its maximum velocity 76 feet from the end of the weapon and slows until it reaches the target.

AIR RESISTANCE (DRAG)

5-103. Air resistance, or drag, immediately produces a slowing effect on a bullet.

ALTITUDE OR AIR DENSITY

5-104. The greater the altitude, the thinner the air and the longer the bullet will travel (with a correspondingly flatter trajectory). Each 5,000-foot elevation will raise the strike of the bullet ½ to 1 minute of angle (MOA).

TEMPERATURE

5-105. Deviation from standard daytime temperature (59 degrees Fahrenheit/15 degrees Celsius) affects bullet trajectory.

Cold Temperatures

5-106. Cold air is denser than warm air; the bullet must travel through more tightly packed air particles. This causes the bullet to lose velocity, causing the bullet to impact lower than intended. Cooler air also causes lower chamber pressure, which reduces the initial velocity.

Hot Temperatures

5-107. Warm or hot temperatures cause the strike of the round to move up.

TRAJECTORY

5-108. When a projectile exits the weapon's muzzle, it drops from the line of departure, otherwise known as the center/bore line. As the projectile travels downrange, air drag decreases the velocity, giving way to the inevitable force of gravity. This effect creates trajectory.

Line of Sight

5-109. The line of sight is an imaginary straight line extending from the firer's eye through the telescopic sight, or rear and front sight, to the target.

Line of Departure

5-110. The line of departure is an imaginary straight line extending from the center of the barrel to infinity.

Zero Range

5-111. Zero range is where the projectile intersects the line of sight. It occurs twice—once on the way up and once on the way down.

Apex

5-112. Otherwise known as midrange trajectory, the apex is the point where the projectile is at its highest in relation to the line of sight.

Bullet Path

5-113. The bullet path is the relationship of a projectile and the line of sight at any given range (normally expressed in inches).

WIND

5-114. External factors influence the trajectory relative to the point of aim, such as wind, altitude, temperature, humidity, and barometric pressure. Wind is by far the most significant. Consider the following effects of wind:

- Because the bullet is moving through the air, the air moves the bullet. Wind deflection is always in the same direction the wind is moving. A wind blowing from the left will move the bullet to the right.
- Deflection decreases as the angle of the wind to the line of flight decreases.

5-115. Effectively reading and correcting for wind effects takes practice, especially at longer ranges where accuracy in correcting is more critical. To shoot accurately in the wind, a firer must know the wind velocity, the wind direction, and the value of deflection at the range at which he is shooting.

NOTE: See paragraph 5-74 for more information on the effects of wind.

ANGLES

5-116. Firing uphill or downhill normally causes the bullet to hit high relative to a horizontal trajectory. If the firer is firing on an angle up or down at a slanted range of 100 meters, the point of impact will be higher than it would be for a level shot of 100 meters. The height depends on the angle.

5-117. Gravity acts on a bullet only during the horizontal component of its flight (the distance from the firer to the target measured as if they were both at the same level). Since the horizontal component will always be less than the slanted range, gravity will not pull the bullet down as far as it would if the range were level.

5-118. Firing uphill or downhill causes the wind to affect the shot over the entire slant range. The firer should aim at the target as if it were 25 meters away and correct for wind as if it were 400 meters away. The correct method for shooting uphill or downhill is to adjust elevation based on the horizontal range and correct for wind deflection based on the slanted range.

TERMINAL BALLISTICS

5-119. Terminal ballistics deals with what happens to the bullet when it comes in contact with the target.

5-120. Bullet penetration depends on the range, velocity, bullet characteristics, and target material. Greater penetration does not always occur at close range with certain materials, since the high velocity of the 5.56-millimeter bullet causes it to disintegrate soon after impact.

BULLET DISPERSION AT RANGE

5-121. Instructors/trainers must have a working knowledge of the effects of bullet dispersion and accuracy at various ranges.

MINUTE OF ANGLE

5-122. An MOA is the standard unit of measurement used in adjusting a weapon's sights and other ballistic-related measurements. It is also used to indicate the accuracy of a weapon.

5-123. A circle is divided into 360 degrees. Each degree is further divided into 60 minutes; therefore, a circle contains 21,600 minutes. An MOA is an angle beginning at the muzzle that covers 2.54 centimeters at a distance of 91.4 meters (Figure 5-36).

5-124. When the range is increased to 182.8 meters, the angle covers twice the distance, or 5.08 centimeters. The rule applies as range increases—7.62 centimeters at 274.2 meters, 10.16 centimeters at 365.6 meters, and so on.

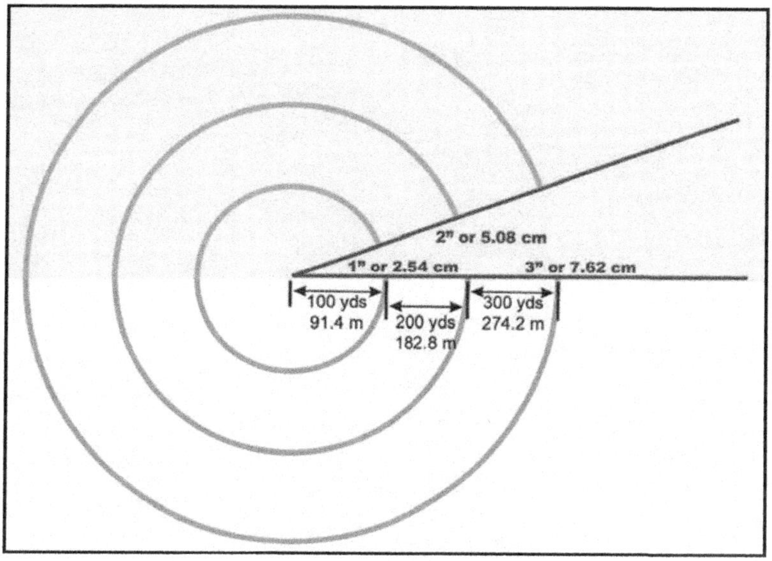

Figure 5-36. Minute of angle.

INCREASE OF SHOT GROUP SIZE

5-125. Just as the distance covered by an MOA increases each time the range increases, a shot group can be expected to do the same. If there are 2.54 centimeters between bullets on a 25-meter target, there will be an additional 2.54 centimeters of dispersion for each additional 25 meters of range. A 2.54-centimeter shot group at 25 meters (about 3.5 MOA) is equal to a 25.4-centimeter shot group at 250 meters (Figure 5-37).

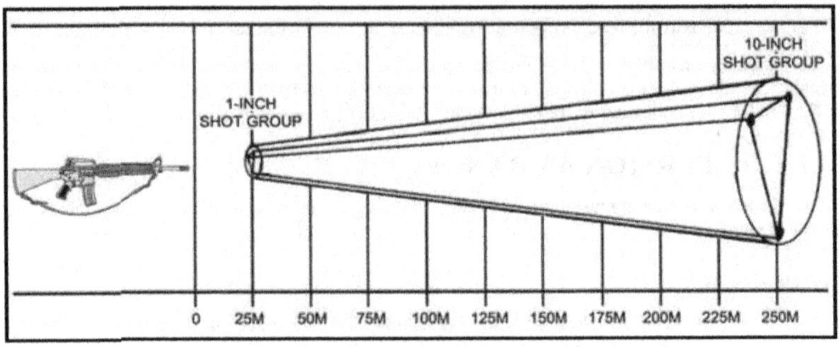

Figure 5-37. Increase in shot group size as range increases.

Chapter 6

Field Fire

Field firing is part of the continued progression in the development of combat shooting skills. This begins the Soldier's critical transition from unstressed firing at single, known distance targets to targets at various ranges for short exposures. It also requires the Soldier to practice and refine previously taught skills.

This chapter introduces the techniques for scanning the range for targets, estimating range, and firing quickly and accurately.

NOTE: See Table 6-1 for the current training program.

Table 6-1. Field Fire I and II training program.

FIELD FIRE I AND II
Instructional Intent
• Reinforce PMI and downrange feedback by detecting and engaging single and multiple timed targets with the M16-/M4-series weapon.
Special Instructions
Ensure that— • The rear sight is on the proper setting (M16A2/3=8/3; M16A4 and M4=6/3 flush; M16A1=the unmarked aperture, short-range). • The rear sight aperture is set on 300, not 800. • The small aperture is being used. • The range consists of targets at 75, 175, and 300 meters.
Observables
• Coaches are used to analyze the firer's adherence to the fundamentals—not as scorers. • Soldiers detect and achieve 22 target hits out of 36 timed target exposures (Field Fire I). • Soldiers detect and achieve 27 target hits out of 44 timed target exposures (Field Fire II). • Soldiers that don't achieve the standard receive remedial training prior to refiring.

SECTION I. TARGET DETECTION

For most Soldiers, finding the target is a greater problem than hitting it. Target detection is the process of locating, marking, prioritizing, and determining the range to combat targets. Target detection must be conducted as part of individual training and tactical exercises, and must be integrated into day and night LFXs.

NOTE: Refer to TC 25-8 for construction of a target detection range.

LOCATING TARGETS

6-1. The ability to locate a combat target depends on the observer's—

- Position.
- Skill in scanning.
- Ability to observe the area and recognize the type of indicators made by the target.

Selection of a Position

6-2. A good position is one that offers maximum visibility of the area, while affording cover and concealment. In this case, the word "position" refers to both the observer's location on the ground and the position of his body at that location.

NOTE: Instructors must continuously refer to and emphasize the importance of the observer's position when conducting practical exercises.

6-3. Depending on the situation, the individual Soldier may or may not select his own position.

- In most defensive situations, the Soldier is told where to prepare his position.
- Some situations (for example, the attack and reorganization on the objective) require the Soldier to select his own position.

Scanning

6-4. To scan the area, Soldiers use three methods of search:

- Self-preservation method.
- 50-meter overlapping strip method.
- Maintaining observation of the area.

Self-Preservation Method of Search

6-5. When moving into a new area, Soldiers use the self-preservation method of search. To perform this method, use the following techniques:

(1) For approximately 30 seconds, quickly scan the area for enemy activity that may be of immediate danger.

(2) Make quick glances at specific points throughout the area, rather than just sweeping the eyes across the terrain in one continuous panoramic view.

NOTE: The eyes are sensitive to slight movements that occur within the area the eyes are focused on; panoramic views do not allow the eyes to detect the slight movements of a concealed target.

50-Meter Overlapping Strip Method of Search

6-6. If the Soldier fails to locate the enemy during the initial search, he must begin a systematic examination known as the 50-meter overlapping strip method of search. To perform this method—

(1) Begin the search with the area offering the greatest potential danger, the terrain nearest to your position.

(2) Beginning at either flank, systematically search the terrain to your front in a 180-degree arc, 50 meters in depth.

NOTE: Become familiar with the terrain as you search it. Take advantage of peripheral vision by focusing the eyes on specific points as you search from one flank to the other. Make mental notes of prominent terrain features and areas that may offer cover and concealment to the enemy.

---

(3) After reaching the opposite flank, search a second 50-meter strip farther out, but overlapping the first strip by approximately 10 meters.

(4) Continue in this manner until the entire area has been searched.

NOTE: This method should also be used as part of maintaining observation of the area and when the observer has been distracted from his area of responsibility.

Maintaining Observation of the Area

6-7. After completing his detailed search, the Soldier may be required to maintain observation of the area. To perform this method—

- Glance quickly at various points throughout the entire area, focusing the eyes on specific features.
- Always search the area in the same manner to ensure complete coverage of all terrain.

NOTE: Since this quick search may fail to detect the initial movement of an enemy, the observer should periodically repeat the procedures outlined in the 50-meter overlapping strip method of search.

TARGET INDICATORS

6-8. A target indicator is anything that a Soldier (friendly or enemy) does or fails to do that reveals his position. Since these indicators apply equally to both sides of the battlefield, Soldiers must learn to use target indicators to locate the enemy and to prevent the enemy from using the same indicators to locate them. For instructional purposes, these indicators can be grouped into three general areas:

- Sound.
- Movement.
- Improper camouflage.

Sound

6-9. Sounds, such as footsteps, coughing, or equipment noises, provide only a direction and general location, making it difficult to pinpoint a target by sound alone. However, detection of a sound alerts the observer, greatly increasing the possibility that he will eventually locate the target through other target indicators.

Movement

6-10. The degree of difficulty in locating moving targets depends primarily on the speed of movement. Slow, deliberate movements are much more difficult to notice than those that are quick and jerky.

Improper Camouflage

6-11. The lack or improper use of camouflage or concealment reveals the majority of targets detected on the battlefield; alert observers easily notice indicators such as light reflecting from shiny surfaces or a contrast with the background. Three general indicators may reveal a camouflaged or concealed target:

- Shine.
- Regularity of outline.
- Contrast with the background.

Shine

6-12. Metal objects, such as belt buckles, reflect light and act as a beacon to the wearer's position. This is as true at night as it is during the day.

Regularity of Outline

6-13. Humans and most types of military equipment cast outlines that are familiar to all Soldiers. The outlines of rifles, helmets, and vehicles are all easily identified. The reliability of this indicator depends on the visibility and the experience of the observer. On a clear day, most Soldiers can easily identify enemy riflemen or equipment if a distinctive outline is presented. At night or during other periods of poor visibility, seeing outlines is not only more difficult, but inexperienced troops will frequently mistake stumps and rocks for enemy Soldiers. This is an additional reason for Soldiers to become completely familiar with the terrain during periods of good visibility.

Contrast with the Background

6-14. If a Soldier wearing a dark uniform moves into a position in front of a snow bank, the contrast between the white snow and the dark uniform makes him clearly visible. However, if he was wearing a white or light-colored uniform, he would be more difficult to see.

6-15. Contrast with the background is the most difficult target indicator for a Soldier to avoid. During operations in which the Soldier is moving, he is usually exposed to numerous background colors. Since no single type of personal camouflage blends in with all areas, a moving Soldier must be continually aware of the surrounding terrain and vegetation.

MARKING TARGETS

6-16. A Soldier observes two enemy riflemen moving into completely concealed positions, one behind a bush and the other into a depression.

- By selecting a point of aim on the bush, the Soldier should hit the enemy rifleman even though he can't see him. If the target cannot be engaged, the point of aim also allows for quick and accurate engagement once the target is re-exposed.
- The enemy rifleman who moved into the depression provides no distinguishable point of aim, so the Soldier must select a nearby feature as a reference point and determine its distance and general direction from the depression. A reference point provides a general point of aim on a concealed target.

6-17. Of the two, a point of aim is usually the more effective means of delivering accurate fire. The difficulty in using reference points to mark targets moving from one location to another depends on the following factors:

- Number of targets.
- Exposure time to target.
- Spacing of targets.
- Good and poor points of aim.

Number of Targets

6-18. If several targets appear and disappear at approximately the same time, it is very difficult to note each target's point of disappearance.

Exposure Time of Target

6-19. Usually, moving targets are exposed for only a short period of time, so the observer must be alert to note the point of disappearance for all of the targets. In such situations, the Soldier should mark the location of as many targets as possible before engaging any of them. By doing so, he will know the location of several targets and can engage each of them in rapid succession.

Spacing of Targets

6-20. The greater the interval between targets, the more difficult it is to note each target's movements. When there is considerable distance between targets, the observer should accurately locate and mark the one nearest to his position and note the general area of the others.

Good and Poor Points of Aim

6-21. Good points of aim are easily distinguishable in the surrounding terrain. Targets disappearing behind good points of aim, such as manmade objects and large terrain features, can be easily marked for future reference.

6-22. Poor points of aim are not easily distinguishable within the surrounding terrain. Targets disappearing behind poor points of aim are difficult to mark accurately and are easily lost.

6-23. If two targets offer the same degree of danger to the Soldier, but one disappears behind a good point of aim and the other behind a poor point of aim, the Soldier should mark the location of the target behind the good point of aim and engage the other target first.

RANGE DETERMINATION

6-24. Range determination is the process of finding the distance between two points. In most situations, one of these points will be the Soldier's own position. The other may be a target or prominent feature.

6-25. To accomplish the mission, combat riflemen must be able to accurately determine range. Not only does this affect his combat marksmanship proficiency, it is also required to report information and adjust artillery and mortar fire.

6-26. The methods of range estimation used during this period are—

- 100-meter unit of measure method.
- Appearance of objects method.
- Front sightpost method.

100-Meter Unit of Measure Method

6-27. To perform this method, the Soldier must be able to visualize a distance of 100 meters on the ground.

- For ranges up to 500 meters, he determines the number of 100-meter increments between the two points.
- Beyond 500 meters, the Soldier selects a point halfway to the target, determines the number of 100-meter increments to the halfway point, and then doubles it to find the range to the target.

6-28. During training exercises, the Soldier must become familiar with the effect that sloping ground has on the appearance of a 100-meter increment.

- Ground that slopes upward gives the illusion of greater distance, and observers tend to underestimate a 100-meter increment.
- Ground that slopes downward gives the illusion of shorter distance, and observers tend to overestimate a 100-meter increment.

6-29. Proficiency in the 100-meter unit of measure method requires constant practice. While training this technique, comparisons should be made continually between the range as determined by the Soldier and the actual range as determined by pacing or other more accurate means of measurement.

NOTE: The best training technique is to require the Soldier to pace the range after he has visually determined it. In this way, he discovers the actual range for himself, which makes a much greater impression than if he is simply told the correct range.

6-30. The greatest limitation of the 100-meter unit of measure method is that its accuracy is directly related to the amount of terrain visible to the observer. This is particularly true at longer ranges. If a target appears at a range of 500 meters or more and the observer can see only a portion of the ground between himself and the target, it becomes very difficult to use the 100-meter unit of measure method with any degree of accuracy.

APPEARANCE OF OBJECTS METHOD

6-31. The appearance of objects method is a means of determining range by the size and other details of the object observed. This is a common method of determining distances and is used by most people in their everyday living. For example, a motorist attempting to pass another car must judge the distance of oncoming vehicles based on his knowledge of how vehicles appear at various distances. Suppose the motorist knows that, at a distance of one mile, an oncoming vehicle appears to be 1 inch wide and 2 inches high. Then, any time he sees other oncoming vehicles that fit these dimensions, he knows they are about one mile away.

6-32. The rifleman can use this technique to determine ranges on the battlefield. If he knows the characteristics, size, and detail of personnel and equipment at known ranges, he can compare these characteristics to similar objects at unknown ranges. When the characteristics match, so do the ranges.

6-33. To use the appearance of objects method with any degree of accuracy, the Soldier must be thoroughly familiar with the details of objects as they appear at various ranges. For example, the Soldier should study the appearance of a man standing at a range of 100 meters. He fixes the man's appearance firmly in his mind, carefully noting details of size and the characteristics of uniform and equipment. Next, he studies the same man in a kneeling position and in a prone position. By comparing the appearance of Soldiers in these positions at known ranges from 100 to 500 meters, the Soldier can establish a series of mental images that will help him determine range on unfamiliar terrain.

NOTE: Training should also be conducted in the appearance of other familiar objects, such as weapons or vehicles.

6-34. Because the successful use of this method depends upon visibility, anything that limits visibility (for example, weather, smoke, or darkness) will limit the effectiveness of this method.

FRONT SIGHTPOST METHOD

6-35. Using the front sightpost as a scale is another method of estimating range. This method can be used for a quick on-the-spot estimation and engagement.

- Generally, if a man-sized target is ½ of the width of the front sightpost, the target is approximately 300 meters away.
- If the target is ¼ of the width of the front sightpost, the target is approximately 600 meters away.

SECTION II. FIELD FIRE TRAINING

Field fire training provides the transition from unstressed, slow firing at known distances or feedback targets to engaging pop-up silhouettes 50 to 300 meters away. Two basic types of field firing exercises are single and multiple target timed engagements, which use 75-, 175-, and 300-meter targets. Pop-up targets are used to add stress and simulate the short exposure times of combat targets. During field fire training, the firer learns to quickly detect targets, apply SPORTS, and apply the four fundamentals simultaneously.

CONDUCT OF A FIELD FIRE RANGE

6-36. On field fire ranges, Soldiers fire (from the supported and prone unsupported firing positions) at F-type silhouettes 75 meters away and E-type silhouettes 175 and 300 meters away. Initial training begins with single exposed targets and increased time for target exposures (Field Fire I). As Soldiers become proficient, multiple target engagements with shorter exposure times are introduced (Field Fire II).

NOTE: There are two types of electronic pop-up targets used on a field fire range: those that rise to the upright position from the back and those that rise to the upright position from the side. When using targets that rise from the side, Soldiers should be instructed to wait until the target is fully raised before engaging it. If the target is engaged as it is rising, the computer will not register the hit even though the target may fall.

Remedial Training

6-37. Soldiers who miss most targets should be removed from the firing line for remedial training if their problem cannot be corrected. A Soldier who fires at a 300-meter target 10 times and misses it 10 times is not learning, instead he is losing confidence in his ability. The typical Soldier should hit the 300-meter target at least 7 out of 10 times.

Peer Coaching

6-38. Peer coaches should assist Soldiers in observing the strike of rounds and identifying firing problems. If the target is missed and the coach cannot observe the bullet strike, the coach should instruct the Soldier to aim lower for the next shot, expecting to see the strike of the bullet in the ground. With this information, the coach can instruct the Soldier where to aim to hit the target.

Organization

6-39. Live-fire training can be organized in several ways. A unit is divided into two or more firing orders based on the number of personnel to be trained:

- The first order is the firer.
- The second order is the coach.
- The third order is the scorer (if required).

6-40. At the conclusion of each exercise, positions rotate until all orders have fired. Standard field fire scenarios have been developed to provide several target exposures. Although they are recommended for IET, local commanders can develop any variety of more challenging target sequences. Ammunition is allocated based on one round for each target.

Recording

6-41. During live-fire, the Soldier's hit-and-miss performance is recorded to facilitate the instructor/trainer's critiques or to indicate where more training, closer supervision, or remedial training is needed. Two methods are used to record firing performance:

- Manually marked scorecards.
- Automated computer printouts.

Manual Recording

6-42. When manual recording is used, the unit provides Soldiers for recording information on either DA Form 3601-R (Single Target Field Fire I Scorecard) or DA Form 5241-R (Single and Multiple Targets Field Fire II Scorecard).

NOTE: See the end of this publication for blank reproducible copies of these forms.

Automated Recording

6-43. When firing exercises are conducted using automated field fire (AFF) ranges, a computer printout is provided for each firing order. At the conclusion of each firing order, the range noncommissioned officer in charge (NCOIC) completes the printout and ensures that the Soldier identification is matched with each firing point. He adds the Soldier's numbered code to the top of each lane/firing point data column. Based on a one-round allocation for each target exposure, data should be collected on hits, misses, no-fires, and repeated shots to assist the instructor/trainer in assessing firing proficiency.

FIELD FIRE I (SINGLE TIMED TARGET)

6-44. Field Fire I is broken down into four firing tables (Table 6-3).

CONCEPT

6-45. Firing Table 1 helps the firer practice shooting skills and develop the sense of timing and rhythm required to make the transition from KD to field fire. This builds confidence prior to firing the exercises in Firing Tables 2, 3, and 4, and identifies Soldiers who are having difficulty and need reinforcement.

6-46. When firing the exercises in Firing Tables 2, 3, and 4, each Soldier demonstrates his ability to apply the fundamentals of marksmanship during the integrated act of firing by successfully detecting and engaging single timed targets.

CONDUCT

NOTE: When firing record fire, each Soldier must wear the proper uniform: the helmet, LBE, and IBA with all SAPI plates (if available). No other armor is required.

6-47. Each firer receives 54 rounds of 5.56-millimeter ball ammunition, with two 18-round magazines (one magazine each for Firing Tables 1 and 2) and two 9-round magazines (one magazine each for Firing Tables 3 and 4). Table 6-2 depicts the number of rounds that must be fired from each position. Each Soldier must achieve 22 hits out of 36 timed target exposures in Firing Tables 2, 3, and 4. Table 6-3 shows the number of target exposures, target ranges, and exposure times for each firing table.

Table 6-2. Number of rounds that must be fired from each position during Field Fire I.

POSITION	NUMBER OF ROUNDS FIRED
Supported firing position	18
Supported firing position	18
Unsupported firing position	9
Kneeling	9

Table 6-3. Field Fire I firing tables.

FIRING TABLE 1			FIRING TABLE 2			FIRING TABLE 3			FIRING TABLE 4		
ROUND	RANGE (m)	TIME (sec)	ROUND	RANGE (m)	TIME (sec)	ROUND	RANGE (m)	TIME (sec)	ROUND	RANGE (m)	TIME (sec)
1	75	6	1	75	6	1	75	6	1	75	6
2	175	8	2	175	8	2	175	8	2	175	8
3	300	10	3	300	10	3	300	10	3	75	6
4	175	8	4	175	8	4	175	8	4	175	8
5	75	6	5	75	6	5	75	6	5	75	6
6	300	10	6	300	10	6	300	10	6	175	8
7	300	10	7	300	10	7	300	10	7	75	6
8	75	6	8	75	6	8	75	6	8	75	6
9	175	8	9	175	8	9	175	8	9	175	8
10	175	8	10	175	8						
11	300	10	11	300	10						
12	175	8	12	175	8						
13	75	6	13	75	6						
14	300	10	14	300	10						
15	175	8	15	175	8						
16	75	6	16	75	6						
17	300	10	17	300	10						
18	75	6	18	75	6						

FIELD FIRE II (MULTIPLE OR SINGLE TIMED TARGETS)

6-48. Field Fire II consists of three firing tables.

CONCEPT

6-49. Firing Table 1 helps the firer practice shooting skills and develop the sense of timing and rhythm required to make the transition from single timed targets to multiple or single timed fleeting combat targets. This builds confidence prior to firing the exercises in Firing Tables 2 and 3, and identifies Soldiers who are having difficulty and need reinforcement.

6-50. When firing the exercises in Firing Tables 2 and 3, each Soldier demonstrates his ability to apply the fundamentals of marksmanship during the integrated act of firing by successfully detecting and engaging multiple and single timed targets.

CONDUCT

NOTE: When firing record fire, each Soldier must wear the proper uniform: the helmet, LBE, and IBA with all SAPI plates (if available). No other armor is required.

6-51. Every firer receives 54 rounds of 5.56-millimeter ball ammunition, with 10 rounds loaded into one magazine (for Firing Table 1) and 44 rounds loaded into two separate magazines, with 22 rounds each (for Firing Tables 2 and 3). Table 6-4 depicts the number of rounds that must be fired from each position. Each Soldier must achieve 27 hits out of the 44 timed target exposures in Firing Tables 2 and 3. Table 6-5 shows the number of target exposures, target distance, and exposure times for each firing table.

Table 6-4. Number of rounds that must be fired from each position during Field Fire II.

POSITION	NUMBER OF ROUNDS FIRED
Supported firing position	10
Supported firing position	22
Unsupported firing position	22

Table 6-5. Field Fire II firing tables.

FIRING TABLE 1			FIRING TABLE 2			FIRING TABLE 3		
ROUND	RANGE (m)	TIME (sec)	ROUND	RANGE (m)	TIME (sec)	ROUND	RANGE (m)	TIME (sec)
1	75	5	1	175	7	1	75	6
2	175	7	2	75	10	2	175	8
3	75	11	3	300		3	75	13
4	300		4	75	9	4	300	
5	75	9	5	175		5	75	11
6	175		6	300	9	6	175	
7	75	10	7	75	9	7	75	12
8	300		8	175		8	300	
9	175	11	9	175	11	9	175	13
10	300		10	300		10	300	
			11	75	9	11	75	11
			12	175		12	175	
			13	175	11	13	175	8
			14	300		14	75	6
			15	75	5	15	75	11
			16	175	11	16	175	
			17	300		17	75	12
			18	75	9	18	300	
			19	175		19	75	11
			20	75	10	20	175	
			21	300		21	175	13
			22	175	7	22	300	

SECTION III. RECORD QUALIFICATION

The objective of record fire is to access and confirm the individual proficiency of firers and the effectiveness of the training program. Important statistical data, such as qualification ratings and first-time GO rates, are obtained from record fire. These data provide goals for the Soldier and aid the commander in identifying the quality of his training.

PRACTICE RECORD FIRE I AND II

6-52. Although the Soldier receives a practice rating based on the number of target hits, practice record fire should also be considered a valuable training exercise. When practice record fire is correctly conducted, all Soldiers gain valuable experience and become more confident in engaging combat targets.

NOTE: See Table 6-6 for the current training program.

Table 6-6. Practice Record Fire I and II training program.

PRACTICE RECORD FIRE I AND II
Instructional Intent
• Reinforce PMI and KD firing and apply the techniques of target detection by engaging a more difficult course of fire with single and multiple pop-up targets and increased time constraints.
Special Instructions
Ensure that— • The rear sight is on the proper setting (M16A2/3=8/3; M16A4 and M4=6/3 flush; M16A1=the unmarked aperture, short-range). • The rear sight aperture is set on 300, not 800. • The small aperture is being used. • Peer coaching is stressed (Practice Record Fire I only).
Observables
• Soldiers apply all aspects of BRM. • Soldiers hit 23 out of 40 target exposures. • Soldiers that do not meet the standard receive remedial training before refiring. • Practice record fire should be conducted on a different range than record fire.

CONCEPT

6-53. Practice Record Fire I and II should closely resemble all aspects of actual qualification.

CONDUCT

NOTES: 1. When firing record fire, each Soldier must wear the proper uniform: the helmet, LBE, and IBA with all SAPI plates (if available). No other armor is required.

2. Before engaging targets, target detection is accomplished with a dry-fire scenario.

3. If possible, Soldiers should fire Practice Record Fire I and II on different ranges. Soldiers firing Practice Record Fire II on the same range as Practice Record Fire I must fire on a different lane.

6-54. Each firer receives 40 single or multiple target exposures at ranges from 50 to 300 meters and 40 rounds of 5.56-millimeter ball ammunition. Table 6-7 depicts the number of rounds that must be fired from each position. Each Soldier must hit a minimum of 23 out of 40 target exposures.

Table 6-7. Number of rounds that must be fired from each position during Practice Record Fire I and II.

POSITION	NUMBER OF ROUNDS FIRED
Prone supported firing position or (at the unit commander's discretion) the foxhole supported firing position	20
Prone unsupported firing position	10
Kneeling unsupported firing position	10

6-55. Soldiers adhere to the following guidelines:

- Based on the total number of hits achieved in each table, Soldiers are critiqued on the practice record fire score. If a firer consistently misses targets or experiences problems with target detection and range estimation, coaches should point out the shooting error to help correct it.
- Exposure times are three to seven seconds at ranges of 50 to 300 meters. Since it requires one to two seconds for the manually activated target mechanism to raise the target, timing begins when

the target is fully exposed, rather than when the tower operator activates the target switch. When practice record fire is conducted on automated record fire (ARF) ranges, these factors are included in the computer program.

NOTE: Practice Record Fire I allows peer coaching and the use of dummy ammunition. Practice Record Fire II does not allow peer coaching, and dummy ammunition will not be used.

ALIBI FIRING

6-56. Alibi firing should be conducted at the end of each firing table IAW tower operator commands. Alibis are provided during practice record fire for three reasons:

- Malfunction of the weapon.
- Malfunction of the target mechanism.
- Faulty ammunition.

RANGE TRAINING AREAS

6-57. The three range training areas are as follows:

- Orientation area.
- Ready area.
- Retired area.

Orientation Area

6-58. The orientation area is located so that firers cannot see the firing area. Practice record fire orientation includes instructions on the conduct of fire, safety, and range operations, including the procedures used in ready and retired areas.

Ready Area

6-59. The ready area is located near the firing range, but is positioned so that firers cannot see targets on the range. While in this area, the firer blackens the weapon sights, lubricates the weapon, and checks for defects that might cause malfunctions.

Retired Area

6-60. The retired area is about 100 meters behind the ready area. Soldiers completing practice record fire move to the retired area to clean their weapons and be critiqued on their firing performance.

PRACTICE RECORD FIRE STANDARDS

6-61. A firer who fails to qualify on his first try should refire the practice record fire range after his problem has been diagnosed and remedial training has been provided. Practice qualification ratings are shown in Table 6-8.

Table 6-8. Qualification ratings for Practice Record Fire I and II.

QUALIFICATION RATINGS	NUMBER OF TARGETS HIT
Expert	36 to 40
Sharpshooter	30 to 35
Marksman	23 to 29

RECORD OF PERFORMANCE

6-62. Accurate performance data are critical. The firer's score is manually recorded using DA Form 3595-R (Record Fire Scorecard) or automatically documented using a computer printout provided on the automated range. Based on the data recorded, an AAR can be performed by range and firing position to discuss firing performance.

NOTE: See Appendix B for a sample of a completed DA Form 3595-R and the end of this publication for a blank, reproducible copy.

RECORD FIRE

6-63. The intent of record fire is to facilitate the commander's evaluation of several individual tasks and integrated marksmanship skill performances, and to provide unit readiness indicators. The qualification standards are specifically related to a prescribed procedure for conducting record fire. Individual performance must be evaluated IAW two components:

- What record fire test was used (standard, KD, or alternate scaled)?
- How were individual performances distributed (first-time GO rates, percent in each qualification rating)?

NOTE: See Table 6-9 for the current training program.

Table 6-9. Record Fire training program.

RECORD FIRE
Instructional Intent
• Reinforce all phases of BRM. • Allow Soldiers to practice and refine critical marksmanship skills. • Measure the Soldier's complete understanding of BRM.
Special Instructions
Ensure that— • The rear sight is on the proper setting (M16A2/3=8/3; M16A4 and M4=6/3 flush; M16A1=the unmarked aperture, short-range). • The rear sight aperture is set on 300, not 800. • The small aperture is being used. • All targets are operational. • Each Soldier has one 20-round magazine and two 10-round magazines.
Observables
• Soldiers apply all aspects of BRM. • Soldiers hit 23 out of 40 target exposures. • Soldiers that do not meet the standard receive remedial training before refiring.

CONCEPT

6-64. Since all Soldiers must fire the record fire course at least once a year for qualification, the course can provide excellent firing performance evaluations. It also provides excellent diagnostic information for instructor/trainers who are concerned with scheduling training to overcome the most serious firing weaknesses. The standard record fire course should be used for all Soldiers, but there are times when qualification exercises must be conducted on alternate courses.

DEVELOPMENT OF STANDARDS

6-65. Testing and development indicates that the Soldier should hit at least 39 of 40 targets if he applies the marksmanship fundamentals correctly (assuming that target mechanisms have been checked and are functioning). This probability of hit (PH) is provided as a guide to consider the capability of the typical weapon, ammunition, and Soldier firing a standard course (Table 6-10).

Table 6-10. Probability of hits.

RANGE (m)	PH	NUMBER OF TARGETS
50	1.0	6
100	1.0	8
150	1.0	11
200	.99	7
250	.95	5
300	.90	3

6-66. When the IET BRM POI or an adequate unit training program is conducted, the following PH can be expected (Table 6-11).

Table 6-11. Results from an adequate unit training program.

RANGE (m)	TARGETS	LOW PH	AVERAGE PH	HIGH PH
50	6	.80	.95	.98
100	8	.70	.90	.95
150	11	.65	.90	.95
200	7	.45	.70	.90
250	5	.35	.60	.85
300	3	.25	.50	.80
		23 hits	32 hits	37 hits

TARGET FUNCTION

6-67. The first task on a standard record fire course is to ensure that all targets function properly. When in doubt, a lane should be fired to ensure that a bullet strike will activate each target. Sometimes slapping a target with a cleaning rod can cause it to activate, but a bullet impact will not. Hot plastic targets may allow a 5.56-millimeter bullet to pass through without causing sufficient vibration to activate the mechanism, resulting in a requirement to change targets more often, use double targets, or use different silhouettes.

CONDUCT

NOTE: When firing record fire, each Soldier must wear the proper uniform: the helmet, LBE, and IBA with all SAPI plates (if available). No other armor is required.

6-68. Each firer receives 40 single or multiple target exposures at ranges from 50 to 300 meters and 40 rounds of 5.56-millimeter ball ammunition. Table 6-12 depicts the number of rounds that must be fired from each position. Each Soldier must hit a minimum of 23 out of 40 target exposures.

Table 6-12. Number of rounds that must be fired from each position during Record Fire.

POSITION	NUMBER OF ROUNDS FIRED
Prone supported firing position or (at the unit commander's discretion) the foxhole supported firing position	20
Prone unsupported firing position	10
Kneeling unsupported firing position	10

6-69. Soldiers adhere to the following guidelines:

- Credit for targets hit should not be given when rounds are "saved" from difficult targets to be used on easier targets (for example, not firing at the 300-meter target so an additional round can be fired at the 150-meter target).
- When double targets are exposed, the Soldier should fire two rounds. If the first target is missed, he may fire at that same target with the second round.
- Soldiers engage the target that poses the greatest threat first (normally assumed to be the closer target). No scoring distinction is made between near targets and far targets or the sequence in which they are engaged.
- Credit is not given if unused ammunition from one 20-round table is added to a magazine provided for the next table.

Alibi Firing

6-70. Alibi firing is reserved for Soldiers who encounter a malfunctioning target, ammunition, or weapon. A Soldier will not be issued more than 20 rounds for Table 1, 10 rounds for Table 2, or 10 rounds for Table 3. Soldiers who fire 20 rounds, despite a target malfunction, will not be issued additional alibi rounds. There are no alibis for Soldier-induced weapon malfunctions or for targets missed during the application of immediate action procedures.

NOTE: The ammunition allocation and alibi procedures for practice record fire and record fire are conducted the same. The only exception is that coaching is authorized for practice record fire.

6-71. If a weapon or target malfunction occurs—

- The Soldier must apply immediate action procedures and continue to fire the exercise.
- After firing, the Soldier notifies the NCOIC to determine if the ammunition was faulty or if the target malfunctioned.
- The NCOIC verifies the malfunction.
- The Soldier is permitted to fire at that target(s) with the exact number of rounds equal to the target malfunctions. For example, a Soldier had two confirmed target malfunctions at 250 meters. Although he may have had five rounds left from the overall exercise, the Soldier receives only two rounds to engage the two 250-meter target exposures, if repaired, or the nearest target. He is not allowed to fire all five remaining rounds at the two 250-meter target exposures.
- On a computerized range, the tower operator confirms that the target malfunctioned and indicates the number of malfunctions that occurred.

6-72. Inoperable weapons are uncorrectable malfunctions such as the following:

- A broken firing pin.
- Jam caused by a double feed, not by the Soldier.
- Failure to extract due to a broken extractor.
- Round in the bore.

6-73. The Soldier must apply correct immediate action procedures to eliminate stoppages. If a stoppage is determined to be correctable (for example, the Soldier did not apply correct immediate action procedures and, as a result, the Soldier did not engage the required number of targets), the Soldier is at fault.

Troubleshooting Performance

6-74. Onsite observation, detailed analysis and evaluation of individual results, and unit performance identify weaknesses such as the following:

- Unserviceable weapons could cause poor zeroes or failures to fire and, therefore, failures to qualify.
- Some Soldiers may not qualify because of a lack of understanding of immediate action procedures or weapon and magazine maintenance procedures.

- Soldiers who miss targets are not applying the four fundamentals or are not accurately zeroing the weapon.
- Soldiers who do not fire at exposed targets during qualification might be experiencing:
  - Failure to scan the designated area.
  - Lack of ability to detect targets.
  - Lack of ability to shift from one target to another.
  - Failure to manage ammunition.
  - A stoppage.

6-75. Training can then focus on combat tasks, skills, or other factors that address these weaknesses.

Refire

6-76. Qualified weapons personnel or the NCOIC must verify weapon malfunctions before the Soldier can refire the course. Soldiers who erroneously claim a malfunction on the firing line are considered unqualified and refire as a second-time firer. Soldiers who fail to qualify on the first attempt should be given appropriate remedial training and allowed to refire in a few days. When a Soldier refires the course—

- He remains unqualified if he hits 22 targets or less.
- A rating of marksman is awarded for a score of 23 to 40 target hits.
- If automated scoring procedures (if available) allow the Soldier's performance to be stored and retrieved before a weapon malfunction, his performance is added to the score of his first attempt after weapon repair and refire.
- If a Soldier's weapon becomes inoperable and his performance before a malfunction precludes qualification, he is considered unqualified and must refire.

Qualification Ratings

6-77. One point is awarded for each round within or touching some part of the silhouette facing. Qualification ratings are shown in Table 6-13.

Table 6-13. Qualification ratings for Record Fire.

QUALIFICATION RATINGS	NUMBER OF TARGETS HIT
Expert	36 to 40
Sharpshooter	30 to 35
Marksman	23 to 29

Record of Performance

6-78. The record fire range is fired and recorded IAW DA Form 3595-R.

NOTE: See Appendix B for a sample completed scorecard and the end of this publication for a blank, reproducible copy.

SECTION IV. ALTERNATE QUALIFICATION COURSES

Units should conduct weapon qualification on a standard record fire range. Convenience and comfort should not be the prime consideration when choosing a range. Authorized alternate record fire courses are—

- KD record fire range.
- 25-meter scaled target alternate course.
- 15-meter scaled target alternate course.

NOTES: 1. The official records of personnel who use an alternate qualification course are noted to distinguish alternate qualification ratings from standard record fire course ratings. For example, official personnel records are annotated as follows:

JONES, John Q. 000-00-0000 Expert 36 (record fire [RF])

JONES, John Q. 000-00-0000 Expert 38 (known distance alternate course [KDAC])

JONES, John Q. 000-00-0000 Expert 38 (alternate course [AC])

2. The uniform for all alternate qualification courses is a helmet, LBE, and IBA with all SAPI plates (if available). No other armor is required.

3. Firers should engage targets from left to right, from nearest to the farthest (50-meter, 100-meter left, 100-meter center, 100-meter right, 150-meter left, 150-meter right, 200-meter left, 200-meter right, 250-meter, and 300-meter). This ensures that firers do not forget which targets they engaged during qualification. It also alleviates the possibility of shooting each target more than the prescribed number of times.

KNOWN DISTANCE RECORD FIRE RANGE

NOTE: The KD record fire range is used by all components of the U.S. Army, U.S. Army Reserve, and Army National Guard when a record fire range is not available.

6-79. The KD record fire range allows Soldiers to engage targets at range while experiencing time constraints, feedback, and the effects of wind and gravity.

CONDUCT

6-80. To complete this course, Soldiers fire three tables. Table 6-14 depicts these three tables and provides related information, such as time constraints, number of rounds that must be fired, type of target that must be used, and the distance from the firer that the target must be placed.

NOTE: Before firing the course, all Soldiers confirm the zero of their assigned weapons by assuming the prone position and firing six rounds from the 300-meter line. Zero rounds do not count for score.

Table 6-14. Known distance record fire range firing tables and related information.

TABLE	POSITION	TIME CONSTRAINTS	NUMBER OF ROUNDS	TYPE OF TARGET	DISTANCE (m)
Table 1	Prone supported firing position	2 min	20	E-type silhouette target	300
Table 2	Prone unsupported firing position	60 sec	10	E-type silhouette target	200
Table 3	Prone unsupported firing position	60 sec	10	F-type silhouette target	100

QUALIFICATION RATINGS

6-81. Scoring is conducted in the pits, with the results provided after each firing table. One point is awarded for each round within or touching some part of the silhouette facing. Qualification ratings for the KD record fire range are shown in Table 6-15.

Table 6-15. Qualification ratings for the known distance record fire range.

QUALIFICATION RATINGS	NUMBER OF TARGETS HIT
Expert	38 to 40
Sharpshooter	33 to 37
Marksman	26 to 32
Unqualified	25 and below

RECORD OF PERFORMANCE

6-82. The KD record fire range is fired and recorded IAW DA Form 5789-R (Record Fire Scorecard—Known Distance Course).

NOTE: See Appendix B for a sample completed form and the end of this publication for a blank, reproducible copy.

25-METER SCALED TARGET ALTERNATE COURSE

NOTE: The 25-meter scaled target alternate course is used when a standard record fire or KD range is unavailable for weapon qualification.

6-83. The 25-meter scaled target alternate course enables units to test a Soldier's weapon marksmanship proficiency, and firing at scaled silhouettes allows Soldiers to engage targets with time constraints and feedback.

CONDUCT

6-84. To complete this course, Soldiers fire three tables. Table 6-16 depicts these three tables and provides related information, such as time constraints, number of rounds that must be provided, number of silhouettes that must be engaged, and the distance from the firer that the target must be placed.

NOTES:

1. Soldiers should not receive training on target detection or the effects of wind and gravity by engaging targets at 25 meters. These skills are trained by firing at longer distances.
2. If zeroing/grouping exercises are not performed on the day of record fire, all Soldiers confirm the zero of their assigned weapons by firing six rounds of training/sustainment ammunition from the 25-meter line before firing the course. Zero rounds do not count for score.

Table 6-16. 25-meter scaled target alternate course firing tables and related information.

TABLE	POSITION	TIME CONSTRAINTS	NUMBER OF ROUNDS	NUMBER OF SILHOUETTES	ADDITIONAL INFORMATION
Table 1	Prone supported firing position or foxhole supported firing position	120 sec	20-round magazine, two rounds for each silhouette if firing from the prone supported firing position or foxhole supported firing position	10 silhouettes on the same target sheet	No more than two hits for each silhouette will be scored for this table.
Table 2	Prone unsupported firing position	60 sec	10-round magazine, one round for each silhouette	10 silhouettes on the same target sheet	No more than one hit for each target will be scored for this table.
Table 3	Kneeling unsupported firing position	60 sec	10-round magazine, one round for each silhouette	5 silhouettes on the same target sheet (50 to 150 m)	No more than two hits for each target will be scored for this table.

Time Between Firing Positions

6-85. The time between each firing position is not specified, but enough time should be allotted to allow the firer to clear his weapon, quickly change firing positions, and reload before beginning the next firing table.

Duties of Range Personnel

6-86. The following personnel perform range duties:

- Officer in charge (OIC).
- Range safety officer (RSO).
- Firing line safety crew.

Officer In Charge

6-87. The OIC briefs all Soldiers on the proper scoring procedures.

Range Safety Officer

6-88. To facilitate the timely flow of the record fire qualification table, the RSO ensures that enough time is given between firing positions.

Firing Line Safety Crew

6-89. Firing line safety crew personnel—

- Perform as scorers.
- Inform the chief range officer of crossfires.
- Inform the chief range officer of allowable alibis.
- Accurately count hits and misses.
- Count only four hits for each silhouette for score.
- Complete the scorecard.
- Assist the Soldier with target repair.
- Total, sign, and return the completed scorecard to the chief range officer.

SCORING

6-90. One hit is awarded for each round that strikes within or touches some part of the silhouette. If a bullet hole does not touch some part of the scaled silhouette, it is counted as a miss. Ricochets are counted as hits or misses.

6-91. A maximum of four hits for each silhouette on the same target sheet are scored. The same target sheet is used for every 40-round qualification table that a firer completes.

6-92. DA Form 5790-R (Record Fire Scorecard—Scaled Target Alternate Course) is used to score alternate course record fire qualifications.

NOTE: See Appendix B for a sample completed form and the end of this publication for a blank, reproducible copy.

6-93. The NSNs for scaled silhouette targets are—

- 25-meter (NSN 6920-01-1671398).
- 15-meter (NSN 6920-01-1396).

RATINGS

6-94. Qualification ratings for the 25-meter scaled target alternate course are shown in Table 6-17.

Table 6-17. Qualification ratings for the 25-meter scaled target alternate course.

QUALIFICATION RATINGS	NUMBER OF TARGETS HIT
Expert	38 to 40
Sharpshooter	33 to 37
Marksman	26 to 32
Unqualified	25 and below

15-METER SCALED TARGET ALTERNATE COURSE

NOTE: Units are permitted to use the 15-meter scaled alternate course only when standard record fire and KD ranges, and 25-meter scaled target alternate courses are unavailable.

6-95. The 15-meter scaled target alternate course is conducted on a 50-foot indoor range using a .22-caliber rimfire adapter (RFA). Qualification is conducted using the 15-meter alternate course C target (NSN 6920-01-167-1396).

NOTES:
1. See Appendix A for more information about the RFA.
2. Prior to qualification, all Soldiers battlesight zero their weapons using the 15-meter battlesight zeroing target (NSN 6920-01-167-1393).
3. The conduct of fire, scoring, scorecard, and qualification ratings are the same as those used for the 25-meter scaled target alternate course.

Chapter 7

Advanced Rifle Marksmanship

The procedures and techniques for implementing the ARM training program are based on all Soldiers understanding common firing principles, being proficient marksmen, and being confident in applying their firing skills in combat. This chapter concentrates on the advanced techniques and procedures that Soldiers need to participate in collective training during unit live-fire training exercises. This chapter describes advanced firing positions, combat firing techniques, CBRN firing, unassisted night fire, moving target engagements, SRM training, and SDM training.

NOTE: Unit METL and STRAC allocation determine which ARM tasks will be trained.

SECTION I. ADVANCED FIRING POSITIONS

After mastering the four marksmanship fundamentals in the basic firing positions, Soldiers master the four fundamentals while firing from a variety of advanced firing positions. The firer's position may change, but the three remaining fundamentals never change. Ultimately, any firing position that aids the firer in applying the fundamentals is acceptable, as long as the firer applies it consistently to avoid changing his sight picture.

With minor modifications, the dry-fire exercises taught during PMI can effectively train and evaluate a Soldier's ability to apply the fundamentals while in advanced firing positions. Repetitive training (muscle memory) will teach the Soldier the corrections needed to keep the same point of aim while in different firing positions.

NOTE: The act of assuming different firing positions while keeping the same point of aim increases first-time target hits and Soldier survivability.

KNEELING SUPPORTED FIRING POSITION

7-1. This position allows the Soldier to obtain the height necessary to observe many target areas while taking advantage of available cover. This position is the same as the kneeling unsupported firing position, except the Soldier uses some form of support to stabilize his body. Solid cover that can support any part of the body or weapon assists in firing accuracy.

NOTE: See Chapter 4 for more information about the kneeling unsupported firing position.

7-2. To assume the kneeling supported firing position (Figure 7-1)—

(1) Drop to the knee toward the inside of the covered position.

(2) Place the foot toward the outside of cover, with the toes pointing toward the engagement area.

(3) Place the nonfiring hand on the edge of the cover to support the weapon.

(4) Place the firing hand on the pistol grip, with the weapon's buttstock between the SAPI plate and the bicep to stabilize the weapon and absorb recoil.

NOTE: When firing from the strong side, the firer can place the elbow of the firing side on the outside knee to provide stability. When firing from the weak side, the firer should cant the weapon approximately 45 degrees and thrust the hips forward to minimize exposure to the enemy. When placing the weapon against the cover, the firer should take care to prevent the ejection port from becoming obstructed.

Figure 7-1. Kneeling supported firing position.

STANDING UNSUPPORTED FIRING POSITION

7-3. While the standing position provides the least stability, it can be assumed quickly while moving and is a good position for target area observation. Support for any portion of the body or weapon improves stability.

7-4. To assume the standing unsupported firing position (Figure 7-2)—

(1) Face the target.

(2) Step toward the target with the foot closest to the target.

(3) Spread the feet a comfortable distance apart.

(4) Place the firing hand on the pistol grip and the nonfiring hand on the upper handguard.

(5) Place the weapon's buttstock between the SAPI plate of the IBA and the bicep.

NOTE: This action stabilizes the weapon and absorbs recoil.

---

(6) Shift the feet until aiming naturally at the target.

(7) Evenly distribute body weight.

NOTE: More stability can be obtained by adjusting the ammunition pouch to support the nonfiring-side elbow. This allows the weapon's magazine to rest in the nonfiring hand.

Figure 7-2. Standing unsupported firing position.

STANDING SUPPORTED FIRING POSITION AROUND OBSTACLES

7-5. To assume the standing supported firing position when firing around obstacles (Figure 7-3)—

(1) Face the target.

(2) Execute a facing movement to the firing side.

(3) Spread the feet a comfortable distance apart.

(4) Place the firing hand on the pistol grip and the nonfiring hand on the upper handguard.

(5) Place the weapon's buttstock between the SAPI plate and the bicep.

NOTE: This action helps stabilize the weapon and absorbs recoil.

---

(6) Lean into the wall or obstacle, with the nonfiring-side forearm, shoulder, and thigh touching the obstacle for support.

(7) Shift the feet until aiming naturally at the target.

(8) Evenly distribute body weight.

NOTE: When firing from the left side of any obstacle, the firer should take care to prevent the ejection port from becoming obstructed.

Figure 7-3. Standing supported firing position around obstacles.

MODIFIED FIRING POSITIONS

7-6. Soldiers should be encouraged to modify positions by—

- Taking advantage of available cover.
- Using anything that helps to steady the weapon.
- Making changes that allow them to hit more combat targets.

7-7. To provide maximum stability, Soldiers should use prone and supported positions when firing M16- or M4-series weapons in the automatic or burst fire mode. If the weapon is equipped with the ARS, the Soldier should use the vertical pistol grip to further increase control of the weapon. Optional modifications include the following;

- Maximize use of artificial support.
- Grip the weapon firmly, and pull it into the shoulder securely. This helps offset the progressive displacement of weapon/target alignment caused by recoil.
- Use sandbags to support the weapon.
- Position the nonfiring hand on the weapon wherever it provides the most stability and flexibility. The goal is to maintain weapon stability and minimize recoil.
- Form a 5-inch loop with the sling at the upper sling swivel. Grasp this loop with the nonfiring hand, and pull down and to the rear while firing.
- Grasp the small of the stock with the nonfiring hand, and apply pressure down and to the rear while firing.
- Assume the modified supported prone firing position (Figure 7-4). This position uses sandbags to support the handguard and frees the nonfiring hand to hold the magazine steady.

Figure 7-4. Modified supported prone firing position.

URBAN OPERATIONS FIRING POSITIONS

7-8. Although the same principles of rifle marksmanship apply, the selection of firing positions during urban operations (UO) requires some special considerations. During UO, Soldiers may be required to fire—

- Over rooftops.
- Around obstacles.
- From windows.

FIRING OVER ROOFTOPS

7-9. Long-range observation may require Soldiers to occupy positions that are high above the ground. Figure 7-5 shows a Soldier firing over a rooftop, exposing only the parts of his body necessary to engage a target.

Figure 7-5. Firing over a rooftop.

FIRING AROUND OBSTACLES

7-10. In the urban environment, Soldiers may encounter man-made and natural obstacles. Figure 7-6 shows a Soldier firing around an urban obstacle.

NOTE: Firing around corners could require the Soldier to fire from the opposite shoulder to avoid exposing himself to enemy fire.

Figure 7-6. Firing around an obstacle.

FIRING FROM WINDOWS

7-11. When firing from windows, Soldiers should stay in the shadows and make sure that the weapon's muzzle does not protrude out of the opening (Figure 7-7).

Figure 7-7. Firing from a window.

SECTION II. COMBAT FIRE TECHNIQUES

Combat is the ultimate test of a Soldier's ability to apply the fundamentals of marksmanship and firing skills. Soldiers must apply the marksmanship skills mastered during training, practice, and record fire exercises to many combat situations (for example, attack, assault, ambush, or UO). Although these situations present problems, basic techniques and fundamentals require only two modifications: changes to the rate of fire and alterations in weapon/target alignment.

NOTE: The necessary changes are significant and must be thoroughly taught and practiced before performing LFXs.

RAPID SEMIAUTOMATIC FIRE

7-12. The most important firing technique during fast-moving, modern combat is rapid semiautomatic fire. It is the most accurate technique of placing a large volume of fire on poorly defined targets or target areas, such as short exposure, multiple, or moving targets. To apply rapid semiautomatic fire, the Soldier intentionally fires a quick series of shots into the target area to ensure a high probability of a hit.

NOTE: Increased speed and volume should be sought only after the Soldier has demonstrated expertise and accuracy during slow semiautomatic fire.

EFFECTIVENESS AND CONTROL OF RAPID SEMIAUTOMATIC FIRE

7-13. With proper training, Soldiers can select the appropriate mode of fire: semiautomatic fire, rapid semiautomatic fire, or automatic/burst fire.

NOTE: Leaders must ensure that Soldiers apply proper fire discipline at all times. Even in training, unaimed fire must never be tolerated, especially unaimed automatic fire.

7-14. While Soldiers sacrifice some degree of accuracy to deliver a greater volume of fire, it is surprising how devastatingly accurate rapid semiautomatic fire can be. At ranges beyond 25 meters, rapid semiautomatic fire is superior to automatic fire in all measures: shots per target, trigger pulls per hit, and time to hit. Proper training and repeated practice increases the degree of accuracy.

7-15. Rapid application of the four fundamentals will result in a well-aimed shot every one or two seconds. This technique of fire allows a unit to place the most effective volume of fire in a target area while conserving ammunition. It is the most accurate means of delivering suppressive fire.

MODIFICATIONS FOR RAPID SEMIAUTOMATIC FIRE

7-16. Trainers must consider the impact of the increased rate of fire on the Soldier's ability to properly apply the fundamentals of marksmanship and other combat firing skills, such as immediate action procedures.

Marksmanship Fundamentals

7-17. The following paragraphs describe the modifications necessary for Soldiers to apply the four fundamentals when firing in the rapid semiautomatic fire mode.

Steady Position

7-18. Consider the following modifications to achieve a steady position:

- Make sure that the weapon is well-supported to improve accuracy and reduce recovery time between shots.
- Grip the handgrip tightly to reduce recovery time and rapidly shift or distribute fire to subsequent targets.
- When possible, pivot the weapon where the nonfiring hand meets the support.
- Avoid changing the position of the nonfiring hand on the support; it is awkward and time-consuming when rapidly firing a series of shots.

Aiming

7-19. Consider the following recommendations to properly aim the weapon:

- Do not change sighting and stock weld during rapid semiautomatic fire. Keep the cheek on the stock for every shot, align the firing eye with the rear aperture, and focus on the front sightpost.
- When using slow semiautomatic fire, seek a stable sight picture.
- In the fast-moving situations that require rapid semiautomatic fire, accept target movement and unsteady sight picture, and keep firing into the target area until the target is down or there is no chance of a hit.
- Aim every shot.

Breath Control

7-20. Breath control must be modified because the Soldier does not have time to take a complete breath between shots. Consider the following modifications to achieve proper breath control:

- Hold your breath at some point in the firing process.
- Take shallow breaths between shots.

Trigger Squeeze

7-21. To maintain the desired rate of fire, the Soldier has a brief period of time to squeeze the trigger. The firer must cause the weapon to fire in about half of a second or less and still not anticipate the precise moment of firing. Consider the following modifications to achieve proper trigger squeeze:

- Apply initial trigger pressure as soon as a target is identified and while the front sightpost is being brought to the desired point of aim.
- When the front sightpost reaches the point of aim, apply final pressure to cause the weapon to fire almost at once. Apply this additional pressure, also known as final trigger squeeze, without disturbing the lay of the weapon.
- Increase the firing rate by firing, releasing enough trigger pressure to reset the sear, and then immediately firing the next shot. This technique is called rapid trigger squeeze. It eliminates the time used in fully releasing pressure on the trigger and allows the firer to rapidly deliver subsequent rounds.

NOTE: Training and practice sessions are required for Soldiers to become proficient in the technique of rapid trigger squeeze.

7-22. Repeated dry-fire training using simulators, such as the EST 2000 and LMTS, and live-fire practice ensure that the Soldier can squeeze the trigger and maintain a rapid rate of fire consistently and accurately.

Immediate Action Procedures

7-23. To maintain an increased rate of suppressive fire, Soldiers must apply immediate action quickly. Repeated dry-fire practice using blanks or dummy rounds, followed by live-fire training and evaluation, ensures that Soldiers can rapidly apply immediate action procedures while other Soldiers initiate fire.

RAPID SEMIAUTOMATIC FIRE TRAINING

NOTE: Soldiers should be well-trained in all aspects of slow semiautomatic firing before attempting any rapid semiautomatic fire training. Those who display a lack of knowledge of fundamental marksmanship skills should not advance to rapid semiautomatic fire training until these skills are learned and mastered.

7-24. Initial training should focus on the modifications to the fundamentals and other basic combat skills necessary during rapid semiautomatic firing.

NOTE: See Table 7-1 for the current training program.

Table 7-1. Rapid semiautomatic fire training program.

RAPID SEMIAUTOMATIC FIRE TRAINING PROGRAM
Instructional Intent
• Soldiers learn to engage targets using rapid semiautomatic fire and practice rapid magazine changes.
Special Instructions
Ensure that— • The M16A2/A3/A4 rifle's or M4 carbine's rear sight is set on the 0-2 aperture. • The M16A1's rear sight is set on the unmarked aperture. • Soldiers use a 25-meter alternate course C qualification target. • Each Soldier is given four 5-round magazines of 5.56-millimeter ball ammunition. • Soldiers use rapid semiautomatic fire to engage targets. • Each Soldier fires one round at each of the 10 silhouettes on the alternate course C qualification target. • Each Soldier does a rapid magazine change after each magazine is fired. • The first iteration of 10 rounds is fired within a time limit of 40 seconds. • The second iteration of 10 rounds is fired within a time limit of 30 seconds. • Each target is inspected, and the results are posted after each iteration.
Observables
• Coaches continuously analyze the firer's application of the fundamentals. • Each Soldier obtains 14 hits out of 20 silhouette target exposures.

Conduct

7-25. Each Soldier receives four 5-round magazines of 5.56-millimeter ball ammunition. Using rapid semiautomatic fire, the Soldier fires one round at each of the 10 silhouettes on the alternate course C qualification target. Soldiers fire two iterations, performing a rapid magazine change after each magazine is fired. The targets are inspected, and the results are posted after each iteration. Each Soldier must obtain 14 hits out of 20 silhouette target exposures.

7-26. Table 7-2 depicts the two iterations and provides related information, such as time constraints, number of rounds that must be fired, type of target that must be used, and the distance away from the firer that the target must be placed.

Table 7-2. Rapid semiautomatic fire training and related information.

ITERATION	TIME CONSTRAINTS	NUMBER OF ROUNDS	TYPE OF TARGET	DISTANCE (m)
1	40 sec	10	25-m alternate course C qualification target	25
2	30 sec	10	25-m alternate course C qualification target	25

Dry-Fire Exercises

7-27. Repeated dry-fire exercises are the most efficient means to ensure that Soldiers can apply modifications to the fundamentals. Multiple dry-fire exercises are needed, emphasizing a rapid shift in position and point of aim, followed by breath control and fast trigger squeeze.

NOTES:
1. Blanks or dummy rounds may be used to train rapid magazine changes and immediate action procedures.
2. The Soldier should display knowledge and skill during dry-fire exercises before attempting LFXs.

Live-Fire Exercises

7-28. There are two types of LFXs:

- Individual.
- Collective.

Individual

7-29. To conduct an individual LFX—

- Ensure that the emphasis is on each Soldier maintaining a heavy volume of accurate fire.
- Keep weapon downtime (during immediate action and rapid magazine changes) to a minimum.
- Begin by firing at shorter ranges, progressing to longer ranges as Soldiers display increased proficiency.
- Shorten exposure or engagement times and increase the number of rounds to simulate the need for a heavy volume of fire.
- Provide downrange feedback to determine the accuracy of fire.

Collective

7-30. Rapid semiautomatic fire should be the primary means of delivering fire during a collective LFX. To conduct a collective LFX, ensure that the emphasis is on performing staggered rapid magazine changes, maintaining a continuous volume of fire, and conserving ammunition.

AUTOMATIC OR BURST FIRE

NOTE: Automatic or burst fire should be trained only after the Soldier has demonstrated expertise during slow and rapid semiautomatic fire.

7-31. When applying automatic or burst fire, Soldiers deliver the maximum number of rounds (one to three rounds per second) into a designated target area while rapidly applying the four fundamentals. This specialized technique of delivering suppressive fire may not apply to most combat engagements.

NOTE: The M16A1/A3 rifle and M4A1 carbine have fully automatic settings. The M16A2/A4 rifle and M4 carbine use a three-round burst capability.

Effectiveness and Control of Automatic or Burst Fire

7-32. Automatic or burst fire is inherently less accurate than semiautomatic fire. The first fully automatic shot fired may be on target, but recoil and a high cyclic rate of fire often combine to place subsequent rounds far from the desired point of impact. Even controlled (three-round burst) automatic or burst fire may place only one round on the target. Because of these inaccuracies, it is difficult to evaluate the effectiveness of automatic or burst fire, and even more difficult to establish absolute guidelines for its use.

Factors for Use of Semiautomatic Versus Automatic or Burst Fire

7-33. Trainers must ensure that Soldiers understand the capabilities and limitations of automatic or burst fire. They must know when it should and should not be used.

Semiautomatic Fire

7-34. M16 rifles and M4 carbines should normally be employed in the semiautomatic fire mode.

7-35. Depending on the tactical situation, Soldiers should employ the semiautomatic fire mode in the following conditions:

- Ammunition is in short supply, or resupply may be difficult.
- Single targets are being engaged.
- Widely spaced multiple targets are being engaged.
- The target is located more than 50 meters away.
- The effect of bullets on the target cannot be observed.
- Artificial support is not available.
- Targets may be effectively engaged using semiautomatic fire.

Automatic or Burst Fire

7-36. In some combat situations, the use of automatic or burst fire can improve survivability and enhance mission accomplishment. Clearing buildings, final assaults, FPF, and ambushes may require limited use of automatic or burst fire.

7-37. Depending on the tactical situation, Soldiers should employ automatic or burst fire in the following conditions:

- Ammunition is readily available, and there are no problems with resupply.
- Closely spaced multiple targets are located 50 meters away or less.
- Maximum fire is immediately required at an area target.
- Tracers or some other means can be used to observe the effect of bullets on the target.
- Leaders can maintain adequate control over weapons firing in the automatic fire mode.
- Good artificial support is available.
- The initial sound of gunfire disperses closely spaced enemy targets.

Modifications for Automatic or Burst Fire

7-38. Automatic or burst fire is inherently less accurate than semiautomatic fire. Trainers must consider the impact of recoil and the high cyclic rate of fire on the Soldier's ability to properly apply the fundamentals of marksmanship and other combat firing skills, such as immediate action procedures and rapid magazine changes.

Marksmanship Fundamentals

7-39. The following paragraphs describe the modifications necessary for Soldiers to apply the four fundamentals when firing in the automatic fire mode.

Steady Position

7-40. Consider the following modifications to achieve a steady position:

- Make sure that the weapon is well-supported.
- Grip the weapon a little more firmly and pull it into the shoulder a little tighter than when in the semiautomatic fire mode.

NOTE: This support and increased grip help offset the progressive displacement of weapon/target alignment caused by recoil.

- To provide maximum stability, assume the modified supported prone firing position (Figure 7-4).

NOTE: If the weapon is equipped with the ARS, use the vertical pistol grip to further increase control of the weapon.

Aiming

7-41. Consider the following recommendations to properly aim the weapon:

- Do not change sighting and stock weld during automatic or burst fire. Keep the cheek on the stock for every shot, align the firing eye with the rear aperture, and focus on the front sightpost.
- Although recoil may disrupt this process, try to apply the aiming techniques throughout recoil.

Breath Control

7-42. Breath control must be modified because the Soldier does not have time to take a complete breath between shots. Consider the following modifications to achieve proper breath control:

- Hold your breath at some point in the firing process.
- Take shallow breaths between shots.

Trigger Squeeze

7-43. Training and repeated dry-fire practice aid the Soldier in applying proper trigger squeeze during automatic firing. LFXs enable him to improve this skill.

M16A2/3/4 Rifles and M4 Carbines

7-44. Until the weapon fires, trigger squeeze is applied in the normal manner. To use the burst fire mode—

(1) Hold the trigger to the rear until three rounds are fired.
(2) Release pressure on the trigger until it resets.
(3) Reapply pressure for the next three-round burst.

NOTES: 1. Do not slap or jerk the trigger. Squeeze it, and then quickly release pressure.

2. Depending on the position of the burst can when the selector is moved to the burst fire mode, the weapon may fire one, two, or three rounds when the trigger is held to the rear for the first time. If the weapon fires only one or two rounds, quickly release pressure on the trigger and squeeze again, holding it to the rear until a three-round burst is completed.

M16A1 Rifles

7-45. Until the weapon fires, trigger squeeze is applied in the normal manner. Because three-round bursts are the most effective rate of fire, pressure on the trigger should be released as quickly as possible. To use the burst fire mode, keep the index finger on the trigger, but quickly release pressure to prevent an excessive number of rounds from being fired in one burst. With much dry-fire practice, the Soldier can become proficient at delivering three-round bursts with the squeeze/release technique.

Immediate Action

7-46. To maintain an increased rate of suppressive fire, Soldiers must apply immediate action quickly. Repeated dry-fire practice using blanks or dummy rounds, followed by live-fire training and evaluation, ensures that Soldiers can rapidly apply immediate action procedures.

Rapid Magazine Changes

7-47. Rapid magazine changes are vital in maintaining automatic or burst fire. Rapid magazine changes must be correctly taught and practiced during dry-fire and live-fire exercises until the Soldier becomes proficient.

Automatic or Burst Fire Training

NOTE: Soldiers should be well-trained in all aspects of slow semiautomatic firing before attempting any automatic training. Those who display a lack of knowledge of fundamental skills should not advance to automatic or burst fire training until these skills are learned.

7-48. Initial training should focus on the modifications to the fundamentals and other basic combat skills necessary during automatic firing.

7-49. Unit training is vital to properly applying this technique. Soldiers must be taught the advantages and disadvantages of automatic and burst firing so they know when it should be used. Without this knowledge, Soldiers tend to switch to the automatic or burst fire mode in life-threatening situations.

NOTE: See Table 7-3 for the current training program.

Table 7-3. Automatic or burst fire training program.

AUTOMATIC OR BURST FIRE TRAINING PROGRAM
Instructional Intent
• Soldiers learn the advantages and disadvantages of automatic or burst fire.
Special Instructions
Ensure that—
• The M16A2/A3/A4 rifle's or M4 carbine's rear sight is set on the 0-2 aperture.
• The M16A1's rear sight is set on the unmarked aperture.
• Soldiers use a 25-meter alternate course C qualification target.
• Each Soldier is in a proper modified automatic/burst firing position.
• Each Soldier is given two 15-round magazines of 5.56-millimeter ball ammunition.
• Each Soldier fires one 3-round burst at each of the 10 silhouettes on the alternate course C qualification target.
• Each Soldier does a rapid magazine change after each magazine is emptied.
Observables
• Each Soldier obtains five target hits.
• Soldiers demonstrate control of the weapon in the automatic/burst fire mode.

Conduct

7-50. Each Soldier receives two 15-round magazines of 5.56-millimeter ball ammunition. Each Soldier fires one 3-round burst at each of the 10 silhouettes on the alternate course C qualification target, performing a rapid magazine change after each magazine is emptied. Each Soldier must obtain five target hits.

7-51. Table 7-4 depicts automatic or burst fire training and provides related information, such as number of rounds that must be fired, type of target that must be used, and the distance away from the firer that the target must be placed.

Table 7-4. Automatic or burst fire training and related information.

FIRING POSITION	NUMBER OF ROUNDS	TYPE OF TARGET	DISTANCE (m)
Modified automatic/burst firing position	30, one 3-round burst at each of the 10 silhouettes	Alternate course C qualification target	25

Dry-Fire and Live-Fire Exercises

7-52. Repeated dry-fire exercises are the most efficient means to ensure that Soldiers can apply modifications to the fundamentals. Multiple dry-fire exercises are needed, emphasizing a stable position and point of aim, followed by breath control and appropriate trigger squeeze.

NOTES: 1. Blanks or dummy rounds may be used to train trigger squeeze, rapid magazine changes, and immediate action procedures.

2. The Soldier should display knowledge and skill during dry-fire exercises before attempting LFXs.

SUPPRESIVE FIRE

7-53. Suppressive fire is precisely aimed at a definite point or area target. Some situations may require a Soldier to place suppressive fire into a wide area (for example, wood line, hedgerow, or small building) while, at other times, the target may be a smaller area (for example, a bunker or window). Suppressive fire is used to control the enemy and the area he occupies. It is employed to kill the enemy or to prevent him from observing the battlefield, effectively using his weapons, or moving.

EFFECTIVENESS AND CONTROL OF SUPPRESSIVE FIRE

7-54. Many Soldiers have difficulty delivering effective suppressive fire when they cannot see a definite target, only likely locations or general areas where the enemy is known to exist. Even though definite targets cannot be seen, most suppressive fire should be well-aimed.

7-55. When controlling suppressive fires, two factors must be considered:

- Point of aim.
- Rate of fire.

Point of Aim

7-56. Suppressive fire should be well-aimed, sustained, semiautomatic fire. Although lacking a definite target, the Soldier must be taught to control and accurately deliver fire within the limits of the suppressed area. As when engaging a point target, the weapon sights are used, with the front sightpost placed so each shot impacts within the desired area.

Rate of Fire

7-57. During most phases of live-fire training (for example, grouping, zeroing, qualifying), shots are delivered using slow semiautomatic fire (one round every 3 to 10 seconds). During training, this allows a slow and precise application of the fundamentals. Successful suppressive fire requires a faster, but sustained, rate of fire. Soldiers may need to fire full automatic or bursts (13 rounds per second) for a few seconds to gain initial fire superiority. Rapid semiautomatic fire (one round every one or two seconds) allows the firer to sustain a large volume of accurate fire while conserving ammunition.

MODIFICATIONS FOR SUPPRESSIVE FIRE

7-58. The tactical situation dictates the most useful rate of fire, but the following must be considered:

- Marksmanship fundamentals.
- Rapid magazine changes.
- Ammunition conservation.

Marksmanship Fundamentals

7-59. As the stress of combat increases, some Soldiers may fail to apply the fundamentals of marksmanship. This factor contributes to reduced accuracy and effectiveness. While some modifications are appropriate, the basic fundamentals should be applied and emphasized—regardless of the rate of fire or combat stress. Strategies to enhance marksmanship skills under combat stress include shooting in the prone position, as opposed to standing.

7-60. Factors that contribute to combat stress are:

- Environmental.
- Operational.

Environmental

7-61. Environmental stressors have been shown to degrade marksmanship accuracy up to 20 percent. Such stressors include—

- Heat.
- Altitude.

Operational

7-62. Operational stressors have been shown to degrade marksmanship accuracy from 17 percent to 136 percent. Such stressors include—

- MOPP gear.
- Tasks that require carrying rucksacks, litter patients, and other equipment on the body.
- Sleep deprivation.

Rapid Magazine Changes

7-63. One of the keys to sustained suppressive fire is reloading the weapon rapidly. Rapid magazine changes must be correctly taught and practiced during dry-fire and live-fire exercises until the Soldier becomes proficient. Small-unit training exercises must be conducted so Soldiers who provide suppressive fire practice staggered magazine changes.

Ammunition Conservation

7-64. Automatic or burst fire should be used sparingly and only to gain initial fire superiority. Depending on the tactical situation, the rate of fire should be adjusted so that a minimum number of rounds are expended. Accurate fire conserves ammunition, while preventing the enemy from placing effective fire on friendly positions.

Suppressive Fire Training

NOTE: See Table 7-5 for the current training program.

Table 7-5. Suppressive fire training program.

SUPPRESSIVE FIRE TRAINING PROGRAM
Instructional Intent
• Soldier learns to suppress targets using suppressive fire.
Special Instructions
Ensure that— • The M16A2/A3/A4 rifle's or M4 carbine's rear sight is set on the 0-2 aperture. • The M16A1's rear sight is set on the unmarked aperture. • Soldiers use a 25-meter scaled landscape target. • Each Soldier is given two 9-round magazines and one 12-round magazine of 5.56-millimeter ball ammunition. • Each Soldier is in a proper supported firing position. • Each Soldier fires 9 rounds at the open window area of the target using rapid semiautomatic fire with the first 9-round magazine. • Each Soldier fires 12 rounds at the fence or hedgerow area of the target using rapid semiautomatic fire with the 12-round magazine. • Each Soldier fires three 3-round bursts at the tank turret area of the target using the automatic/burst fire mode with the second 9-round magazine.
Observables
• Each Soldier achieves 5 hits inside the open window area within 18 seconds. • Each Soldier achieves 10 hits inside the dotted lines surrounding the fence or hedgerow area within 24 seconds. • Each Soldier achieves 3 hits inside the tank turret area within 24 seconds.

7-65. Figure 7-8 shows a landscape target suitable for suppressive fire training. When this type of target is used, trainers must develop a firing program to include areas of engagement and designated target areas. At 25 meters, this target provides the firer with an area to suppress without definite targets to engage.

Figure 7-8. Landscape target.

Conduct

7-66. Each Soldier receives two 9-round magazines and one 12-round magazine of 5.56-millimeter ball ammunition. The Soldier engages three areas of a 25-meter scaled landscaped target: the open window area, the fence or hedgerow area, and the tank turret area. Each Soldier achieves 5 hits inside of the open window area, 10 hits inside of the dotted lines surrounding the fence or hedgerow area, and 3 hits inside of the tank turret area.

7-67. Table 7-6 depicts suppressive fire training and provides related information, such as number of rounds that must be fired, type of target that must be used, and the distance away from the firer that the target must be placed.

Table 7-6. Suppressive fire training and related information.

FIRING POSITION	TYPE OF TARGET	AREA OF TARGET ENGAGED	NUMBER OF ROUNDS	TYPE OF FIRE	TIME CONSTRAINTS
Supported firing position	25-m scaled landscape target	Open window	9	Rapid semiautomatic	18 sec
		Fence or hedgerow	12	Rapid semiautomatic	24 sec
		Tank turret	9, in three 3-round bursts	Automatic/burst	24 sec

QUICK FIRE

7-68. The two main techniques of directing fire with a rifle or carbine are—

- Aim using the sights.
- Use weapon alignment, instinct, bullet strike, or tracers to direct the fire.

7-69. The preferred technique is to use the sights, but sometimes quick reflex action is required. Quick fire, also known as instinctive firing or quick kill, is a technique used to deliver fast, effective fire on surprise personnel targets 25 meters away or less.

EFFECTIVENESS AND CONTROL OF QUICK FIRE

7-70. Quick fire techniques are appropriate when Soldiers are presented with close, suddenly appearing, surprise enemy targets; or when close engagement is imminent.

NOTE: Fire may be delivered in the SEMIAUTO or AUTOMATIC/BURST fire mode. For example, a point man in a patrol may carry the weapon on AUTOMATIC/BURST. This may also be required when clearing a room or bunker. Initial training should be in the SAFE mode.

7-71. Two techniques of delivering quick fire are:

- Aimed.
- Pointed.

7-72. The difference in the speed of delivery of these two techniques is small. Pointed quick fire can be used to fire a shot about one-tenth of a second faster than aimed quick fire. The difference in accuracy, however, is more pronounced:

- A Soldier well-trained in pointed quick fire can hit an E-type silhouette target at 15 meters, although the shot may strike anywhere on the target.
- A Soldier well-trained in aimed quick fire can hit an E-type silhouette target at 25 meters, with the shot or burst striking 5 inches from the center of mass.

7-73. This variance of target hit for this type of engagement reinforces the need for well-aimed shots.

7-74. Pointed and aimed quick fire should be used only when a target cannot be engaged fast enough using the sights in a normal manner. These techniques should be limited to targets appearing at 25 meters or less. Modern short-range combat (SRC) techniques emphasize carrying the weapon with the buttstock high so that the weapon sights can be brought into display as quickly as firing a hasty unaimed shot. In extremely dangerous moments, special reaction teams (SRTs) commonly advance with weapons shouldered, aiming as they advance.

Aimed

7-75. When using this technique, a Soldier can accurately engage a target at 25 meters or less in one second or less.

7-76. To use aimed quick fire (Figure 7-9)—

(1) Bring the weapon to the shoulder.

(2) With the firing eye, look through or just over the rear sight aperture.

(3) Use the front sightpost to aim at the target.

(4) Quickly fire a single shot.

Figure 7-9. Aimed quick fire.

Pointed

7-77. When using this technique, a Soldier can engage a target at 15 meters or less in less than one second.

7-78. To use pointed quick fire (Figure 7-10)—

(1) Keep the weapon at your side.

(2) Keeps both eyes open, and use instinct and peripheral vision to line up the weapon with the target.

(3) Quickly fire a single shot or burst.

Figure 7-10. Pointed quick fire.

FACTORS FOR USE OF CONTROLLED PAIRS VERSUS BURST FIRE

7-79. Tactical considerations dictate whether controlled pairs or burst fire is most effective in a given situation.

MODIFICATIONS FOR QUICK FIRE

7-80. Trainers must consider the impact of the increased rate of fire on the Soldier's ability to properly apply the fundamentals of marksmanship and other combat firing skills.

Marksmanship Fundamentals

7-81. Quick fire techniques require major modifications to the four fundamentals of marksmanship. Initial training in these differences, followed by repeated dry-fire exercises, will be necessary to prepare the Soldier for live-fire.

Steady Position

7-82. The quickness of shot delivery prevents the Soldier from assuming a stable firing position. Consider the following modifications:

- Fire from the present position when the target appears.
- If moving, stop.
- Do not make adjustments for stability and support before the round is fired.

Aimed

7-83. Consider the following modifications:

(1) Pull the weapon's buttstock into the pocket of the shoulder as the cheek comes in contact with the stock.
(2) Firmly grip the weapon with both hands, applying rearward pressure.
(3) Place the firing eye so that it looks through or just over the rear sight aperture.
(4) Place the sight on the target.

Pointed

7-84. Consider the following modifications:

- Pull the weapon into the side.
- Firmly grip the weapon with both hands, applying rearward pressure.

Aiming

7-85. This fundamental must be highly modified because the Soldier may not have time to look through the rear sight, find the front sight, and align it with the target.

NOTE: When using either aiming technique, bullets may tend to impact above the desired location. Repeated live-fire practice is necessary to determine the best point of aim or the best focus. Such practice should begin with the Soldier using a center of mass aim.

Aimed

7-86. Consider the following modified procedure:

(1) Initially focus on the target.

(2) Place the firing eye so that it looks at the target through or just over the rear sight aperture.

(3) Using peripheral vision, locate the front sightpost and bring it to the center of the target.

NOTE: Focus remains on the front sightpost throughout the aiming process.

---

(4) When the front sightpost is in focus, fire a controlled pair.

Pointed

7-87. Consider the following modifications:

- Place the focus on or slightly below the center of the target as you align the weapon with it, and fire the weapon.
- Use your instinctive pointing ability and peripheral vision to aid in proper alignment.

Breath Control

7-88. This fundamental has little application to the first shot of quick fire. The round must be fired before a conscious decision can be made about breathing. If subsequent shots are necessary, breathing must not interfere with the necessity of firing quickly. When possible, use short, shallow breaths.

Trigger Squeeze

7-89. Consider the following modifications:

(1) Apply initial pressure as weapon alignment is moved toward the target.

(2) Exert trigger squeeze so when weapon/target alignment is achieved, the rounds are fired at once.

7-90. Perfecting rapid trigger squeeze requires much training and practice.

QUICK FIRE TRAINING

NOTE: Only Soldiers in basic training will conduct quick fire training. SRM will be conducted at the unit level. See Section VI of this chapter for more information about SRM training.

7-91. The key to the successful employment of both quick fire techniques is practice. Both pointed and aimed quick fire must be repeatedly practiced during dry-fire training. LFXs provide further skill enhancement and illustrate the difference in accuracy between the two techniques.

NOTE: See Table 7-7 for the current training program.

Table 7-7. Quick fire training program.

QUICK FIRE TRAINING PROGRAM
Instructional Intent
• Soldiers learn how to engage targets using the quick fire techniques.
Special Instructions
Ensure that— • The M16A2/A3/A4 rifle's or M4 carbine's rear sight is set on the 0-2 aperture. • The M16A1's rear sight is set on the unmarked aperture. • Each Soldier is given two 10-round magazines. • Each Soldier engages 10 target exposures of 2 seconds each at 15 meters using the first 10-round magazine. • Each Soldier moves to the 25-meter line and engages 10 target exposures of 2 seconds each at 25 meters using the second 10-round magazine.
Observables
• Each Soldier achieves 7 hits out of 10 target exposures at 15 meters. • Each Soldier achieves 5 hits out of 10 target exposures at 25 meters.

Conduct

7-92. Each Soldier receives two 10-round magazines. Each Soldier must achieve 7 target hits out of 10 target exposures at 15 meters and 5 target hits out of 10 target exposures at 25 meters.

7-93. Table 7-8 depicts quick fire training and provides related information, such as the number of target exposures, distance from the firer, number of rounds that must be fired, and time constraints.

Table 7-8. Quick fire training and related information.

NUMBER OF TARGET EXPOSURES	DISTANCE	NUMBER OF ROUNDS	TIME CONSTRAINTS
10	15	10	2 sec per target exposure
10	15	10	2 sec per target exposure

SECTION III. CHEMICAL, BIOLOGICAL, RADIOLOGICAL, AND NUCLEAR FIRING

All Soldiers must effectively fire their weapons to accomplish combat missions in a CBRN environment. With proper training and practice, Soldiers gain confidence in their ability to effectively hit targets in full MOPP equipment. MOPP firing proficiency must be part of every unit's training program.

MISSION-ORIENTED PROTECTIVE POSTURE EQUIPMENT FIRE

7-94. Firing weapons is only part of overall CBRN training. Soldiers must be familiar with CBRN equipment, its use, and proper wear before they progress to learning the techniques of MOPP firing.

MODIFICATIONS FOR MISSION-ORIENTED PROTECTIVE POSTURE FIRE TRAINING

7-95. Trainers must consider the impact of MOPP equipment (for example, hood or mask, gloves, overgarments) on the Soldier's ability to properly apply the fundamentals of marksmanship and combat firing skills.

Operation and Function

7-96. Many actions are affected by MOPP equipment: handling the weapon, performing operation and function checks, loading and unloading, and cleaning. Consider the following modifications:

- Movements are slowed; tasks take longer to complete and often require more effort.
- Vision is impaired.
- Care is needed to avoid damaging MOPP equipment and possible exposure to lethal agents.

7-97. Because of the great differences between MOPP Level 1 and MOPP Level 4, Soldiers must be trained in all aspects of weapon operation and maintenance while practicing at the highest MOPP level. Only through repeated training and practice can the Soldier be expected to perform tasks efficiently.

Immediate Action

7-98. Consider the following modifications:

- Under normal conditions, a Soldier should be able to clear a stoppage in 3 to 5 seconds. Under increased MOPP levels, however, this may take as long as 10 seconds to successfully complete.
- Mask (with or without hood) and gloves must be worn. Care must be taken not to snag or damage the gloves or dislodge the hood or mask during movements.
- Vision is limited to what can be seen through the mask lenses or faceplate. Peripheral vision is severely restricted. The lenses or faceplate may be scratched or partly fogged, further restricting vision.

NOTE: Soldiers requiring corrective lenses must be issued insert lenses before training.

- Scanning movements may be restricted by the hood or mask.

7-99. These factors could adversely affect the Soldier's ability to quickly and accurately detect targets. Dry-fire practice under these conditions is necessary to reduce time and streamline actions. Until Soldiers can instinctively apply immediate action to stoppages, they should practice using dummy or blank ammunition.

Marksmanship Fundamentals

7-100. Although the four marksmanship fundamentals remain valid during MOPP firing, some modifications may be needed to accommodate the equipment.

Steady Position

7-101. Consider these modifications:

- Due to the added bulk of the overgarment, adjust firing positions for stability and comfort.
- Stand, crouch, or squat during dry- and live-fire to reduce bodily contact with contaminated ground or foliage.
- Understand that a consistent spot or stock weld is difficult to maintain due to the shape of the protective mask. This requires the firer to hold his head in an awkward position to place the eye behind the sight.

Aiming

7-102. Wearing a protective mask may force firers to rotate (cant) the weapon to see through the rear aperture. This ideal aiming procedure (Figure 7-11) should be the initial procedure taught and practiced. If this cannot be achieved, a canted sight picture may be practiced.

Ideal Aiming Procedure

7-103. To perform the ideal aiming procedure—

(1) Rotate the weapon as little possible to see through and line up the sights.

(2) Place the center tip of the front sightpost on the ideal point of aim.

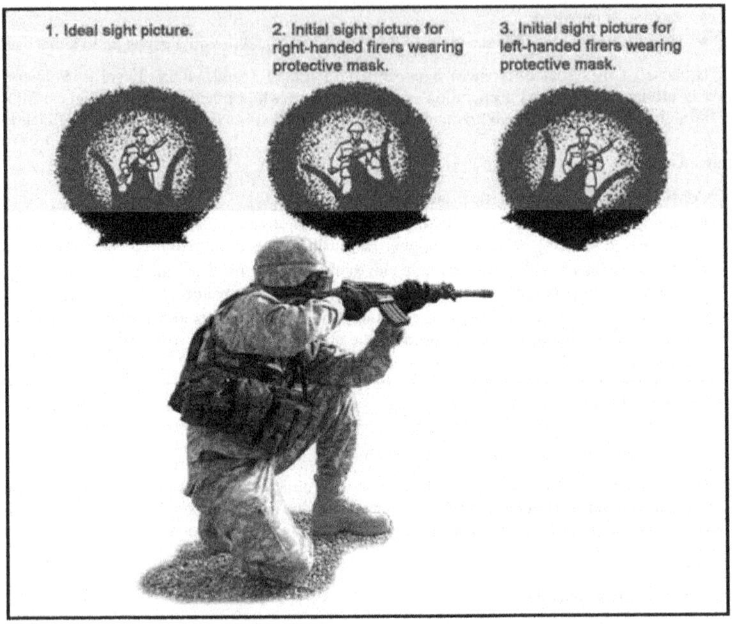

Figure 7-11. Sight picture when canting the rifle while wearing a protective mask (75-meter target).

Canted Sight Picture

NOTE: The normal amount of cant has a limited influence on rounds fired at ranges of 75 meters or less. Rifle ballistics cause the strike of the bullet to impact low in the direction of the cant at longer ranges.

7-104. Due to a shift in bullet strike and the many individual differences in sight alignment when wearing a protective mask, it is important to conduct downrange feedback training at ranges beyond 75 meters on KD ranges. This allows Soldiers to determine the aiming adjustments needed to hit the center of the target.

7-105. Figure 7-12 shows what might be expected for a right-handed firer engaging a target at 175 meters with a certain amount of cant; it depicts the adjustment in point of aim needed to move the bullet strike to the center of the target. Figure 7-13 shows what might be expected for a right-handed firer engaging a 300-meter target.

NOTE: The adjustments in point of aim for left-handed firers are the opposite of those shown in Figures 7-12 and 7-13.

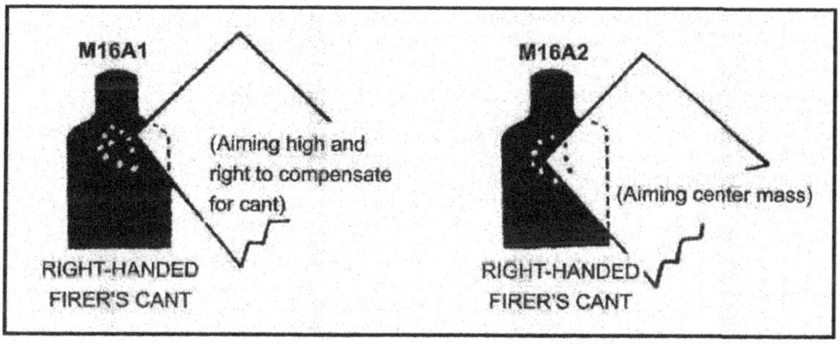

Figure 7-12. Engagement of 175-meter target.

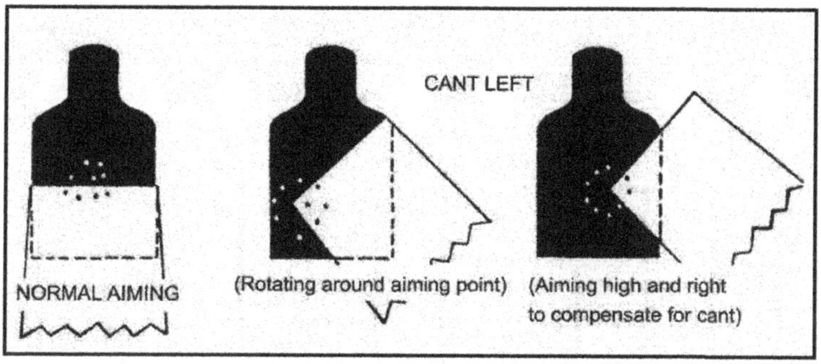

Figure 7-13. Engagement of 300-meter target.

7-106. Although bullet strike is displaced when using a cant, individual differences vary; center-of-mass aiming should be used until the individual knows what aiming adjustment is needed. When distant targets are missed, a right-handed firer should usually adjust his point of aim to the right and high; a left-handed firer should adjust to the left and high. Actual displacement of the point of aim must be determined by using downrange feedback targets positioned more than 75 meters from the firer.

Breath Control

7-107. Moving when encumbered by MOPP equipment requires more physical effort, and breathing is restricted while wearing the protective mask. Physical exertion can produce labored breathing and make settling into a normal breathing rhythm much more difficult, which, in turn, can increase the breath rate. All of these factors make holding and controlling the breath to produce a well-aimed shot more energy- and time-consuming. Emphasis must be placed on rapid target engagement during the limited amount of time a firer can control his breath.

Trigger Squeeze

7-108. Consider the following modifications:

- MOPP gloves complicate the act of grasping the pistol grip and squeezing the trigger with the index finger. The action of the trigger finger is restricted, and the fit of the glove may require the release of the swing-down trigger guard.
- Because the trigger feels different, control differs from that used in barehanded firing. This difference cannot be accurately predicted.

7-109. Dry-fire training using dime/washer exercises or simulators such as the EST 2000 or LMTS may be necessary to ensure that the firer knows the changes he will encounter during live-fire.

MISSION-ORIENTED PROTECTIVE POSTURE EQUIPMENT TRAINING

7-110. Repeated dry-fire training and LFXs are the most efficient means to prepare a Soldier wearing MOPP Level 4 equipment for successful target engagements at any range. The Soldier must follow these procedures and applications to be combat-effective in a CBRN environment.

Conduct

NOTE: See Table 7-9 for the current training program.

Table 7-9. Chemical, Biological, Radiological, and Nuclear fire training program.

CBRN FIRE TRAINING PROGRAM
Instructional Intent
• Develop the Soldier's confidence and ability to engage targets while wearing any level of MOPP equipment.
Special Instructions
Ensure that—
• The M16A2/A3/A4 rifle's or M4 carbine's rear sight is set on the 0-2 aperture.
• The M16A1's rear sight is set on the unmarked aperture.
• Soldiers have inserted lenses before firing, if required.
• Soldiers have a proper seal on the mask to prevent fogging and loss of visibility.
• Each Soldier is issued 20 rounds of ammunition to be loaded in two magazines (10/10).
• Each Soldier engages 20 targets at 50 meters.
• Target exposures consist of 10 from the right and 10 from the left from the foxhole supported firing position using Table 1 of the Record Fire Qualification firing table (DA Form 3595-R).
Observables
• Each Soldier obtains 11 hits out of 20 target exposures.

Dry-Fire Exercises

7-111. As with all marksmanship training, Soldiers must start at the basics in order to become proficient at CBRN fire. Modified fundamentals can be taught anywhere, and Soldiers must learn them before participating in an LFX. The dry-fire exercises used during CBRN training are the same as those used during initial rifle marksmanship (dime/washer exercise, target box, SPORTS, EST 2000, and LMTS). Soldiers can also practice by using MILES equipment during force-on-force training.

NOTE: Soldiers should wear MOPP Level 4 when participating in dry-fire exercises; training at the highest degraded level allows Soldiers to adjust their shooting technique to increase their marksmanship ability in a CBRN environment.

Downrange Feedback

7-112. CBRN downrange feedback gives Soldiers the confidence, knowledge, and skills required to consistently deliver accurate, well-aimed fire against combat targets as far away as 300 meters. On a KD range, the Soldier will perform the following scenario:

- The Soldier will be issued six magazines.
  - The first and second magazine will have 5 rounds each.
  - The third and fourth magazine will have 10 rounds each.
  - The fifth and sixth magazine will have 5 rounds each.

- The Soldier engages the appropriate targets.
  - The Soldier engages the 75-meter target with the first and second 5-round magazine from the foxhole, standing, or kneeling supported position.
  - The Soldier engages the 175-meter target with both 10-round magazines from the foxhole, standing, or kneeling supported firing position.
  - The Soldier engages the 300-meter target with the remaining two 5-round magazine from the foxhole, standing, or kneeling supported firing position.
- In order to receive a GO, the Soldier must obtain a minimum of—
  - 8 hits out of 10 shots on the 75-meter target.
  - 14 hits out of 20 shots on the 175-meter target.
  - 5 hits out of 10 shots on the 300-meter target.

NOTES: 1. DA Form 5789-R is used to record the firer's results.

2. Soldiers can use ammunition allocated by DA PAM 350-38 for advanced skill training for the CBRN downrange feedback scenario.

50-Meter Live-Fire Exercise

7-113. The basic CBRN LFX allows all Soldiers to gain confidence in their CBRN firing abilities. Practice and proficiency firing can be conducted on any range; however, when a Remote Electronic Target System (RETS) range is used for this exercise, the two 50-meter mechanisms are used. For CBRN LFXs, Soldiers perform the following scenario.

- Each Soldier is issued 20 rounds of ammunition to be loaded in two 10-round magazines.
- On the command "GAS–GAS–GAS," each Soldier engages twenty 50-meter targets. Target exposures consist of 10 from the right and 10 from the left from the foxhole supported or the kneeling firing position.
- Each Soldier must achieve 11 hits out of 20 target exposures.
- This is a GO/NO GO exercise.

Alternate Fire Exercise

7-114. The CBRN alternate fire course uses the 25-meter scaled silhouette timed-fire target. This silhouette can be used on any 25-meter range, and the target provides feedback to the firer on where the strike of the round impacts the target. This exercise is conducted in the same manner as the 25-meter alternate course record fire.

- Each Soldier will be in MOPP Level 4.
- Each Soldier will be issued 20 rounds of ammunition to be loaded two 10-round magazines.
- Each Soldier will engage each silhouette with two rounds from the foxhole supported or kneeling position using Table I of DA Form 5790-R.
- Each Soldier must achieve 11 hits out of 20 shots.
- This is a GO/NO GO exercise.

NOTE: At the commander's discretion, Table II of DA Form 3595-R may be used as an alternate course of fire. This table allows Soldiers to engage targets from 50 to 300 meters.

SECTION IV. NIGHT FIRE TRAINING

Soldiers should be able to fire their weapons effectively in total darkness, in bright sunlight, and under all conditions between these two extremes. Scheduled marksmanship training should provide a variety of night and limited visibility conditions. The battlefield may be illuminated by ground flares, handheld flares, M203 flares, mortar and artillery illumination, aerial flares, searchlights, exploding rounds, and burning vehicles, or obscured by smoke, fog, and various environmental conditions. The well-trained Soldier should have experienced a number of these conditions and be confident that he can effectively employ his weapon when required. This section provides guidance on training Soldiers to be effective in total darkness without using iron sights and in limited visibility when using iron sights.

NOTE: See Appendix C for more detailed information about night fighting.

UNASSISTED NIGHT FIRE TRAINING

7-115. All Soldiers must be able to effectively employ their weapons during limited visibility. Soldiers must experience the various conditions of night combat—from total darkness to the many types of artificial illumination.

NOTE: See Table 7-10 for the current training program.

Table 7-10. Unassisted night fire training program.

UNASSISTED NIGHT FIRE TRAINING PROGRAM
Instructional Intent
• Develop the Soldier's confidence in his ability to hit targets when he cannot see through his weapon sights and does not have night vision capability.
Special Instructions
Ensure that— • The M16A2/A3/A4 rifle's or M4 carbine's rear sight is set on the 0-2 aperture. • The M16A1's rear sight is set on the unmarked aperture. • Each Soldier is given two 15-round magazines with 10 rounds of ball ammunition and 5 rounds of tracer ammunition in each magazine. • Each Soldier engages the 50-meter F-type silhouette target from the prone supported or foxhole supported fighting position with one magazine. • Each Soldier engages the 50-meter F-type silhouette target from the prone unsupported fighting position or kneeling position with the second magazine.
Observables
• Each Soldier achieves 7 hits out of 30 target exposures.

MODIFICATIONS FOR NIGHT FIRE

7-116. During limited visibility, a firer cannot effectively use his iron sights in most situations, and without artificial illumination, the sights block his field of vision. Trainers must consider the impact of limited visibility on the Soldier's ability to properly apply combat firing skills and the fundamentals of marksmanship.

Operation and Maintenance of the Weapon

7-117. Many actions are affected by nighttime conditions: handling the weapon, performing operation and function checks, loading and unloading, and maintenance. Consider the following modifications:

- Because combat conditions and enforcement of noise and light discipline restrict the use of illumination, Soldiers must be trained to operate, service, and clean their weapons in total darkness.
- Movements are slower, tasks take longer to complete, vision is impaired, and equipment is more easily misplaced or lost.

NOTE: Although initial practice of these tasks should occur during daylight to facilitate control and error correction, repeated practice during actual nighttime conditions should be integrated with other training. Only through repeated practice and training can the Soldier be expected to perform all tasks efficiently.

Immediate Action

7-118. Under normal conditions, a Soldier should clear a stoppage in three to five seconds. After dark, this task usually takes longer. Identifying the cause of the stoppage may be difficult and frustrating for the Soldier. A dry-fire practice (applying SPORTS) under these conditions using dummy or blank rounds is necessary to reduce time and build confidence. Once the Soldier is confident in applying immediate action in darkness, he can perform such actions rapidly during actual firing.

NOTES: 1. Training should be practiced first during daylight for better control and error correction by the trainer.

2. To learn the hands-only technique of identifying a stoppage, the firer must practice applying immediate action with his eyes closed. This is a technique that Soldiers should be able to master with practice.

Firing Positions

7-119. The firing position recommended for use during limited visibility is the supported firing position.

NOTE: This unassisted night fire supported position differs slightly from the supported position taught in earlier periods of instruction because the firer cannot use his sights during limited visibility.

7-120. To effectively engage targets during limited visibility—

(1) Assume a supported firing position.

(2) Establish a raised stock weld (look 2 to 3 inches above the sights, level with the barrel).

(3) Point the weapon at the target.

(4) Fire in the semiautomatic fire mode.

NOTE: To obtain optimum results, keep the eyes open, and move the head, arms, and weapon as one unit.

Marksmanship Fundamentals

7-121. The four marksmanship fundamentals apply to night firing, but some modifications are needed depending on the conditions.

Steady Position

7-122. When applying unassisted night fire, the Soldier must change his head position or stock weld, especially when using weapon/target alignment techniques. Consider the following modifications:

- Position the head high so that the weapon is aligned on the target.
- Look just over the iron sights.
- Keep the cheek in contact with the stock.

7-123. Repeated dry-fire practice, followed by live-fire training, is necessary to learn and refine these modifications and still achieve the steadiest position.

Aiming

7-124. Modifications to the aiming process vary. Consider these modifications:

- When firing unassisted, use off-center vision instead of pinpoint focus.
- Open both eyes to gather the maximum available light.
- Focus both eyes downrange.

Breathing

7-125. This fundamental is not affected by unassisted night fire conditions.

Trigger Squeeze

7-126. This fundamental is not affected by unassisted night fire conditions. The objective is to not disrupt weapon's alignment with the target when squeezing the trigger.

Target Detection

7-127. Trying to detect a target during the day is difficult enough, but at night, it becomes even more so. The firer must detect and engage targets without artificial illumination or NVDs. Consider these modifications:

- Scan potential target areas and when a target is detected, engage it using a modified quick fire position.
- Take a few seconds to improve weapon/target alignment by pointing slightly low to compensate for the usual tendency to fire high (Figure 7-14).
- Apply the three principles of night vision:
  - Dark adaptation.
  - Off-center vision.
  - Scanning.

7-128. Tracer ammunition may provide feedback on the line of trajectory and facilitate any adjustments in the weapon/target alignment.

Figure 7-14. Lower weapon—target alignment.

Dark Adaptation

7-129. This process conditions the eyes to see under low levels of illumination. The eyes of the average person take about 30 minutes to acquire 98 percent night vision in a completely darkened area. Moving from illuminated to darker areas will decrease night vision until the eyes have adjusted to the surrounding area again.

Off-Center Vision

7-130. When an individual looks at an object during the daytime, he looks directly at it. However, if he did this at night he would only see the object for a few seconds. To see this object for any length of time, he must look 6 to 10 degrees from this object (Figures 7-15 and 7-16) while concentrating his attention on the object. This allows the light sensitive area of the eye to detect faint light sources or reflection.

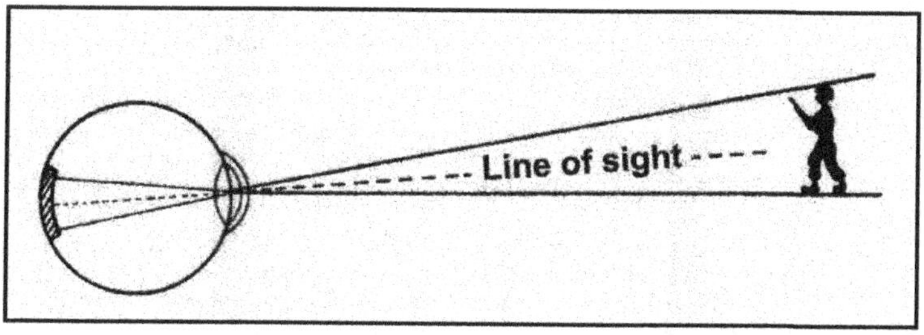

Figure 7-15. Daytime field of view using pinpoint focus.

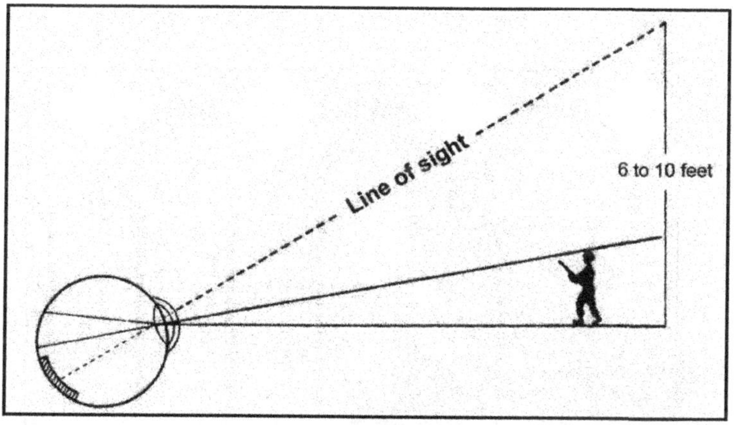

Figure 7-16. Nighttime field of view using off-center vision.

Scanning

7-131. The act of scanning relates to the short, abrupt, irregular movement of the firer's eyes every 4 to 10 seconds around an object or area. Be aware that scanning ranges vary according to levels of darkness.

NIGHT FIRE TRAINING

7-132. All units must include basic unassisted night fire training annually in their unit marksmanship programs.

7-133. Combat units should conduct tactical night fire training at least quarterly. This tactical training should include MILES, during force-on-force training, as well as live-fire training.

Live-Fire Exercises

7-134. The basic unassisted LFX allows all Soldiers to apply night fire principles and to gain confidence in their ability to effectively engage targets out to 50 meters. Practice and proficiency firing can be conducted on any range equipped with mechanical lifters and muzzle flash simulators (Figure 7-17). The muzzle flash simulator provides the firer with a momentary indication that a target is presenting itself for engagement. When a RETS range is used for this exercise, the two 50-meter mechanisms are used. For the unassisted night LFX, the Soldier will perform the following scenario:

- Each Soldier will be issued two 15-round magazines with a tracer and ball combination.
  - The Soldier engages the F-type silhouette target at 50 meters while in the prone supported or foxhole supported firing position. The Soldier uses one magazine of 15 rounds (10 rounds ball; 5 rounds tracer). The Soldier will detect and engage 15 target exposures at 50 meters.
  - The Soldier engages the F-type silhouette target at 50 meters while in the prone unsupported position or kneeling position. The Soldier uses a second magazine of 15 rounds (10 rounds ball; 5 rounds tracer). The Soldier will detect and engage 15 target exposures at 50 meters.
- Each Soldier must achieve 7 hits out of 30 target exposures.
- When the automated range is used, the Soldier's performance is recorded in the tower. If automatic scoring is not available, a coach can observe and score the number of target hits the firer achieves using NVDs.

Figure 7-17. Night-fire target.

ARTIFICIAL ILLUMINATION TRAINING

7-135. Artificial illumination allows for better target detection and long-range accuracy than unassisted night vision. When the artificial light is gone, time must be spent regaining night vision and adapting. Only when the level drops enough so that the target cannot be seen through the iron sights should the firer resume short-range scanning, looking just over the sights.

NOTE: See Table 7-11 for the current training program.

Table 7-11. Artificial illumination training program.

ARTIFICIAL ILLUMINATION TRAINING PROGRAM
Instructional Intent
• Develop the Soldier's confidence in his ability to locate, mark, prioritize and engage targets at night using artificial illumination.
Special Instructions
Ensure that— • M16A2/A3/A4 or M4-series weapon's rear sight is set on the 0-2 aperture. • M16A1 rear sight is set on the unmarked aperture. • Each Soldier is given two 15-round magazines with 10 rounds of ball and 5 rounds of tracer ammunition. • Each Soldier engages 15 target exposures from the foxhole supported fighting position with the first 15-round magazine. • Each Soldier engages 15 target exposures from the prone unsupported fighting position with the second 15-round magazine.
Observables
• Each Soldier achieves 15 hits out of 30 target exposures.

EFFECTIVENESS AND CONTROL OF ARTIFICIAL ILLUMINATION

7-136. When artificial illumination is used, the eyes lose most of their night adaptation, and off-center vision is no longer useful. Artificial illumination allows the firer to use the iron sights with the 0-2 rear sight aperture, as he does during the day.

7-137. To preserve night vision while artificial illumination is being used, the Soldier closes his firing eye and scans his sector for enemy targets with his nonfiring eye. This allows the Soldier to have night vision in at least one eye to keep scanning his sector for enemy targets after the artificial illumination has burned out. However, keeping one eye closed to preserve its night vision results in a drastically altered sense of perception when both eyes are opened.

NOTE: Repeated dry-fire training and target detection practice are the keys to successful engagement of targets out to 250 meters or more during live-fire under artificial illumination.

LIVE-FIRE EXERCISE

7-138. The unassisted LFX with artificial illumination allows all Soldiers to apply night fire principles and to gain confidence in their abilities to effectively detect and engage targets 150 meters away and beyond when using artificial illumination. Soldiers use the night record fire scenario:

- Each Soldier is issued two 15-round magazines with the appropriate tracer and ball combination.
  - During the night, each Soldier engages 15 E-type silhouette target exposures from 50 to 250 meters with one magazine of 15 rounds (10 rounds ball, 5 rounds tracer) while in the foxhole supported or prone supported firing position.
  - During the night, each Soldier engages 15 E-type silhouette target exposures from 50 to 250 meters with the second magazine of 15 rounds (10 rounds ball, 5 rounds tracer) while in the prone unsupported or kneeling firing position.
- Each Soldier must achieve 15 hits out of 30 shots.

SECTION V. MOVING TARGET ENGAGEMENTS

In combat situations, enemy Soldiers do not stand still; they rush from one covered or concealed position to another. While making the rush, enemy Soldiers present rapidly moving targets. Moving targets are open to aimed fire at two points in the rush: as the target begins to gain speed (at the beginning) and as it slows down to a new position.

NOTE: See Table 7-12 for the current training program.

Table 7-12. Moving target engagement training program.

MOVING TARGET ENGAGEMENT TRAINING PROGRAM
Instructional Intent
• Soldiers learn to detect and engage moving and stationary targets with the M16 rifle or M4 carbine.
Special Instructions
Ensure that—
• The M16A2/A3/A4 rifle's or M4 carbine's rear sight is set on the 0-2 aperture.
• The M16A1's rear sight is set on the unmarked aperture.
• Soldiers get into a proper semisupported firing position.
• Soldiers understand and apply lead guidance rules.
• Each Soldier receives two magazines with 25 rounds each of 5.56-millimeter ball ammunition.
• Each Soldier engages 34 moving target exposures at ranges from 35 to 185 meters, and 15 stationary target exposures at ranges from 50 to 300 meters.
Observables
• Each Soldier achieves 18 hits out of 50 target exposures.

MODIFICATIONS FOR MOVING TARGET ENGAGEMENTS

7-139. Soldiers in combat do not know if their next target will be stationary or moving; they must fire immediately at whatever target presents itself. Trainers should consider the following when conducting moving target engagement instruction:

- More dispersion and erratic shots are expected when Soldiers are trained to hit moving targets.
- Considering the environment and the variables of the weapon and ammunition, well-trained Soldiers should be able to hit 300-meter stationary silhouette targets.
- When the target is moving laterally, well-trained Soldiers may only hit 150-meter targets 7 out of 10 times. This is considered an acceptable performance.

7-140. Further, trainers should consider modifications to the fundamentals for engaging stationary targets:

- Steady position.
- Aiming.
- Breath control.
- Trigger squeeze.

STEADY POSITION

7-141. When firing at moving targets, firers should assume the standard supported firing position, but be flexible so that they can track any target in the sector. When a target is moving directly at the firer, directly away from the firer, or at a slight angle, the Soldier engages the target without changing his firing position. Consider the following aspects of firing at moving targets:

- When targets are moving laterally, only minor changes are needed to allow effective target engagement.
- Most moving targets are missed in the horizontal plane (firing in front of or behind the target) and not in the vertical plane (firing too low or too high).

7-142. The Soldier must make other adjustments, as highlighted in Table 7-13.

Table 7-13. Modifications for a steady position when firing at moving targets.

PART OF BODY	MODIFICATION
Nonfiring hand	Grip the weapon more tightly with the nonfiring hand, and apply more pressure to the rear. This helps to maintain positive control of the weapon and steady it for rapid trigger action.
Nonfiring-side elbow	Lift the nonfiring-side elbow from the support position only to maintain a smooth track.
Firing hand	Apply more rearward pressure to the pistol grip to steady the weapon during trigger squeeze.
Firing-side elbow	Lift the firing-side elbow from support only to help maintain a smooth track. NOTE: The weapon pocket in the shoulder and the stock weld are the same as for stationary targets.

AIMING

7-143. When aiming at moving targets, Soldiers must apply precise lead rules, and in turn, Soldiers must accurately estimate speed, angle, and range to the target to apply precise lead rules. Then, he must apply the single-lead rule in order to place effective fire on combat targets.

7-144. The procedures used to engage moving targets vary as the angle and speed of the target vary. For example, when a target is moving directly at the firer, stationary target procedures apply. However, for a close, fast-moving target at a 90-degree angle, the weapon and firer's entire upper body must be free from support so the target can be tracked.

Lead Requirements

7-145. Aiming directly at a 300-meter target moving 8 miles per hour at a 90-degree angle would result in missing it; this type of target covers 4 ½ feet while the bullet is traveling toward him. To hit the target, the Soldier must apply target lead (Figure 7-18) and understand how target lead and bullet speed relate to the range, angle, and speed of the target. To hit the target depicted, the Soldier must aim and fire at position D when the target is at position A.

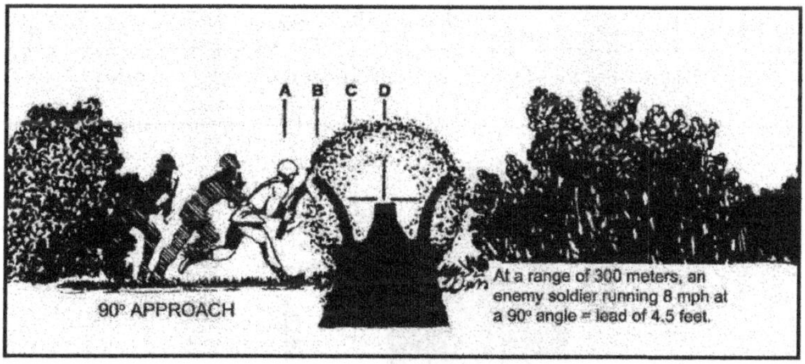

Figure 7-18. Lead requirement based on distance and approach angle.

Single-Lead Rule

7-146. The single-lead rule says: To hit a target moving laterally, place the trailing edge of the front sightpost at the target's center (Figure 7-19). This rule also provides that the lead increases as the range to the target increases (Figure 7-20).

NOTE: At 100 meters, the rule begins to break down for targets moving at slight and large angles.

EXAMPLE

As Figure 7-20 depicts, the front sightpost covers about 1.6 inches at 15 meters and about 16 inches at 150 meters. Since the center of the front sightpost is the actual point of aim, placing the trailing edge of the front sightpost at the target's center provides a .8-inch lead on a 15-meter target and an 8-inch lead on a target at 150 meters.

This provides a dead-center hit on a 15-meter target moving at 7 miles per hour at a 25-degree angle because the target moves .8 inches between the time that the weapon is fired and the time that the bullet arrives at the target. A 150-meter target moving at 7 miles per hour at a 25-degree angle moves 8 inches between the time the weapon is fired and the bullet arrives.

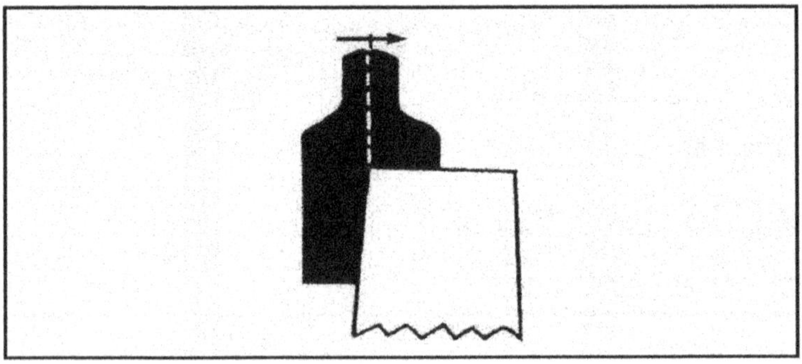

Figure 7-19. Sight-target relationship for the single-lead rule.

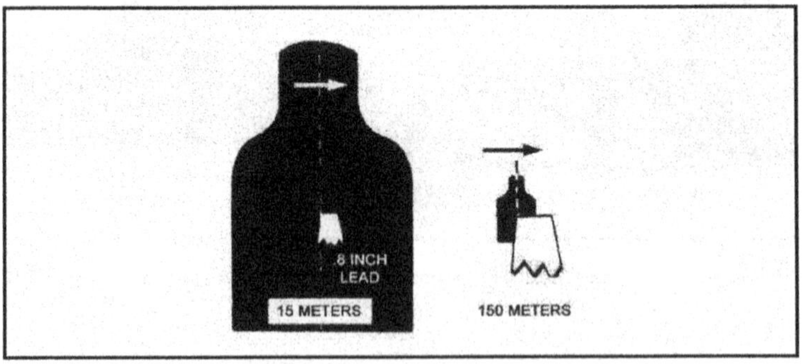

Figure 7-20. Lead increasing at greater ranges.

Target Speed

7-147. Figure 7-21 reflects the differences in lateral speeds for various angles of target movement for a target traveling at 8 miles per hour at a distance of 150 meters from the firer. The angle of target movement is the angle between the target/firer line and the target's direction of movement. An 8-mile-per-hour target moves 24 inches during the bullet's flight time. If the target is moving on a 15-degree angle, it moves 6 inches (the equivalent of 2 miles per hour).

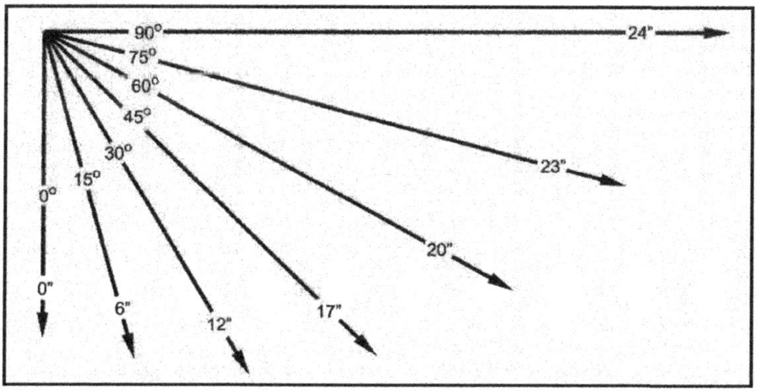

Figure 7-21. Target movement (distance) at various angles.

7-148. Since the target lead is half the perceived width of the front sightpost, at 100 meters, the standard sight provides 5.4 inches of lead for M16A1/2/3/4 rifles and M4 carbines (Table 7-14).

Table 7-14. Angle of target movement.

ANGLE OF TARGET MOVEMENT (Degrees)	RANGE: 100 m		
	(STANDARD SIGHT) TARGET SPEED		
	4 mph	6 mph	8 mph
5	+4.9"	+4.5"	+4.3"
10	+4.1"	+3.5"	+2.7"
15	+3.5"	+2.5"	+1.5"
20	+2.8"	+1.5"	+.2"
25	+2.2"	+.7"	-1.0"
30	+1.7"	-.2"	-2.0"
35	+1.1"	-1.1"	-3.2"
40	+.6"	-1.9"	-4.3"
45	0"	-2.7"	-5.4"
50	-.4"	-3.3"	-6.2"
55	-.8"	-4.0"	-7.0"
60	-1.2"	-4.5"	-7.7"
65	-1.5"	-4.9"	-8.4"
70	-1.7"	-5.3"	-8.8"
75	-1.9"	-5.6"	-9.2"
80	-2.0"	-5.9"	-9.6"
85	-2.1"	-5.9"	-9.7"
90	-2.1"	-6.0"	-9.8"
NOTE: Plus (+) indicates bullet strike in the direction of movement; minus (-) indicates bullet strike behind the target's center.			

Target Distance

7-149. The front sightpost covers only a small part of close-in targets, providing hits on close targets moving at any angle and any speed. However, if the lead rule is applied on more distant targets moving at a slight angle—for example, 5 degrees at 100 meters—the bullet strikes about 4 inches forward of the target's center. Soldiers must be taught to fire at targets as though they are stationary until lateral movement is observed (15 degrees).

7-150. The rule provides for many speed/angle combinations that place the bullet within 2 inches of the target's center (Table 7-14). Since the Soldier is expected to fire a 12-inch group on moving targets at 100 meters, the rule provides for hits on the majority of targets. Even the worst case (a 90-degree target moving at 8 miles per hour) would result in the shot group's center being located 9.8 inches behind the target's center. If bullets were evenly distributed in a 12-inch group, this would result in hitting the target 40 percent of the time.

7-151. Soldiers should be taught to increase their lead if they miss the target, which increases their probability of hitting all targets. For example, if target exhibits much lateral movement and the Soldier feels that he has missed the target by applying the lead rule and firing fundamentals, he should increase his lead.

7-152. The training program must be simple and provide Soldiers with information relevant to improving their performance in combat. It should address the following topics:

- Soldiers should understand and apply the single-lead rule in the absence of more information.
- Soldiers should engage moving targets coming toward them or on a slight angle (0 to 15 degrees) as stationary targets.
- Information should be presented and practice allowed on applying additional lead to targets for Soldiers who demonstrate this aptitude.

Target Angle

7-153. The single-lead rule does not apply to targets moving at small and large angles (Table 7-15). A walking enemy Soldier at 250 meters is hit dead center when he is moving at 40 degrees. Hits can be obtained if he is moving on any angle between 15 and 75 degrees. When he is running, a center hit is obtained when the target is on an angle of 18 degrees; misses occur when he exceeds an angle of 30 to 35 degrees.

7-154. The information provided in Figure 7-21 and Table 7-15 is designed to enhance instructor understanding so proper concepts are presented during instruction. For example, a target at 100 meters moving at 6 miles per hour receives a center hit when moving at 29 degrees. When moving at an angle less than 29 degrees, the bullet strikes somewhat in front of the target's center. When moving at an angle of more than 29 degrees, the bullet strikes somewhat behind the target's center.

Table 7-15. Target angle when dead center; hits occur using the single-lead rule.

STANDARD SIGHT			
RANGE (m)	4 mph	6 mph	8 mph
25	48°	30°	22°
50	47°	30°	22°
100	45°	29°	21°
150	44°	28°	20°
200	41°	27°	19°
250	40°	26°	18°
300	33°	21°	16°
350	38°	24°	18°
400	35°	22°	17°
450	33°	21°	16°

Tracking

7-155. Tracking is a more accurate technique of engaging targets used by experienced firers. It involves establishing and maintaining a point of aim in relationship to the target and moving with the target to maintain that sight picture, while squeezing the trigger. As the target moves, this technique puts the firer in position for a second shot if the first one misses.

Trapping

7-156. Trapping involves setting up a point of aim forward of the target and along the target path. The trigger is squeezed as the target comes into the sights. This technique works on targets with slow lateral movement. It does not require tracking skills, but the firer must know precisely when the weapon is going to fire.

NOTE: Soldiers who can squeeze the trigger without reacting to the weapon firing may fire better using this technique.

BREATH CONTROL

7-157. This fundamental is unchanged.

TRIGGER SQUEEZE

7-158. To use proper trigger squeeze—

- Apply rearward pressure on the handguard and pistol grip to hold the weapon steady while applying pressure to the trigger.
- Squeeze the trigger quickly (almost a controlled jerk).
- Apply heavy pressure on the trigger (at least half of the pressure it takes to make the weapon fire) before squeezing.

MOVING TARGET LIVE-FIRE EXERCISE

7-159. Soldiers engage in a firing scenario once for practice and then for qualification. Soldiers who fail to qualify on the initial day of qualification receive only one refire on the same day.

SECTION VI. SHORT-RANGE MARKSMANSHIP TRAINING

SRM training allows Soldiers to quickly and effectively engage targets at ranges less than 50 meters. Although normally associated with UO, SRM techniques are also used during operations in restrictive terrain, such as clearing a trench line, the final assault across an objective during an attack or raid, or when fighting in dense vegetation or during periods of limited visibility. SRM instruction consists of four components:

- Phase I—Reflexive Fire Training (blank-fire day and night).
- Phase II—Target Discrimination (blank-fire day and night).
- Phase III—Short-Range Marksmanship Qualification (day and night live-fire) and Barricade Transition Qualification.
- Phase IV—Shotgun and Automatic or Burst Firing Familiarization.

CONDUCT OF SHORT-RANGE MARKSMANSHIP TRAINING

7-160. SRM requires individual Soldiers to be trained to standard on reflexive firing, target discrimination, and all necessary BRM fundamentals prior to semiannual qualification.

NOTE: An explanation of the base level proficiency requirements is provided with each course of fire.

7-161. At a minimum, Soldiers should be qualified on their individual weapon within the previous six months. Shotgun and automatic firing is required for annual familiarization only. Reflexive MILES dry-fire drills are an essential part of the training process and should be conducted by the team leader or squad leader during troop-leading procedures and before any SRC or SRM training.

NOTE: See Table 7-16 for the current training program.

Table 7-16. Short-range marksmanship training program.

SHORT-RANGE MARKSMANSHIP TRAINING PROGRAM
Instructional Intent
• Soldiers gain confidence and knowledge in SRM fundamentals.
Special Instructions
Ensure that— • The M16A2/A3/A4 rifle's or M4 carbine's rear sight is set on the 0-2 aperture. • The M16A1's rear sight is set on the unmarked aperture. • Each Soldier is given one 20-round magazine of 5.56-millimeter ball ammunition for marksmanship qualification. • Each Soldier is given one 28-round magazine and three 4-round magazines of 5.56-millimeter ball ammunition for barricade transition firing. NOTE: The round must impact within the lethal zone to be scored a hit.
Observables
• Each Soldier achieves 20 target hits during the day marksmanship qualification. • Each Soldier achieves 16 target hits during the night marksmanship qualification. • Each Soldier achieves 14 target hits during the day marksmanship qualification while wearing a protective mask. • Each Soldier achieves 12 target hits during the night marksmanship qualification while wearing a protective mask.

NOTE: All SRC and SRM training should begin with a review of the principles of safe weapon handling: Assume the weapon is always loaded, and never point the weapon at anything that you do not intend to shoot.

FUNDAMENTALS OF SHORT-RANGE MARKSMANSHIP

7-162. During SRC, there is little margin for error. Too slow a shot at the enemy, too fast a shot at a noncombatant, or inaccurate shots can all be disastrous for the Soldier. Further, the risk of fratricide or noncombatant casualties is greatest during SRC. To survive and accomplish missions in close quarters, Soldiers must master four fundamentals of SRM:

- Firing stance, weapon ready positions, and movement techniques.
- Aiming technique.
- Point of aim.
- Trigger manipulation.

Firing Stance, Weapon Ready Positions, and Movement Techniques

7-163. Regardless of the weapon ready position used, Soldiers must always assume the correct firing stance to ensure stability and accuracy when engaging targets.

Firing Stance

7-164. To assume the correct fighting stance—

- Keep the feet approximately shoulder-width apart.
- Point the toes straight to the front (direction of movement).
- Stagger the firing-side foot slightly to the rear of the nonfiring-side foot.
- Bend the knees slightly.

- Lean the upper body slightly forward.
- Square the shoulders and pull them back; don't roll them over or slouch.
- Keep the head up and both eyes open.

7-165. When engaging targets—

- Hold the weapon with the buttstock firmly against the shoulder.
- Hold the firing-side elbow close against the body.

7-166. Although short-range engagements generally take place from the standing position, a Soldier may be required to engage targets from the kneeling position. The kneeling position is generally used when correcting a weapon malfunction.

Weapon Ready Positions

7-167. There are two weapon ready positions: high ready and low ready.

High Ready Position

7-168. The high ready position (Figure 7-22) is best suited for the lineup outside of a building, room, or bunker entrance.

7-169. To hold the weapon in the high ready position—

(1) Hold the weapon's buttstock under the armpit, with the barrel pointed slightly up so that the top of the front sightpost is just below the line of sight, but still within peripheral vision.

(2) Grasp the handguards toward the front sling swivel with the nonfiring hand. Keep the trigger finger outside of the trigger well and the thumb of the firing hand on the selector lever.

7-170. To engage a target from the high ready position—

(1) Push the weapon forward, as if to bayonet the target.

(2) Bring the buttstock firmly against the shoulder as it slides up the body.

Figure 7-22. High ready position.

Low Ready Position

7-171. The low ready position (Figure 7-23) is best suited for movement inside of buildings.

7-172. To hold the weapon in the low ready position—

(1) Place the weapon's buttstock firmly in the pocket of the shoulder, with the barrel pointed down at a 45-degree angle.

(2) Grasp the handguards toward the front sling swivel. Keep the trigger finger outside of the trigger well and the thumb of the firing hand on the selector lever.

7-173. To engage a target from the low ready position, bring the weapon up until the proper sight picture is achieved.

Figure 7-23. Low ready position.

Movement Techniques

7-174. Soldiers must practice moving with their weapons up until they no longer look at the ground, but concentrate on their sectors of responsibility and move without stumbling over their own feet.

NOTE: The low ready position is the best position to use when moving or turning.

7-175. To engage a target to the left—

(1) Step toward the target with the left foot.

(2) Move the right foot even with the left, assuming a proper standing firing position.

7-176. To engage a target to the right—

(1) Step toward the target with the right foot.

(2) Move the left foot even with the right, assuming a proper standing firing position.

7-177. To turn to the rear—

(1) Position the firing-side foot forward.

(2) Place the body weight on the firing-side foot.

(3) Pivot the body, similar to the drill movement "rear march."

AIMING TECHNIQUES

7-178. Four aiming techniques are used during SRC:

- Slow aimed fire.
- Rapid aimed fire.
- Aimed quick kill.
- Instinctive fire.

7-179. Each has advantages and disadvantages, and the Soldier must understand when, how, and where to use each technique.

Slow Aimed Fire

7-180. Slow aimed fire is the slowest, but most accurate, technique. When using this technique, Soldiers take a steady position, properly align the sight picture, and squeeze off rounds. This technique should only be used to engage targets more than 25 meters away, when good cover and concealment is available, or when the need for accuracy overrides the need for speed.

Rapid Aimed Fire

7-181. The rapid aimed fire technique utilizes an imperfect sight picture. When using this technique, the Soldier focuses on the target and raises his weapon until the target is obscured by the front sightpost. Elevation is less critical than windage when using this technique. This aiming technique is extremely effective on targets 0 to 15 meters away.

Aimed Quick Kill

7-182. The aimed quick kill technique is the quickest and most accurate method of engaging targets up to 12 meters away. As Soldiers become more experienced at using this technique, they may use it at greater ranges. When using this technique, the Soldier aims over the rear sight, down the length of the carrying handle, and places the top ½ to ¾ of an inch of the front sightpost on the target.

Instinctive Fire

7-183. Instinctive fire is the least accurate technique and should only be used in emergencies. It relies on instinct, experience, and muscle memory. To use this technique, the firer concentrates on the target and points the weapon in the general direction of the target. While gripping the handguards with the nonfiring hand, he extends the index finger to the front, automatically aiming the weapon on a line toward the target.

POINT OF AIM

7-184. Most short-range engagements will be decided by who hits his target with the first round. During this type of engagement, it is more important to put the target down as quickly as possible than it is to kill him immediately.

7-185. Soldiers must aim at the lethal zone (center of mass) of the body. Although shots to the center of the body may prove to be eventually fatal, they may not immediately incapacitate the target. A shot that does not immediately incapacitate the target may be no better than a clean miss. Because of this, and the possible presence of military equipment or protective vests, Soldiers must also be able to engage targets with incapacitating shots.

Lethal Shot Placement

7-186. The target's lethal zone (Figure 7-24) is its center of mass, between the waist and the chest. Shots in this area maximize the hydrostatic shock of the shot pellets. Due to the nature of SRC, Soldiers must continue to engage targets until they go down.

Figure 7-24. Lethal zone.

Incapacitating Shot Placement

7-187. Only one shot placement guarantees immediate and total incapacitation: roughly centered in the face, below the middle of the forehead and the upper lip, and from the eyes in. Shots to the side of the head should be centered between the crown of the skull and the middle of the ear opening, from the center of the cheekbones to the middle of the back of the head (Figure 7-25).

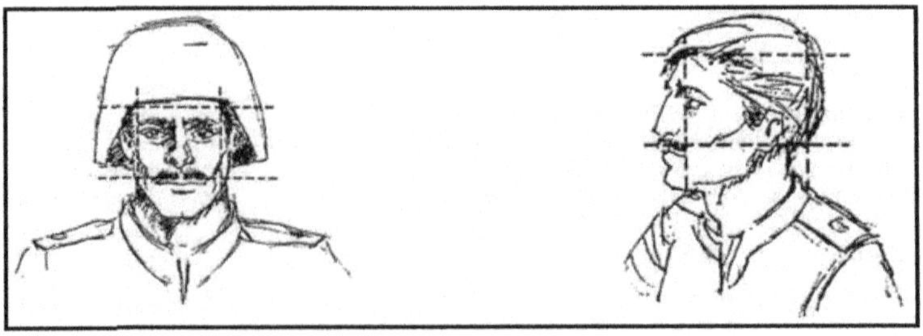

Figure 7-25. Incapacitation zone.

Trigger Manipulation

7-188. SRC engagements are usually quick, violent, and deadly. Due to the reduced reaction time, imperfect sight picture, and requirement to effectively place rounds into targets, Soldiers must fire multiple rounds during each engagement in order to survive. Multiple shots may be fired either through the use of a controlled pair or automatic weapon fire.

Controlled Pair

7-189. A controlled pair is two rounds fired in rapid succession. Controlled pairs should be fired at single targets until they go down. When multiple targets are present the Soldier must fire a controlled pair at each target, and then reengage any targets left standing. To fire a controlled pair—

(1) Fire the first round, and allow the weapon to move in its natural arc without fighting the recoil.

(2) Rapidly bring the weapon back on target, and fire a second round.

7-190. Soldiers must practice firing the controlled pair until it becomes instinctive.

Automatic Fire

7-191. While rapid, aimed, semiautomatic fire is the most accurate method of engaging targets during SRC and controlled three-round bursts are better than automatic fire, automatic weapon fire may be necessary to maximize violence of action or gain fire superiority when gaining a foothold in a room, building, or trench. When properly trained, Soldiers should be able to fire six rounds (two three-round bursts) in the same time it takes to fire a controlled pair. With practice, the accuracy of engaging targets can be equal to that of semiautomatic fire at 10 meters.

NOTE: The key to burst or automatic firing is to squeeze the trigger, not jerk it.

7-192. For the majority of Soldiers, fully automatic fire is rarely effective and can lead to unnecessary noncombatant casualties or fratricide. Not only is fully automatic fire inaccurate and difficult to control, but it also rapidly empties ammunition magazines. A Soldier who finds himself out of ammunition with an armed, uninjured enemy Soldier during SRC will become a casualty unless a fellow Soldier intervenes.

Failure Drill

7-193. To make sure that a target is completely neutralized, Soldiers should be trained to execute the failure drill. The firer will fire a controlled pair into the lethal zone, followed by a third round placed into the incapacitation zone. This type of target engagement is particularly useful when engaging targets wearing body armor.

PRELIMINARY SHORT-RANGE MARKSMANSHIP INSTRUCTION

7-194. As with all other forms of marksmanship training, preliminary SRM instruction must be conducted to establish a firm foundation. Soldiers must be taught, and must understand, the fundamentals of SRM. Blank-fire drills are conducted to ensure a thorough understanding of the fundamentals, as well as to provide the trainers with valuable feedback about each Soldier's level of proficiency.

NOTE: To maximize safety during training and in combat situations, it is important to emphasize muzzle awareness and selector switch manipulation during preliminary SRM instruction. The risk of fratricide or noncombatant casualties is greatest during SRC.

7-195. Table 7-17 outlines the tasks that preliminary SRM instruction should include (at a minimum).

Table 7-17. Preliminary SRM tasks and explanation.

TASK	EXPLANATION
Weapon ready positions and firing stance	Ensure that each Soldier understands and can properly carry his weapon.
Magazine changing drills	Have Soldiers perform magazine changes using the two methods: • Parallel method. • L-shaped method. NOTE: See the paragraph below for more information about magazine changing drills.
Moving with a weapon	Ensure that Soldiers can walk, run, and turn left, right, and to the rear, as well as move from the standing to kneeling firing position and the kneeling back to the standing firing position.
Weapon malfunction drills	Ensure that Soldiers instinctively drop to the kneeling firing position, clear a malfunction using SPORTS, and continue to engage targets. This drill can be performed by issuing each Soldier a magazine loaded with six to eight rounds of blank ammunition with one expended blank round.
Target engagement drills	Teach Soldiers to move from the weapon ready position to the firing stance, emphasizing speed and precision. Soldiers must be observed to ensure that the finger is outside of the trigger well and that the selector switch remains on SAFE until the weapon is raised to the firing position. NOTE: This is a force protection issue and must be drilled until all Soldiers can perform to standard.

MAGAZINE CHANGING DRILLS

NOTE: Before performing magazine changing drills, have Soldiers configure their LBE or magazine racks to allow for the easiest access for the magazine change. The time limit to perform any of the magazine changing drills is 8 seconds.

7-196. There are two methods of changing the magazine:

- Parallel method.
- L-shaped method.

Parallel Method

7-197. When shooting from the right side, perform the right-side parallel magazine changing method (Figure 7-26):

(1) Engage the target until the magazine is empty.

(2) Take a knee.

(3) Place the weapon at a 45-degree angle, with the magazine well facing in, the muzzle pointed in a safe direction, and the buttstock in the crease of the right elbow.

(4) Slide the nonfiring hand down the handguard to the receiver, and press the magazine release catch.

(5) Secure a full magazine with two fingers and the thumb of the nonfiring hand.

(6) Bring the full magazine (top of the magazine toward the magazine well) next to and parallel with the empty magazine.

(7) Slide the empty magazine out with the nonfiring hand and insert the full magazine.

(8) Hit the bolt release.

(9) Stow the empty magazine in the ammunition pouch.

Figure 7-26. Right-side parallel magazine changing method.

7-198. When shooting from the left side, perform the left-side parallel magazine changing method (Figure 7-27):

(1) Engage the target until the magazine is empty.

(2) Take a knee.

(3) Place the weapon at a 45-degree angle, with the magazine well facing in, the muzzle pointed in a safe direction, and the buttstock in the crease of the left elbow.

(4) Secure a full magazine with two fingers and the thumb of the nonfiring hand.

(5) Bring the full magazine (top of the magazine toward the magazine well) next to and parallel with the empty magazine.

(6) Press the magazine release catch with the trigger finger.

(7) Slide the empty magazine out with the nonfiring hand, and insert the full magazine.

(8) Hit the bolt release.

(9) Stow the empty magazine in the ammunition pouch.

Figure 7-27. Left-side parallel magazine changing method.

L-Shaped Method

7-199. When shooting from the right side, perform the right-side L-shaped magazine changing method (Figure 7-28):

(1) Engage the target until the magazine is empty.

(2) Take a knee.

(3) Place the weapon at a 45-degree angle, with the magazine well facing in, the muzzle pointed in a safe direction, and the buttstock in the crease of the right elbow.

(4) Slide the nonfiring hand down the handguard to the receiver, and press the magazine release catch.

(5) Secure a full magazine with two fingers and the thumb of the nonfiring hand.

(6) Bring the full magazine (top of the magazine facing to the rear) next to and perpendicular to the empty magazine, forming an L-shape.

(7) Slide the empty magazine out with the nonfiring hand, rotate the full magazine so the top of the magazine faces the magazine well, and insert the full magazine.

(8) Hit the bolt release.

(9) Stow the empty magazine in the ammunition pouch.

Figure 7-28. Right-side L-shaped magazine changing method.

7-200. When shooting from the left side, perform the left-side L-shaped magazine changing method (Figure 7-29):

(1) Engage the target until the magazine is empty.

(2) Take a knee.

(3) Place the weapon at a 45-degree angle, with the magazine well facing in, the muzzle pointed in a safe direction, and the buttstock in the crease of the left elbow.

(4) Press the magazine release catch with the trigger finger.

(5) Secure a full magazine with two fingers and the thumb of the nonfiring hand.

(6) Bring the full magazine (top of the magazine facing to the rear) next to and perpendicular to the empty magazine, forming an L-shape.

(7) Slide the empty magazine out with the nonfiring hand, rotate the full magazine so the top of the magazine faces the magazine well, and insert the full magazine.

(8) Hit the bolt release.

(9) Stow the empty magazine in the ammunition pouch.

Figure 7-29. Left-side L-shaped magazine changing method.

PHASE I—REFLEXIVE FIRE TRAINING

7-201. Reflexive fire training involves the practical application of all four of the fundamentals of SRM. All Soldiers must receive a GO on the task, Conduct Reflexive Firing, before proceeding with training. This is a perishable skill that must be constantly reinforced, and reflexive fire training should be conducted frequently as refresher training to ensure that the Soldiers' skills are always at the highest level possible.

REFLEXIVE FIRING TARGETS

7-202. Targets can be purchased locally (FBI style) or manufactured by the unit (bowling pin targets). E-type silhouettes may be painted as shown in Figure 7-30.

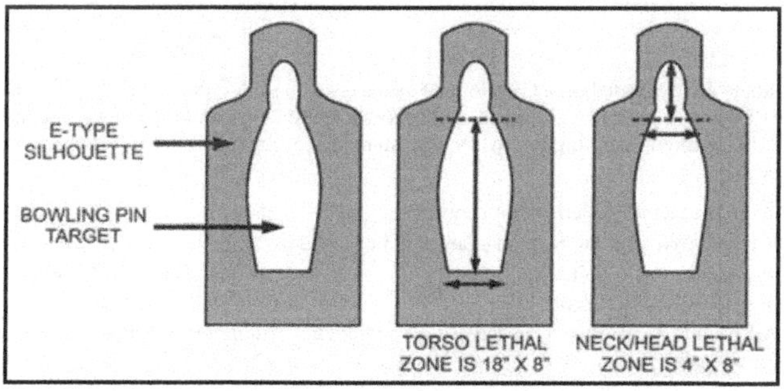

Figure 7-30. Dimensions and placement of bowling pin targets.

RANGE SETUP

7-203. The range must be at least 25 meters in length, with identification marks at the 5-, 10-, 15-, and 25-meter distances. Each lane should be marked in a way that prevents crossfiring between lanes. A lane safety coach is assigned to each lane to observe and evaluate the Soldiers' performance, as well as ensure the safe conduct of firing. All firing cues are given by the tower or line safety.

CONDUCT OF TRAINING

NOTE: Unit commanders should conduct training continually to first establish and then sustain levels of proficiency in reflexive firing.

7-204. Each Soldier conducts a dry-fire exercise and a blank-fire exercise prior to conducting the LFX. The dry-fire and blank-fire exercises give the Soldier the repetition needed to successfully engage targets quickly and accurately.

7-205. During the dry-fire exercise, Soldiers identify and engage the proper targets at ranges from 5 to 25 meters from the stationary position, while turning and walking. The tower or line safety gives all firing cues. All tables are fired at night, with and without protective masks, using automatic fire for familiarization, and while using NVDs. The dry-fire exercise proceeds as follows:

(1) The Soldier starts at the 25-meter line at the low ready position, facing the targets.

(2) The Soldier is then told the engagement position (for example, facing left, turns right) and, once in position, is given the cue to fire.

(3) On cue, the Soldier assumes the proper firing position and stance, places the selector lever on SEMI, uses the correct aiming technique for the target's distance, and engages the target.

(4) After engaging the target, the Soldier continues to cover the target to reinforce firing until the threat is eliminated.

NOTE: If Soldiers will be engaging targets using lasers, optics, or the protective mask, they should complete all steps using the same equipment. Do not have Soldiers familiarize with iron sights and then fire the LFX with optics.

Evaluation

7-206. Soldiers are evaluated on a GO/NO GO basis, based on the standards in the training and evaluation outline (T&EO) and scoring table. Soldiers must score a GO on the familiarization firing tables (Table 7-18 and Table 7-19) before attempting to qualify. The standard for protective mask firing is 60 percent day and 50 percent night.

7-207. Use the following guidelines for scoring:

- Rounds fired after the time standard will be scored as a miss.
- The number of rounds fired after the time standard will be subtracted from the total number of hits the Soldier has scored.
- All rounds must impact on the E-type silhouette. Hits are defined as being in the lethal zone (bowling pin).

Table 7-18. Familiarization (stationary).

POSITION	ROUNDS FIRED	DISTANCE (m)	METHOD	TIME STANDARD	LETHAL ZONE HIT STANDARD
Straight ahead	4	25	Single shot	None	3
Straight ahead	4	10	Single shot	None	3
Straight ahead	4	25	Controlled pair	None	3
Straight ahead	4	10	Controlled pair	None	3

Table 7-19. Familiarization (moving).

POSITION	ROUNDS FIRED	DISTANCE (m)	METHOD	TIME STANDARD	LETHAL ZONE HIT STANDARD
Facing left; turning right	4	25	Controlled pair	None	3
Facing right; turning left	4	25	Controlled pair	None	3
Straight ahead walking	4	10, start at 15	Controlled pair	None	3
Straight ahead	4	5	Controlled pair	None	3

PHASE II—TARGET DISCRIMINATION TRAINING

7-208. During SRC, there is little or no margin for error. Too slow a shot at the enemy, too fast a shot at a noncombatant or friendly Soldier, or inaccurate shots can all be disastrous for the Soldier. Target discrimination is the act of distinguishing between threat and nonthreat targets during SRC. Target discrimination is an inescapable responsibility and must be stressed in all situations regardless of mission.

NOTE: It is essential that target discrimination training be aimed at instilling fire control and discipline in Soldiers. The first priority is always the safety of a Soldier.

TARGET DISCRIMINATION TARGETS

7-209. Targets used to conduct target discrimination training include—

(1) Two or more E-type silhouettes with bowling pins painted on each side of the silhouette (for example, brown side and green side).

(2) Multiple E-type silhouettes with different painted shapes (for example, squares, triangles, and circles).

(3) A series of 3-inch circles on E-type silhouettes.

(4) Pop-up targets (for example, electrical or pull targets).

(5) Target's hands and center of the uniform.

Two or More E-Type Silhouettes with Bowling Pins Painted on Each Side

7-210. Target discrimination is best taught using two or more E-type silhouettes with bowling pins painted on each side of the silhouette. To conduct training using this type of target—

(1) The instructor calls out a color for the firer to identify.

(2) On the command "READY, UP" or at the whistle blast, the firer quickly scans all targets for the color and engages the correct target using a controlled pair.

7-211. This exercise trains Soldiers to accomplish missions under the expected ROE. During force-on-force training, the OPFOR wears distinctive uniforms; this exercise prepares Soldiers to eliminate threats based on enemy uniforms and reduces the chances of a Soldier hesitating and becoming a casualty. Using realistic targets displaying threat and nonthreat personnel is another variation.

Multiple E-Type Silhouettes with Different Painted Shapes

7-212. To conduct training using this type of target—

(1) The instructor calls out a shape for the firers to identify.

(2) On the command "READY, UP" or at a whistle blast, the firers quickly scan all targets, searching for the shape, and engage the correct target using the controlled pair technique.

7-213. This is repeated until one shape is mastered. Subsequently, a sequence of shapes is announced, and the firers engage accordingly.

Series of 3-Inch Circles on E-Type Silhouettes

7-214. To conduct training using this type of target—

(1) Instructors call out which circle to engage (for example, top left).

(2) Firers react accordingly.

7-215. Marksmanship is emphasized using this technique.

Target's Hands and Center of the Uniform

7-216. If a target is a threat, the first and most obvious indicator is a weapon in the target's hands. A good technique for teaching Soldiers target discrimination is to have them focus on the target's hands and the center of the uniform. To conduct this exercise, the Soldier must visualize the entire target because an armed target could be a fellow Soldier or other friendly. This level of target discrimination should not be trained until Soldiers are thoroughly proficient in basic SRC and SRM tasks.

Range Setup

7-217. The range must be at least 25 meters in length, and each lane should be at least 5 meters wide. Each lane should have target holders and should be marked in a way that prevents crossfiring between lanes. A coach/safety is assigned to each lane to observe and control the Soldier's performance. The tower, lane safety, or senior instructor gives all firing commands.

Conduct

7-218. Regardless of the type of target used, the exercise proceeds as follows:

NOTE: Each Soldier must complete a dry-fire exercise and a blank-fire exercise before moving to the live-fire portion.

---

(1) The Soldier faces away from the target.

NOTE: This action requires the Soldier to identify and discriminate targets, and reinforces skills used during reflexive firing training.

---

(2) The Soldier is given a target description.

(3) On the command "READY," the Soldier begins to scan for the target.

(4) On the command "UP," voice command, or whistle blast, the Soldier turns toward and engages the target.

NOTES: 1. Instructors should vary commands and targets so that the Soldier does not fall into a pattern. Intermixing "NO FIRE" commands will add to realism.

2. The first priority is always the safety of the Soldier.

7-219. All Soldiers must receive a GO on this task before SRM qualification. Targets must be scored and marked after each firing distance. A Soldier will be scored as a NO GO if he fails to engage a target or engages a target other than the one called for by the instructor. Soldiers will also receive a NO GO if, at any time, they point their weapon at another Soldier or they fail to keep their weapon on SAFE before acquiring and engaging the targets.

NOTE: Initial training and sustainment training may be conducted by changing the uniform in the standards statement.

EVALUATION

7-220. Table 7-18 will be used to score this exercise.

PHASE III—SHORT-RANGE MARKSMANSHIP QUALIFICATION

7-221. Soldiers should conduct SRM qualification semiannually. In addition to qualification, commanders should conduct familiarization using the same qualification standards while altering the conditions. Firing the qualification tables in protective masks and during periods of limited visibility should be included; Soldiers should train as they fight—with all assigned equipment.

NOTE: Although the qualification is intended to be fired with open sights only, iterations using laser aiming devices, CCOs, and NVDs are highly encouraged.

RECORD AND PRACTICE FIRE

Conduct

NOTES:
1. Soldiers must complete a blank-fire iteration of the qualification tables before conducting live-fire qualification.
2. Each Soldier will have a coach to ensure that he acquires the target, keeps the weapon on SAFE until time to engage the target and is then places it back on SAFE, and maintains muzzle awareness throughout the exercise.
3. If Soldiers will be engaging targets with lasers, optics, or the protective mask, they should complete all steps using the same equipment. Do not have Soldiers familiarize with iron sights and fire the exercise while wearing the protective mask.

7-222. Each Soldier engages the target IAW the firing table (Table 7-20).

Table 7-20. Record and practice fire.

POSITION	ROUNDS FIRED	DISTANCE (m)	METHOD	TIME STANDARD
Straight ahead	2	4	Controlled pair	3 sec from command "UP"
Left turn	2	7	Controlled pair	3 sec from command "UP"
Right turn	2	10	Controlled pair	3 sec from command "UP"
Walk straight	2	5, start at 10	Controlled pair	3 sec from command "UP"
Walk straight	2	10, start at 15	Controlled pair	3 sec from command "UP"
Run, stop, and shoot	2	10, start at 20	Controlled pair	3 sec from command "UP"
Straight ahead	2	25	Controlled pair	3 sec from command "UP"
Run, stop, kneel, and shoot	2	15, start at 25	Controlled pair	3 sec from command "UP"
Walk laterally right	2	7	Controlled pair	3 sec from command "UP"
Walk laterally left	2	7	Controlled pair	3 sec from command "UP"

Evaluation

> **NOTE:** For scoring purposes, a hit is a round that impacts within the lethal zone.

7-223. Each Soldier scores 16 hits day and night. The standard when wearing a protective mask is 14 day and 12 night. In addition to achieving a qualifying score, all 20 rounds must hit the E-type silhouette in order to qualify.

BARRICADE TRANSITION QUALIFICATION FIRING

Range Setup

7-224. The range must be at least 50 meters in length, with tall barricades at the 35- and 50-meter distances and short barricades at the 25- and 40-meter distances. Each lane should be marked in a way that prevents crossfiring between lanes. A lane safety coach is assigned to each lane to observe and evaluate the Soldier's performance, as well as ensure that firing is conducted safely. The tower or line safety gives all firing cues.

Conduct

> **NOTE:** Soldiers must complete a dry-fire and a blank-fire iteration of the qualification tables before conducting live-fire qualification. The dry-fire and blank-fire exercises gives the Soldier the repetition needed to successfully engage targets quickly and accurately.

7-225. The exercise proceeds as follows:

(1) The Soldier receives one 28-round and three 4-round magazines.
(2) The Soldier loads one 28-round magazine.
(3) The Soldier assumes the low ready position at the 50-meter line, facing the targets.
(4) The Soldier engages the targets as shown in Table 7-21.
(5) On cue, the Soldier assumes the proper firing position and stance, places the selector lever on SEMI, uses the correct aiming technique for the target's distance, and engages the target.

> **NOTE:** If a Soldier is having difficulty during the blank-fire exercise, he must be retrained before continuing with the qualification.

Evaluation

7-226. Use the following guidelines for scoring:

- All rounds must impact on the E-type silhouette.
- Hits are defined as being in the lethal zone (bowling pin).

Table 7-21. Barricade transition fire.

BARRICADE/DISTANCE	POSITION	ROUNDS FIRED	METHOD
Start with 28-round magazine in the open.			
Open/50 m	Standing	2	Controlled pair
	Kneeling	2	Controlled pair
	Prone	2	Controlled pair
	Sitting	2	Controlled pair
Tall/50 m	Left	2	Controlled pair
	Right	2	Controlled pair
Short/40 m	Left	2	Controlled pair
	Top	2	Controlled pair
	Right	2	Controlled pair
Tall/35 m	Left	2	Controlled pair
	Right	2	Controlled pair
Short/25 m	Left	2	Controlled pair
	Top	2	Controlled pair
	Right	2	Controlled pair
Transition back to the tall 35-m barricade. Perform parallel or L-shaped magazine change. Insert 4-round magazine.			
Tall/35 m	Left	2	Controlled pair
	Right	2	Controlled pair
Transition back to short 40-m barricade. Perform parallel or L-shaped magazine change. Insert 4-round magazine.			
Short/45 m	Left	2	Controlled pair
	Right	2	Controlled pair
Transition back to tall 50-m barricade. Perform parallel or L-shaped magazine change. Insert 4-round magazine.			
Tall/50 m	Left	2	Controlled pair
	Right	2	Controlled pair

PHASE IV—SHOTGUN AND AUTOMATIC OR BURST FIRING FAMILIARIZATION

7-227. Shotgun and automatic or burst firing familiarization is no different for SRM than for BRM.

NOTE: Publications for shotgun firing familiarization are currently being developed.

SECTION VII. SQUAD DESIGNATED MARKSMAN TRAINING

Advances in technology have led to the development of weapon systems that are increasingly more accurate and able to engage targets at much longer ranges. Conversely, today's rifleman is trained to engage targets only out to 300 meters. This 300-meter limit is well short of the weapon/ammunition combination's capability. Snipers engage targets at 600 meters and beyond. The SDM engages targets with direct small arms fire in the gap between the engagement range of the average combat Soldier and the sniper. Possessing the ability to estimate range, detect targets, and place effective, well-aimed fire on intermediate range targets, the SDM plays a vital role on the modern battlefield.

MISSION OF THE SQUAD DESIGNATED MARKSMAN

7-228. The SDM program provides the squad with a designated marksman that has been trained to engage targets from 300 to 500 meters. He will operate and maneuver as a rifleman, but will have the added responsibility of engaging targets out to 500 meters with effective, well-aimed fires. The SDM is a vital member of his individual squad, not a squad sniper. The SDM has neither the equipment nor training to engage targets at extended ranges with precision fires while operating individually or in a small team. He can also be used to help direct the fires of other squad members into enemy positions. Due to the increased skill level required for his position, the SDM must maintain a high level of proficiency through continued training of the required skills. The SDM must possess a thorough understanding and mastery of—

- The fundamentals of rifle marksmanship.
- Ballistics.
- Elevation and windage.
- Hold-off (adjusted points of aim).
- Sight manipulation.
- Range estimation.

SELECTION

7-229. The platoon sergeant and squad leaders must take special consideration in selecting the SDM. The SDM must have—

- A solid marksmanship performance.
- A clear understanding of the fundamentals.
- The ability to apply these fundamentals consistently during dry-fire and live-fire training.

SQUAD DESIGNATED MARKSMAN SKILLS PROGRESSION

7-230. The skills progression program assesses the Soldier's ability to apply the fundamentals of marksmanship and trains and assesses the Soldier's proficiency in several key areas. Firing events will serve to both reinforce and assess these areas.

CONDUCT

7-231. While conducting the skills progression program, instructors-trainers will adhere to the following guidelines:

- The skills progression program for the SDM is based on the M16-/M4-series weapon system and a 98-rounds-per-man ammunition requirement.
- Soldiers will use their assigned weapon during the training.
- The firing events will be conducted with the iron sights or BUISs only.
- The firing events will be conducted on a KD range that enables firing out to 600 meters at a minimum.

QUALIFICATION

7-232. SDM qualification requires the completion of five phases:

(1) Position evaluation.
(2) Dry-fire training.
(3) Range estimation and sight manipulation.
(4) Hold-off.
(5) Field fire.

7-233. Each phase stresses marksmanship fundamentals and specific skill areas required to perform as an SDM. To continue training, Soldiers must receive a GO in each phase. Soldiers who fail in any area should be removed from training.

NOTE: If an optic is issued for use, the phases dealing with hold-off and field record fire will be removed and relevant optics training and testing will be substituted.

PHASE I—POSITION EVALUATION

7-234. Phase I of the training consists of demonstrating the ability to consistently assume proper firing positions. The foxhole supported and prone unsupported firing positions will provide the Soldier with the smallest target exposure to the enemy and will be used during this training cycle. The prone supported position can be substituted for the foxhole supported position dependent on range configurations. Prior to this phase of training, trainers ensure that the—

- Weapon is cleared and that no ammunition is loaded prior to training.
- Weapon is zeroed prior to training.
- Soldier is able to assume a steady firing position.

Foxhole Supported

7-235. The Soldier must be able to successfully assume a proper supported position while firing from a foxhole. The trainer must ensure that the Soldier has a good steady position. After the Soldier has assumed a good supported position in the foxhole—

(1) The trainer uses DA Form 7650-R (Squad Designated Marksman—Position Evaluation) to evaluate his position and take notes on all of the following characteristics:

NOTE: See the end of this publication for a blank, reproducible copy of DA Form 7650-R.

- Eye relief.
- Trigger finger.
- Elbows.
- Nonfiring hand.
- Legs.

NOTE: The main areas that will differ between the foxhole supported and the prone supported positions are in the placement of the elbows, legs, and nonfiring hand. These body positions will be similar to those of the prone unsupported position.

(2) After all characteristics have been noted, the trainer has the Soldier lay his weapon down, relax, and then assume another supported position in the foxhole.

(3) The trainer evaluates this position by comparing his notes from the original supported position. The Soldier should maintain the same characteristics in the second evaluation as he did in the first.

(4) Once the trainer is satisfied that the Soldier has demonstrated the proper position and is able to show it in two consecutive attempts, the Soldier moves to the unsupported prone position.

Eye Relief

7-236. To evaluate the Soldier's eye relief, ensure that the Soldier—

- Demonstrates a consistent eye relief by checking the placement of the Soldier's cheek on the weapon's buttstock.
- Places his eye is the same distance from the rear sight each time he is evaluated.

Trigger Finger

7-237. To evaluate the placement of the Soldier's trigger finger, ensure that the Soldier—

- Uses his own style; not all Soldiers place their finger on the trigger in the same place.
- Places his finger on the trigger the same way each time he is evaluated.

Elbows

7-238. To evaluate the placement of the Soldier's elbows, ensure that the Soldier—

- Places his elbows firmly a comfortable distance apart on the outside edge of the foxhole.
- Uses a sandbag, and not his arms, to support the weapon's weight.
- Assumes a stable position each time he is evaluated by slightly nudging him.

Nonfiring Hand

7-239. To evaluate the placement of the Soldier's nonfiring hand, ensure that the Soldier—

- Places the nonfiring hand in a position that is comfortable and provides the best weapon stability and support. Show the Soldier different ways this can be done.
- Is supporting the weapon properly by nudging him after the weapon has been stabilized.

Legs

7-240. To evaluate the placement of the Soldier's legs, ensure that the Soldier—

- Places the legs inside the foxhole while firing.
- Plants the legs firmly for a stable position while firing. Slightly nudge the Soldier to make sure that his legs are firmly planted in the foxhole.

Prone Unsupported

7-241. The Soldier must be able to successfully assume a proper unsupported firing position. The trainer must ensure that the Soldier has a good steady position. After the Soldier has assumed a good unsupported firing position —

(1) Taking special care to observe the positioning of the elbows, the nonfiring hand, and the legs, the trainer uses DA Form 7650-R (Squad Designated Marksman—Position Evaluation) to evaluate the same characteristics as with the supported firing position with the exception of the—

- Elbows.
- Nonfiring hand.
- Legs.

NOTE: See the end of this publication for a blank, reproducible copy of DA Form 7650-R.

---

(2) Once the trainer has noted the Soldier's position, he has the Soldier lay his weapon down, stand up, relax, and then get back down into another unsupported prone position.

(3) The trainer evaluates this position by comparing his notes from the original position. The Soldier should maintain the same characteristics in the second evaluation as he did in the first evaluation.

(4) The trainer lets the Soldier hold this firing position for approximately 15 seconds to check for shaking. If the Soldier starts to shake, have him relax and reposition himself.

(5) Once the trainer is satisfied that the Soldier has demonstrated the proper position and is able to accomplish it in two consecutive attempts, the Soldier moves on to the next phase of training.

Elbows

7-242. To evaluate the placement of the Soldier's elbows, ensure that the Soldier—

- Places the elbows a comfortable distance apart on the ground.
- Uses the bone, not the muscles, to support the weapon's weight. This will prevent any unnecessary muscle fatigue and will allow for a steadier firing position.
- Assumes a stable position. Slightly nudge the Soldier to ensure that his position is stable.

Nonfiring Hand

7-243. To evaluate the placement of the Soldier's nonfiring hand, ensure that the Soldier—

- Places his nonfiring hand in a comfortable position on the handguards.
- Does not support his nonfiring hand on the ground, sandbag or anything that would create a supported position.

Legs

7-244. To evaluate the placement of the Soldier's legs, ensure that the Soldier—

- Positions his legs in such a way that he has a stable position; not all Soldiers position their legs the same way while shooting from the prone position.
- Spreads his legs a comfortable distance apart, with the heels on the ground or as close as possible without causing strain.

Zero Confirmation

7-245. After completing Phases I and II, the Soldier conducts a firing event (Table 7-22) to zero or confirm the zero on his weapon and reinforce the fundamentals of marksmanship. This firing event will be conducted on a 25-meter range. If the Soldier cannot zero within 18 rounds, the trainer recommends re-training, retesting, or possible removal from the course. After the weapon is zeroed, any additional rounds will be fired and the coach will observe the Soldier for deficiencies in his marksmanship fundamentals.

Table 7-22. Zero/zero confirmation firing event.

FIRING EVENT	ROUNDS	TARGET RANGE (m)
Zero/Zero Confirmation	18	25

PHASE II—DRY-FIRE TRAINING

7-246. SDMs must have a solid grasp on the fundamentals to successfully engage targets at longer ranges. During this phase of training, the Soldier must demonstrate that he can apply the fundamentals of marksmanship correctly. If the Soldier does not receive a GO in this phase of training, he will be dropped from the course. The components of this phase of training are—

- Follow-through.
- Borelight exercise.
- Target box exercise.
- Dime/washer drill.
- Zero confirmation.

7-247. Prior to this phase of training trainers ensure that the—

- Weapon is cleared and no ammunition is loaded prior to training.
- Weapon is zeroed prior to training.
- Soldier is able to consistently apply the fundamentals of marksmanship.

Follow-Through

7-248. Follow-through involves applying all of the marksmanship fundamentals while and after the weapon fires. A good follow-through ensures that the weapon is allowed to fire and recoil naturally. The Soldier/weapon combination reacts as a single unit to such actions. It consists of the following:

- Keeping the cheek in firm contact with the stock (stock weld).
- Keeping the finger on the trigger all the way to the rear.
- Continuing to look through the rear aperture.
- Keeping muscles relaxed.
- Avoiding reaction to recoil or noise.
- Releasing the trigger only after the recoil has stopped.

Borelight Exercise

7-249. The borelight dry-fire exercise provides evaluation of the Soldier throughout the integrated act of firing.

NOTE: If a borelight is not available, the target box exercise will be used.

7-250. To perform a borelight exercise—

(1) The trainer attaches a 25-meter zero target to a flat surface.

(2) The trainer positions the Soldier 10 meters away, facing the target.

(3) The Soldier assumes a good prone supported firing position with the borelight inserted in the weapon's barrel and with the weapon placed in the dry-fire mode.

(4) The trainer uses DA Form 7650-R (Squad Designated Marksman—Position Evaluation) to evaluate the Soldier's position:

NOTE: See the end of this publication for a blank, reproducible copy of DA Form 7650-R.

---

(5) The Soldier aims at the silhouette's center of mass on the 25-meter zero target and squeezes the trigger.

(6) The borelight is activated as the trigger is fired. The laser is seen on the 25-meter zero target.

(7) The trainer marks the 25-meter zero target exactly were the borelight laser hit the target.

(8) The Soldier gets out of position and then back into a prone supported firing position.

7-251. This process will be done until a three-round shot group has been achieved. The Soldier will do the same from the prone unsupported firing position. To receive a GO, the Soldier must place a three-round shot group in a 3-centimeter circle from both prone positions.

Target Box Exercise

7-252. The target box exercise checks the consistency of aiming and placement of three-round shot groups in a dry-fire environment. To conduct the exercise—

(1) The target man places the silhouette on a plain sheet of paper 25 meters away from the firer and moves the silhouette target as directed by the Soldier.

(2) When the Soldier establishes proper aiming, he squeezes the trigger.

(3) The target man marks through the silhouette with a pen or pencil at the target's center of mass.

(4) The target man moves the silhouette to another spot on the paper and tells the firer to repeat the process twice more to obtain a shot group.

NOTE: A simulated shot group covered within a 1/2-centimeter circle indicates consistent aiming.

7-253. Since no weapon or ammunition variability is involved and shot groups do not have to be placed in a certain location, a 1/2-centimeter standard may be compared to obtaining a 4-centimeter shot group on the 25-meter live-fire zero range.

Dime/Washer Drill

7-254. The dime/washer drill is an effective way of measuring the Soldier's trigger squeeze. To conduct the exercise—

(1) The Soldier takes aim and squeezes the trigger.

(2) If the dime or washer remains in place, he has successfully squeezed the trigger

7-255. The Soldier must successfully obtain five out of five consecutive shot groups within 1 centimeter and without allowing the dime or washer to drop. The trainer evaluates the Soldier's performance and gives the Soldier a GO or NO GO. If the Soldier receives a NO GO, the trainer recommends retraining, retesting, or possible removal from the course.

Zero Confirmation

7-256. Zero confirmation is conducted as shown in Table 7-6.

PHASE III—RANGE ESTIMATION AND SIGHT MANIPULATION 100 TO 500 METERS

7-257. SDMs must use range estimation methods to determine the distance between their position and the target. Trainers ensure that the—

- Weapon is cleared and no ammunition is loaded prior to training.
- Weapon is zeroed prior to training.
- Soldier knows how to adjust for wind and gravity.
- Soldier can manipulate the rear sight for different ranges.

7-258. The trainer sets up a range estimation course using E-type silhouettes at ranges from 100 meters to 700 meters. Soldiers practice on this course until they find the method that works best for them.

7-259. Once the Soldiers have had time to practice, trainers test their ability to estimate range. The Soldier is given six targets. He must estimate the range within 50 meters of the actual range to receive a GO. The Soldier must estimate range correctly six out of six targets to move on to the next portion of this phase.

Methods of Range Determination

7-260. SDMs can use five different methods of range determination:

- 100-meter unit-of-measure method.
- Range card method.
- Front sightpost method.
- Appearance of objects method.
- Combination method.

100-Meter Unit-of-Measure Method

7-261. To use this method, the SDM must be able to visualize a distance of 100 meters on the ground.

- For ranges up to 500 meters, the SDM determines the number of 100-meter increments between the two objects he wishes to measure.
- Beyond 500 meters, he must select a point halfway to the object, determine the number of 100-meter increments to the halfway point, and then double the number.

NOTES: 1. See Chapter 6 for more information about this method of range determination.

2. For example, terrain with much dead space limits the accuracy of the 100-meter method.

Range Card Method

7-262. SDMs use a range card to quickly determine ranges throughout the target area. Once a target is detected, the SDM determines its location on the card and then reads the proper range to the target.

Front Sightpost Method

7-263. Using the front sightpost as a scale is another method of estimating range. This method can be used for a quick on-the-spot estimation and engagement.

- If a man-sized target is approximately ½ of the width of the front sightpost, he is approximately 300 meters away.
- If the target is 1/2 the width of the front sightpost, then the target is approximately 600 meters away.

NOTE: See Chapter 6 for more information about this method of range determination.

Appearance of Objects Method

7-264. This method of range determination is based on the size and visible characteristics of an object. To use this method with any degree of accuracy, the SDM must be familiar with the appearance and visible detail of an object at various ranges. Some common guidelines can be used to determine the range of a human target:

- At 200 meters, a human target is clear and details can be seen.
- At 300 meters, the target is still clear, but no details can be seen.
- At 400 meters, the target's outline is clear; however, the target itself is blurry.
- At 500 meters, the body tapers and the head disappears.
- At 600 meters, the body resembles a wedge shape.

NOTE: See Chapter 6 for more information about this method of range determination.

Combination Method

7-265. In a combat environment, perfect conditions rarely exist. Therefore, only one method of range estimation may not be enough for the SDM's specific mission. By using a combination of two or more methods to determine an unknown range, an experienced SDM should arrive at an estimated range close to the true range.

Factors Affecting Range Estimation

7-266. Three factors affect range estimation:

- Nature of the target
- Nature of the terrain.
- Light conditions.

Nature of the Target

7-267. The nature of the target affects its perceived range:

- An object with a regular outline, such as a house, appears closer than one with an irregular outline, such as a clump of trees.
- A target that contrasts with its background appears to be closer than it actually is.
- A partly exposed target appears more distant than it actually is.

Nature of the Terrain

7-268. The contour of the terrain affects the observer's ability to estimate range:

- As the observer's eye follows the contour of the terrain, he tends to overestimate distant targets.
- Observing over smooth terrain, such as sand, water, or snow, causes the observer to underestimate distant targets.
- Looking downhill, the target appears farther away.
- Looking uphill, the target appears closer.

Light Conditions

7-269. Light conditions affect range estimation:

- The more clearly a target can be seen, the closer it appears.
- When the sun is behind the observer, the target appears to be closer.
- When the sun is behind the target, the target is more difficult to see and appears to be farther away.

Elevation Knob Training

7-270. Elevation knob training involves nothing more than being able to adjust the rear elevation knob (front sightpost on the M16A4) for the various ranges that the SDM must engage. With this knowledge, he can better determine his range settings for the different distances between the 100-meter adjustments.

7-271. The rear elevation knob adjusts the point of aim—

- From 300 to 800 meters on the M16A2.
- From 300 to 600 meters on the M16A4 and M4.

Conduct

7-272. During elevation knob training, the Soldier determines the number of adjustments (clicks) between the different range settings on his rear elevation adjustment knob (Tables 7-23 and 7-24). Once the Soldier understands how to set the proper point of aim for his target using his rear elevation knob, the instructor/trainer has him conduct another range estimation course. This time, the instructor/trainer has him estimate the range and set the rear elevation for the range that he has estimated. The Soldier must estimate range and set his rear elevation knob properly six out of six times to receive a GO. If the Soldier receives a NO GO, the trainer recommends retraining, retesting, or possible removal from the course.

7-273. Once the Soldier has an understanding of range estimation and sight manipulation, he is able to begin the live-fire training exercise (Table 7-25). The Soldier will be given 20 rounds in which to engage 20 targets at ranges from 100 to 500 meters using mechanical sight adjustments.

Table 7-23. Elevation knob, M16A2/3 and front sightpost, M16A4.

DISTANCE (m)	DISTANCE ONE CLICK WILL ADJUST THE POINT OF IMPACT		
	FRONT SIGHTPOST	WINDAGE KNOB	ELEVATION WHEEL
25	.83 cm (3/8 in)	.33 cm (1/8 in)	.5 cm (1/4 in)
50	1.50 cm (5/8 in)	.5 cm (1/4 in)	1.5 cm (1/2 in)
75	2.50 cm (1 in)	1.0 cm (3/8 in)	2.0 cm (3/4 in)
100	3.50 cm (1 3/8 in)	1.5 cm (1/2 in)	2.75 cm (1 in)
150	5.00 cm (2 in)	2.0 cm (3/4 in)	4.0 cm (1 1/2 in)
175	6.00 cm (2 3/8 in)	2.25 cm (7/8 in)	5.0 cm (2.0 in)
200	6.50 cm (2 5/8 in)	2.5 cm (1 in)	5.5 cm (2 1/4 in)
250	8.50 cm (3 3/8 in)	3.5 cm (1 1/4 in)	7.0 cm (2 3/4 in)
300	10.0 cm (4 in)	4.0 cm (1 1/2 in)	8.5 cm (3 1/4 in)
400	13.5 cm (5 3/8 in)	5.5 cm (2 1/4 in)	11.0 cm (4 1/2 in)
500	17.0 cm	6.5 cm (2 1/2 in)	14.0 cm (5 1/2 in)
600	20.5 cm	8.0 cm (3 1/8 in)	16. 75 cm (6 1/2 in)
700	24.0 cm	9.0 cm (3 5/8 in)	19.5 cm (7 1/2 in)
800	27.5 cm	10.5 cm (4 1/8 in)	22.5 cm (8 3/4 in)
NOTE: All values were rounded off.			

Table 7-24. Elevation knob, M4/M4A1 and windage, M16A4.

DISTANCE (m)	DISTANCE ONE CLICK WILL ADJUST THE POINT OF IMPACT		
	FRONT SIGHTPOST	WINDAGE KNOB	ELEVATION WHEEL
25	1.2 cm (1/2 in)	.5 cm (1/4 in)	.5 cm (1/4 in)
50	2.4 cm (1 in)	1.5 cm (1/2 in)	1.5 cm (1/2 in)
75	3.6 cm (1 1/2 in)	2.0 cm (3/4 in)	2.0 cm (3/4 in)
100	4.8 cm (1 7/8 in)	2.75 cm (1 in)	2.75 cm (1 in)
150	7.2 cm (2 7/8 in)	4.0 cm (1 1/2 in)	4.0 cm (1 1/2 in)
175	8.4 cm (3 3/8 in)	5.0 cm (2.0 in)	5.0 cm (2.0 in)
200	9.6 cm (3 3/4 in)	5.5 cm (2 1/4 in)	5.5 cm (2 1/4 in)
250	12.0 cm (4 3/4 in)	7.0 cm (2 3/4 in)	7.0 cm (2 3/4 in)
300	14.4 cm (5 3/4 in)	8.5 cm (3 1/4 in)	8.5 cm (3 1/4 in)
400	19.2 cm (7 1/2 in)	11.0 cm (4 1/2 in)	11.0 cm (4 1/2 in)
500	24.0 cm (9 1/2 in)	14.0 cm (5 1/2 in)	14.0 cm (5 1/2 in)
600	28.8 cm (11 1/4 in)	16. 75 cm (6 1/2 in)	16.75 cm (6 1/2 in)
NOTE: All values were rounded off.			

Table 7-25. Known distance (mech. adj.) firing event.

FIRING EVENT	ROUNDS	TARGET RANGE (m)
Known Distance (Mech. Adj.)	20	100 to 500

PHASE IV—HOLD-OFF 100 TO 500 METERS

7-274. Hold-off is shifting the point of aim to achieve a desired point of impact. Hold-off is used only when the SDM does not have time to change his sight setting. Certain situations, such as multiple targets at varying ranges and rapidly changing winds, do not allow proper windage and elevation adjustments. Familiarization and practice of elevation and windage hold-off techniques prepares the SDM to meet these situations. The SDM rarely achieves pinpoint accuracy when holding off; a minor error in range determination or a lack of a precise point of aim might cause the bullet to miss the desired point.

7-275. Prior to this phase of training, trainers ensure that the—

- Weapon is cleared and that no ammunition is loaded prior to training.
- Weapon is zeroed prior to training.
- Soldier knows how to adjust for wind and gravity.
- Soldier can manipulate the rear sight for different ranges.

Elevation

7-276. When a Soldier aims directly at a target at ranges greater than the set range, his bullet will hit below the point of aim. At lesser ranges, his bullet will hit higher than the point of aim. If the SDM understands this and knows about trajectory and bullet drop, he will be able to hit the target at ranges other than that for which the weapon was adjusted.

7-277. For example, the SDM adjusts the weapon for a target located 500 meters downrange, and another target appears at a range of 600 meters. The hold-off would be 25 inches; that is, the SDM should hold off 25 inches above the center of visible mass in order to hit the center of mass of that particular target. If another target were to appear at 400 meters, the SDM would aim 14 inches below the center of visible mass in order to hit the center of mass.

7-278. The chart in Figure 7-31 shows the projectile's trajectory when fired from the M4 carbine and the M16A2 rifle. This demonstrates the drop of the round at various ranges.

NOTE: This diagram will assist the trainer in teaching vertical hold-off during this phase.

7-279. As the chart in Figure 7-31 shows, the hold-off at 400 meters is about half the height of the standard E-type silhouette; to hold-off at 400 meters, the firer aims half the height of the target over the target to hit it. The drop at 500 meters is considerably larger, so holding off will not be practical. The firer will have to adjust his rear elevation knob to get the proper point of aim for that distance.

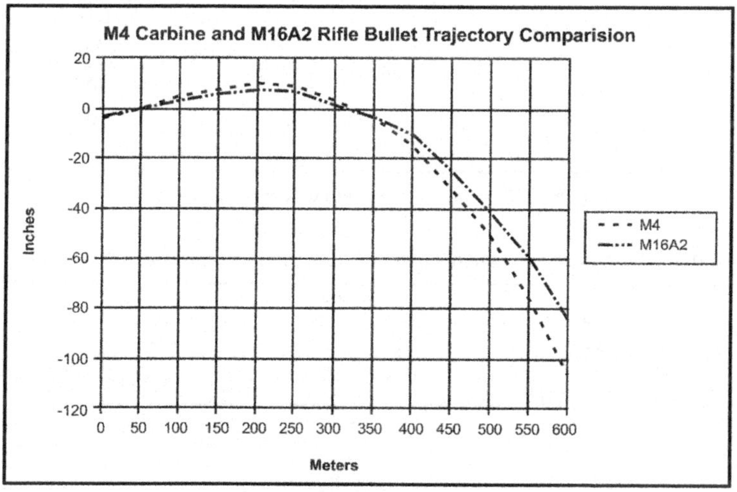

Figure 7-31. Bullet trajectory comparison.

Windage

7-280. When firing during windy conditions and there is no time to make sight adjustments, the SDM must use hold-off to adjust for windage (Figure 7-32).

- When holding off, the SDM aims into the wind. If the wind is moving from the right to left, his point of aim is to the right. If the wind is moving from left to right, his point of aim is to the left.
- If the SDM misses the target and the point of impact of the round is observed, he notes the lateral distance of his error and re-fires, holding off that distance in the opposite direction.

7-281. Table 7-26 shows calculated adjusted points of aim based on wind speed.

7-282. Table 7-27 shows the drift for a 10-mph wind using 5.56-millimeter M855 ball ammunition fired in a M16A2 rifle with a 300-meter battlesight zero.

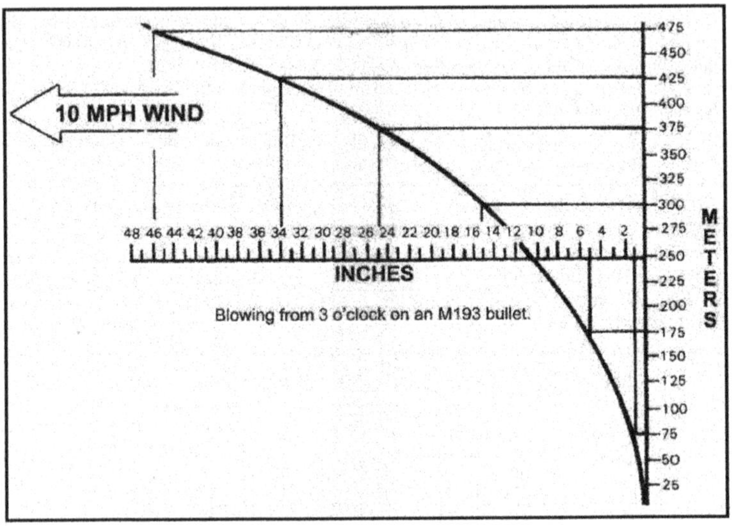

Figure 7-32. Windage effects of a 10-mph crosswind.

Table 7-26. Calculated adjusted point of aim based on wind speed (full value).

WIND SPEED	RANGE (m)								
	25	50	75	100	150	175	200	250	300
5 mph	1/4 in	3/8 in	1/2 in	1 in	2 in	2.5 in	3.5 in	5 in	7.5 in
10 mph	1/2 in	3/4 in	1 in.	2 in	4 in	5 in	7 in	10 in	15 in
15 mph	3/4 in	1 1/8 in	1.5 in	3 in	6 in	7.5 in	10.5 in	15 in	22.5 in

Table 7-27. Drift for 10-mph wind using M855 ammunition when fired from M16A2 rifle with 300-meter battlesight zero.

RANGE (m)	VELOCITY (fps)	TRAJECTORY (in)	DROP (in)	DRIFT (in)
0	3,100	-2.5	0.0	0.0
100	2,751	4.4	-2.3	1.1
200	2,420	5.8	-10.2	4.9
300	2,115	0.0	-25.3	11.8
400	1,833	-15.0	-49.5	22.4
500	1,569	-42.9	-86.7	38.0
600	1,323	-88.2	141.3	59.5
700	1,106	-156.1	-220.9	88.4
800	1,010	-267.7	-339.2	124.9

7-283. Firers use the M15 sighting device to demonstrate that they understand holding off. To do so, the firer aligns the sights on the silhouette on the proper adjusted point of aim.

NOTE: See Appendix A for information about the M15 sighting device.

7-284. Once the firer has an understanding of elevation and windage hold-off, he can begin the live-fire training exercise (Table 7-28). The firer will be given 20 rounds in which to engage 20 targets at ranges from 100 to 500 meters using elevation and windage hold-off.

Table 7-28. Firing event, known distance (hold off).

FIRING EVENT	ROUNDS	TARGET RANGE (m)
Known Distance (Hold Off)	20	100 to 500

PHASE V—FIELD FIRE 100 TO 500 METERS

7-285. The field fire events (Table 7-29) will test the individual's marksmanship, range estimation, and target detection skills. Field fire will consist of both a Record Fire I and a Record Fire II course. The Record Fire I course requires the individual to use mechanical elevation and windage adjustments. The Record Fire II course requires the individual to use elevation and windage hold-off (adjusted points of aim).

Conduct

7-286. To complete this course, Soldiers fire two firing events: Record Fire I and Record Fire II.

7-287. During each event, each Soldier will engage a total of 20 targets with 20 rounds. Soldiers must attain a total of 14 hits out of 20 targets on each record fire to pass. Table 7-29 depicts these two events and provides related information, such as number of rounds that must be fired, position that must be used, and the distance away from the firer that the target must be placed.

Table 7-29. Firing event, Record Fire I and II.

FIRING EVENT	POSITION	ROUNDS	NUMBER OF TARGETS	TARGET RANGE (m)
Record Fire I	Foxhole supported or prone supported position and the prone unsupported firing position	20	20	100 to 500
Record Fire II	Foxhole supported or prone supported position and the prone unsupported firing position	20	20	100 to 500

NOTE: If the SDM is issued an optic, the Record Fire II course will substitute use of that optic instead of using adjusted points of aim.

7-288. Prior to training, ensure that—

- The weapon is zeroed prior to training.
- Each Soldier assumes a steady firing position.
- Each Soldier consistently applies the fundamentals of marksmanship.
- Each Soldier knows how to adjust for wind and gravity.
- Each Soldier manipulates the rear sight for different ranges.

NOTE: See Appendix B for a sample completed form and the end of this publication for a blank, reproducible copy.

CERTIFICATION

7-289. Once the firer has successfully completed the SDM program, he is designated as an SDM and will be able to perform all duties and responsibilities set forth by these guidelines.

NOTE: SDM skills are highly perishable, and sustainment training should be conducted to ensure retention of the skills. At a minimum, sustainment training should be conducted semiannually.

Chapter 8

Advanced Optics, Lasers, and Iron Sights

BRM teaches Soldiers how to effectively engage targets with their weapons using the iron sights. ARM adds other marksmanship situations that a combat Soldier may encounter. This chapter discusses how to enhance marksmanship skills through proper training, using the Army's newest optics and lasers to ensure that Soldiers can fight as well at night as they can during the day. This chapter implements new night qualification standards to complement current Army training strategies.

SECTION I. BORELIGHT

The borelight is an accurate means of zeroing weapons and most aided-vision equipment without the use of ammunition. The time and effort required to ensure a precise boresight will, in turn, save time and ammunition. Table 8-1 outlines weapon/aided-vision device combinations that can be zeroed using the borelight.

NOTE: See Table 8-2 for the current training program.

Table 8-1. Weapon/aided-vision device combinations.

WEAPON / AIDED-VISION DEVICE	M16A2	M16A3/A4	M4/MWS
IRON SIGHT	X	N/A	N/A
BUIS	N/A	X	X
AN/PAQ-4B/C	X	X	X
AN/PEQ-2A/B	X	X	X
AN/PAS-13B/C/D	X	X	X
M68 CCO	X	X	X
ACOG	X	X	X

NOTES: 1. Precise boresighting of a laser will allow direct engagement of targets without a 25-meter zero. If a borelight is not available, a 25-meter zero must be conducted to zero the laser.

2. All optics must be 25-meter zeroed; a borelight only aids in zeroing.

Table 8-2. Borelight training program.

BORELIGHT
Instructional Intent
• Soldiers align the optic, laser, or iron sight to the bore of the weapon to reduce or eliminate the time and ammunition it requires to live-fire zero.
Special Instructions
Ensure that Soldiers— • Zero the borelight. • Use only approved 10-meter boresighting targets from Picatinny Arsenal. • Use the proper 10-meter boresighting target for weapon configuration. • Boresight 10 meters from the end of the barrel. • Stabilize both the weapon and the target (or the boresight will not be accurate). • Install filters for aiming lasers to reduce blooming.
Observables
Soldiers confirm that the— • Borelight spins on itself when zeroed at 10 meters. • Weapon configuration is boresighted using the official and proper targets only. • Target and weapon do not move during the boresighting procedure. • Borelight is centered on the circle on the target. • Aiming device is aiming at the center of the crosshair on the offset.

CONCEPT

8-1. Boresighting is a simple procedure that saves time and ammunition. The visible laser of the borelight is aligned with the barrel of a designated weapon. Then, using a 10-meter boresighting target, the weapon can be boresighted with any optic, laser, or iron sight that the Soldier is assigned to fire.

ZEROING THE BORELIGHT

WARNINGS

Before using the borelight, ensure that the weapon is clear and on SAFE, and that the bolt is locked in the forward position.

When rotating the borelight to zero it, ensure that the mandrel is turning counterclockwise (from the firer's point of view) to avoid loosening the borelight from the mandrel.

DANGERS

DO NOT STARE INTO THE VISIBLE LASER BEAM.

DO NOT LOOK INTO THE VISIBLE LASER BEAM THROUGH BINOCULARS OR TELESCOPES.

DO NOT POINT THE VISIBLE LASER BEAM AT MIRROR-LIKE SURFACES.

DO NOT SHINE THE VISIBLE LASER BEAM INTO OTHER INDIVIDUALS' EYES.

8-2. Before boresighting the weapon system, the borelight must first be zeroed to the weapon. To zero the borelight to the weapon—

(1) Stabilize the weapon by placing it in a rifle box rest or by laying two rucksacks side by side and placing another rucksack on top of the weapon.

NOTE: The weapon does not have to be perfectly level with the ground when boresighting.

---

(2) Align the visible laser with the weapon's barrel.

CAUTION

Do not over-adjust the laser. Do not point the laser at Soldiers or reflective material.

(3) Attach the 5.56-millimeter mandrel to the borelight.

(4) Insert the mandrel into the weapon's muzzle.

NOTE: The borelight is seated properly when the mandrel cannot be moved any further into the muzzle and the mandrel spins freely.

---

(5) Measure 10 meters with the 10-meter cord that comes with the borelight, or pace off eleven paces.

(6) Draw a zeroing mark (small dot) on a piece of paper or tree bark, or use the borelight reference point on the 10-meter boresighting target (Figure 8-1).

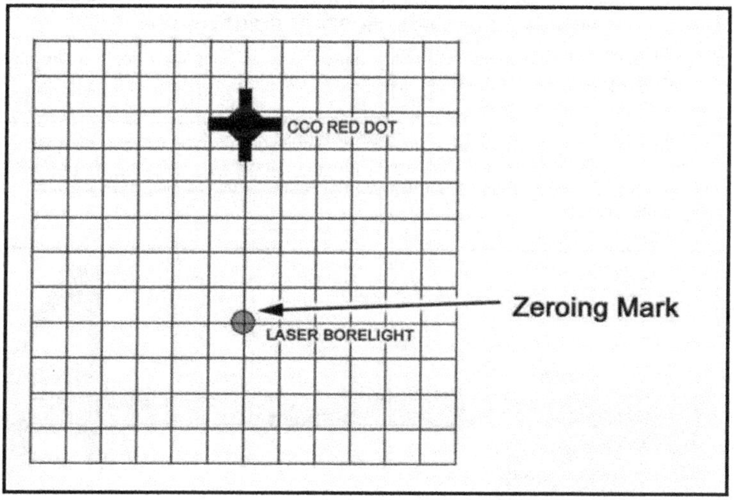

Figure 8-1. Example of a zeroing mark.

(7) Rotate the borelight until the battery compartment is facing upward and the adjusters are on the bottom (Figure 8-2).

NOTE: Once the Soldier performs step (7), the borelight's position and the place where the visible laser is pointing are identified as the start point. The command "START POINT" is given to ensure clear communication between the Soldier at the weapon and the Soldier at the boresighting target.

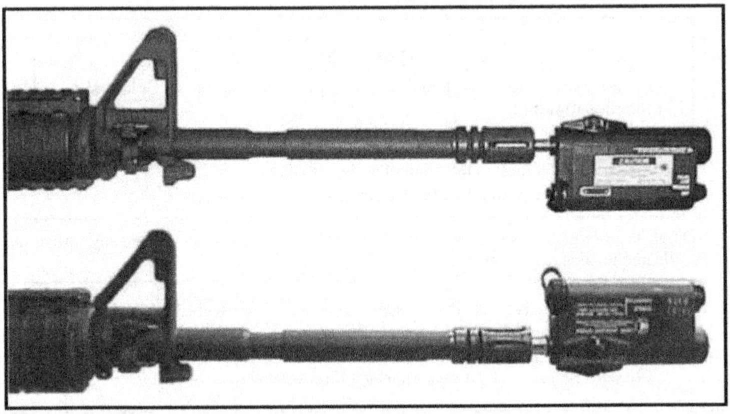

Figure 8-2. Borelight in the START POINT position.

(8) Rotate the borelight until the battery compartment is facing downward and the adjusters are on top to allow for easy access to the adjusters and help with communication and stabilization of the weapon (Figure 8-3).

NOTE: Once the Soldier performs step (8), the borelight's position and the place where the visible laser is pointing are identified as the half-turn position. The command "HALF-TURN" is given to ensure clear communication between the Soldier at the weapon and the Soldier at the boresighting target.

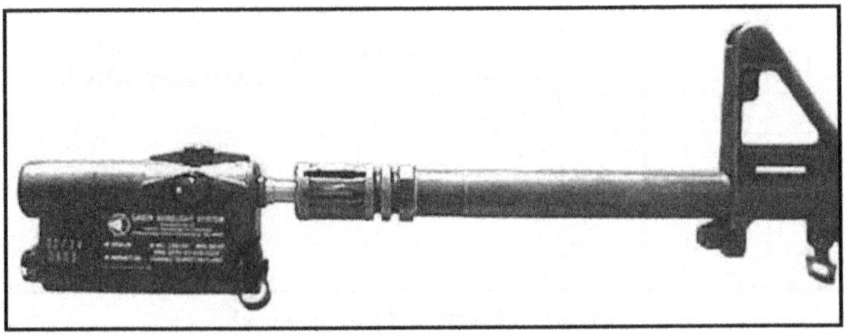

Figure 8-3. Borelight in the HALF-TURN position.

(9) Identify the point approximately halfway between the start point and the half-turn point. This is the reference point (Figure 8-4).

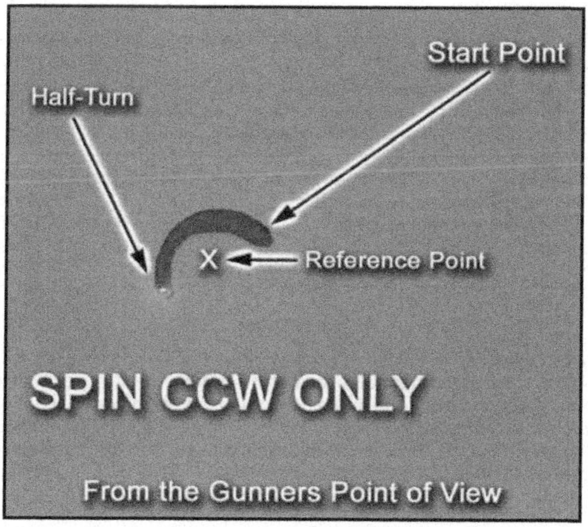

Figure 8-4. Example of a start point, half-turn, and reference point.

(10) Turn the borelight on and spin it until it is in the start point position.

(11) Place the zeroing mark approximately 10 meters from the end of the barrel so that the visible laser strikes the zeroing mark.

(12) Slowly rotate the borelight 180 degrees while watching the visible laser made by the borelight.

NOTE: If the visible laser stops on the zeroing mark, the borelight is zeroed to the weapon. If the borelight does not stop on the zeroing mark, elevation and windage adjustments must be made to the borelight.

---

(13) From the start point, realign the zeroing mark with the visible laser, rotate the borelight 180 degrees to the half-turn position, and identify the reference point.

(14) Using the adjusters on the borelight, move the visible laser to the reference point.

(15) Rotate the borelight back to the start point; move the zeroing mark to the visible laser.

NOTE: If the visible laser cannot be located when the borelight is spun to the half-turn position, start this procedure at 2 meters instead of 10 meters. When the visible laser is adjusted to the reference point at 2 meters, restart the procedure at 10 meters.

---

(16) Repeat steps (13) through (15) until the visible laser spins on itself.

NOTE: Every barrel is different, so steps (10) through (15) must be performed with every weapon to ensure that the borelight is zeroed to that barrel. If the borelight is zeroed, go directly to the boresighting procedures.

BORESIGHTING

8-3. Two Soldiers are required to properly boresight a weapon: a firer and a target holder. Their duties are as follows:

- The firer's primary duty is to zero the borelight and make all adjustments on the aided-vision device being used.
- The target holder secures the 10-meter boresighting target straight up and down 10 meters from the borelight and directs the firer in making necessary adjustments to the aiming device. The target holder must wear night vision goggles when boresighting IR aiming lasers.

NOTE: Appendix F shows the most current 10-meter boresighting target. The 10-meter boresighting target grids are 1x1-centimeter squares, unlike those on 25-meter zeroing targets. Contact the proponent of this publication (C Co, 2/29 IN, Fort Benning, GA) for information about the availability of boresighting targets.

CAUTION

Weapon stability is crucial in boresighting. The weapon should be in the bolt forward position and must not be canted left or right during boresighting procedures. If the weapon is boresighted using field-expedient methods (for example, sandbags, rucksacks) and the weapon is laid on its side for stability, ensure that the boresighting target is also oriented in the same manner.

Boresighting the Weapon Using the Backup Iron Sights

8-4. The BUIS can be boresighted to a new user to expedite 25-meter zeroing. To boresight the weapon using the BUIS—

(1) Align the BUIS with the Canadian bull on the 10-meter boresighting target.

(2) Make adjustments to the windage and elevation of the BUIS until the borelight is centered with the circle on the boresighting target.

Boresighting the Weapon Using the M68 Close Combat Optic

NOTES: 1. Before boresighting, ensure that the borelight has been zeroed to the weapon. The more accurate the boresight of the M68 CCO to the assigned weapon, the closer to a battlesight zero the weapon will be.

2. 25-meter zeroing must be conducted to ensure that the M68 CCO is properly zeroed.

3. The M68 CCO is a parallax-free sight beyond 50 meters; boresighting is conducted at 10 meters. To get a solid boresight, ensure that the red dot is centered within both the CCO and the target's center of mass.

8-5. To boresight the weapon using the M68 CCO—

(1) Select the proper 10-meter boresighting target for the weapon/M68 configuration.

(2) With the help of an assistant, place the boresighting target 10 meters in front of the weapon.

(3) Turn the M68 CCO to the desired setting by turning it clockwise to position 10, turning it counterclockwise one click at a time until you can no longer see the dot, and then turning it clockwise two clicks.

NOTE: The desired setting may need to be adjusted throughout the day depending on ambient light conditions.

(4) Get behind the weapon in a stable supported firing position, and look through the M68 CCO.

(5) Aim the red dot of the M68 CCO on the crosshair located on the 10-meter boresighting target.

(6) Make adjustments to the M68 CCO until the visible laser of the borelight is centered on the borelight circle on the 10-meter boresighting target.

(7) Turn the borelight off.

(8) Move the weapon off of the crosshair, realign the M68 CCO's red dot on the crosshair, and turn the borelight back on.

(9) If the borelight is on the circle and the M68 CCO's red dot is on the crosshair, the weapon system is boresighted.

(10) Turn the laser off, and carefully remove the borelight and the mandrel from the weapon so that the borelight device is not damaged.

NOTE: If the firer does not get the same sight picture after the second realignment, he more than likely has a fundamentals problem with his firing position and sight picture. To save time on the range, a coach should troubleshoot the Soldier before continuing to boresight the M68 CCO.

Boresighting the Advanced Combat Optical Gunsight

NOTES: 1. Before boresighting, ensure that the borelight has been zeroed to the weapon. The more accurate the boresight of the ACOG to the assigned weapon, the closer to a battlesight zero the weapon will be.

2. 25-meter zeroing must be conducted to ensure that the ACOG is properly zeroed.

8-6. To boresight the ACOG—

NOTE: Use the 10-meter boresighting target that is used for boresighting the M68 CCO.

(2) With the help of an assistant, place the boresighting target 10 meters in front of the weapon.

(4) Get behind the weapon in a stable supported firing position, and look through the ACOG.

(5) Aim the tip of the 300-meter post on the crosshair located on the 10-meter boresighting target.

(6) Make adjustments to the ACOG until the visible laser of the borelight is centered on the borelight circle on the boresighting target.

(7) Turn the borelight off.

(8) Move the weapon off of the crosshair, realign the tip of the 300-meter post on the crosshair, and turn the borelight back on.

(10) If the borelight is on the circle and the tip of the 300-meter post is on the crosshair, the weapon system is boresighted.

(10) Turn the laser off, and carefully remove the borelight and the mandrel from the weapon so that the borelight device is not damaged.

BORESIGHTING THE THERMAL WEAPON SIGHT

NOTES: 1. Before boresighting the TWS, ensure that the borelight has been zeroed to the weapon. The more accurate the boresight of the TWS to the assigned weapon, the closer to a battlesight zero the weapon will be.

2. 25-meter zeroing must be conducted to ensure that the TWS is properly zeroed.

3. Both the narrow field of view (NFOV) and wide field of view (WFOV) must be boresighted and zeroed.

8-7. To boresight the TWS—

(1) Select the proper 10-meter boresighting target for the weapon/TWS configuration.
(2) With the help of an assistant, place the boresighting target 10 meters in front of the weapon.
(3) Ensure that the weapon's reticle is displayed.
(4) Get behind the weapon in a stable supported firing position, and look through the TWS.
(5) Place a finger on each oval on the 10-meter boresighting target.
(6) Aim between the fingers with the 300-meter point of aim.
(7) Make adjustments to the TWS until the visible laser of the borelight is centered on the borelight circle on the 10-meter boresighting target.
(8) Move off of the aiming block, realign the TWS to the center of the heated block, and then turn the borelight back on.
(9) If the boresight is properly aligned, the weapon system is boresighted; otherwise, you will need remedial training on your sight picture.
(10) Change the field of view (FOV) on the sight by rotating the FOV ring, and repeat steps (1) through (9).
(11) Turn the laser off, and carefully remove the borelight and the mandrel from the weapon so that the borelight device is not damaged.

BORESIGHTING THE AN/PAQ-4B/C

NOTE: Before boresighting the AN/PAQ-4B/C, make sure that the borelight has been zeroed to the weapon. The more accurate the boresight of the AN/PAQ-4B/C to the assigned weapon, the closer to a battlesight zero the weapon will be.

DANGERS

DO NOT STARE INTO THE IR LASER BEAM WITH THE NAKED EYE OR THROUGH BINOCULARS OR TELESCOPES.

DO NOT POINT THE IR LASER BEAM AT MIRROR-LIKE SURFACES OR OTHER INDIVIDUALS' EYES.

ALTHOUGH THIS LASER IS EYE-SAFE, IT IS A SAFE PRACTICE TO TREAT ALL LASERS AS NOT EYE-SAFE.

3x EXTENDERS MAGNIFY THE LASER AIMING LIGHT. WHEN USING THE 3x EXTENDERS, THE AN/PAQ-4B/C IS NOT CONSIDERED EYE-SAFE AT ANY DISTANCE.

DO NOT STORE THE AN/PAQ-4B/C WITH THE BATTERIES INSTALLED.

8-8. To boresight the AN/PAQ-4B/C —

(1) Select the proper 10-meter boresighting target for the weapon/AN/PAQ-4B/C configuration.

(2) With the help of an assistant, place the boresighting target 10 meters in front of the weapon.

(3) Install the borelight filter, and turn the AN/PAQ-4B/C on.

(4) Align the 10-meter boresighting target with the visible laser of the borelight.

(5) Adjust the adjusters on the AN/PAQ-4B/C until the IR laser is centered on the crosshair located on the 10-meter boresighting target.

NOTE: For windage and elevation, one click equals 1 centimeter at 25 meters. A click is defined as the sound or feel of the positive detent movement.

CAUTION

Do not turn the adjustment screws too much. Over-turning the adjustment screws will cause them to break.

NOTES: 1. The boresighting target and zeroing mark must be kept stable during the boresighting procedure.

2. Regardless of the mounting location, the adjuster that is on top or bottom will always be the elevation adjuster, and the one on the side will always be the windage adjuster.

BORESIGHTING THE AN/PEQ-2A/B

NOTES: 1. Before boresighting the AN/PEQ-2A/B, ensure that the borelight has been zeroed to the weapon. The more accurate the boresight of the AN/PEQ-2A/B to the assigned weapon, the closer to a battlesight zero the weapon will be.

2. The boresighting target and zeroing mark must be kept stable during the boresighting procedure.

DANGERS

THE AN/PEQ-2A/B EMITS INVISIBLE LASER RADIATION. AVOID DIRECT EXPOSURE TO THE BEAM.

DO NOT STARE INTO THE IR LASER BEAM WITH THE NAKED EYE OR THROUGH BINOCULARS OR TELESCOPES.

DO NOT POINT THE IR LASER BEAM AT MIRROR-LIKE SURFACES OR OTHER INDIVIDUALS' EYES.

EYE DAMAGE CAN OCCUR IF CARELESS HANDLING OF THE LASER OCCURS. EYE-SAFE DISTANCE IS BEYOND 25 METERS (IN DUAL LO MODE) IN TRAINING MODE AND BEYOND 220 METERS IN TACTICAL MODE.

3x EXTENDERS MAGNIFY THE LASER AIMING LIGHT. WHEN USING THE 3x EXTENDERS, THE AN/PEQ-2A/B IS NOT CONSIDERED EYE-SAFE AT ANY DISTANCE.

DO NOT STORE THE AN/PEQ-2A/B WITH THE BATTERIES INSTALLED.

8-9. To boresight the AN/PEQ-2A/B—

(1) Select the proper 10-meter boresighting target for the weapon/AN/PEQ-2A/B configuration.
(2) With the help of an assistant, place the boresighting target 10 meters in front of the weapon.
(3) Install the filter on the aiming laser, and turn the AN/PEQ-2A/B on.
(4) Align the 10-meter boresighting target with the visible laser of the borelight.
(5) Adjust the adjusters on the AN/PEQ-2A/B until the IR laser is centered on the crosshair located on the 10-meter boresighting target.

NOTE: For windage and elevation, one click equals 1 centimeter at 25 meters. A click is defined as the sound or feel of the positive detent movement.

CAUTION

Do not turn the adjustment screws too much. Over-turning the adjustment screws will cause them to break.

NOTES: 1. Each click of elevation and windage is 1 centimeter and each square of the 25-meter zeroing target is .9 centimeter. For ease of use, round up to one square per click of elevation. This will have a negligible effect on the zero unless a large adjustment is needed.

2. Regardless of the mounting location, the adjuster that is on top or bottom will always be the elevation adjuster, and the one on the side will always be the windage adjuster.

(6) Adjust the illuminator in the same manner.
(7) Turn the laser off, and carefully remove the borelight and the mandrel from the weapon so that the borelight device is not damaged.

Boresighting the AN/PVS-4

NOTES: 1. Before boresighting, ensure that the borelight has been zeroed to the weapon. The more accurate the boresight of the AN/PVS-4 to the assigned weapon, the closer to a battlesight zero the weapon will be.

2. 25-meter zeroing must be conducted to ensure that the AN/PVS-4 is properly zeroed.

(1) Select the proper 10-meter boresighting target for the weapon/AN/PVS-4 configuration.
(1) With the help of an assistant, place the boresighting target 10 meters in front of the weapon.
(2) Ensure that the M16 reticle is displayed.
(3) Get behind the weapon in a stable supported firing position, and look through the AN/PVS-4.
(4) Turn the borelight laser on.
(5) Align the borelight laser with the circle on the 10-meter target offset.
(6) Keeping the laser in place, adjust the windage and elevation until the reticle of the AN/PVS-4 is aligned with the circular crosshair.

NOTE: If there is not enough ambient light to see the 10-meter target offset circular crosshair, shine a flashlight indirectly at the target to provide ambient light.

(7) Turn the borelight off.

(8) Move the reticle off of the circular crosshair, and then realign on the target.

(9) Turn the borelight laser back on.

(9) If the borelight is in the circle, then the AN/PVS-4 is boresighted.

(10) Turn the laser off, and carefully remove the borelight and mandrel from the weapon so that the borelight device is not damaged.

SECTION II. TRAINING STRATEGIES AND QUALIFICATION STANDARDS

Before beginning a night marksmanship program, Soldiers must qualify on their assigned weapons during the day, as outlined in the previous chapters of this manual. Commanders should follow these training strategies and abide by the qualification standards set forth. Although some courses of fire may seem redundant or inappropriate, numerous tests show that these training strategies work, and the qualification standards are achievable if the strategies are followed.

BACKUP IRON SIGHT

8-10. The BUIS is a semipermanent flip-up sight equipped with a rail-grabbing base. The BUIS provides a backup capability effective out to at least 600 meters and can be installed on M16A4 rifles and M4 carbines.

NOTE: See Table 8-3 for the current training program.

Table 8-3. Backup iron sights training program.

BACKUP IRON SIGHTS
Instructional Intent
• Soldiers zero and qualify with the BUIS.
Special Instructions
Ensure that Soldiers— • Apply the marksmanship fundamentals. • Place the BUIS in the full vertical position and lock them prior to firing. • Install the plastic insert in the BUIS during boresighting and zeroing.
Observables
• Soldiers attain the same day standards for zeroing and qualification as with the standard iron sights (See Chapter 5).

CONCEPT

8-11. The BUIS is adjusted for a 300-meter battlefield zero to provide backup in the event an optic or laser device fails to function. The BUIS is zeroed on the M4/M4A1 target on the backside of the M16A2 zeroing target (NSN 6920-01-395-2949).

NOTE: The 25-meter zeroing procedures are the same as for conventional rear sight assembly on M16 rifles and M4 carbines.

CONDUCT OF TRAINING

NOTES: 1. The BUIS training strategy is the same as the iron sight training strategy.

2. All procedures for the BUIS are the same as with standard iron sights.

8-12. BUIS equipment training should familiarize the Soldier with the proper operation and characteristics of the BUIS.

Boresight the Backup Iron Sights

8-13. Optional.

Zero

8-14. The zeroing standards for the BUIS are the same as with iron sights.

- To zero the BUIS for M4 carbines, set the range selector to 300 meters.
- To zero the BUIS to an M16A4, place the range selector on the white line below the 300-meter mark.

Target Detection

8-15. Target detection procedures for the BUIS are the same as with standard iron sights.

Practice Qualification

8-16. A practice qualification must always precede an actual qualification. Practice qualification allows the Soldier to practice and refine the skills needed to succeed during qualification. Practice qualification standards for the BUIS are the same as with standard iron sights. If the Soldier qualifies during the practice qualification, it may be counted as the record qualification.

Record Qualification

8-17. Qualification with the BUIS is conducted on a standard record fire range, and the standards for qualification are the same as the record fire day standards.

M68 CLOSE COMBAT OPTIC

8-18. The M68 CCO is a reflex (nontelescopic) sight. It uses a red aiming reference (collimated dot) and is designed for the two eyes open method of sighting. The dot follows the horizontal and vertical movement of the firer's eye, while remaining fixed on the target. No centering or focusing is required.

NOTE: See Table 8-4 for the current training program.

Table 8-4. M68 close combat optic training program.

M68 CLOSE COMBAT OPTIC
Instructional Intent
• Soldiers qualify with the M68 CCO.
Special Instructions
Ensure that Soldiers— • Are proficient with the M68 CCO. • Install the half-moon spacer when using the M68 CCO with the M4, M16A4, and MWS. • Do not install the half moon spacer when using the M68 CCO with the M16A1/A2. • Use the proper offset during boresighting procedures. • Confirm 10-meter boresight with a 25-meter zero. • Retighten the rail grabber after the initial 3 rounds are fired. • Sets the M68 CCO's dot for best sight picture. • Apply the marksmanship fundamentals. • Zero and qualify with the same sight picture (one eye or two eyes open method). • Zero on the M16A2 25-meter zeroing target. Ensure that the designated impact zone is 1.4 centimeters down from the center of mass of the 300-meter silhouette on the 25-meter zeroing target.
Observables
• Soldiers zero the M68 CCO to the same standards as with iron sights. • Rounds impact in the 4x4-centimeter square designated impact zone. • Soldiers achieve the same practice qualification and qualification standards as with day record fire.

CONCEPT

8-19. Soldiers must qualify on their assigned weapons during daylight conditions, as outlined in this manual. The integrated act of firing with the M68 CCO is identical to the iron sights except for the change in the sight picture.

CONDUCT OF TRAINING

NOTES: 1. The M68 CCO training strategy is the same as the iron sight training strategy.

2. All procedures for the M68 are the same as with standard iron sights.

8-20. M68 equipment training should familiarize the Soldier with the proper operation and characteristics of the M68 CCO IAW TM 9-1240-413-13&P.

MODIFIED FUNDAMENTALS

8-21. M68 CCO operation requires modifications to the fundamentals of marksmanship:

- Steady position.
- Aiming.
- Breath control.
- Trigger squeeze.

Steady Position

8-22. When operating the M68 CCO, Soldiers no longer need a good stock weld to get a good sight picture after the M68 is zeroed at 25 meters. The M68's reflexive sight allows the Soldier to fire the weapon with his cheek in a comfortable position, but since the CCO is parallax-free beyond 50 meters only, the Soldier must zero and fire using the same cheek position.

Aiming

8-23. When using the M68 CCO, the two eyes open method is the preferred method of aiming.

NOTE: The aiming method used to zero must also be used to engage targets. When using the M68 CCO, the weapon must not be canted during aiming or firing.

Two Eyes Open Method (Preferred)

8-24. This method allows a much greater field of view and makes scanning for targets much easier; however, getting accustomed to the two eyes open method takes practice. The Soldier must keep the rifle and M68 in a vertical alignment each time he fires.

8-25. To use the two eyes open method—

(1) Position the head so that one eye can focus on the red dot and the other eye can scan downrange.

(2) Place the red dot on the target's center of mass, and engage.

One Eye Open Method

8-26. To use the one eye open method—

(1) With the nonfiring-side eye closed, look through the M68 CCO to ensure that the red dot can be seen clearly.

(2) Place the red dot on the target's center of mass, and engage.

Breath Control

8-27. This fundamental does not change.

Trigger Squeeze

8-28. This fundamental does not change.

ZEROING WITHOUT A BORELIGHT

8-29. To zero without a boresight—

(1) Begin with a securely installed and live-fire zeroed BUIS.

(2) Mount the CCO to the front of the receiver rail or to the top ARS, as preferred.

(3) Adjust windage and elevation on the reflex sight until the center of the aiming dot is at the tip of the front sightpost when viewed through the BUIS while assuming a normal firing position.

25-METER ZEROING PROCEDURES

NOTE: Conduct zeroing only on the M16A2 25-meter target.

8-30. When zeroing the CCO at 25 meters, a designated point of impact zone must be identified on the 25-meter zeroing target (Figure 8-5). To zero the CCO at 25 meters—

(1) Starting from center of mass of the 300-meter silhouette on the 25-meter zeroing target, count down 1½ squares or 1.4 centimeters. This is now the point of impact when zeroing the M68 CCO.

(2) Continue to aim at the center of mass of the 300-meter silhouette, and make adjustments to the M68 CCO so that the rounds impact in the secondary 4x4-centimeter circular box, 1½ squares or 1.4 centimeters down from the point of aim.

8-31. Other procedures are the same as standard iron sight procedures:

NOTE: A click is defined as the sound or feel of the positive detent movement.

- For windage and elevation, two clicks equal 1 centimeter at 25 meters.
- For elevation, one clockwise click moves the bullet strike down.
- For windage, one clockwise click moves the bullet strike left.

NOTES: 1. At ranges of 50 meters and beyond, the effects of parallax are minimal. However, at ranges of 50 meters and closer, parallax exists and the firer must ensure that the red dot is centered while zeroing.

2. The aiming method (two eyes open or one eye open) used to zero must be used to engage targets.

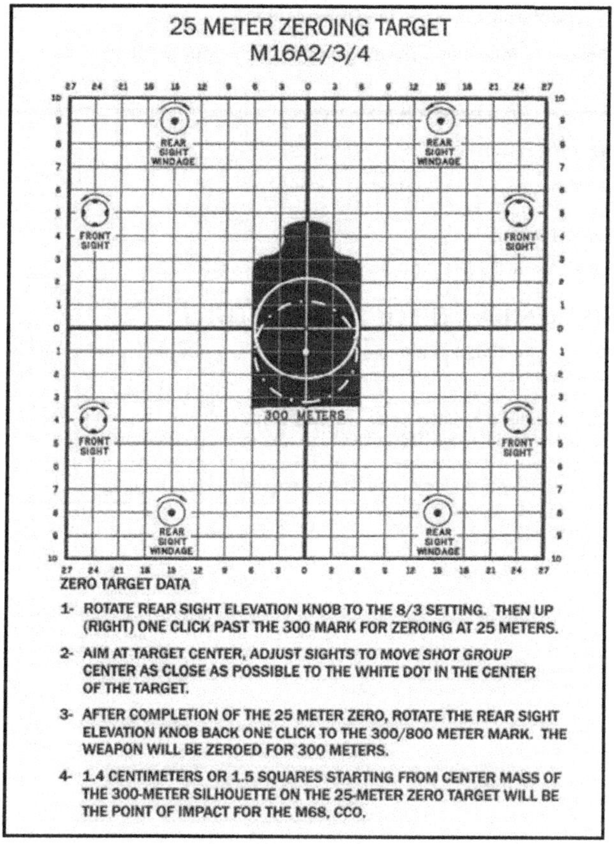

Figure 8-5. Close combat optic, 25-meter zeroing target.

TARGET DETECTION

8-32. Target detection procedures for the M68 are the same as with standard iron sights.

WARNING

In Positions 4 and above, the red dot is visible through the front of the sight. For night vision operations—

- **Close the front lens cover before turning the rotary switch clockwise to Positions 2 and 3.**
- **Check the light for proper intensity before opening the front lens cover.**
- **Close the front lens cover before turning the rotary switch counterclockwise to the OFF position.**

Failure to follow this warning could reveal your position to the enemy.

PRACTICE QUALIFICATION

8-33. The procedures are the same as standard iron sight procedures.

RECORD QUALIFICATION

8-34. The procedures are the same as standard iron sight procedures.

ADVANCED COMBAT OPTICAL GUNSIGHT

8-35. The ACOG scope is designed to provide enhanced target identification and hit probability for the M16/M4 series weapons out to 800 meters. It is designed with dual illuminated technology using fiber optics for daytime illumination and tritium for night and low-light use. The ACOG scope is a lightweight rugged, fast and accurate 3.5 power optic scope. It is internally adjustable to allow the shock of rough handling to be carried by the scope body and not the adjustment mechanism.

NOTE: See Table 8-5 for the current training program.

Table 8-5. Advanced combat optical gunsight training program.

ADVANCED COMBAT OPTICAL GUNSIGHT
Instructional Intent
• Soldiers qualify with the ACOG.
Special Instructions
Ensure that Soldiers— • Are proficient with the ACOG. • Use the proper offset during boresighting procedures. • Confirm the 10-meter boresight with 100-meter zero (or 25-meter zero if the 100-meter range is not available). • Retighten the rail grabber after the initial 3 rounds are fired. • Apply the marksmanship fundamentals. • Zero and qualify with the same sight picture. Ensure that— • The point of aim and point of impact are the same at 100 meters. • The designated impact zone is 1 centimeter down from the center of mass of the 300-meter silhouette on the 25-meter zeroing target.
Observables
• Soldiers zero the ACOG to the same standards as with iron sights. • Soldiers achieve the same practice qualification and qualification standards as with day record fire.

CONCEPT

8-36. Soldiers must qualify on their assigned weapons during the day, as outlined in this manual. The integrated act of firing with the ACOG is identical to the iron sights except for the change in sight picture.

CONDUCT OF TRAINING

NOTES: 1. The ACOG training strategy is the same as the iron sight training strategy.

2. All procedures for the ACOG are the same as with standard iron sights.

8-37. ACOG equipment training should familiarize the Soldier with its proper operation and characteristics.

OPERATION

8-38. The ACOG scope is internally adjustable. Adjustment is made using the adjuster mechanisms located inside the adjuster caps on the top and right-hand side of the scope. Adjustment can be made with a small screwdriver, coin, or other hard object that fits the adjustment screws. The caps are very tight to ensure a waterproof seal with the O-rings inside. The caps should only be off the scope when adjustments are being made.

CAUTION

The ACOG contains an internal adjustment mechanism to allow zeroing. Adjustments to the extreme ends of the range can result in damage to the internal prism assembly. Do not continue to adjust the windage and elevation mechanisms if you encounter resistance.

8-39. The ACOG scope is shipped with a pre-centered setting. Normally this means that only small adjustments are necessary. Do not adjust the scope to the extremes. It is possible that over-adjustment will damage the precise alignment of the prism assembly inside the rifle scope.

8-40. The ACOG scope is adjusted at the factory to be as parallax-free as possible at 100 meters. This in no way affects the accuracy of the scope.

8-41. The reticle patterns in the scope have been designed to provide many features while retaining simplicity of operation. The user does not need to make any manual adjustments between shots at different ranges. Ranging capability is built into the reticle patterns.

8-42. The outside legs of the chevron reticle in the ACOG correspond to 19 inches (average width of a man's shoulders) at 300 meters.

8-43. The widths of the horizontal hash marks on the BDC reticles in all ACOG scopes correspond to the width of a .5 meter (19 inches) silhouette (man-size) at that range (Figure 8-6).

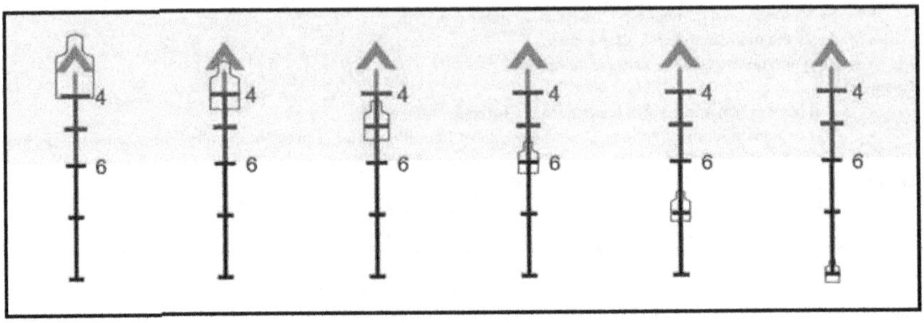

Figure 8-6. Width of horizontal hash marks.

ZEROING THE ADVANCED COMBAT OPTICAL GUNSIGHT

8-44. The adjustment increments for the ACOG are 1/4 inch per click at 100 meters. A click is defined as the sound or feel of the positive detent movement. This means that 4 clicks are required to move the bullet approximately one inch on the target at 100 meters. At 25 meters, 16 clicks move the bullet one inch. Turning the screw in the direction of the arrow moves the bullet impact in the direction marked (up/right).

NOTE: To ensure a consistent zero, tap the scope with the palm of your hand to stabilize the adjustment mechanism after an adjustment has been made, and then fire a three-shot group on the target.

8-45. The ACOG scope with chevron reticle is designed to be zeroed at 100 meters, using the tip of the chevron reticle for POA/POI (Figure 8-7).

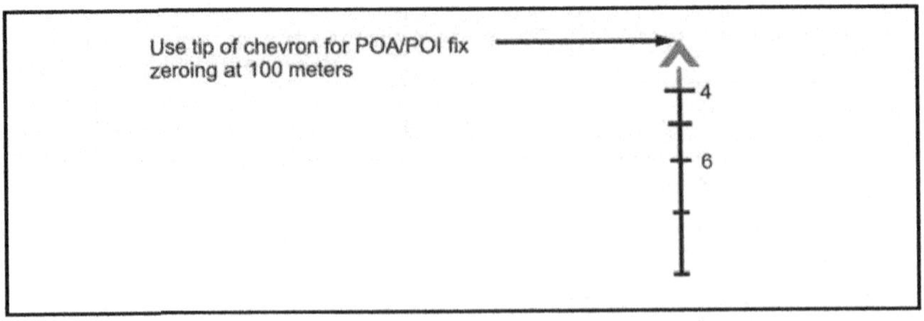

Figure 8-7. Advanced combat optical gunsight reticle point of aim at 100 meters.

8-46. For a quick combat zero, the scope may be zeroed at 25 meters, using the tip of the 300-meter post as POA/POI. A 25-meter zero is less precise than a 100-meter zero and should be verified at longer distances once time and a range is available (Figure 8-8).

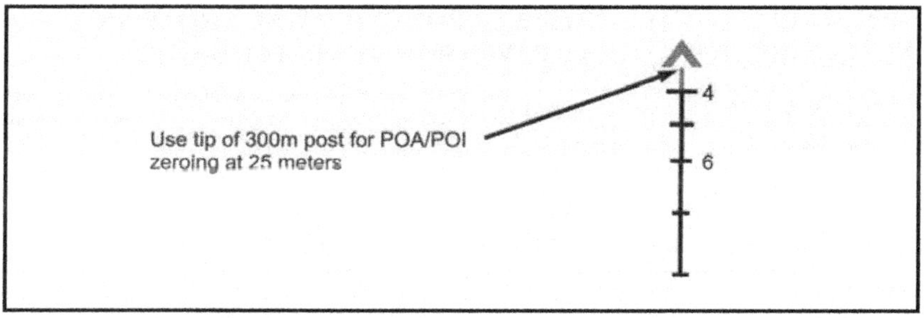

Figure 8-8. Advanced combat optical gunsight reticle point of aim at 25 meters.

MODIFIED FUNDAMENTALS

8-47. ACOG operation requires modifications to the following fundamentals of marksmanship:

- Steady position.
- Aiming.
- Breath control.
- Trigger squeeze.

Steady Position

8-48. This fundamental does not change.

Aiming

8-49. When the TA11F ACOG is attached to the carrying handle of an M16A1/A2/A3 rifle, the BDC reticle will be off slightly at extended distances. Beginning at 400 meters, the BDC reticle will be approximately one MOA off to the extent of the BDC reticle, once properly zeroed.

Example: At 500 meters, the impact will be 5 inches high; at 600 meters, the impact will be 6 inches high.

Once the target is ranged beyond 400 meters, hold slightly low.

Breath Control

8-50. This fundamental does not change.

Trigger Squeeze

8-51. This fundamental does not change.

TARGET DETECTION

8-52. Target detection procedures for the ACOG are the same as with standard iron sights.

PRACTICE QUALIFICATION

8-53. The procedures are the same as standard iron sight procedures.

RECORD QUALIFICATION

8-54. The procedures are the same as standard iron sight procedures.

AN/PAS-13B/C/D (V1) LIGHT WEAPON THERMAL SIGHT AND AN/PAS-13B/C/D (V3) HEAVY WEAPON THERMAL SIGHT

8-55. The AN/PAS-13B/C/D (V1, V3) TWS is an IR imaging sensor used for target acquisition under conditions of low visibility. IR light is received through the telescope, detected by an IR sensor, converted to digital data, processed, and displayed for the user.

NOTE: See Table 8-6 for the current training program.

Table 8-6. AN/PAS-13B/C/D thermal weapon sight training program.

AN/PAS-13B/C/D TWS
Instructional Intent
• Soldiers qualify with the ANPAS-13B/C/D TWS.
Special Instructions
Ensure that Soldiers— • Are proficient with the TWS. • Use the spacer with the M4, M16A4, and MWS. • Use the proper 10-meter boresighting target during boresighting procedures. • Boresight both FOVs. • Confirm the 10-meter boresight with a 25-meter zero. • Use the M16A2 zeroing target with a 4x4-centimeter square cut out of the center of the silhouette. • Thermalize the zero range and qualification range (inspect for targets that aren't thermalized). • Use every other lane during zero and qualification.
Observables
• Soldiers zero the TWS to the same standards as with iron sights. • Soldiers achieve the same practice qualification and qualification standards as with day record fire.

WARNING

Ensure that the weapon is not loaded and is on SAFE before installing the TWS on the weapon. A loaded weapon may accidentally discharge, causing severe injury or death.

CONCEPT

8-56. The AN/PAS-13B/C/D TWS training strategy is much the same as that of aiming lights. The TWS does not require the use of NVDs. The course of fire for the TWS is the same scenario as the day qualification tables, with the same standards of fire as for current day standards. Qualification standards are the same for day and night.

CONDUCT OF TRAINING

8-57. AN/PAS-13B/C/D equipment training should familiarize the Soldier with the proper operation and characteristics of the TWS IAW the TM.

MODIFIED FUNDAMENTALS

8-58. AN/PAS-13B/C/D TWS operation requires modifications to the fundamentals of marksmanship:

- Steady position.
- Aiming.
- Breath control.
- Trigger squeeze.

Steady Position

8-59. This fundamental slightly changes due to the height of the sight. Consider the following modifications:

- Soldiers must adjust their position so they can properly look through the sight.
- In most cases, the cheek-to-stock weld no longer exists.

Aiming

8-60. To properly aim with the TWS, Soldiers must ensure that the correct reticle is selected in the sight.

NOTE: Refer to TM 11-5855-312-10 for reticle selection and point of aim for use with the TWS.

Breath Control

8-61. This fundamental is not affected by night firing conditions when using the TWS.

Trigger Squeeze

8-62. This fundamental of marksmanship does not change during night firing.

25-METER ZEROING PROCEDURES

NOTE: Refer to TM 11-5855-312-10 for target preparation.

8-63. To zero the AN/PAS-13B/C/D TWS at 100 meters, use the same procedures and standards as with iron sights, along with the following:

- At the 25-meter range, each incremental adjustment to the azimuth or elevation setting moves strike of the round as follows:
  - 1 ½ centimeters for LWTS on WFOV.
  - ½ centimeter for LWTS on NFOV.
  - ¾ centimeter for HWTS on WFOV.
  - ¼ centimeter for HWTS on NFOV.
- Retighten the rail grabber after firing the first three rounds.
- Zero both FOVs (Figure 8-9).

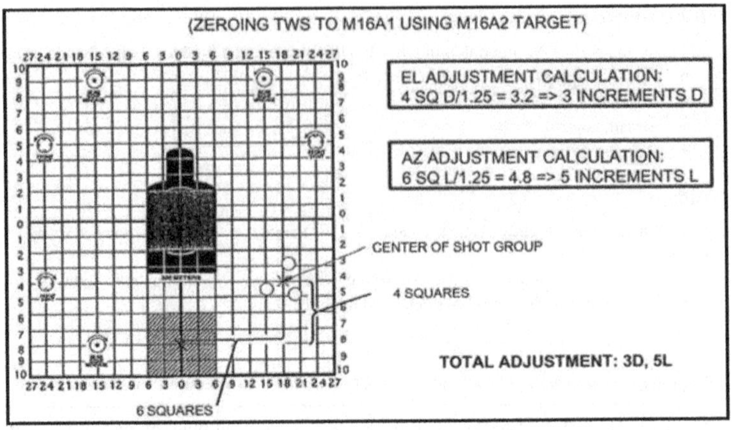

Figure 8-9. Example of thermal weapon sight zeroing adjustments.

TARGET DETECTION

8-64. With NVDs, the field of view is much smaller, scanning becomes much more deliberate, and, with the TWS, camouflage becomes less of a factor. Even though NVDs greatly enhance the Soldier's ability to acquire a target at night, increased awareness of target detection must be trained to allow the Soldier to recognize the visual cues of IR imagery.

Select Position

8-65. The TWS is a large device; selecting a position that allows for good fields of view, but does not silhouette the Soldier and his equipment, might be a challenge. Since the TWS detects thermal energy (heat) emitted from an object, a position near an object emitting a vast amount of thermal energy (for example, a vehicle with the engine running or a fire) may affect the Soldier's ability to acquire a target.

Scanning

8-66. With earlier versions of the TWS, scanning too fast causes a stuttering on the screen, which might cause the Soldier to miss or overlook a target. With these versions, Soldiers must scan slowly in order to maintain a good thermal image on the screen.

8-67. With the newer version, this stuttering is not as obvious. One advantage of the TWS is its two FOVs—wide and narrow. Each has advantages and disadvantages:

- The NFOV increases magnification but decreases the field of view.
- The WFOV decreases magnification but increases the field of view.

Target Indicators

8-68. While scanning the sector or lane with the TWS, the Soldier should recognize thermal cues that allow him to detect and identify targets. The engine compartment, exhaust, and tires of a vehicle that has been moving are all examples of thermal cues. Adjusting the brightness, contrast, and polarity helps enhance the thermal cues of a target, allowing for quicker detection and identification.

Sound

8-69. Use the same techniques outlined in day and night target detection.

Movement

8-70. Thermal cues become much more obvious on a moving object than on an object standing still. A good example is a vehicle's tires. When the vehicle is not moving, the tires are cold. On a moving vehicle, the friction between the road and the tires causes the tires to heat up and become prominent when observed through the TWS. The same is true with the human body—a moving person generates more heat than someone standing still.

Camouflage

8-71. The TWS allows Soldiers to see thermal signatures not completely masked by camouflage, such as paint, foliage, and camouflage netting, thereby increasing both day and night target detection abilities.

> **WARNING**
>
> **If the TWS is operated with the eyecup removed, light emitting from the eyepiece may be visible to the enemy's NVDs.**

PRACTICE QUALIFICATION

8-72. Practice qualification with the TWS is the same as day practice qualification with iron sights. Dry-fire exercises are performed to allow Soldiers to make adjustments to the TWS. Every other firing lane should be used so that the Soldier engages only the targets in his lane.

RECORD QUALIFICATION

8-73. Record qualification with the TWS is the same as day record qualification with iron sights.

NOTES:
1. Record qualification with the TWS can be done day and or night. Regardless of the qualification, the standard day record fire for the iron sights will be used. The standards for qualification with the TWS, either day or night, are 23 out of 40.
2. During practice qualification and qualification, the Soldier chooses his own polarity and FOV.

AN/PAQ-4B/C AND AN/PEQ-2A/B INFRARED AIMING LASERS

8-74. The newest IR aiming lasers greatly increase the night firing accuracy of all infantry weapons. IR aiming lasers complete the transition from day optics to night optics. Their effectiveness is limited by the capability of the image-intensifier (I2) sight with which they are used.

NOTE: See Table 8-7 for the current training program.

Table 8-7. AN/PAQ-4B/C or AN/PEQ-2A/B infrared aiming laser training program.

AN/PAQ-4B/C OR AN/PEQ-2A/B INFRARED AIMING LASERS
Instructional Intent
• Soldiers qualify with the AN/PAQ-4B/C or AN/PEQ-2A/B IR aiming laser.
Special Instructions
Ensure that Soldiers— • Are proficient with the AN/PAQ-4B/C or AN/PEQ-2A/B IR aiming laser. • Use the proper 10-meter boresighting target during boresighting procedures. • Use the borelight filter. • Set the AN/PEQ-2A/B IR aiming laser to AIM LO. • Boresight the illuminator on the AN/PEQ-2A/B. • Use the M16A2 25-meter zeroing target for 25-meter zero. Ensure that a 3x3-centimeter hole is cut in the center of the 25-meter zeroing target and E-type silhouette.
Observables
• Soldiers conduct either 10-meter boresight or a 25-meter zero. • Soldiers display good scanning, IR discipline, and IR walking technique. • Soldiers achieve at least 17 target hits out of 40 target exposures.

CONCEPT

8-75. Two training strategies have been devised to adequately train Soldiers in the use of AN/PAQ-4B/C and AN/PEQ-2A/B IR aiming lasers: the night initial training strategy and the night sustainment training strategy. Units should always review the night initial training strategy prior to the night sustainment training strategy.

Night Initial Training Strategy

8-76. The night initial training strategy is used for Soldiers who have little or no previous experience with night vision goggles, or for units beginning a night training program.

Night Sustainment Training Strategy

8-77. The night sustainment training strategy is for Soldiers who are familiar with night vision goggles, and for units that have already implemented a night training program.

CONDUCT OF TRAINING

8-78. AN/PAQ-4B/C and AN/PEQ-2A/B IR aiming laser equipment training should familiarize Soldiers with the proper operation and characteristics of AN/PAQ-4B/C and AN/PEQ-2A/B IR aiming lasers IAW the TM.

MODIFIED FUNDAMENTALS

8-79. NVD operation requires modifications to the fundamentals of marksmanship:

- Steady position.
- Aiming.
- Breath control.
- Trigger squeeze.

Steady Position

8-80. Consider the following modifications:

- Realize that a good cheek-to-stock weld is not possible with NVGs mounted on the head.
- Ensure that the weapon's buttstock is firmly pulled into the pocket of the shoulder to prevent the laser from wobbling.
- When ready to fire, plant the elbows firmly on the ground to prevent the laser from wobbling excessively.

Aiming

8-81. Consider the following modifications:

- Practice raising the head just enough to clear the weapon with the NVGs.
- Acquire a good sight picture by walking the laser onto the target and then aiming at the center of mass.

Breath Control

8-82. This fundamental is not modified for night firing conditions.

Trigger Squeeze

8-83. Do not disrupt the laser/target alignment by jerking the trigger.

25-METER ZEROING PROCEDURES

8-84. If the borelight is not available, a 25-meter zero must be conducted. A 25-meter zeroing target is shown in Figure 8-10.

AN/PAQ-4B/C Infrared Aiming Laser

8-85. The 25-meter zeroing procedures for the AN/PAQ-4B/C IR aiming laser are as follows:

NOTE: The zero standards are the same as with iron sights.

---

(1) Set the adjusters to their zero preset position.

NOTE: See TM 11-5855-301-12&P for more information about setting the adjusters to their zero preset position.

---

(2) Prepare a 25-meter zeroing target by cutting a 3x3-centimeter square out of the center of the silhouette.

(3) Adjust for windage and elevation.

- For windage and elevation, one click equals 1 centimeter at 25 meters.
- For elevation, one clockwise click moves the bullet strike up.
- For windage, one clockwise click moves the bullet strike left.

(4) Retighten the rail grabber after the first three rounds are fired.

NOTE: When cutting the 3x3-centimeter square out of the target, some of the strike zone may be cut out. Take care when annotating the impact of the rounds. When the weapon is close to being zeroed, some of the shots may be lost through the hole in the target.

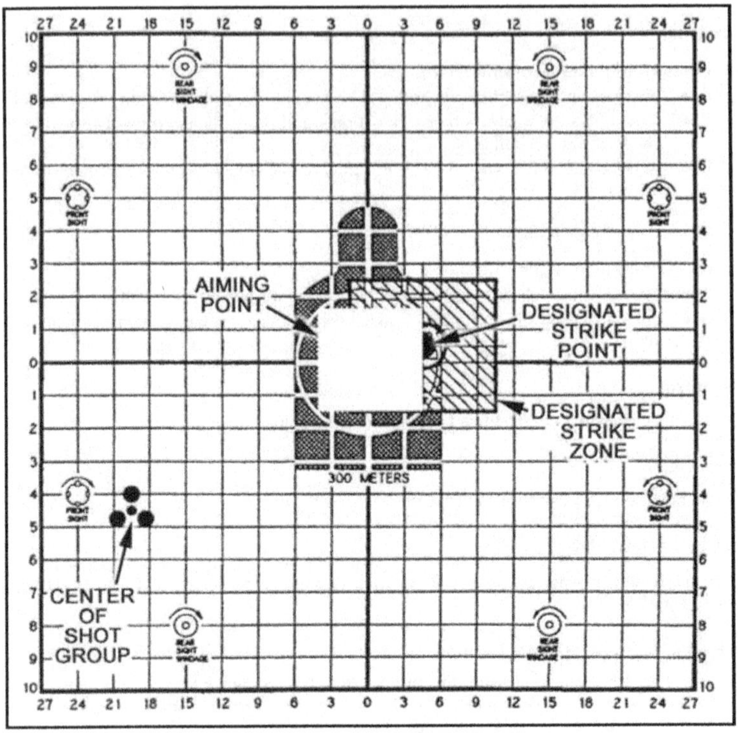

Figure 8-10. Example of shot group adjustment with strike zone.

AN/PEQ-2A/B Infrared Aiming Laser

8-86. The 25-meter zeroing procedures for the AN/PEQ-2A/B IR aiming laser are as follows:

NOTE: The zero standards are the same as with iron sights.

---

(1) Set the adjusters to their zero preset position.

NOTE: See TM 11-5855-308-12&P for more information about setting the adjusters to their zero preset position.

---

(2) Prepare the 25-meter zeroing target by cutting out a 3x3-centimeter square in the center of the target and E-type silhouette.

(3) Turn the aiming beam on in the low power setting (AIM LO).

(4) Install the aim point filter to eliminate excessive blooming.

(5) Adjust the AN/PEQ-2A/B (top-mounted) aiming point as follows:

- For windage and elevation, one click equals 1 centimeter or one square at 25 meters.
- For elevation, one clockwise click moves the bullet strike up.
- For windage, one clockwise click moves the bullet strike right.

(6) Adjust the AN/PEQ-2A/B (top-mounted) target illuminator as follows:
- For windage and elevation, one click equals 1 centimeter or one square at 25 meters.
- For elevation, one clockwise click moves the bullet strike down.
- For windage, one clockwise click moves the bullet strike right.

(7) Retighten the rail grabber and the AN/PEQ-2A/B.

(8) Once the aiming beam is zeroed, rotate the selector knob to the DUAL LO, DUAL LO/HI, or DUAL HI/HI mode to observe both aiming and illumination beams.

(9) Rotate the illumination beam adjusters to align the illumination beam with the aiming beam.

NOTES: 1. Failure to fully tighten the mounting brackets and AN/PEQ-2A/B thumbscrew may cause zero retention problems. Confirm that equipment is tight prior to zeroing.

2. To retain zero, remove the TPIAL and rail grabber as a whole assembly and place back onto the same notch as removed.

TARGET DETECTION

8-87. Soldiers should receive in-depth instruction on the proper use and fit of night vision goggles, to include characteristics and capabilities, maintenance, and mounting procedures. Extensive testing has proven that the average Soldier does not properly use NVDs. Unit leaders must be proficient in the train-the-trainer strategy. At night, Soldiers should conduct a terrain walk to become more familiar and build confidence using the night vision goggles.

Scanning for Targets

8-88. NVDs have a 40-degree field of view, which causes the average firer to miss easy targets of opportunity. Soldiers must be trained to aggressively scan their sectors of fire for targets.

8-89. The art of target detection at night is only as good as the Soldier practices. Regular blinking during scanning relieves some of the eyestrain that Soldiers experience when trying to spot distant targets. Regular blinking must be reinforced during training. After Soldiers have mastered the art of scanning, they will find that targets are more easily detected by acknowledging the flicker or movement of a target.

Infrared Discipline

8-90. Soldiers must be taught that what they can see downrange or on the battlefield through NVGs, the enemy can also see. Soldiers must train to activate the laser at the base of the target and engage the target as soon as the target is detected. After the target has been engaged, the laser is deactivated.

8-91. When a Soldier uses proper IR discipline while scanning for targets, he must keep his weapon oriented within his sector of fire. When the target is detected, the Soldier orients his weapon around the base of the target, activates his laser, and walks the laser to the target's center of mass for engagement.

FIELD FIRE

8-92. During dry-fire exercises, Soldiers acquire a sight picture on all exposed silhouette targets before conducting the field fire scenario. This allows Soldiers to focus on the targets at range. The procedures for field fire include the following:

- Conduct a dry-fire exercise prior to conducting a live-fire.
- Conduct Field Fire in the same manner as Field Fire II.
- Engage targets at 50, 150, and 250 meters.
- Fire 36 rounds:
  - 18 rounds from the supported firing position.
  - 18 rounds from the prone unsupported firing position.

PRACTICE QUALIFICATION

8-93. The procedures for practice qualification include the following:

- Conduct a dry-fire exercise.
- Use coaches.
- Fire 40 rounds:
  - 20 rounds from the prone supported firing position.
  - 10 rounds from the unsupported firing position.
  - 10 rounds from the kneeling firing position.
- Engage targets from 50 to 250 meters.
- Meet the standards (17 hits out of 40 target exposures).

RECORD QUALIFICATION

8-94. The procedures for record qualification include the following:

- Conduct a dry-fire exercise.
- Fire 40 rounds:
  - 20 rounds from the prone supported firing position.
  - 10 rounds from the unsupported firing position.
  - 10 rounds from the kneeling firing position.
- Engage targets from 50 to 250 meters.
- Meet the standards (17 hits out of 40 target exposures).

AN/PVS-4 NIGHT VISION DEVICE

8-95. The AN/PVS-4 NVD is a portable, battery-operated electro-optical instrument used for observation and aimed fire of weapons at night. It amplifies reflected light, such as moonlight, starlight, and sky glow, so that the viewed scene becomes clearly visible to the operator. It can be mounted on the M16A2/A3/A4 rifle and M4/M4 MWS. Mounting brackets are provided for each type of weapon.

NOTE: See Table 8-8 for the current training program.

Table 8-8. AN/PVS-4 night vision device training program.

AN/PVS-4 NIGHT VISION DEVICE
Instructional Intent
• Soldiers qualify with the AN/PVS-4 NVD.
Special Instructions
Ensure that Soldiers— • Are proficient with the AN/PVS-4 NVD. • Install the spacer and Picatinny rail grabber when mounting on the MWS. • Use the proper 10-meter boresighting target during boresighting procedures. • Use the proper reticle. • Confirm 10-meter boresight with a 25-meter zero.
Observables
• Soldiers zero the AN/PVS-4 to the same standard as with the iron sight. • Soldiers achieve the same practice and qualification standards as with day record fire.

CONCEPT

8-96. Training strategy for the AN/PVS-4 is much the same as that for aiming lights, and the course of fire and qualifications standards are the same as those for aiming lasers.

CONDUCT OF TRAINING

8-97. This training should familiarize the Soldier with the proper operation and characteristics of the AN/PVS-4 IAW the TM.

MODIFIED FUNDAMENTALS

8-98. NVD operation requires modifications to the fundamentals of marksmanship:

- Steady position.
- Aiming.
- Breath control.
- Trigger squeeze.

Steady Position

8-99. This fundamental slightly changes due to the height of the sight. Consider the following modifications:

- Adjust your position so you can properly look through the sight.
- Recognize that, in most cases, the cheek-to-stock weld no longer exists.

Aiming

8-100. Consider the following modifications:

- To properly aim the AN/PVS-4, ensure that the proper reticle is inserted in the sight.
- Place the aiming point on the target's center of mass.

NOTE: Refer to TM 11-5855-213-10 for more information about inserting the proper reticle.

Breath Control

8-101. This fundamental of marksmanship does not change during night firing.

Trigger Squeeze

8-102. This fundamental of marksmanship does not change during night firing.

25-METER ZEROING PROCEDURES

8-103. To zero the AN/PVS-4 at 25 meters, use the same procedures and standards as with the iron sights, along with the following:

- At 25-meter range, each increment of azimuth or elevation setting moves the strike of the round .63 centimeters or ¼ mil.
- Two clicks of the windage or elevation will move the strike of the round approximately one square on the M16A2 zeroing target.
- Retighten the thumbscrew on the rail grabber after firing the initial three rounds.

NOTE: If there is not enough ambient light to see the boresight mark at ten meters or the silhouette on the zeroing target during boresighting or zeroing procedures, shine a flashlight indirectly at the target to provide ambient light.

Target Detection

8-104. Target detection with the AN/PVS-4 is very similar to target detection with the night vision goggles. The AN/PVS-4 has a 14.5-degree field of view, causing the average firer to miss easy targets of opportunity, more commonly the 50-meter left or right.

8-105. Soldiers must be trained to aggressively scan their sectors of fire for targets. The art of target detection at night is only as good as the Soldier practices. Regular blinking during scanning relieves some of the eyestrain that Soldiers experience when trying to spot distant targets. Regular blinking must be reinforced during training. After Soldiers have mastered the art of scanning, they will find that targets are more easily detected by acknowledging the flicker or movement of a target.

Field Fire

8-106. During dry-fire exercises, Soldiers acquire a sight picture on all exposed silhouette targets before conducting the field fire scenario. This allows Soldiers to focus on the targets at range. The procedures for field fire include the following:

- Conduct a dry-fire exercise prior to conducting a live-fire.
- Conduct Field Fire in the same manner as Field Fire II.
- Engage targets at 50, 150, and 250 meters.
- Fire 36 rounds:
  - 18 rounds from the supported firing position.
  - 18 rounds from the prone unsupported firing position.

Practice Qualification

8-107. The procedures for practice qualification include the following:

- Conduct a dry-fire exercise.
- Use coaches.
- Fire 40 rounds:
  - 20 rounds from the prone supported firing position.
  - 10 rounds from the unsupported firing position.
  - 10 rounds from the kneeling firing position.
- Engage targets from 50 to 250 meters.
- Meet the standards (17 hits out of 40 target exposures).

Record Qualification

8-108. The procedures for record qualification include the following:

- Conduct a dry-fire exercise.
- Fire 40 rounds:
  - 20 rounds from the prone supported firing position.
  - 10 rounds from the unsupported firing position.
  - 10 rounds from the kneeling firing position.
- Engage targets from 50 to 250 meters.
- Meet the standards (17 hits out of 40 target exposures).

Appendix A

Training Aids, Devices, and Exercises

Training aids and devices must be included in a marksmanship program. This appendix lists those available and provides information on how to obtain them for marksmanship training.

SECTION I. ENGAGEMENT SKILLS TRAINER 2000

The EST 2000 (Figure A-1) supports realistic and comprehensive gated rifle marksmanship instruction, identifies Soldiers' needs by requiring them to satisfy gate requirements in order to progress, and facilitates any necessary remedial training prior to qualification. This unit/institutional, indoor, multipurpose, multilane, small arms, crew-served, and individual antitank training simulator is used to—

- Train and evaluate individual marksmanship training for initial entry Soldiers (BCT/OSUT).
- Provide unit sustainment training for active and reserve components in preparation for qualification on individual and crew small arms live-fire weapons.
- Provide unit collective tactical training for static dismounted infantry, scout, engineer, military police squads, and sustainment elements.
- Simulate training events which lead to live-fire individual and crew weapon qualification.
- Simulate training events that contribute to increased weapon, crew, fire team, and squad combat effectiveness.
- Simulate squad collective, defensive, ambush, gunnery, and tactical tasks.
- Train leaders of fire teams and squads in the command, control, and distribution of fires.
- Save ammunition, travel time, transportation costs, and other range support resources.
- Support functional gunnery training strategies and standards in weapon training.

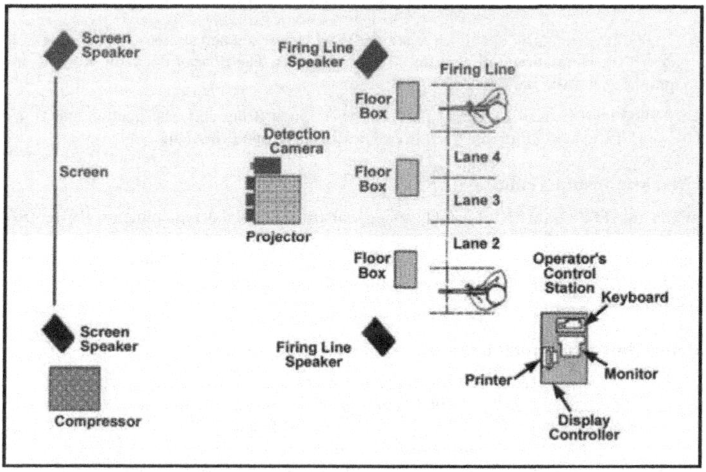

Figure A-1. Engagement skills trainer (five-lane configuration).

GENERAL CHARACTERISTICS

A-1. The EST 2000 replicates eleven weapons, including the rifle, carbine, pistol, grenade launcher, all machine guns, MK19, shotgun, and AT4. It has three modes of training:

- Marksmanship training.
- Tactical collective training.
- Shoot/don't shoot rules of engagement training.

A-2. It also includes a scenario editor for generating or tailoring new scenarios.

MARKSMANSHIP TRAINING

A-3. When used in the marksmanship training mode, the EST 2000—

- Uses Army standard courses of fire for all small arms weapons.
- Accurately simulates live-fire ranges in daylight and limited visibility conditions using precision-scaled targets, high-resolution imagery, and the essential weapon's system accuracy to compensate for errors (drift, parallax).
- Isolates, captures, and displays shots with replay that highlights firer's errors in the application of the fundamentals of marksmanship. Replay of the point of aim trace (before the shot, during the shot, and after the shot) diagnoses firer problems with aiming, breathing, steady hold, trigger control, and shot recovery for on-the-spot corrections. Cant sensors visually indicate firer-induced right or left cant, possibly resulting in missed shots.

Basic Rifle Marksmanship Training

NOTE: The marksmanship core scenarios can be found in the EST 2000 Operator's Manual, TD-07-6910-702-10.

A-4. The EST 2000 begins training the fundamentals of marksmanship right from the beginning, before the Soldier has a chance to develop bad habits. Using EST 2000 technology, Soldiers and units can reduce their rate of marksmanship failures and increase Soldiers' confidence in being able to fire their assigned weapons.

A-5. EST 2000 is particularly useful for teaching BRM, where a gated strategy requires the Soldier to pass certain gates in the simulation before firing live ammunition. The Soldier does not proceed or pass a gate scenario until he meets the standard.

A-6. EST 2000 marksmanship training provides basic range firing and qualification and is accomplished in 5-, 10-, or 15-lane configurations. Each firer is restricted to one firing lane.

Remedial Marksmanship Training

A-7. While the EST 2000 BRM gated strategy often reduces the requirements for remedial live-fire training, it is highly useful in diagnosing and correcting problems before the Soldier fires live rounds. Using the EST 2000 technology of rifle cant, trigger pressure, and before-the-shot, during-the-shot, and after-the-shot AARs, trainers can quickly identify and correct problems, thus raising confidence and first-time qualifications.

Individual and Unit Sustainment Training

A-8. Sustainment training and prequalification refresher training can be conducted with the EST 2000. During individual and unit sustainment training, grouping, zeroing, and Field Fire I and II should be conducted to standard in the EST 2000 prior to live-fire qualification.

NOTE: EST 2000 will not be used in lieu of live-fire qualification except for those outlined in DA Pam 350-38.

TACTICAL COLLECTIVE TRAINING

A-9. When used in the tactical collective training mode, the EST 2000—

- Presents tactical collective exercises that train squad, team, and element leaders in fire distribution and control.
- Provides fully articulated interactive targets with variable outcomes based on a squad's action or inaction.
- Uses realistic 3D-modeled battlefield terrain and variable environmental effects (for example, day/night and dawn/dusk, variable weather conditions, and illumination).
- Uses other special effects, such as a weapon's effects, explosions, and vehicle damage, to enhance the static eye point of the battlefield.
- Allows trainers and units to build scenarios as they would fight.
- Provides shot-by-shot feedback that is tied to each firer's lane of fire.
- Will soon include an entry-level indirect fire capability (product improvement).

NOTE: Tactical collective training core scenarios can be found in the EST 2000 Operator's Manual, TD-07-6910-702-10.

A-10. Tactical collective training is conducted on two networked five-lane subsystems. This configuration can support up to 11 weapons, including tandem weapons.

NOTE: The tandem weapons capability is available in collective training only. This capability allows the use of an extra weapon connected to the fifth lane (port 6) in the third floor box, allowing a firer to manage two weapons. For further instructions, refer to the EST 2000 Operator's Manual.

SHOOT/DON'T SHOOT RULES OF ENGAGEMENT TRAINING

A-11. When used in the shoot/don't shoot rules of engagement training mode, the EST 2000—

- Can be configured to enhance special operations and counterterrorism training.
- Is the premier training simulation for stability and support operations training.
- Uses video-based graphic overlays with multiple escalation or de-escalation points that require the firer to justify his actions based on his situational awareness.

NOTE: Shoot/don't shoot core scenarios can be found in the EST 2000 Operator's Manual, TD-07-6910-702-10.

A-12. Shoot/don't shoot rules of engagement training is conducted on a five-lane subsystem. This training uses video-based graphic overlays that provide important clues (for example, facial expressions and body language) for the firer to cue on. Multiple escalation or de-escalation points require the firer to justify his actions based on his situational awareness.

SCENARIO EDITOR

A-13. The scenarios currently available in the EST 2000 meet 90 percent of a unit's training requirements. As a unit's mission changes, weapons and TTPs are changed, or additional training requirements occur, the unit can use the scenario editor to generate or tailor new scenarios. The scenario editor can be used to enhance the individual Soldier's skills and, collectively, the squad's ability to engage and destroy an enemy threat.

NOTE: Detailed instructions on how to create and modify scenarios can be found in the EST 2000 training support package.

WEAPON SAFETY

A-14. The following general safety precautions should be adhered to:

- Fire simulated weapons only if they are pointed downrange.
- Post warning signs at all entry doors.
- Do not allow personnel to stand downrange from the firing line.

WARNING

No one should be allowed beyond the firing line.

- Instruct weapon handlers never to look directly into a barrel.
- Take the weapon off-line for testing and service at the first indication of malfunction and refer to the troubleshooting procedures.

DANGER

Each simulated weapon has the same appearance as a fully functioning weapon, with the exception of the trainer-peculiar umbilical cable. Under certain circumstances, especially in the subdued light of a training room, it is possible to mistake a live firearm for a simulated weapon. This creates the potential for personal injury or damage to property. To avoid confusion, no live or blank ammunition and no live weapons should be allowed in the training room.

Simulated weapons will not accept live or blank ammunition. Any attempt to chamber a live or blank round may damage the simulated weapon and create an unsafe situation.

LASER SAFETY

A-15. The lasers used in the simulated weapons meet ANSI Standard Z136.1-1993 Class I Standards for single laser pulse power. This classification is commonly referred to by the industry rating of "eye-safe." However, even eye-safe lasers may be dangerous under extraordinary circumstances. To ensure personnel safety, weapon handlers should not stare directly down a simulated weapon's barrel. Serious eye injury could result if a laser malfunctioned while a user was staring into the weapon's muzzle.

WARNING

Laser light is used in the operation of this equipment. Injury may result if personnel fail to observe safety precautions.

Never stare into the laser beam, look down the barrel of the simulated weapon, or directly view the laser beam with optical instruments.

Avoid direct eye exposure.

The instructor should ensure that all persons entering the training room are aware that laser radiation is present.

SECTION II. LASER MARKSMANSHIP TRAINING SYSTEM

The LMTS supports training with a Soldier's own weapon without the use of live ammunition. It is not designed to replace live-fire training or to eliminate the need for knowledgeable instructors. Major components include a battery-powered laser transmitter mounted to a mandrel inserted in the rifle barrel or affixed to the front sight with a mounting bracket, and a variety of laser-sensitive targets. The exercise is performed in the same manner as an LFX, except that the ammunition is a laser beam. The target senses shot locations, which are shown on a laptop screen. Unit commanders should expect and require the following:

- A training process that focuses on the four fundamentals of marksmanship (steady position, sight alignment and picture, breath control, and trigger squeeze).
- Opportunity for experienced marksmen to "test out" and serve as peer trainers or return to other duties.
- Real-time feedback.
- All-season training.
- Soldiers trained on their assigned weapon throughout the process.

Using LMTS technology, units can consistently reduce the rate of first-time marksmanship failures and increase new Soldiers' confidence in their ability to fire their basic weapon.

EQUIPMENT

A-16. Software enhancements continue to optimize the training process and minimize computer requirements by enabling an instructor to control up to 10 targets with only one computer. This feature reduces overall system costs and provides maximum throughput with a minimum number of instructors.

A-17. Minimum LMTS systems consist of a basic laser transmitter with a rod to fit the weapon and a laser target. Systems can be expanded to include a variety of components.

NOTE: Table A-1 provides a complete component list.

Table A-1. Laser marksmanship training strategy parts list.

PART	INCLUDES
110 system (110v or 220v)	TR-700 target, LT100C laser, transmitter rod, mask set, AC power adapter, and user's manual
330A system (110v or 220v)	TR-900 target, LT100C laser, transmitter rod, cable, software, AC power adapter, and user's manual
360 system (available in 3-, 4-, and 5-target array)	TR-900 target, LT100 laser, transmitter rod, cable, software, AC power adapter, control box, and user's manual
430 system (110v or 220v)	TR-900 target, MP400 laser, 556C rod, software, AC power adapter, and user's manual
Mini-range	TR-700 target, LT100C laser, transmitter rod, transceiver unit with RS-232 cable, software, E-tat, AC power adapter, and user's manual
Sound and recoil replicator system	M16A2 upper receiver M4 upper receiver CO_2-powered weapon simulator (M4 or M16)
Borelight kit	MP400 laser, LTA-556C transmitter rod, and carrying case
Targets	TR-700 electronic TR-900 electronic 25-m zero reflective (small) 25-m zero reflective (large) Other reflective targets available
Laser transmitters	LT-100 laser transmitter LT-500 in-barrel laser transmitter

Table A-1. Laser marksmanship training strategy parts list (continued).

PART	INCLUDES
System software	330A system 360 system Mini-range system
Instruction booklets	110 system 330A system 360 system
Laser transmitter rods	LTA-1200 Type B 12-gauge LTA-190 Type B Cal .177 LTA-220 Type B Cal .22 LTA-240 Type B Cal .25 LTA-310-2 Type A Cal .30, 2" barrel LTA-380 Type A Cal .38 and .357 LTA-380-2 Type A Cal .38 and .357, 2" barrel LTA-410 Type A Cal .40 and .41 LTA-440 Type A Cal .44 LTA-450-2 Type A Cal .45, 2" barrel LTA-500 Type A Cal .50 LTA-556C Cal 5.56-mm (special order) LTA-762C Cal 7.62-mm (special order)
Mask sets for TR-700 targets	Various E-type silhouettes; military mask set to simulate 300 m, 450 m, and 600 m for training conducted at 25 m
System components and accessories	Training vest for TR-700 target Sound and Recoil Replicator Consumable—5.56-mm, 7.62-mm, and 9-mm Notebook computer Safety rod CO2 4 oz., 7.5 oz., and 20 lbs. AC-600 100v adapter for TR-700 target AC-610 220v adapter for TR-700 target AC-910 110v adapter for TR-900 target AC-920 220v adapter for TR-900 target RC-260 remote cable BL-265 daisy chain cable BL-640 long cable BL-654 daisy chain cable RS-232 15-m cable extension CB-440 control boxTR-700 RF control (E-tag) Mini-range transceiver unit (tag-receiver)
Carrying cases	110 system, 330A system, and 360 system (per 3 targets) M16/M4 sound and recoil replicator Borelight kit

MARKSMANSHIP TRAINING

A-18. For initial skill development (for example, IET), Exercises 1 through 4 should be conducted sequentially. After grouping and zeroing standards, the Soldier moves to the LMTS alternate course C target where the course of fire replicates the live-fire course, except that the ammunition is a laser beam. Failure to meet the standards for this course of fire identifies the Soldier as a candidate for remedial training.

REMEDIAL TRAINING

A-19. Failure to achieve the standards set forth in this manual identifies the Soldier as a candidate for remedial training. Using the LMTS technology, trainers can quickly identify and correct problems, significantly raising qualification rates after subsequent attempts at qualification. After remedial training, the Soldier moves to the LMTS alternate course C target where the course of fire replicates that of the live-fire course, except that the ammunition is a laser beam.

SUSTAINMENT TRAINING

A-20. The training model in the exercises provides commanders and unit trainers with a sustainment training system that can be employed year-round, ideally as integrated concurrent training to cause the least disruption to other planned training.

A-21. Soldiers are administered a skill test at a regular frequency (current training guidance recommends quarterly). The results of this test allow commanders to focus training efforts on those Soldiers least able to demonstrate the minimum skills required.

A-22. For quarterly sustainment training, Soldiers should be pretested to determine the extent of training required. The pretest should begin with the grouping exercise (from Exercise 3) followed by the electronic alternate course C or mini-RETS (Exercise 4). Soldiers unable to meet pretest standards are given refresher training in the four fundamentals of rifle marksmanship, followed by completion of Exercises 1 through 4.

EXERCISES

A-23. LMTS exercises define procedures for using LMTS equipment to train and sustain basic marksmanship fundamentals. They may be conducted as independent stations or combined on a single station as appropriate for the training scenario. Trainers should employ LMTS equipment in a manner that accounts for:

- Space and time available at the training site.
- Unit size and composition.
- Remedial training requirements.
- Equipment availability.

NOTE: Check the LMTS operator's manual for specific information about equipment setup and operation.

A-24. Training in Exercises 1 through 3 should be conducted using the Soldier's own service rifle in the dry-fire mode. Exercise 4 may be conducted in the dry-fire mode, but an optional sound and recoil replicator should be employed for added realism.

NOTE: Sound and recoil replicators provide nearly 100 percent of the recoil felt with full rifle function. They require the Soldier to properly load magazines and enable the trainer to cause the rifle to misfeed or misfire to verify a Soldier's ability to perform immediate action procedures to reduce a stoppage.

A-25. If LMTS training immediately precedes a live-fire grouping and zeroing exercise and time permits, trainers may wish to use the system's prezeroing capability during Exercise 3 by using calibrated or "spun" lasers Using the calibrated lasers, Soldiers make adjustments to their own rifle sights during Exercise 3, resulting in a savings of time and ammunition on the grouping and zeroing range. All LMTS-based zeros must be confirmed by live-fire. If no live-firing is planned, calibrated lasers need not be used, and adjustments are made to the laser in Exercise 3.

NOTE: See the LMTS operator's manual for a description of the laser calibration process.

Exercise 1: Reflective Target Exercise

Table A-2. Action, conditions, and standards for a reflective target exercise.

ACTION	Demonstrate the four fundamentals of rifle marksmanship while using the LMTS reflective zeroing target.
CONDITIONS	Given an M16-/M4-series weapon, laser transmitter with mandrel, and reflective target.
STANDARDS	Demonstrate the four fundamentals of marksmanship by: • Achieving a good steady position. • Applying the proper sight alignment and sight picture. • Applying proper breath control. • Applying proper trigger squeeze.

A-26. Exercise 1 introduces Soldiers to the four fundamentals of marksmanship, outlines how to diagnose and correct firer problems, and reinforces proper application of the fundamentals.

A-27. This exercise requires a high degree of instructor involvement, but one instructor may effectively train up to 20 lanes. Decreased trainer/firer ratio will result in decreased efficiency and effectiveness. One trainer per 10 lanes is the optimum ratio.

A-28. The exercise requires little time to complete; to retain group integrity, it should be combined with Exercise 2 to allow more advanced firers to progress while problem firers receive remedial training.

A-29. A reflective zeroing target with MP400 laser/mandrel provides a simple but effective tool for remedial training during LFXs. Problem firers should be sent to a remedial station for a quick check of the application of the fundamentals and remedial training.

Exercise Performance

A-30. To perform Exercise 1, use the following procedures:

(1) The Soldier assumes a proper supported position using sandbags. The trainer inserts the MP-400/LTA-556C assembly into the rifle barrel and uses laser windage and elevation adjustments to achieve a bold sight adjustment with laser spot on front sight (Figure A-2). Then, the Soldier turns the laser to the ON position.

NOTE: Soldiers should become familiar with both supported and unsupported firing positions.

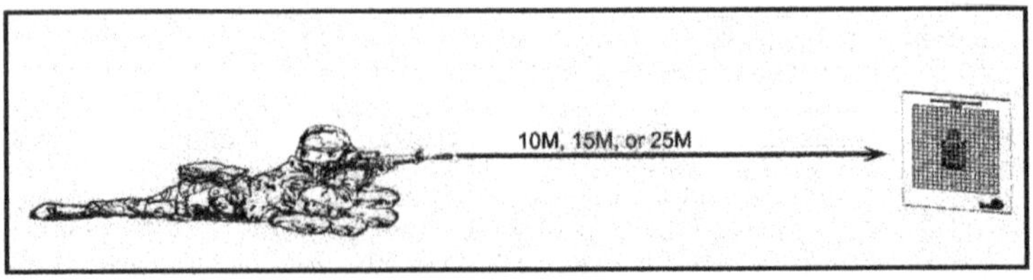

Figure A-2. Exercise 1.

(2) Under trainer supervision, the Soldier establishes a proper sight alignment and sight picture on a reflective zeroing target set at 10 meters, 15 meters, or 25 meters. With the MP-400 laser turned to ON, the trainer or coach blocks the beam with his finger.

NOTE: Use the appropriate target with the corresponding distance.

---

(3) When the Soldier is confident with the sight alignment and picture, the trainer removes his finger and observes the location of the red laser dot on the target.

(4) If the laser dot is in the 4-centimeter circle, proceed to Step 5. If the laser dot is outside of the 4-centimeter circle, the trainer instructs the Soldier regarding correct aiming techniques to bring the dot inside the circle, and repeats Step 2.

NOTE: If the trainer is reasonably certain that the laser and sights are aligned, the visible laser dot may be used to help the firer understand correct sight picture and alignment. The firer should be instructed to bring the laser dot to the target silhouette's center of mass, and then observe the relationship of the front and rear sights to the target.

---

(5) With the MP-400 in constant ON mode, the trainer or coach uses the red dot trace to confirm steady hold and proper breathing and trigger control.

NOTE: This trace can also be used to show the effects of improper steady position breath control and trigger control, and reinforce proper techniques.

---

(6) The trainer or coach turns the MP-400 to the training (TRN) mode and instructs the Soldier to fire six shots into the target's center of mass. The trainer or coach observes the laser hits to confirm proper application of the four fundamentals of rifle marksmanship.

NOTE: Failure to achieve this standard provides an early indication of the need for more intense instruction in the fundamentals of marksmanship or remedial training.

EXERCISE 2: INTERACTIVE DRY-FIRE EXERCISE

Table A-3. Action, conditions, and standards for an interactive dry-fire exercise.

ACTION	Demonstrate the integrated act of firing while using the LMTS 130-target system.
CONDITIONS	Given an M16-/M4-series weapon, laser transmitter with mandrel, and TR-700 targets with military masks.
STANDARDS	• Achieve 8 hits out of 10 shots two times on an open-faced target from the prone unsupported position. • Achieve 8 hits out of 10 shots two times on a 300-meter masked target from the supported position.

A-31. This exercise provides Soldiers with an opportunity for practicing the four fundamentals of rifle marksmanship in the integrated act of firing and may easily be conducted concurrently with Exercise 1 on the same station. TR-700 targets may be used both indoors and outdoors in various environments and arrangements to meet the training requirement.

Exercise Performance

A-32. To perform Exercise 2, use the following procedures:

(1) The Soldier assumes a proper firing position using sandbags for a supported position (Figure A-3).

NOTE: Sleeping mats should be used on hard floors.

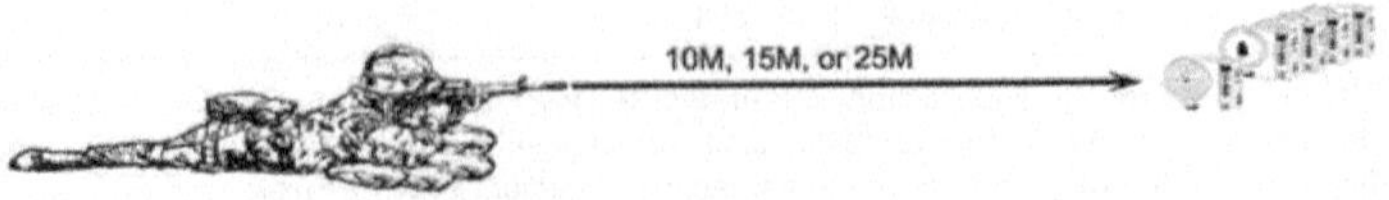

Figure A-3. Exercise 2.

(2) The Soldier applies the four fundamentals of marksmanship to engage a TR-700 open-faced target with 10 shots from the prone unsupported position. The Soldier cocks the rifle after each shot, forcing a break and reestablishing a proper stock weld to build muscle memory.

NOTE: The TR-700 open-faced target at 25 meters equals a doublewide E-silhouette target at 300 meters.

(3) The trainer inspects the target score for the number of hits. If the number is less than eight, the trainer should perform a visual laser/sight alignment check. If the laser/sight alignment is correct, the trainer reconfirms the Soldier's understanding of the four fundamentals of marksmanship and directs the Soldier to repeat Step 2. If the number of hits is less than eight after several tries, the Soldier reports for remedial training. If the number of hits is eight or more, the Soldier repeats Step 2 to confirm, and then proceeds to Step 4.

(4) After the Soldier completes Step 3, a 300-meter scaled E-silhouette mask is installed over the face of a TR-700 target. The Soldier repeats Step 2 from the supported position and repeats Step 4 to confirm.

(5) OPTIONAL: As time allows, the Soldier increases the number of shots to 20 and/or installs smaller masks for an additional skill challenge. Additional firing positions may also be reinforced, if needed.

NOTES:
1. The largest mask presents a 300-meter E-target size scaled for 25 meters.
2. The middle mask presents a 300-meter E-target size scaled for 15 meters or a 450-meter E-target size scaled for 25 meters.
3. The smallest mask presents a 300-meter E-target size scaled for 10 meters, a 450-meter E-target size scaled for 15 meters, or a 600-meter E-target size scaled for 25 meters.

EXERCISE 3: GROUPING AND ZEROING EXERCISE

Table A-4. Action, conditions, and standards for a grouping and zeroing exercise.

ACTION	Group and zero an M16-/M4-series weapon using the TR-900 target system with military mask.
CONDITIONS	Given an M16-/M4-series weapon, laser transmitter with mandrel, and TR-900 target system with military mask.
STANDARDS	From the supported firing position— • Grouping. Fire up to 27 shots (dry-fire) in three-round shot groups and achieve two consecutive shot groups within a 4-centimeter circle (25 meters), 2.4-centimeter circle (15 meters), or 1.6-centimeter circle (10 meters). • Battlesight Zero. Adjust the sights so that five out of six rounds in two consecutive shot groups strike within the zeroing circle in the silhouette on the zeroing target.

A-33. This exercise evaluates a Soldier's ability to apply the four fundamentals of rifle marksmanship in the integrated act of firing through shot grouping. The exercise is conducted in the same manner as live-fire grouping and zeroing exercises and can make those exercises more efficient and effective. All normal range commands should be used to reinforce training in proper range procedures. Up to 10 targets may be grouped together for scoring on one computer. This exercise is most efficient with one trainer to run the control and scoring console, plus one trainer for every five lanes. Training distance must correspond to the distance used in Exercise 4.

NOTE: Whenever this exercise is conducted prior to an LFX, calibrated lasers should be used to support prezeroing. Adjustments to the rear sight of an M16A2/A3/A4 or M4-series weapon, associated optics, and the front sight of an M16A1 must be made when training at 10 meters or 15 meters to compensate for parallax error. See the LMTS operator's manual for a detailed description of these adjustments.

Exercise Performance

A-34. To perform Exercise 3, use the following procedures:

(1) From the supported firing position (Figure A-4), the Soldier fires three-round shot groups at the target overlay's center of mass, continuing until two consecutive groups fall within a 4-centimeter circle anywhere on the target (maximum 27 shots). Trainers should provide feedback to the Soldier between each shot group. If the Soldier is unable to achieve the standard within 27 shots, the trainer attempts remedial actions or sends the Soldier to the remedial training station.

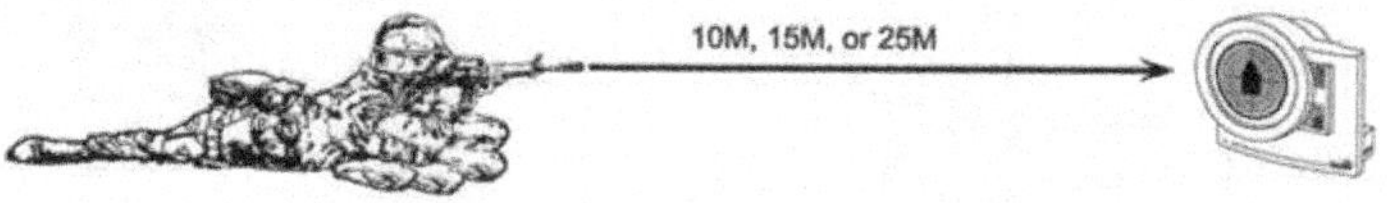

Figure A-4. Exercise 3.

(2) When the grouping standard is met, the Soldier makes appropriate sight changes as instructed by the trainer, who begins a new session for the zeroing process. The Soldier continues to fire three-round shot groups, adjusting the sights as instructed by the trainer to bring the shot groups (maximum 18 shots) within the zeroing circle on the target silhouette. When a shot group falls within the zeroing circle, the Soldier fires an additional shot group for confirmation. Five of six shots must fall within the zeroing circle.

NOTE: Failure to achieve the standard identifies the Soldier as a candidate for remedial training. The Soldier should not progress to Exercise 4 until the standard is met.

Exercise 4: Laser Marksmanship Training Strategy Prequalification Exercise

Table A-5. Action, conditions, and standards for a laser marksmanship training strategy prequalification exercise.

ACTION	Engage 10-, 15-, or 25-meter alternate course C scaled silhouettes with an M16-/M4-series weapon.
CONDITIONS	Given an M16-/M4-series weapon, laser transmitter with mandrel, and electronic alternate C target system with ten 10-, 15-, or 25-meter scaled silhouettes. Engage each silhouette with two shots from the supported position and two shots from the prone unsupported position.
STANDARDS	Without assistance, the Soldier engages 10 target silhouettes using the M16-/M4-series weapon with laser transmitter and achieves a minimum of 30 hits out of 40 shots.

A-35. Exercise 4 is used as a skill test to determine the need for training or the results of training, and serves as an accurate predictor of live-fire alternate course C performance.

NOTE: Soldiers failing to meet the standards of this exercise should receive remedial training prior to live-fire qualification.

A-36. The exercise may be conducted in the dry-fire mode, with Soldiers recocking the rifle between shots. When performing the dry-fire method, Soldiers should use a magazine with the follower and spring removed. Another option for the dry-fire mode involves removing the charging handle and attaching a piece of cord (looped on both ends with the free end about 3 inches from the end of the stock) to the bolt. The coach can recock the rifle between shots by pulling the cord directly to the rear.

A-37. The sound and recoil replicator options include the M16A2 Blazer and the alternate laser-mounting bracket used with a standard blank firing attachment (BFA). Both options offer added realism by providing full rifle function, sound, and recoil. When using the special safe Blazer blanks, the Blazer option may be employed indoors without hearing protection. Standard M200 blanks may only be used with the BFA while outdoors and while wearing hearing protection. When available, the LMTS mini-RETS range should be employed for added training realism and to prepare Soldiers for firing on pop-up targets.

Exercise Performance

A-38. To perform Exercise 4, use the following procedures:

(1) The Soldier assumes the proper supported firing position using sandbags (Figure A-5).

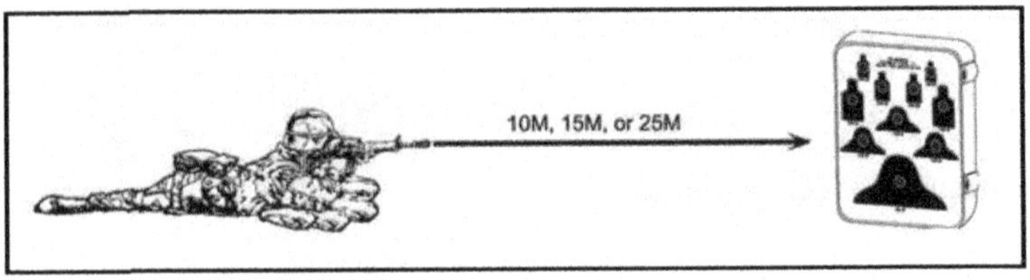

Figure A-5. Exercise 4.

(2) The trainer prepares the electronic 25-meter alternate C target and computer control station for the prequalification skill test. The course may be conducted at 25 meters, 15 meters, or 10 meters using the appropriate target overlay.

NOTE: The trainer should use appropriate range commands and enforce all range safety procedures.

---

(3) To fire the alternate course C, the Soldier applies the four fundamentals of rifle marksmanship using the service rifle (laser-zeroed) during Exercise 3.

(4) The computer automatically times the test and can print a score sheet.

SOUND AND RECOIL REPLICATOR

A-39. To add realism to the training, a special upper receiver sound and recoil replicator provides full live-fire functionality (without the projectile). It supplies nearly 100 percent of the recoil with 50 percent of the noise using a specially-designed nontoxic theatrical blank.

SECTION III. OTHER TRAINING DEVICES AND EXERCISES

This section provides the classification and nomenclature for training aids, devices, and targets.

TRAINING SUPPORT CENTERS

A-40. Training support centers (TSCs) are located throughout the world to provide training aids and devices. Each TSC provides training aid services to customers (for example, active Army units and schools, reserve components, and ROTC units) in their geographic area of support.

NOTE: For more information concerning TSC operations, write:

Commander
United States Army Training Support Center
ATTN: ATIC-DM
Fort Eustis, VA 23604

SELECTION OF TRAINING AIDS AND DEVICES

A-41. After training requirements have been established, appropriate training aids and devices can be selected from the TSC. Table A-6 lists many of those aids and devices available and their identification numbers.

Table A-6. Training aids and devices.

TYPE/NOMENCLATURE	IDENTIFICATION NUMBER
Rifle Marksmanship Trainer (Weaponeer)	DVC 07-57/DVC 07/57B
M15A2 Blank Firing Attachment	Supply Item (see TM)
Chamber Block (M16A1/A2)	TAD-0001
M16 Sighting Device (A1 0r A2) (Left and Right)	DVC-T 7-84
Target Box Paddle	DVC-T 7-86
Riddle Device	DVC-T 7-87
M16 Rifle Brass Deflector	DVC-T 23-30
M15A1 Aiming Card	DVC-T 07-26
M16A1 Display Mat (canvas)	TAD-0034 (locally)
Rifle Rest (for target box exercise)	TAD-12 (locally)
Front and Rear Sight, M16 Rifle	TAD-26 (locally)
Front and Rear Sight, M16A2 Rifle	TAD-0026A
Graphic Training Aids (GTA)	
M16A1 Disassembly Mat (paper)	GTA 09-06-43
Rifle, 5.56-mm, M16A1 Mechanical	GTA 7-1-26
Rifle, M16 Disassembly (M16A1)	GTA 9-6-43
M16A1 Rifle Malfunction	GTA 9-6-44
M16A1 Rifle Maintenance Card	GTA 21-1-3
Training Films	
*Rifle, M16A1 Part I, Care, Cleaning, Lubrication	TF 21-3907
*Rifle, M16A1 Part II Field Expedients	TF 21-3908
*Also available in videotape.	
Videotapes	
Engagement of Moving Personnel Targets with the M16A1 Rifle Team from the Foxhole Position	2E/010-071-1271-B
Cycle of Functioning M16A1 Rifle	2E/010-071-0444-B
Overview of BRM Training	2E/010-071-0086-B
TVT 7-13 (Feb 87)	2E/010-071-0725-B
TVT 7-1 Teaching Rifle Marksmanship: Part I	
TVT 7-2 Teaching Rifle Marksmanship: Part II	

TARGET ORDERING NUMBERS

A-42. Table A-7 lists the description and NSN to use when ordering marksmanship targets.

Table A-7. Target ordering numbers.

DESIGNATION	DESCRIPTION	NSN
D Prone	Full-length face with V through two scoring areas	6920-00-922-7450
D Prone	Repair center with V through two scoring areas	6920-00-922-7451
E-Silhouette	Full-length face, solid color paper	6920-00-600-6874
E-Silhouette	Full-length, pop-up, solid color plastic	6920-00-071-4780
E-Silhouette	Full-length face, cardboard, kneeling	6920-00-079-1806
F-Silhouette	Short-length face, solid color paper	6920-00-610-9086
F-Silhouette	Short-length, pop-up, solid color plastic	6920-00-071-4589
F-Silhouette	Short-length face, pasteboard	6920-00-795-1807
25-Meter Alternate Course Scaled Qualification Target	50- to 300-meter scaled silhouette target	6920-01-167-1398
15-Meter Battlesight Zeroing Target (.22 Caliber RFA)	250-meter scaled silhouette target (50-foot indoor range)	6920-01-167-1393
15-Meter Alternate Course C (.22 Caliber RFA)	50- to 300-meter scaled silhouette target (50-foot indoor range)	6920-01-167-1396
25-Meter M16A1 Zeroing Target	250-meter scaled silhouette target	6920-01-167-1392
25-Meter M16A2 Zeroing Target	300-meter scaled silhouette target	6920-01-253-4005
25-Meter M16A1 Slow Fire Target	75- to 300-meter scaled silhouette target	6920-01-167-1391
25-Meter M16A1 Timed Fire Target	50- to 300-meter scaled silhouette target	6920-01-167-1397
75-Meter M16A1 Feedback Target	75-meter scaled F-type silhouette	6920-01-169-6921
75-Meter M16A2 Feedback Target	75-meter scaled F-type silhouette	6920-01-253-4006
175-Meter M16A1 Feedback Target	175-meter scaled E-type silhouette	6920-01-167-1395
175-Meter M16A2 Feedback Target	175-meter scaled E-type silhouette	6920-01-167-1395
Pasters, Black		6920-00-165-6354
Pasters, Buff		6920-00-172-3572
Landscape target		6920-00-713-8253
Spindle, Target Spotter, Wood		6920-00-713-8257
Spotters, 1 1/2 inches in diameter		6920-00-789-0869
Spotters, 3 inches in diameter		6920-00-713-8255
Spotters, 5 inches in diameter		6920-00-713-8254
Thermal Blankets		6920-01-516-9912

TRAINING DEVICES

A-43. Several marksmanship training devices are available to aid in PMI and sustainment training. They are beneficial when ammunition is limited for training or during practice exercises, such as field firing on the EST 2000 or zeroing and qualifying with short-range training ammunition (SRTA). Some training devices are complex, costly, and in limited supply, while others are relatively simple, cheap, and in large supply. Individuals or squads can sustain and practice basic marksmanship skills and fundamentals using devices and aids alone or in combinations.

AIMING CARD

A-44. The M15A1 aiming card (Figure A-6) determines if the Soldier understands how to aim at a target's center of mass, how to adjust the point of aim, how to allow for gravity, and how to engage a moving target.

Figure A-6. M15A1 aiming card.

A-45. To use the aiming card—

(1) The card is misaligned, the Soldier is instructed to establish the correct point of aim, and a trainer checks it. Each Soldier demonstrates six out of six of the points of aim.

NOTE: The sight/target relationship on the card is the same visual perception that the Soldier should have when he is zeroing on a standard silhouette target.

---

(2) The Soldier shows the side alignment technique three times—place the front sightpost on the left or right edge of the target and bring the front sightpost to the target's center of mass.

(3) The Soldier shows the bottom-up alignment technique—place the front sightpost at the bottom of the target and bring the front sightpost to target's center of mass.

Riddle Sighting Device

A-46. The Riddle sighting device (Figure A-7) indicates if the Soldier understands the aiming process while using the rifle. It is a small plastic plate with a magnet and a drawing of an E-type silhouette target. A two-man team is required for its use.

A-47. To use the Riddle sighting device—

- (1) The Soldier assumes a supported or prone firing position. The assistant places the Riddle device on the front sight assembly and adjusts the plastic plate in the direction of the firer until he reports the proper sight picture.
- (2) Without disturbing the plastic plate, the trainer or coach aims through the sights to determine if the Soldier has aligned the target and sight properly.

A-48. Many sightings are conducted, and the trainer may include variations to ensure that the Soldier understands the process. Each Soldier demonstrates six out of six points of aim, starting with the plastic plate offset to the front sightpost.

NOTE: This device has a small metal clip that slips over the front sight assembly to allow a smooth surface for attachment of the magnet. The device may also be used without the metal clip.

This device is provided with a small metal clip that slips over the front sight assembly. It all a smoother surface for attachment of the magnet; however, the device can be used without the metal clip.

Figure A-7. Riddle sighting device.

M16 Sighting Device

A-49. The M16 sighting device (Figure A-8) is made of metal, with a tinted square of glass placed at an angle. When the device is attached to the rear of the M16A1 carrying handle, an observer can look through the sight to see what the firer sees.

NOTE: The M16 sighting device can be mounted on the M16A2 rifle, but the charging handle must be pulled to the rear first. Then, the M16 sighting device is mounted on the rear of the carrying handle, and the charging handle is returned forward.

A-50. When using the M16 sighting device, observe the following recommendations:

- The M16 sighting device can be used in a dry-fire or live-fire environment, but a brass cartridge deflector must be used during live-fire.
- The observer must practice with the sight to be effective. For example, if the observer looks at a reflected image and the Soldier is aiming to the right, it appears left to the observer.
- The device must be precisely positioned on the rifle (it may need to be bent to stay on).
- The observer's position must remain constant.
- The observer must talk with the firer to ensure a correct analysis of the aiming procedures.

A-51. The Soldier must achieve six out of six proper sight alignment drills.

NOTE: The M16 sighting device is made for left- and right-handed firers.

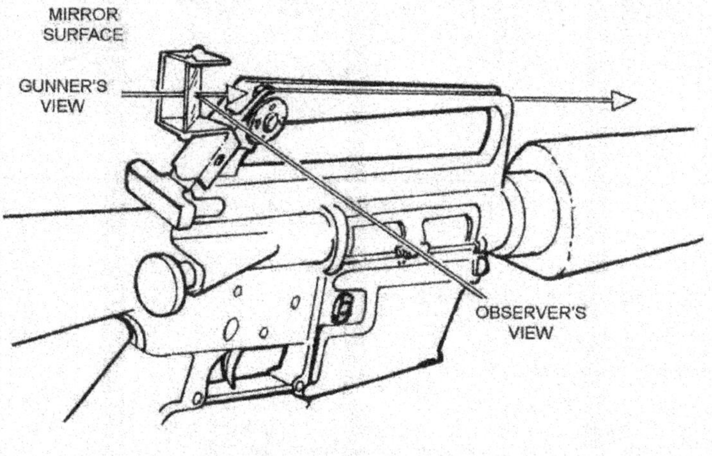

The M16 sighting device is made for left- and right-handed firers, and is available for the M16A2.

Figure A-8. M16 sighting device.

M15A2/M23 Blank Firing Attachment

A-52. The M15A2/M23 BFA (Figure A-9) attaches to the muzzle of M16-/M4-series weapons. It is designed to keep sufficient gas in the weapon's barrel to allow for semiautomatic, automatic, or burst firing with blank ammunition (M200 only).

A-53. When using the M15A2/M23 BFA, observe the following recommendations:

- After firing 50 rounds, check the attachment for a tight fit.
- Continuous blank firing results in a carbon buildup in the bore, gas tube, and carrier key. If this occurs, follow the cleaning procedures outlined in TM 9-1005-249-10.

NOTE: The M15A2 is painted red and is used on the M16-series weapons. The M23 is painted yellow and is used on the M4-series weapons. For identification, the M23 is stamped "M4 Carbine Only."

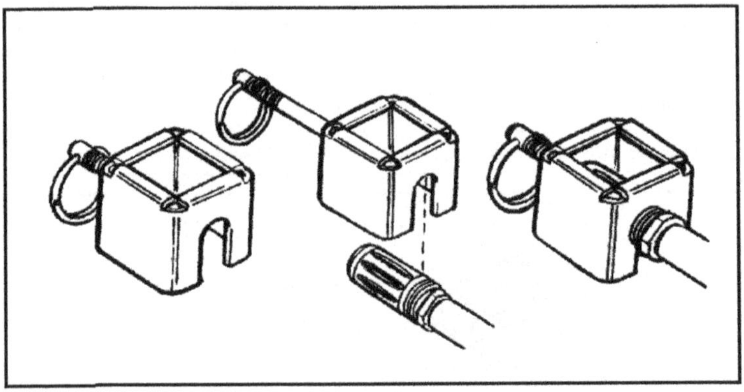

Figure A-9. Blank firing attachment.

CAUTION

Hand tighten the M15A2/M23 only.

Location of Miss-and-Hit System

A-54. The LOMAH system is a range aid used during downrange feedback exercises. The device uses acoustical triangulation to compute the exact location of a supersonic bullet as it passes through a target. The bullet impact is displayed instantly on a video monitor at the firing line. More importantly, it shows the location of a bullet miss, allowing the firer to make either a sight adjustment or a hold-off for subsequent shots.

NOTE: LOMAH, like other devices, is only an aid. When the Soldier uses LOMAH, he must understand the weapon and firing techniques and have a coach/instructor.

A-55. LOMAH ranges have been fielded in USAREUR and Korea. In locations where KD ranges are not available and restrictions prohibit walking downrange, the LOMAH system is a practical alternative to essential downrange feedback.

NOTE: Requests for LOMAH devices should be sent to:

Commander
US Army Training Support Center
ATIC-DM
Fort Eustis, VA 23604.

M261 .22-Caliber Rimfire Adapter

A-56. The .22-caliber RFA can contribute to a unit's marksmanship program when 5.56-millimeter ammunition is not available or when ranges that allow firing 5.56-millimeter ammunition are not available. The RFA can be useful for marksmanship training such as night fire, quick fire, and assault fire, but is not recommended for primary marksmanship training.

Training Considerations

A-57. When service ammunition is in short supply, the RFA can be used to complement a unit's training program.

Rifle Performance

A-58. The RFA cannot be depended on to fire in the same place as 5.56-millimeter ammunition. The RFA and its .22-caliber ammunition cannot replicate the exact ballistics of the 5.56-millimeter ammunition, but efforts to match RFAs with specific rifles can result in reasonable replication.

NOTE: Under ideal training conditions, the RFA should be used with dedicated rifles. It is not necessary for the Soldier to use his own weapon during RFA training.

A-59. Finding the right match of RFA and rifle can eliminate some variability. A trial-and-error technique can match RFAs to rifles, which results in weapons that fire well.

Rifle Zero

A-60. RFA ammunition differs from 5.56-millimeter ammunition in the following ways:

- RFA ammunition will not usually group in the same location as 5.56-millimeter ammunition at 25 meters and cannot be used for weapon zero.
- RFA ammunition normally fires a slightly larger shot group than 5.56-millimeter ammunition.

A-61. When a Soldier uses an RFA in his rifle, he must be careful not to lose his 5.56-millimeter zero. This can be accomplished by using hold-off while firing .22-caliber ammunition or keeping a record of sight changes so the sights can be moved back.

NOTES:

1. The .22-caliber round approximates the 5.56-millimeter trajectory out to 25 meters.
2. The correct zeroing target or appropriate scaled-silhouette targets can be used for practice firing exercises at 15 meters (50 feet) or 25 meters.

Advantages and Disadvantages

A-62. If the RFA is used as a training aid, the advantages and disadvantages must be considered during training.

Advantages

A-63. The .22-caliber ammunition is cheaper and may be available in larger quantities than 5.56-millimeter ammunition. It can be fired on all approved indoor ranges and in other close-in ranges where 5.56-millimeter ammunition is prohibited. RFA training can be used to sustain marksmanship skills during periods when full-caliber 5.56-millimeter ammunition training cannot be conducted.

Disadvantages

A-64. Some negative training aspects exist because of differences in the weapon's functioning when using the RFA. These differences include—

- The forward assist does not work.
- The bolt does not lock to the rear after the last round is fired.
- More malfunctions can occur with the RFA than with 5.56-millimeter ammunition.
- Immediate action procedures are different.

SHORT-RANGE TRAINING AMMUNITION

A-65. SRTA is a plastic practice cartridge (M862) that enables a unit to conduct realistic firing training at shorter distances with reduced danger areas. The M862 has a maximum range of 250 meters. The blue plastic projectile reduces the risk of overpenetration and ricochet, which makes it ideal for UO training.

A-66. To fire the M862 SRTA from an M16-/M4-series weapon, the standard bolt and bolt carrier must be replaced by the M2 practice bolt. The M2 practice bolt consists of a bolt carrier, which is a fixed bolt. The practice bolt changes the weapon from a gas-operated action to a blow-back action that permits cyclic fire with the lower-powered M862.

NOTES: 1. Because of the design of the M2 practice bolt, standard 5.56-millimeter rounds cannot be fired from the weapon while it is installed.

2. See TM 9-6920-746-12&P for more information on the M862 SRTA and the M2 practice bolt.

WEAPONEER

A-67. The Weaponeer is an effective rifle marksmanship training device that simulates the live-firing of the M16-series rifle. The system can be used for developing and sustaining marksmanship skills, diagnosing and correcting problems, and assessing basic skills.

NOTE: The EST 2000 has replaced the Weaponeer, but some units still use the Weaponeer.

Operation

A-68. Figure A-10 shows the Weaponeer in the standing supported firing position. With the exception of smoke and cartridge ejection, the rifle operates normally and has the same weight and balance as the standard weapon. An IR aiming sensor simulates round trajectory and hit point to an accuracy of better than one MOA. The recoil rod that attaches at the rifle's muzzle end simulates recoil.

NOTE: Recoil is provided in both semiautomatic and automatic modes of fire and is adjustable from no-net force to 30 percent more than that of a live M16.

A-69. Sound is provided through headphones and is adjustable from 115 to 135 decibels. Special magazines are used. One magazine simulates a continuous load; the other is used to train rapid magazine change and can be loaded with 1 to 30 simulated rounds. Selectable misfire can be used to detect gun-shyness and drill immediate action. The front and rear sights are zeroed the same as standard rifles.

Figure A-10. Weaponeer set up in the standing supported position.

A-70. The Weaponeer range can be raised or lowered to accommodate all firing positions. The target assembly contains four targets: a scaled 25-meter zeroing target and three pop-up targets. E-type and F-type silhouettes at ranges from 75 meters can be used on the Weaponeer. Known distance and various other types of targets displayed in fixed or random sequences can be used. Target exposure times may be set to unlimited or from 1 to 30 seconds. The fall-when-hit mode can be selected with the KILL button.

A-71. The operator's console contains the system control buttons, graphics printer, and video feedback monitor. The back of the console has counters that total rounds and hours, and a storage bin for storing magazines, printer paper and ribbon, headphones, two wrenches for assembling the Weaponeer, and a small hex-head wrench for aligning the rifle sensor. A remote control, which attaches to the back of the console, enables a trainer or firer to operate select functions away from the console.

Feedback

A-72. The Weaponeer provides feedback to help trainers to teach and Soldiers to learn marksmanship skills. The tools used for feedback include—

- Fall-when-hit mode.
- Real-time aiming point display.
- Immediate shot impact display.
- Replay.
- Shot groups.
- Printer.

Fall-When-Hit Mode

A-73. Lighting the KILL button enables the fall-when-hit mode. When the button is activated, targets fall when hit. This feedback provides the same hit-or-miss information as a train-fire (RETS) range.

Real-Time Aiming Point Display

A-74. When a firer aims on or near a target, his aiming point relative to the target is continuously displayed on the video screen. The aiming point display allows the trainer to teach and verify aiming techniques, and to continuously monitor the firer's steadiness, techniques, time on target, trigger squeeze, and recovery from recoil.

Immediate Shot Impact Display

A-75. When a shot is fired, its impact relative to the target is immediately displayed on the video screen as a blinking white dot (Figure A-11, left target).

Replay

A-76. After a shot is fired, a real-rate display of how the firer engaged the target can be replayed on the video screen.

A-77. The target to the right in Figure A-11 shows the type of information that can be replayed on the video screen after a series of shots are fired. To show the sequence, the dots have been numbered.

A-78. To show a replay, the firer—

(1) Selects the shot he wishes to replay by operating the EACH SHOT button.

(2) Presses the REPLAY button.

NOTE: Some Weaponeers record and store replays for only the first three shots.

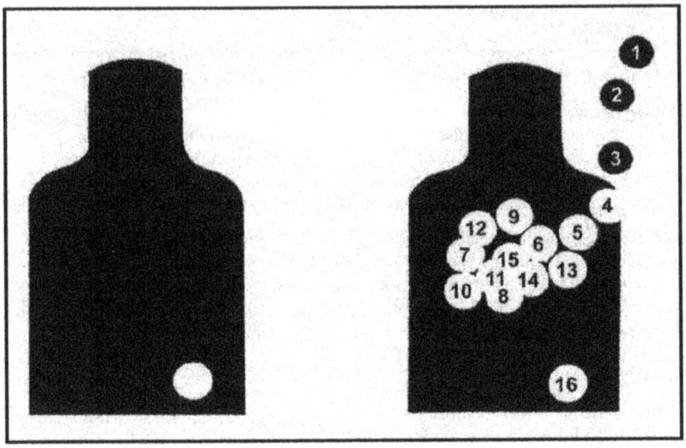

Figure A-11. Replay of shot.

Shot Groups

A-79. The impact location of up to 32 shots is automatically stored in the Weaponeer's memory and displayed on the video screen. Each impact is indicated by a white dot, and the last shot blinks for indication. All 32 shots can be fired and displayed on a single target or split among a combination of targets. The CLEAR button erases all shots from the Weaponeer's memory.

Printer

A-80. A hard copy printer is provided for postfiring analysis, for firer progress tracking, and for record keeping. Pressing the PRINT button causes the target displayed on the video to print (Figure A-12). Some Weaponeers can print the three pop-up targets at the same time by holding in the REPLAY button and pressing the PRINT button.

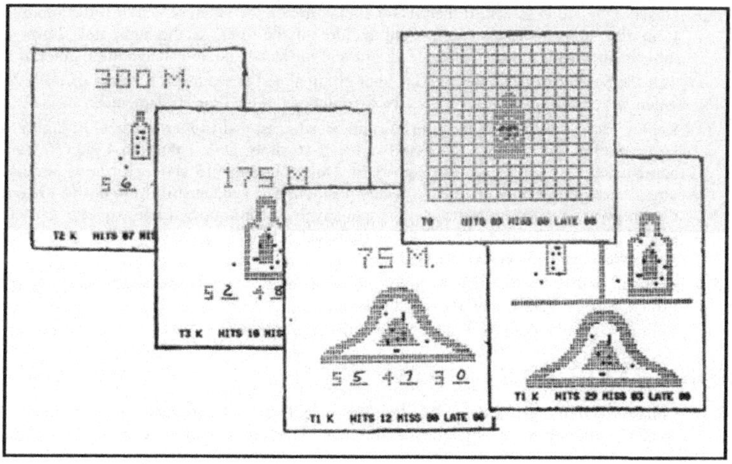

Figure A-12. Weaponeer printouts.

Use of the Weaponeer

A-81. The Weaponeer is used to evaluate the firer's ability to apply the four fundamentals. It is used throughout the program to help diagnose and remediate problems. In the unit, the Weaponeer should be used much like it is used in BRM.

A-82. When training Soldiers on the Weaponeer—

- Proceed at a relaxed pace, and emphasize accuracy before speed.
- If possible, train with small groups, allowing each Soldier several 10- to 15-minute turns on the device.
- For remedial training, try to relax the Soldier. A nervous Soldier will have trouble learning and gaining confidence in his marksmanship skills. For sustainment training, encourage competition between individuals or units.

Diagnosis of Firing Problems

A-83. The Weaponeer assists in the diagnosis of firing problems. The seven-step program is recommended as a guide. These seven steps are designed to diagnose and show the Soldier his firing errors. This could be enough to correct the error.

NOTES: 1. Diagnosis needs to be followed up with remedial exercises with the Weaponeer, target box exercise, or dime/washer exercise.

2. Depending on the extent of the firer's problems and time constraints, the number of shots may be increased.

A-84. To diagnose firing problems—

(1) Tell the Soldier to assume a good firing position, aim at a target, and hold steady (supported and prone unsupported positions).

(2) Visually check the firer's position and correct any gross errors.

(3) Observe the video screen. If there is no aiming dot on the video screen or if the aiming dot is far from the target's center, teach sight picture to the firer. If the light dot shows excessive movement, check and correct the techniques of the steady position and natural point of aim.

(4) Tell the Soldier to fire a three-round shot group aimed at the target's center of mass. Watch the video screen and Soldier as he fires. Note violations of the four fundamentals.

(5) Replay each shot to show the firer his aim, steadiness, and trigger squeeze. In Figure A-11, the target on the right shows a numbered series of 16 shots. Dots 1 through 4 indicate that the firer approached the target from the high right. Dots 5 through 15 show that he is aiming near the target's center, but does not have a steady position. The sudden shift from dot 15 to 16 (dot 16 is the hit point of the shot) indicates that gun-shyness or improper trigger squeeze caused the firer to pull his point of aim down and to the right just before firing. Replay helps the firer understand and correct his firing errors.

(6) Confirm and refine the diagnosis by allowing the Soldier to fire additional three-round shot groups. Use replay to show the firer his faults.

(7) Summarize and record the Soldier's basic firing problems.

TRAINING EXERCISES

A-85. Marksmanship training exercises are designed to aid in PMI and sustainment training. They are beneficial when ammunition is limited for training. Training exercises can be used alone or in combinations.

TARGET BOX EXERCISE

A-86. The target box exercise checks the consistency of aiming and placement of three-round shot groups in a dry-fire environment.

A-87. To conduct the exercise (Figure A-13)—

(1) The target man places the silhouette anywhere on the plain sheet of paper and moves the silhouette target as directed by the firer.

NOTE: The two positions (separated by 15 yards or 25 meters) must have already been established so the rifle is pointed at some place on the paper.

(2) When the firer establishes proper aiming, he signals the target man to "Mark."

NOTE: Only hand signals are used since voice commands would be impractical when training several pairs of Soldiers at one time.

Figure A-13. Target box exercise.

(3) The target man places the pencil through the hole in the silhouette target and makes a dot on the paper. Then, he moves the silhouette to another spot on the paper and indicates to the firer that he is ready for another shot.

(4) When the three shots are completed, the target man triangulates the three shots and labels it "Shot Group 1." The firer and instructor view the shot group.

(5) The Soldier fires several shot groups. After two or three shot groups are completed in one location, the rifle, paper holder, or paper is moved so shots fall on a clean section of the paper.

NOTE: Any movement of the rifle or paper between the first and third shots of a group voids the exercise. Two devices are available to hold the rifle. Rifle-holding devices (Figures A-14 and A-15) are positioned on level ground, or are secured by sandbags or stakes to ensure the rifle does not move during the firing of the three shots. Movement of the paper is eased by using a solid backing (Figure A-16). Any movement of either is reflected in the size of the shot group. Several varieties of wooden target boxes have been locally fabricated.

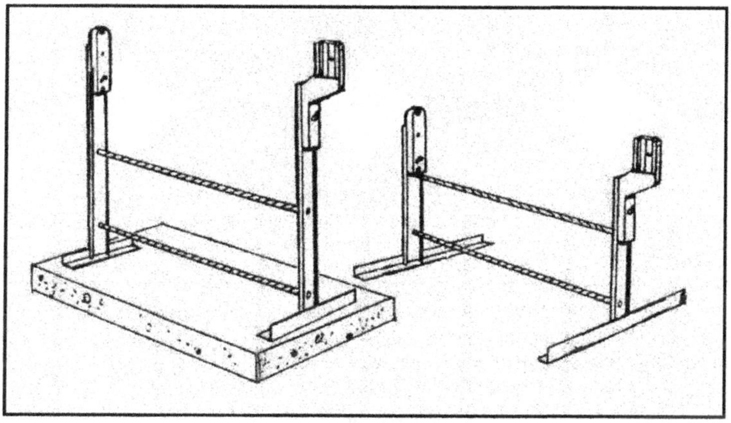

Figure A-14. Rifle-holding device (TA-G-12A).

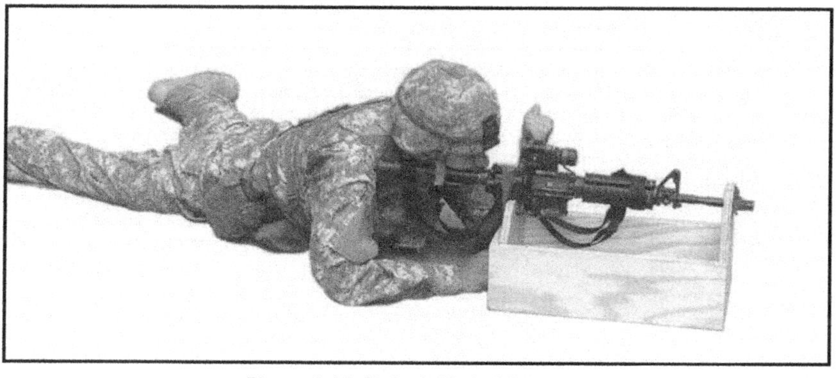

Figure A-15. Staked rifle holding box.

Figure A-16. Paper being placed on a stationary object.

A-88. Each Soldier dry-fires the exercise until he has demonstrated six out of six of the points of aim within the plastic target box paddle's 4-centimeter template. The exercise should be repeated as many times as necessary to achieve two consecutive shot groups that will fit into the same 2-centimeter circle.

A-89. A simulated shot group covered with a 1-centimeter (diameter) circle indicates consistent aiming. Since no rifle or ammunition variability is involved and since there is no requirement to place the shot group in a certain location, a 1-centimeter standard may be compared to obtaining a 4-centimeter shot group on the 25-meter live-fire zero range.

NOTE: The silhouettes on the plastic paddle (Figure A-17) are scaled to represent an E-type silhouette target at 250 meters. The visual perception during the target box exercise is similar to what a Soldier sees while zeroing on a standard zeroing target. The small E-type silhouette is the same scale at 15 yards as the larger silhouette is at the 25-meter range (some training areas are set up at 15 yards; others are set up at 25 meters). While there are some benefits to representing a 250-meter target, the main benefit of this exercise can be obtained at any distance. In place of the paddle, a standard zeroing target can be used at 25 meters by placing a small hole in the center (dot), moving the target sheet over the paper, and marking as previously outlined.

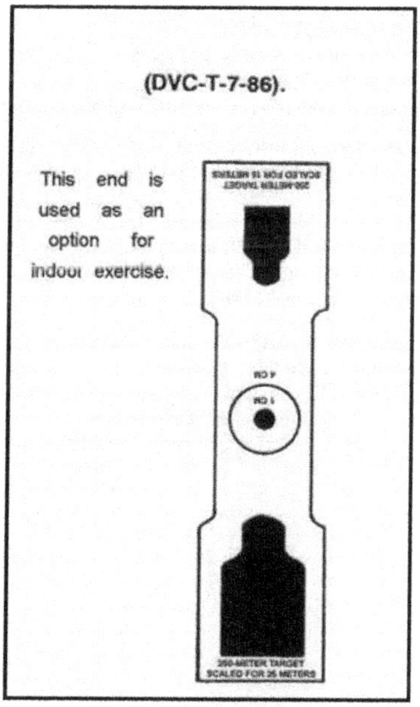

Figure A-17. Target box paddle (DVC-T-7-86).

A-90. The shot group exercise provides a chance for the trainer to critique the Soldier on his aiming procedures, aiming consistency, and placement of shot groups. Assuming the rifle and paper remain stationary and the target man properly marks the three shots, the only factor to cause separation of the dots on the paper is error in the Soldier's aiming procedure. When the Soldier can consistently direct the target into alignment with the sights on this exercise, he should be able to aim at the same point on the zero range or on targets at actual range.

Ball and Dummy Exercise

A-91. This exercise is conducted on a live-fire range and is used to detect if the firer is anticipating the shot or using improper trigger squeeze.

A-92. To perform this exercise—

(1) The coach or designated assistant inserts a dummy round into a magazine of live rounds. The firer must not know when a dummy round is in the magazine.

(2) When the hammer falls on a dummy round (which the firer thought was live), the firer and his coach may see movement. The firer anticipating the shot or using improper trigger squeeze causes this. Proper trigger squeeze results in no movement when the hammer falls.

A-93. The Soldier demonstrates the ability to properly utilize the fundamentals of marksmanship six consecutive times.

Dime/Washer Exercise

A-94. This dry-fire technique is used to teach or evaluate the skill of trigger squeeze and is effective when conducted from an unsupported position. To perform this exercise—

(1) The Soldier cocks the weapon.

(2) The Soldier assumes an unsupported prone firing position.

(3) The Soldier aims at the target. An assistant places a dime or washer on the rifle's barrel between the flash suppressor and front sightpost assembly.

(4) The Soldier tries to squeeze the trigger naturally without causing the dime or washer to fall off.

A-95. Several repetitions of this exercise must be conducted to determine if the Soldier has problems with trigger squeeze. The Soldier is a GO if he can dry-fire six of six consecutive shots without causing the dime or washer to fall.

NOTES: 1. If the dime or washer is allowed to touch the sight assembly or flash suppressor, it may fall off due to the jolt of the hammer. The strength of the hammer spring on some rifles can make this a difficult exercise to perform.

2. When using the M16A2 rifle, the dime/washer exercise is conducted the same, except that a locally fabricated device must be attached to the weapon. A piece of 3/4-inch bonding material is folded into a clothespin shape and inserted in the weapon's flash suppressor and the dime or washer is placed on top of it.

Appendix B
Scorecards

During live-fire events, a Soldier's hit-and-miss performance is recorded to facilitate the instructor/trainer's critiques or to indicate where more training is needed. The following are examples of completed scorecards.

EXAMPLES OF COMPLETED SCORECARDS

B-1. Figures B-1 through B-8 show examples of completed scorecards.

REPRODUCIBLE FORMS

B-2. Blank copies of the following can be found at the end of this publication:

- DA Form 3595-R (Record Fire Scorecard).
- DA Form 3601-R (Single Target—Field Fire I Scorecard).
- DA Form 5239-R (100-, 200-, and 300-Meter Downrange Feedback Scorecard).
- DA Form 5241-R (Single and Multiple Targets—Field Fire II Scorecard).
- DA Form 5789-R (Record Fire Scorecard—Known-Distance Course).
- DA Form 5790-R (Record Fire Scorecard—Scaled Target Alternate Course).
- DA Form 7489-R (Record Night Fire Scorecard).
- DA Form 7649-R (Squad Designated Marksman—Record Fire I and II).
- DA Form 7650-R (Squad Designated Marksman—Position Evaluation).

NOTE: These forms, scorecards, and position evaluation sheets are not available through the normal supply channels. You may reproduce them locally on 8 1/2- x 11-inch paper or download them from the Army Publishing Directorate at http://www.apd.army.mil/.

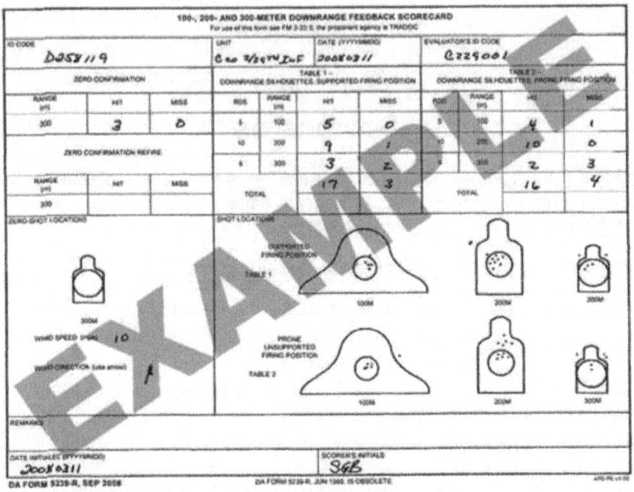

100-, 200- AND 300-METER DOWNRANGE FEEDBACK SCORECARD
For use of this form see FM 3-22.9; the proponent agency is TRADOC

ID CODE	UNIT	DATE (YYYYMMDD)	EVALUATOR'S ID CODE
D258119	Co 3/29TH INF	20080311	C229001

ZERO CONFIRMATION			TABLE 1 – DOWNRANGE SILHOUETTES, SUPPORTED FIRING POSITION				TABLE 2 – DOWNRANGE SILHOUETTES, PRONE FIRING POSITION			
RANGE (m)	HIT	MISS	RDS	RANGE (m)	HIT	MISS	RDS	RANGE (m)	HIT	MISS
300	3	0	5	100	5	0	5	100	4	1
ZERO CONFIRMATION REFIRE			10	200	9	1	10	200	10	0
			5	300	3	2	5	300	2	3
RANGE (m)	HIT	MISS	TOTAL		17	3	TOTAL		16	4
300										

ZERO-SHOT LOCATIONS

300M

WIND SPEED (mph) 10

WIND DIRECTION (use arrow)

SHOT LOCATIONS

SUPPORTED FIRING POSITION

TABLE 1

100M 200M 300M

PRONE UNSUPPORTED FIRING POSITION

TABLE 2

100M 200M 300M

REMARKS

DATE INITIALED (YYYYMMDD) 20080311

SCORER'S INITIALS SGB

DA FORM 5239-R, SEP 2008 DA FORM 5239-R, JUN 1989, IS OBSOLETE

Figure B-1. Example of completed DA Form 5239-R
(100-, 200-, and 300-Meter Downrange Feedback Scorecard).

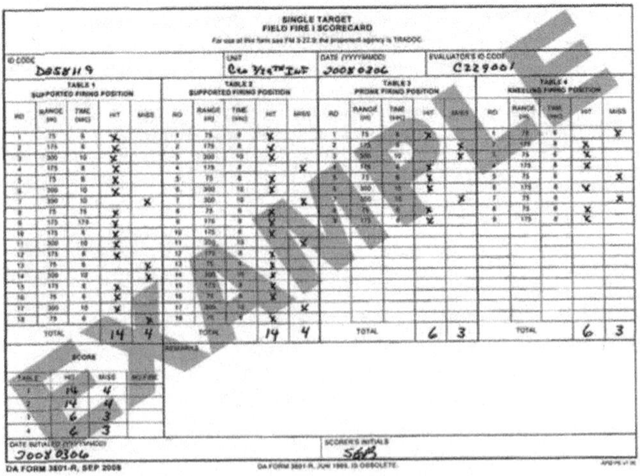

SINGLE TARGET
FIELD FIRE I SCORECARD
For use of this form see FM 3-22.9; the proponent agency is TRADOC

ID CODE	UNIT	DATE (YYYYMMDD)	EVALUATOR'S ID CODE
D258119	Co 3/29TH INF	20080206	C229001

TABLE 1 SUPPORTED FIRING POSITION — TABLE 2 SUPPORTED FIRING POSITION — TABLE 3 PRONE FIRING POSITION — TABLE 4 KNEELING FIRING POSITION

RD, RANGE (m), TIME (sec), HIT, MISS

	TOTAL HIT	TOTAL MISS
Table 1	14	4
Table 2	14	4
Table 3	6	3
Table 4	6	3

REMARKS

SCORE

TABLE	HIT	MISS	NO FIRE
1	14	4	
2	14	4	
3	6	3	
4	6	3	

DATE INITIALED (YYYYMMDD) 20080306

SCORER'S INITIALS SGB

DA FORM 3601-R, SEP 2008 DA FORM 3601-R, JUN 1989, IS OBSOLETE

Figure B-2. Example of completed DA Form 3601-R
(Single Target—Field Fire I Scorecard).

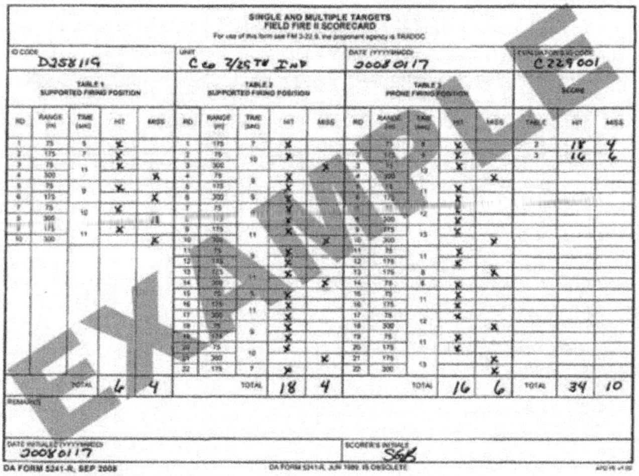

SINGLE AND MULTIPLE TARGETS
FIELD FIRE II SCORECARD
For use of this form see FM 3-22.9; the proponent agency is TRADOC

ID CODE	UNIT	DATE (YYYYMMDD)	EVALUATOR'S ID CODE
D258119	C Co 2/25TH INF	20080117	C229001

TABLE 1 SUPPORTED FIRING POSITION — TOTAL: HIT 6, MISS 4
TABLE 2 SUPPORTED FIRING POSITION — TOTAL: HIT 18, MISS 4
TABLE 3 PRONE FIRING POSITION — TOTAL: HIT 16, MISS 6

SCORE TABLE	HIT	MISS
2	18	4
3	16	6
TOTAL	34	10

REMARKS

DATE INITIALED (YYYYMMDD): 20080117
SCORER'S INITIALS: SGB

DA FORM 5241-R, SEP 2008 — DA FORM 5241-R, JUN 1989, IS OBSOLETE

Figure B-3. Example of completed DA Form 5241-R (Single and Multiple Targets—Field Fire II Scorecard).

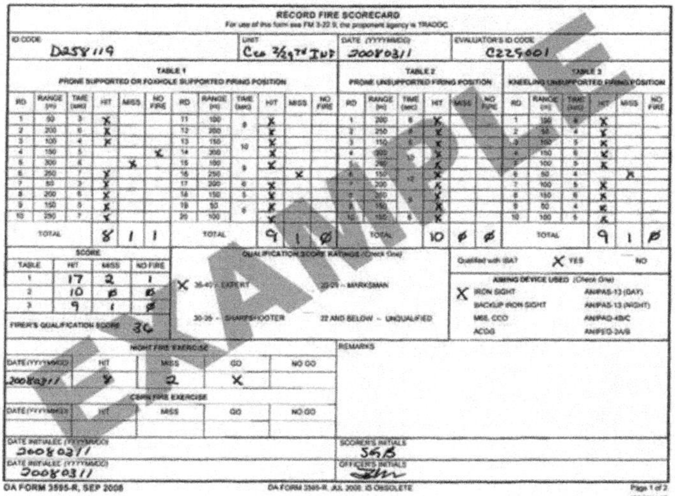

RECORD FIRE SCORECARD
For use of this form see FM 3-22.9; the proponent agency is TRADOC

ID CODE	UNIT	DATE (YYYYMMDD)	EVALUATOR'S ID CODE
D258119	C Co 2/29TH INF	20080311	C225001

TABLE 1 PRONE SUPPORTED OR FOXHOLE SUPPORTED FIRING POSITION — TOTAL: 8 / 1 / 1 and 9 / 1 / Ø
TABLE 2 PRONE UNSUPPORTED FIRING POSITION — TOTAL: 10 / Ø / Ø
TABLE 3 KNEELING UNSUPPORTED FIRING POSITION — TOTAL: 9 / 1 / Ø

SCORE TABLE	HIT	MISS	NO FIRE
1	17	2	1
2	10	Ø	Ø
3	9	1	Ø

FIRER'S QUALIFICATION SCORE: 36

QUALIFICATION SCORE RATINGS (Check One): X 36-40 – EXPERT; 23-29 – MARKSMAN; 30-35 – SHARPSHOOTER; 22 AND BELOW – UNQUALIFIED

Qualified with IBA? X YES — NO

AIMING DEVICE USED (Check One): X IRON SIGHT; BACKUP IRON SIGHT; M68, CCO; ACOG; AN/PAS-13 (DAY); AN/PAS-13 (NIGHT); AN/PAQ-4B/C; AN/PEQ-2A/B

NIGHT FIRE EXERCISE

DATE (YYYYMMDD)	HIT	MISS	GO	NO GO
20080311	8	2	X	

CBRN FIRE EXERCISE

DATE (YYYYMMDD)	HIT	MISS	GO	NO GO

REMARKS

DATE INITIALED (YYYYMMDD): 20080311 — SCORER'S INITIALS: SGB
DATE INITIALED (YYYYMMDD): 20080311 — OFFICER'S INITIALS

DA FORM 3595-R, SEP 2008 — DA FORM 3595-R, JUL 2006, IS OBSOLETE — Page 1 of 2

Figure B-4. Example of completed DA Form 3595-R (Record Fire Scorecard).

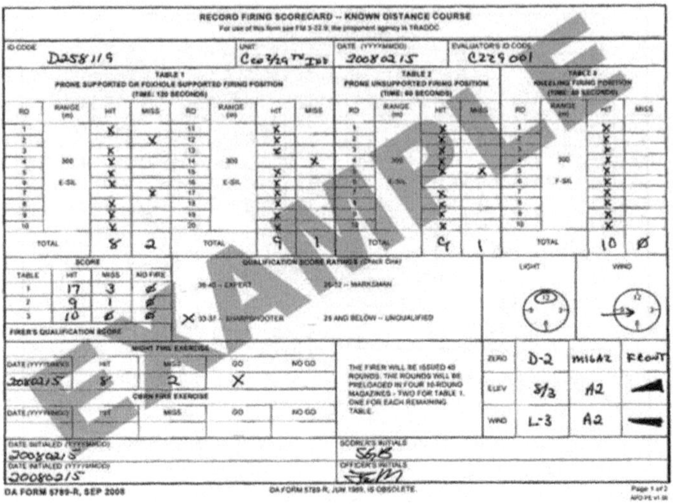

RECORD FIRING SCORECARD -- KNOWN DISTANCE COURSE

For use of this form see FM 3-22.9; the proponent agency is TRADOC

ID CODE	UNIT	DATE (YYYYMMDD)	EVALUATOR'S ID CODE
D258119	C Co 3/29th INF	20080215	C229001

TABLE 1 — PRONE SUPPORTED OR FOXHOLE SUPPORTED FIRING POSITION (TIME: 120 SECONDS)

RD	RANGE (m)	HIT	MISS	RD	RANGE (m)	HIT	MISS
1	300 E-SIL	X		11	300 E-SIL	X	
2			X	12		X	
3		X		13		X	
4		X		14			X
5		X		15		X	
6		X		16		X	
7			X	17		X	
8		X		18		X	
9		X		19		X	
10		X		20		X	
TOTAL		8	2	TOTAL		9	1

TABLE 2 — PRONE UNSUPPORTED FIRING POSITION (TIME: 60 SECONDS); TABLE 3 — KNEELING FIRING POSITION (TIME: 60 SECONDS)

RD	RANGE (m)	HIT	MISS	RD	RANGE (m)	HIT	MISS
1	300 E-SIL	X		1	300 F-SIL	X	
2		X		2		X	
3		X		3		X	
4		X		4		X	
5		X	X	5		X	
6				6		X	
7		X		7		X	
8		X		8		X	
9		X		9		X	
10		X		10		X	
TOTAL		9	1	TOTAL		10	Ø

SCORE

TABLE	HIT	MISS	NO FIRE
1	17	3	Ø
2	9	1	Ø
3	10	Ø	Ø

FIRER'S QUALIFICATION SCORE

QUALIFICATION SCORE RATINGS (Check One): 38-40 -- EXPERT; 26-32 -- MARKSMAN; X 33-37 -- SHARPSHOOTER; 25 AND BELOW -- UNQUALIFIED

LIGHT / WIND

NIGHT FIRE EXERCISE

DATE (YYYYMMDD)	HIT	MISS	GO	NO GO
20080215	8	2	X	

CBRN FIRE EXERCISE

DATE (YYYYMMDD)	HIT	MISS	GO	NO GO

THE FIRER WILL BE ISSUED 40 ROUNDS. THE ROUNDS WILL BE PRELOADED IN FOUR 10-ROUND MAGAZINES - TWO FOR TABLE 1, ONE FOR EACH REMAINING TABLE.

ZERO	D-2	M16A2	FRONT
ELEV	5/3	A2	
WIND	L-3	A2	

DATE INITIALED (YYYYMMDD): 20080215 — SCORER'S INITIALS: SGB

DATE INITIALED (YYYYMMDD): 20080215 — OFFICER'S INITIALS: JEM

DA FORM 5789-R, SEP 2008 — DA FORM 5789-R, JUN 1989, IS OBSOLETE. — Page 1 of 2 APD PE v1.00

Figure B-5. Example of DA Form 5789-R (Record Fire Scorecard—Known Distance Course).

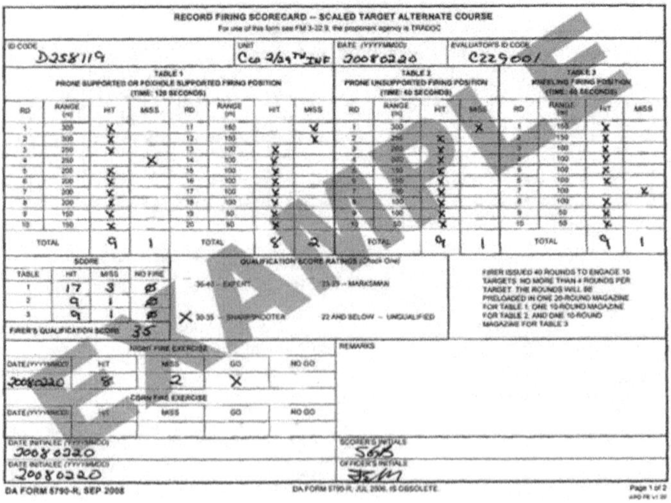

RECORD FIRING SCORECARD -- SCALED TARGET ALTERNATE COURSE

For use of this form see FM 3-22.9; the proponent agency is TRADOC

ID CODE	UNIT	DATE (YYYYMMDD)	EVALUATOR'S ID CODE
D258119	C Co 2/29th INF	20080220	C229001

TABLE 1 — PRONE SUPPORTED OR FOXHOLE SUPPORTED FIRING POSITION (TIME: 120 SECONDS)

RD	RANGE (m)	HIT	MISS	RD	RANGE (m)	HIT	MISS
1	300	X		11	150		X
2	300	X		12	150		X
3	250	X		13	100	X	
4	250		X	14	100	X	
5	200	X		15	100	X	
6	200	X		16	100	X	
7	200	X		17	100	X	
8	200	X		18	100	X	
9	150	X		19	50	X	
10	150	X		20	50	X	
TOTAL		9	1	TOTAL		8	2

TABLE 2 — PRONE UNSUPPORTED FIRING POSITION (TIME: 60 SECONDS); TABLE 3 — KNEELING FIRING POSITION (TIME: 60 SECONDS)

RD	RANGE (m)	HIT	MISS	RD	RANGE (m)	HIT	MISS
1	300		X	1	150	X	
2	250	X		2	150	X	
3	200	X		3	100	X	
4	200	X		4	100	X	
5	150	X		5	100	X	
6	150	X		6	100	X	
7	100	X		7	100		X
8	100	X		8	100	X	
9	100	X		9	50	X	
10	50	X		10	50	X	
TOTAL		9	1	TOTAL		9	1

SCORE

TABLE	HIT	MISS	NO FIRE
1	17	3	Ø
2	9	1	Ø
3	9	1	Ø

FIRER'S QUALIFICATION SCORE: 35

QUALIFICATION SCORE RATINGS (Check One): 36-40 -- EXPERT; 23-29 -- MARKSMAN; X 30-35 -- SHARPSHOOTER; 22 AND BELOW -- UNQUALIFIED

FIRER ISSUED 40 ROUNDS TO ENGAGE 10 TARGETS. NO MORE THAN 4 ROUNDS PER TARGET. THE ROUNDS WILL BE PRELOADED IN ONE 20-ROUND MAGAZINE FOR TABLE 1, ONE 10-ROUND MAGAZINE FOR TABLE 2, AND ONE 10-ROUND MAGAZINE FOR TABLE 3

NIGHT FIRE EXERCISE

DATE (YYYYMMDD)	HIT	MISS	GO	NO GO
20080220	8	2	X	

CBRN FIRE EXERCISE

DATE (YYYYMMDD)	HIT	MISS	GO	NO GO

REMARKS

DATE INITIALED (YYYYMMDD): 20080220 — SCORER'S INITIALS: SGB

DATE INITIALED (YYYYMMDD): 20080220 — OFFICER'S INITIALS: JEM

DA FORM 5790-R, SEP 2008 — DA FORM 5790-R, JUL 2006, IS OBSOLETE. — Page 1 of 2 APD PE v1.00

Figure B-6. Example of completed DA Form 5790-R (Record Fire Scorecard—Scaled Target Alternate Course).

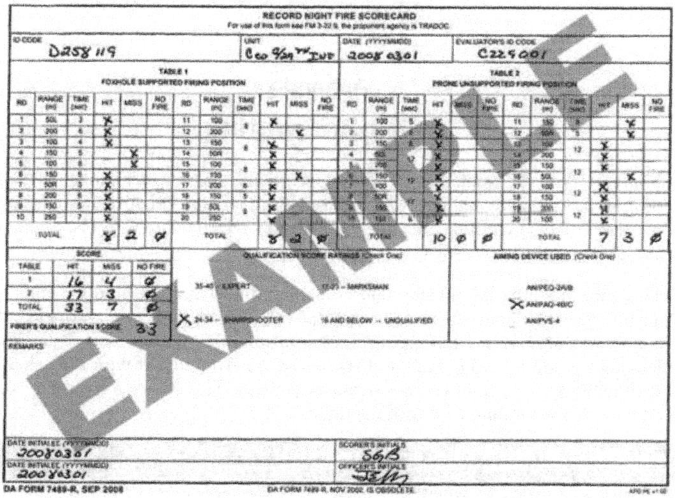

RECORD NIGHT FIRE SCORECARD
For use of this form see FM 3-22.9, the proponent agency is TRADOC

ID CODE	UNIT	DATE (YYYYMMDD)	EVALUATOR'S ID CODE
D258119	C Co 9/29 TH INF	20080301	C229001

TABLE 1 — FOXHOLE SUPPORTED FIRING POSITION

RD	RANGE (m)	TIME (sec)	HIT	MISS	NO FIRE	RD	RANGE (m)	TIME (sec)	HIT	MISS	NO FIRE
1	50L	3	X			11	100	8	X		
2	200	6	X			12	200			X	
3	100	4	X			13	150	8	X		
4	150	5		X		14	50R		X		
5	100	8		X		15	100	8	X		
6	150	5	X			16	150			X	
7	50R	3	X			17	200	6	X		
8	200	6	X			18	150	5	X		
9	150	5	X			19	50L	8	X		
10	250	7	X			20	250		X		
TOTAL			8	2	Ø	TOTAL			8	2	Ø

TABLE 2 — PRONE UNSUPPORTED FIRING POSITION

RD	RANGE (m)	TIME (sec)	HIT	MISS	NO FIRE	RD	RANGE (m)	TIME (sec)	HIT	MISS	NO FIRE
1	100	5	X			11	150	8		X	
2	200	[illegible]	X			12	50R	5		X	
3	150	[illegible]	X			13	[illegible]	12	X		
4	50L	12	X			14	200		X		
5	200		X			15	150	12	X		
6	150	12	X			16	50L			X	
7	100		X			17	100	12	X		
8	50R	12	X			18	150		X		
9	150		X			19	200	12	X		
10	150	6	X			20	100		X		
TOTAL			10	Ø	Ø	TOTAL			7	3	Ø

SCORE

TABLE	HIT	MISS	NO FIRE
1	16	4	Ø
2	17	3	Ø
TOTAL	33	7	Ø

FIRER'S QUALIFICATION SCORE 33

QUALIFICATION SCORE RATINGS (Check One)
36-40 – EXPERT
X 24-34 – SHARPSHOOTER
17-23 – MARKSMAN
16 AND BELOW – UNQUALIFIED

AIMING DEVICE USED (Check One)
AN/PEQ-2A/B
X AN/PAQ-4B/C
AN/PVS-4

REMARKS

DATE INITIALED (YYYYMMDD) 20080301 — SCORER'S INITIALS SGB
DATE INITIALED (YYYYMMDD) 20080301 — OFFICER'S INITIALS JEM

DA FORM 7489-R, SEP 2008 — DA FORM 7489-R, NOV 2002, IS OBSOLETE

Figure B-7. Example of completed DA Form 7489-R (Record Night Fire Scorecard).

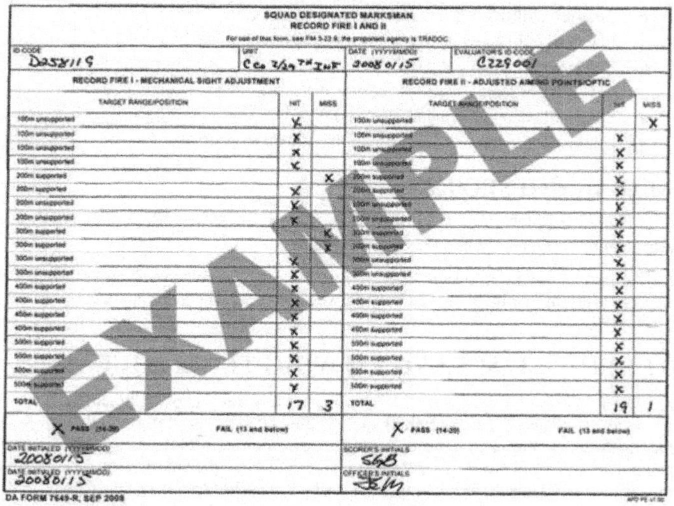

SQUAD DESIGNATED MARKSMAN RECORD FIRE I AND II
For use of this form, see FM 3-22.9, the proponent agency is TRADOC

ID CODE	UNIT	DATE (YYYYMMDD)	EVALUATOR'S ID CODE
D258119	C Co 2/29 TH INF	20080115	C229001

RECORD FIRE I - MECHANICAL SIGHT ADJUSTMENT: TARGET RANGE/POSITION	HIT	MISS	RECORD FIRE II - ADJUSTED AIMING POINTS/OPTIC: TARGET RANGE/POSITION	HIT	MISS
100m unsupported	X		100m unsupported		X
100m unsupported	X		100m unsupported	X	
100m unsupported	X		100m unsupported	X	
100m unsupported	X		100m unsupported	X	
200m supported		X	200m supported	X	
200m supported	X		200m supported	X	
200m unsupported	X		200m unsupported	X	
200m unsupported	X		200m unsupported	X	
300m supported		X	300m supported	X	
300m supported		X	300m supported	X	
300m unsupported	X		300m unsupported	X	
300m unsupported	X		300m unsupported	X	
400m supported	X		400m supported	X	
400m supported	X		400m supported	X	
400m supported	X		400m supported	X	
400m supported	X		400m supported	X	
500m supported	X		500m supported	X	
500m supported	X		500m supported	X	
500m supported	X		500m supported	X	
500m supported	X		500m supported	X	
TOTAL	17	3	TOTAL	19	1

X PASS (14-20) FAIL (13 and below) — X PASS (14-20) FAIL (13 and below)

DATE INITIALED (YYYYMMDD) 20080115 — SCORER'S INITIALS SGB
DATE INITIALED (YYYYMMDD) 20080115 — OFFICER'S INITIALS JEM

DA FORM 7649-R, SEP 2008

Figure B-8. Example of completed DA Form 7649-R (Squad Designated Marksman—Record Fire I and II).

Appendix C

Night Fighting

With the Army's emphasis on night operations, commanders must be sure that riflemen are lethal at night as well as during the day. That lethality depends largely on whether riflemen can fire effectively with today's technology: NVGs, aiming lights, and TWSs.

This appendix provides a better understanding of how eyes adapt to the night, as well as information about night devices. The information has been provided through continuous observation during operational testing, the Army Research Institute, and lessons learned by individual Soldiers across the Army. Be it with the naked eye or with night devices, Soldiers must learn to increase their ability not only to see better at night, but their ability to "own the night."

SECTION I. UNAIDED NIGHT VISION

Although operating at night has definite advantages, it is also difficult. Eyes do not work as well as they do during the day, yet they are crucial to effective performance; 80 percent of sensory input comes through them. Soldiers need to be aware of the constraints that their eyes place upon them at night. These constraints include—

- Reduced ability to see crisp and clear images (visual acuity).
- Inability to distinguish colors under certain conditions.
- Reduced depth perception.
- Difficulty in seeing objects at certain distances due to the night blind spot.
- Lost dark adaptation due to exposure to light.
- Confusion (your eyes may seem to play tricks on you).

NORMAL BLIND SPOTS

C-1. The normal blind spot is always present, day and night. It is caused by the lack of light receptors where the optic nerve inserts into the back of the eye.

C-2. The normal blind spot occurs when you use just one eye. When you close the other eye, objects 12 to 15 degrees away from where you are looking will disappear. When you uncover your eye, the objects will reappear.

NIGHT BLIND SPOTS AND VIEWING TECHNIQUES

C-3. When you stare at an object under starlight or lower levels of illumination, it can disappear or fade away. This is a result of the night blind spot. The night blind spot affects both eyes at the same time and occurs when using the central vision of both eyes. Consequently, when looking directly at an object, Soldiers miss larger objects as the distances increase. A hand grenade 2 meters away might not be seen; an enemy Soldier at 50 meters may be missed. An M1 tank at 300 meters can even be missed.

Avoiding Night Blind Spots

C-4. To avoid night blind spots—

- Look to all sides of objects you are trying to find or follow.
- **DO NOT STARE.**

C-5. Diamond viewing, a technique similar to the off-center vision technique taught in rifle marksmanship, is a good technique for peripheral viewing. To use the diamond viewing technique—

- Move your eyes just slightly, a few degrees, in a diamond pattern around the object you wish to see.
- Do not move your head; use your peripheral vision.

DARK ADAPTATION

C-6. In order for your visual system to work efficiently at night, you need to dark-adapt, or get your eyes used to seeing things under low light conditions. Dark adaptation is similar to walking into a movie theater when it's very dark; you can't see things at first, but as your eyes gradually adapt, you can see better.

C-7. It takes about 30 to 45 minutes to fully dark-adapt when going from a brightly lighted area into the dark. However, people dark-adapt at varying rates. People who are older, people who smoke, or people who are not in great physical shape will take longer to dark-adapt.

C-8. There are three stages of dark adaptation:

- Daylight vision.
- Twilight vision.
- Night vision.

Daylight Vision

C-9. Daylight vision occurs under maximum lighting conditions, such as when the sun is shining or in a well-lit room. Under these conditions, Soldiers have—

- Optimal visual acuity. In daylight conditions, Soldiers use both their central and peripheral vision, which provides optimal visual acuity—20/10, 20/15, and 20/20 vision.
- Optimal color vision. Colors look most vivid under daylight conditions.
- Quickest reaction time.

Twilight Vision

C-10. Twilight vision occurs during many military night operations and when driving around in a car at night. It occurs at dawn and dusk, down to full moonlight; when there is artificial illumination; and when snow is on the ground at night. It can occur in the daytime with several layers of jungle canopy. Under these conditions, Soldiers have—

- Poorer visual acuity. Visual acuity can be as poor as 20/100. Under twilight conditions, optimal visual acuity is between 20/50 and 20/100.
- Poorer color vision. Colors will not be as vivid.
- Slower reaction times.

NIGHT VISION

C-11. Night vision occurs under starlight, as well as on moonless and cloudy nights when there are no stars or cultural lighting. Remember, there is a night blind spot, as discussed earlier. Under these conditions, Soldiers have—

- The worst visual acuity—from 20/200 to 20/400 and possibly much worse. Soldiers can recognize silhouettes, but not details of the objects. This is why knowing the silhouettes of vehicles and critical natural and man-made objects is important.
- Poor color vision. Soldiers cannot see colors, only various shades of gray. The longer wavelengths of light, such as the reds and oranges, are hard to see and appear dark. Unless a dark color is bordered by two lighter colors, it becomes totally invisible. On the other hand, greens and blues appear brighter, but Soldiers may not be able to determine their color.

NOTE: Reds are almost invisible at night. Red crosses are on white backgrounds on tents or vehicles so they can be seen more easily at night.

PROTECTING (BEFORE OPERATION)

C-12. Soldiers must protect their eyes before night operations so they can dark-adapt in an efficient manner. To efficiently dark-adapt, use the following suggestions:

- Don't smoke before night operations. Not smoking four to six hours before night operations will aid in dark adaptation.
- Wear sunglasses if you plan to spend time in the sun. Without sunglasses, it will take longer to dark-adapt.
- Watch what you eat. Maintain adequate levels of Vitamin A.
- Use dim white lighting or red lighting before night operations.

PROTECTING (DURING OPERATION)

C-13. Once a Soldier has dark-adapted, it is important to maintain that dark adaptation. To maintain dark adaptation, use the following suggestions:

- Minimize your use of unnecessary lighting.
- Close one eye before being flashed by flares and other bright lights.

ILLUSION (APPARENT MOVEMENT OF LIGHT)

C-14. The illusion of movement, which a static light exhibits when stared at in the dark, is related to the loss of surrounding visual references that normally serve to stabilize visual perceptions. Consequently, very small eye movements are perceived by the brain as movement of the light. To avoid illusions of movement—

- Begin a scan pattern, and control the eye movement.
- Use large movements, and scan.
- Try to find another light, and shift your gaze back and forth between the lights.

SECTION II. AIDED NIGHT VISION

There are three devices available to dismounted Soldiers that will help increase their lethality at night: night observation devices (NODs), aiming lasers, and TWSs. Each provides the dismounted Soldier with different views of the IR spectrum. Before Soldiers can fully operate these devices, they must receive training on how the systems work within the IR range and the electromagnetic (light) spectrum. Soldiers must also know what constraints and advantages each piece of equipment provides so that they can determine when to employ each device.

ELECTROMAGNETIC (LIGHT) SPECTRUM

C-15. To understand why some night devices cannot be used in conjunction with others, Soldiers must understand the electromagnetic spectrum. The electromagnetic spectrum is a spectrum of energy (light) containing x-rays, gamma rays, radio waves, cosmic rays, and ultraviolet rays. Also within this spectrum is visible light, light visible with the naked eye. Just beyond red visible light is IR light. IR light is broken down into three different ranges: near IR, middle IR, and far IR.

C-16. Two types of night devices will increase the Soldier's vision into the IR range:

- Image-intensification devices.
- Thermal sight.

IMAGE-INTENSIFICATION DEVICES

C-17. Image-intensification devices rely on ambient light and energy within the near IR range, such as light emitted from natural and artificial sources (for example, moonlight or starlight). Image-intensification devices include the AN/PVS-4, PVS-5, PVS-7A/B/C/D, and PVS-14s. The Army also has aiming devices that emit near IR energy in a colliminated beam, such as the AN/PAQ-4B/C and the AN/PEQ-2A/B. Since image-intensification devices and aiming lasers work within the same range of near IR energy, they can be used in conjunction with each other.

THERMAL WEAPON SIGHT

C-18. In the past, thermal technology has been solely reserved for tanks, fighting vehicles, and antiarmor specialists (for example, TOW and Dragon gunners). These devices were bulky, heavy, and impractical for the dismounted Soldier. Now, the Army has a thermal device that can be mounted on a dismounted Soldier's weapon or handheld. The TWS operates within the middle/far IR range. It can detect IR light emitted from friction, combustion, or objects radiating natural thermal energy. Since the TWS and other thermal devices operate within the middle/far IR range, they cannot be used in conjunction with image intensifiers or other image-intensification devices.

IMAGE-INTENSIFICATION DEVICES

C-19. As the name implies, image-intensification devices are designed to amplify light. To be effective, some degree of light must be available. When light enters the image-intensification tube, the light releases electrons, which the tube accelerates until the light is much brighter. Under optimal conditions, second-generation devices, such as the PVS-5-series, intensify ambient light up to about 1,500 times. Third-generation devices, such as the PVS-7/14-series NODs, double that level of intensification.

ADJUSTMENTS

C-20. To acquire and engage a target at night, image-intensification devices must be adjusted properly.

NOTE: Soldiers will not be able to obtain the acuity level that they have during the day. Under optimal night conditions, a Soldier with 20/20 daytime vision can expect no better than 20/50 vision with second-generation NODs and 20/40 vision with third-generation NODs. To approach these levels of acuity, you must be able to adjust your NOD for optimal clarity.

C-21. To adjust the NOD—

(1) Mount the NOD. Mount the head mount or helmet mount IAW the appropriate TM. If using the helmet mount, ensure that the tilt is adjusted until you have a comfortable viewing angle. Use the nape strap to maintain proper acuity with the NODs.

NOTE: If the mounting bracket is permanently attached to the helmet, ensure that the nape strap rear bracket is also permanently attached. (See TM 11-5855-306-10 for more information about attaching the nape strap rear bracket.) Use of the nape strap will prevent the NOD's weight from pulling the helmet downward, causing the NOD to rest on the bridge of your nose. The nape strap allows for proper acuity of the sight and allows you to engage targets with more ease and accuracy.

(2) Set the eye-relief. Move the goggles so that the eyecups cover the eye, but not so close that the eyepiece touches your eyelashes or glasses.

(3) Turn the goggles on.

(4) Set the inner-pupillary distance (AN/PVS-7 series). Move each eyepiece until they are centered over each eye. Close one eye, and make adjustments until the eye that is open is viewing a complete circle and not an oval. Continue to make adjustments to the other eye.

(5) Adjust the diopter focus ring. Close one eye, and with the eye that is open, turn the diopter focus ring in one direction until the diopter is totally out of focus. Then, turn the diopter focus ring the opposite direction until the display is focused to your eye. Follow the same procedures for the other eye if using the AN/PVS-7 series. No further adjustments should have to be made to the diopter focus ring.

NOTE: Adjust the diopter focus ring before adjusting the objective focus ring. The diopter focus ring focuses the display lens to your eye, while the objective focus ring focuses the target. You cannot focus the sight to the target without your eye being focused to the display first.

(6) Adjust the objective focus ring. While looking at an object, turn the objective focus ring until the objective lens is out of focus and then slowly turn the objective focus ring in the opposite direction until the object becomes as clear as possible.

NOTE: Adjustments will have to be made for targets at different ranges using the objective focus ring.

(7) Adjust the variable gain control (AN/PVS-14 only). Turn the variable gain control to the point that both eyes are almost receiving the same amount of light.

NOTE: The AN/PVS-14 has a variable gain control that controls the illumination input to the eye. Keeping the variable gain turned up will cause your brain to form two separate images, one darker and one very bright. With the variable gain turned down to the point that both eyes are almost receiving the same amount of light, the brain will produce one image, making it seem like both eyes are looking through the same sight.

AIMING LASERS

C-22. AN/PAQ-4-series and the AN/PEQ-2A/B aiming lasers operate within the electromagnetic spectrum, specifically near the IR range, and are seen through image-intensification devices. Aiming lasers emit a highly colliminated beam of IR energy that allows for quick "point and shoot" capability at night. Even though aiming lasers provide a quick and easy means of engaging the enemy at night, special attention must be given to the following:

- Proper adjustments to the image intensifiers.
- 10-meter boresighting procedures or 25-meter zeroing procedures.
- Scanning.
- Walking.
- IR discipline.

SCANNING

C-23. NVDs have a 40-degree field of view, leaving the average firer to miss easy targets of opportunity, more commonly targets 50 meters left or right.

C-24. Soldiers must train to aggressively scan their sectors of fire for targets. Training must reinforce regular blinking during scanning, which relieves some of the eyestrain from spotting far targets. After Soldiers master the art of scanning, they find that targets are easier to detect by acknowledging the flicker or movement of a target.

WALKING

C-25. Once a Soldier has located a target, he must be aware of the placement of the aiming laser. If he activates his laser while it is pointing over the target and into the sky, he wastes valuable time trying to locate exactly where the laser is pointing and increases his chances of being detected and fired upon by the enemy. Walking the laser to the target is a quick and operationally secure means of engaging the enemy with the aiming laser. To walk the laser to a target—

(1) Aim the laser at the ground just in front of the target.

(2) Walk the aiming laser along the ground and up to the target's center of mass.

(3) Engage the target.

INFRARED DISCIPLINE

C-26. To exercise proper IR discipline—

- While on the range, actively scan for targets with the laser off.
- Once a target is located, walk the laser to the target, and engage.
- Once a target has been located and engaged with the aiming laser, deactivate the laser.

THERMAL WEAPON SIGHT

C-27. To understand how the TWS is able to convert energy into an image suitable for viewing, Soldiers must understand the electromagnetic spectrum and the range of IR light in which the TWS operates. The TWS absorbs all available light into the lens, and then filters out all light except for middle/far IR (thermal) light. Then, the TWS converts the thermal light into an image and creates a video that is displayed on the raster.

C-28. The TWS can convert thermal energy that is reflected, radiated, or generated from an object. All objects, such as trees, metal, plastic, and living creatures, display a quality that allows them to be seen with thermal technology. How well the objects display these qualities will determine how well they are seen.

Absorption

C-29. During the day, all inanimate objects absorb thermal energy from the sun to varying degrees. Metal objects have a much higher rate of absorption than wood, leaves, or grass; therefore, a metal object sitting in the sun will stand out more than the grass surrounding it when viewed through the TWS.

Exposure

C-30. The amount of time an object is exposed to thermal energy determines how well that object will be seen. Naturally, an object with a long exposure time will have absorbed more thermal energy than an object exposed to the same thermal energy for a shorter period of time.

Emissivity

C-31. Emissivity is the rate at which an object emits the thermal energy it has absorbed or generates. Usually, objects with a high absorption rate will have a high emissivity factor. Although the human body does not have a high absorption rate, it has a high emissivity factor due to the fact that it generates a high amount of thermal energy. An object that has a high emissivity factor will be much hotter, and, therefore, when seen through the TWS, much easier to see and recognize.

Reflection

C-32. Items such as glass and water have virtually no absorption rate. Instead, they reflect thermal energy, which makes it very difficult to see objects through glass and water.

C-33. Snow and ice have the same effect, especially during the day with no clouds present. Snow and ice reflect most of the thermal energy from the sun, so it is difficult to acquire a good thermal image on objects that are close to the ground.

Diurnal Cycle

C-34. There are two times during the day when motionless objects that do not generate their own thermal energy, such as trees, rocks, and man-made objects, become the same temperature as the surrounding air: once in the morning and once in the evening. This is known as the diurnal cycle. The specific times that this cycle will take effect are based on the time of year, but it usually occurs shortly after sunrise and shortly after sunset.

C-35. The diurnal cycle contains crossover points. During the day, a motionless object will absorb thermal energy from the sun; the crossover point is the time when that object stops absorbing thermal energy (day) and starts radiating thermal energy (night). As the night goes on, that same object will come to a point where it stops radiating thermal energy and will once again start absorbing thermal energy (day).

Adjustments

C-36. Rain, snow, fog, smoke, and the diurnal cycle are just a few environmental or combat situations that may affect your thermal image. These conditions may cause objects to be difficult to see, so adjustments must be made to the TWS in order to refine the thermal image.

C-37. To allow Soldiers to maximize the capability of the sight, the TWS is equipped with—

- A diopter focus ring.
- Two FOVs.
- An objective focus ring.
- A brightness knob.
- An auto and manual contrast switch.
- A polarity switch.

Diopter Focus Ring

C-38. When making adjustments to the sight, Soldiers begin with the diopter focus ring. The diopter focus ring focuses the display screen (raster) to the eye. To adjust the diopter focus ring, adjust the diopter focus ring until everything on the display screen is clear and easily read.

NOTE: This is best done with the objective lens cover closed.

C-39. Once you have adjusted the diopter focus to your eye, no other adjustments to the diopter focus ring should be necessary.

Fields of View

C-40. The TWS has two operating FOVs—wide and narrow. The Soldier should be allowed to select the FOV that suits him best. Through use, Soldiers learn to use the appropriate FOV under different combat situations.

NOTE: When selecting an FOV, make sure that the FOV ring is turned completely to the left or to the right. If the FOV ring is turned only halfway, you will not be able to see through the sight.

Wide Field of View

C-41. The WFOV has the least magnification, but a greater FOV and is great for scanning.

Narrow Field of View

C-42. The NFOV has greater magnification, but less degrees of FOV.

Objective Focus Ring

C-43. The objective focus ring focuses the sight to the target. Adjustments to the objective focus ring are based on the range of the object being viewed.

NOTES:
1. Over-adjustment to the objective focus ring will lock the FOV ring to the point that the FOV cannot be changed.
2. Make adjustments to the objective focus ring only after focusing the diopter focus ring.

Brightness Knob

C-44. The dual-function brightness knob is used to turn on the TWS, adjust the brightness of the raster, and refine the thermal image. Used in conjunction with the contrast knob, it helps to combat the effects of the diurnal cycle and other conditions that might require fine-tuned adjustment to the thermal image.

Contrast Switch

C-45. The dual-function contrast switch has automatic and manual contrast modes. Used in conjunction with the brightness knob, the contrast switch allows Soldiers to obtain the best possible thermal image.

Automatic Contrast Mode

C-46. The automatic contrast mode is used under normal operating conditions.

Manual Contrast Mode

C-47. The manual contrast is used under conditions other than normal (for example, during 10-meter boresighting or 25-meter zeroing; during rain, fog, smoke, or snow; during the diurnal cycle; or when trying to obtain as much detail of a target as possible).

Polarity Switch

C-48. The polarity switch has two modes: white hot or black hot. Use of the polarity switch is a user's preference. Through continued use, Soldiers decide which polarity setting works best under different combat or environmental conditions.

White Hot Mode

C-49. When in the white hot mode, hotter objects appear white, while cooler objects have shades of gray to black.

Black Hot Mode

C-50. When using the black hot mode, hotter objects appear black, while the cooler objects are shades of gray to white.

Appendix D

Range Safety and Risk Management

All personnel training on a rifle range should be briefed on the safety and local requirements for that range. This briefing fulfills the minimum requirements for a rifle range safety briefing. Information may be added to conform to local requirements and safety regulations. ARs 385-10 and 385-63 and DA Pam 385-63 should be reviewed by all range personnel before operating any range.

RECOMMENDED BRIEFING

D-1. The first priority on any range is training, but safety must be at the forefront of the training program. The safety program prescribes the safety precautions necessary to minimize accidents when firing and using ammunition during training and range operations. The safety program should identify—

- Surface danger zones (SDZs), as described in AR 385-63.
- The location of medical personnel.
- Left and right limits of the range. Firers never fire outside of these limits.
- The designated smoking area (if applicable).

D-2. Further, the safety program should include the following information:

- Inspect for objects located near the weapon's muzzle before firing, especially during unassisted night fire.
- When not on the firing line, ensure that the weapon's selector lever is on SAFE and the bolt is locked to the rear.
- Ensure that firers enter and exit the firing line at the entry or exit point.
- Before occupying a firing position, inspect it for wildlife or obstructions.
- Always keep the weapon's muzzle pointed downrange when on the firing line, and keep the finger outside of the trigger housing area.
- Never touch a weapon while personnel are downrange or in front of the firing line.
- Load the weapon only on command from the tower or control point.
- Never fire without using hearing protection.
- Ensure that left-handed firers attach left-handed brass deflectors to their weapon before firing.
- When entering or exiting the firing line, clear the weapon with a cleaning rod.
- Consider the rifle loaded at all times, even in break areas. Never point the weapon at anyone.
- Anyone observing an unsafe act will immediately call "CEASE FIRE," place the weapon on SAFE, place it in the V-notched stake or lay it on the sandbags, and give the verbal and visual command of cease fire.
- Once cleared off of the firing line, firers report to the ammunition point and turn in all brass and ammunition.
- No one will leave the range until they have been inspected for live ammunition and brass.
- In case of an electrical storm, personnel will be directed to lock and clear all weapons, ground their equipment (except wet weather gear), and disperse into a predetermined area.
- Eating and drinking are not permitted on the firing line unless the tower operator permits drinking from the canteen.

NOTE: Drink water often to prevent heat injuries.

PERSONNEL AND DUTIES

D-3. To provide a safe and efficient range operation and effective instruction, certain duties may be required of personnel. The personnel may include—

- OIC.
- RSO.
- NCOIC.
- Ammunition detail.
- Unit armorer.
- Assistant instructor.
- Medical personnel.
- Control tower operators.
- Maintenance detail.

OFFICER IN CHARGE

D-4. The OIC is responsible for the overall operation of the range before, during, and after live-firing.

RANGE SAFETY OFFICER

D-5. The RSO—

- Is responsible for the safe operation of the range.
- Conducts a safety orientation before each scheduled LFX.
- Ensures that a brass and ammunition check is made before the unit leaves the range.
- Ensures that all personnel comply with the safety regulations and procedures prescribed for the conduct of an LFX.
- Ensures that all left-handed firers use left-handed firing devices.

NOTE: This officer should not be assigned duties other than those of the safety officer.

NONCOMMISSIONED OFFICER IN CHARGE

D-6. The NCOIC assists the OIC and safety officer by performing duties as required; for example, he might supervise enlisted personnel who are supporting the LFX.

AMMUNITION DETAIL

D-7. This detail is composed of one or more ammunition handlers. The ammunition detail—

- Breaks down, issues, receives, accounts for, and safeguards live ammunition.
- Collects expended ammunition casings and other residue.

UNIT ARMORER

D-8. The unit armorer—

- Repairs rifles.
- Replaces parts.

ASSISTANT INSTRUCTOR

D-9. One assistant instructor is assigned for each one to ten firing points. Each assistant instructor—

- Ensures that all firers observe safety regulations and procedures.
- Assists firers having problems.

MEDICAL PERSONNEL

D-10. Medical personnel provide medical support, as required by regulations governing LFXs.

CONTROL TOWER OPERATORS

D-11. Control tower operators—

- Raise and lower the targets.
- Time the exposures.
- Sound the audible signal.
- Give the fire commands.

NOTE: If possible, two men should be chosen to perform these functions.

MAINTENANCE DETAIL

D-12. This detail should be composed of two segments: one to conduct small arms repair and one to perform minor maintenance on the target-holding mechanisms.

AMMUNITION POSITIONING AND ISSUANCE

D-13. To provide a safe and operational range, the following are recommended procedures for handling ammunition:

- Locate all ammunition for the weapons involved at firing sites outside of the backblast area (when applicable). Store ammunition at a position that minimizes the potential for ignition, explosion, or rapid burning.
- Issue ammunition to firing units immediately before scheduled training exercises. Distribute small arms ammunition to troops only when they are on the ready line or firing line.
- Cover all ammunition to protect it from the elements and direct rays of the sun. For proper ventilation, provide air circulation between the ammunition and cover.
- Limit the unpacking of ammunition at the firing line to the minimum number of rounds needed. Retain packaging material until firing is complete. Units do not burn wooden containers or indiscriminately fire ammunition to preclude return to a storage site.

COMPOSITE RISK MANAGEMENT

D-14. Composite risk management (CRM) is a decision-making process used to mitigate risks associated with all hazards that have the potential to injure or kill personnel, damage or destroy equipment, or otherwise impact mission effectiveness. The guiding principles of CRM are as follows:

- Integrate CRM into all phases of the mission and operations.
- Make risk decisions at the appropriate level.
- Accept no unnecessary risk.
- Apply the process cyclically and continuously.
- Do not be risk averse.

D-15. CRM is a five-step process:

(1) Identify hazards.
(2) Assess hazards to determine risk.
(3) Develop controls, and make risk decisions.
(4) Implement controls.
(5) Supervise and evaluate.

NOTE: Steps 1 and 2 are assessment steps; Steps 3 through 5 are management steps.

Step 1–Identify Hazards

D-16. Hazards may arise from any number of areas. They can be associated with enemy activity, accident potential, weather or environmental conditions, health, sanitation, behavior, and/or materiel or equipment. Mission, enemy, terrain and weather, troops and support available, time available, civil considerations (METT-TC) factors serve as a standard format for identifying hazards, on- or off-duty.

D-17. CRM does not differentiate between the sources of the hazard. The loss of personnel, equipment, or materiel due to any hazard has the same disruptive impact on readiness or mission capabilities, no matter what the source.

Step 2–Assess Hazards to Determine Risk

D-18. During mission analysis, course of action (COA) development, or analysis, rehearsal, and execution steps of the military decision-making process (MDMP), commanders assess hazards and assign risk in terms of probability and severity of adverse impact. During their assessment, they must consider both mission- and non-mission-related aspects that may have an impact. The end result of this assessment is an initial estimate of risk for each identified hazard as determined from the standardized application of the risk assessment matrix.

D-19. There are three substeps in this step:

(1) Assess the probability of an event or occurrence.

(2) Estimate the expected result or severity of an event or occurrence.

(3) Determine the level of risk using the standard risk assessment matrix.

Assess the Probability of an Event or Occurrence

D-20. Probability is the likelihood an event will occur based on prior experience. The probability levels estimated for each hazard are based on the mission, COA, or frequency of a similar event. There are five levels of probability (Table D-1):

- Frequent.
- Likely.
- Occasional.
- Seldom.
- Unlikely.

Table D-1. Five levels of probability.

LEVEL OF PROBABILITY	EXPLANATION
Frequent	Occurs very often, known to happen regularly.
Likely	Occurs several times, a common occurrence.
Occasional	Occurs sporadically, but is not common.
Seldom	Remotely possible, could occur at some time.
Unlikely	Can assume will not occur, but not impossible.

Estimate the Expected Result or Severity of an Event or Occurrence

D-21. Severity is the degree to which an incident will impact combat power, mission capability, or readiness. The degree of severity estimated for each hazard is based on the results of similar events. Severity is addressed in the four levels used on the risk assessment worksheet (Table D-2):

- Catastrophic.
- Critical.
- Marginal.
- Negligible.

Table D-2. Four levels of severity.

LEVEL OF SEVERITY	EXPLANATION
Catastrophic	• Complete mission failure or loss of ability to accomplish a mission. • Death or permanent total disability. • Loss of major or mission-critical systems or equipment. • Major property or facility damage. • Severe environmental damage. • Mission-critical security failure. • Unacceptable collateral damage.
Critical	• Severely degraded mission capability or unit readiness. • Permanent partial disability or temporary total disability exceeding three months time. • Extensive major damage to equipment or systems. • Significant damage to property or the environment. • Security failure. • Significant collateral damage.
Marginal	• Degraded mission capability or unit readiness. • Minor damage to equipment or systems, property, or the environment. • Lost days due to injury or illness, not exceeding three months. • Minor damage to property or the environment.
Negligible	• Little or no adverse impact on mission capability. • First aid or minor medical treatment. • Slight equipment or system damage, but fully functional or serviceable. • Little or no property or environmental damage.

Determine the Level of Risk Using the Standard Risk Assessment Matrix

D-22. Using the standard risk assessment matrix (Table D-3), commanders convert probability and severity for each identified hazard into a specified level of risk. All accepted residual risk must be approved at the appropriate level of command.

NOTE: This assessment is an estimate, not an absolute. It may or may not be indicative of the relative danger of a given operation, activity, or event.

Table D-3. Risk assessment matrix.

RISK ASSESSMENT MATRIX						
Severity		Probability				
		Frequent A	Likely B	Occasional C	Seldom D	Unlikely E
Catastrophic	I	E	E	H	H	M
Critical	II	E	H	H	M	L
Marginal	III	H	M	M	L	L
Negligible	IV	M	L	L	L	L
E – Extremely High			H – High	M – Moderate	L – Low	

D-23. Risk is addressed in the four levels listed in the lower left corner of the matrix (Table D-4):

- Extremely high risk.
- High risk.
- Moderate risk.
- Low risk.

Table D-4. Four levels of risk.

LEVEL OF RISK	EXPLANATION
Extremely High Risk	Loss of ability to accomplish the mission if hazards occur. In the example noted in Table D-3, a frequent or likely probability of catastrophic loss (IA or IB) or a frequent probability of critical loss (IIA) exists. This implies that the risk associated with this mission, activity, or event may have severe consequences. The decision to continue must be weighed carefully against the potential gain to be achieved by continuing this COA.
High Risk	Significant degradation of mission capabilities (in terms of the required mission standard), inability to accomplish all parts of the mission, or inability to complete the mission to standard will result if hazards occur during the mission. Occasional to seldom probability of catastrophic loss (IC or ID) exists. A likely to occasional probability of a critical loss (IIB or IIC) exists. Frequent probability of marginal losses (IIIA) exists. This implies that if a hazardous event occurs, serious consequences will occur. The decision to continue must be weighed carefully against the potential gain to be achieved by continuing this COA.
Moderate Risk	Expected degraded mission capabilities (in terms of the required mission standard) will result if hazards occur during the mission. An unlikely probability of catastrophic loss (IE) exists. The probability of a critical loss is seldom (IID). Marginal losses occur with a likely or occasional probability (IIIB or IIIC). A frequent probability of negligible (IVA) losses exists.
Low Risk	Expected losses have little or no impact on accomplishing the mission. The probability of critical loss is unlikely (IIE), while that of marginal loss is seldom (IIID) or unlikely (IIIE). The probability of a negligible loss is likely or less (IVB through (IVE). Expected losses have little or no impact on accomplishing the mission. Injury, damage, or illness are not expected, or may be minor and have no long-term impact or effect.

Step 3 – Develop Controls, and Make Risk Decisions

D-24. In this step, commanders develop and apply controls, reassess the hazard to determine a residual risk, and make risk decisions. This process continues until an acceptable level of risk is achieved or until all risks are reduced to a level where benefits outweigh the potential cost. This step is accomplished during the COA development, COA analysis, COA comparison, and COA approval of the MDMP.

Step 4 – Implement Controls

D-25. Leaders and staffs ensure that controls are integrated into SOPs, written and verbal orders, mission briefings, and staff estimates. The critical check for this step is to ensure that controls are converted into clear and simple execution orders. This step includes coordination and communication with—

- Appropriate superior, adjacent, and subordinate units, organizations, and individuals.
- Logistics Civil Augmentation Program (LOGCAP) organizations and civilian agencies that are part of the force or may be impacted by the activity, hazard, or its control.
- The media and nongovernmental organizations (NGO) when their presence impacts or is impacted by the force.

Step 5 – Supervise and Evaluate

D-26. This step involves implementing risk controls and enforcing them to standard, and validating the adequacy of selected control measures in supporting the objectives and desired outcomes. This continuous process provides the ability to identify weaknesses and to change or adjust controls based on performance, changing situations, conditions, or events.

RESPONSIBILITIES

D-27. CRM responsibilities are spread across three levels:

- Commander.
- Leaders.
- Individual.

COMMANDER

D-28. During implementation of the CRM process, the commander—

- Ensures that warfighting functions (WFF) are performed to standard to minimize human error, materiel failure, and environmental effects.
- Establishes a force protection policy and publishes a safety philosophy with realistic safety goals, objectives, and priorities.
- Ensures that his training assessment considers the WFF's ability to protect the force. Selects long-, short-, and near-term control actions and ensures implementation to improve force protection.
- Ensures that his staff integrates risk management into the planning and execution of training and operational missions.
- Makes risk decisions. Selects, monitors, and enforces implementation of controls for hazards most likely to result in loss of combat power. After implementing controls, if risk remains above the tolerance level established by higher command, he must elevate the risk decision to the appropriate command level.
- Ensures that the CRM process is evaluated during all AARs.
- Determines if unit performance meets force protection guidance. Determines effectiveness of hazard controls and necessary changes to guidance and controls. Ensures that these changes are fed back into the training management cycle and guidance for operational missions, including unit SOPs.

LEADERS

D-29. During implementation of the CRM process, leaders—

- Enforce METL task performance to standard. Adopt the crawl—walk—run approach in planning and executing training.
- Make use of automated on- and off-duty CRM tools and surveys available from the United States Army Combat Readiness Center (USACRC).
- Execute risk reduction controls selected by the commander by developing and implementing supporting leader level controls. Apply risk management procedures to each.

INDIVIDUAL

D-30. All Soldiers must understand how to use the CRM process to enhance mission success and to reduce or eliminate loss.

D-31. During implementation of the CRM process, Soldiers—

- Support commanders and leaders in the rapid identification and communication of hazards and associated risks that may impact on the mission.
- Provide immediate feedback to the leader as the mission progresses and hazards are encountered. Use short written messages, hand and arm signals, or radio transmissions to communicate first-hand information to leaders.
- In extreme situations, act alone or make risk decisions within the context of orders.

DA FORM 7566 (COMPOSITE RISK MANAGEMENT WORKSHEET)

D-32. DA Form 7566 (Composite Risk Management Worksheet, Figure D-1a and D-1b) provides a starting point to logically track the CRM process. It can be used to document risk management steps taken during planning, preparation, and execution of training and combat missions.

NOTE: For detailed CRM procedures, see FM 5-19.

COMPOSITE RISK MANAGEMENT WORKSHEET

For use of this form, see FM 5-19; the proponent agency is TRADOC.

1. MSN/TASK	2a. DTG BEGIN	2b. DTG END	3. DATE PREPARED *(YYYYMMDD)*

4. PREPARED BY

a. LAST NAME	b. RANK	c. POSITION

5. SUBTASK	6. HAZARDS	7. INITIAL RISK LEVEL	8. CONTROLS	9. RESIDUAL RISK LEVEL	10. HOW TO IMPLEMENT	11. HOW TO SUPERVISE (WHO)	12. WAS CONTROL EFFEC-TIVE?

Additional space for entries in Items 5 through 11 is provided on Page 2.

13. OVERALL RISK LEVEL AFTER CONTROLS ARE IMPLEMENTED *(Check one)*

☐ LOW ☐ MODERATE ☐ HIGH ☐ EXTREMELY HIGH

14. RISK DECISION AUTHORITY

a. LAST NAME	b. RANK	c. DUTY POSITION	d. SIGNATURE

DA FORM 7566, APR 2005

Page 1 of 2

APD V3.00

EXAMPLE

Figure D-1a. Sample DA Form 7566 (Composite Risk Management Worksheet).

ITEMS 5 THROUGH 12 CONTINUED:

5. SUBTASK	6. HAZARDS	7. INITIAL RISK LEVEL	8. CONTROLS	9. RESIDUAL RISK LEVEL	10. HOW TO IMPLEMENT	11. HOW TO SUPERVISE (WHO)	12. WAS CONTROL EFFEC-TIVE?

EXAMPLE

DA FORM 7566, APR 2005

Page 2 of 2
APD V2.00

Figure D-1b. Sample DA Form 7566 (Composite Risk Management Worksheet) (page 2).

D-33. See Table D-5 for instructions on completing DA Form 7566.

Table D-5. Worksheet instructions

ITEM	INSTRUCTION
1 through 4	Self-explanatory.
5	Subtask relating to the mission or task in Block 1.
6	Hazards–Identify hazards by reviewing METT-TC factors for the mission or task. Additional factors include historical lessons learned, experience, judgment, equipment characteristics and warnings, and environmental considerations.
7	Initial Risk Level–Factors include historical lessons learned, intuitive analyses, experience, judgment, equipment characteristics and warnings, and environmental considerations. Determine the initial risk for each hazard by applying the risk assessment matrix. Enter the risk level for each hazard.
8	Controls– For each hazard, develop one or more controls that will eliminate the hazard or reduce the risk of a hazardous incident. Specify who, what, where, why, when, and how for each control. Enter controls.
9	Residual Risk Level–Determine the residual risk for each hazard by applying the risk assessment matrix. Enter the residual risk level for each hazard.
10	How to Implement–Decide how each control will be put into effect or communicated to the personnel who will make it happen (written or verbal instruction; tactical, safety, garrison SOPs; rehearsals). Enter controls.
11	How to Supervise (Who)–Plan how each control will be monitored for implementation (continuous supervision, spot-checks) and reassess hazards as the situation changes. Determine if the controls worked and if they can be improved. Pass on lessons learned.
12	Was Control Effective–Indicate "Yes" or "No." Review during AAR.
13	Overall Risk Level–Select the highest residual risk level and circle it. This becomes the overall mission or task risk level. The commander decides whether the controls are sufficient to accept the level of residual risk. If the risk is too great to continue the mission or task, the commander directs development of additional controls or modifies, changes, or rejects the COA.
14	Risk Decision Authority–Signed by the appropriate level of command.

Appendix E

Range Procedures and Range Operations Checklist

This appendix contains the procedures used to conduct live-fire training exercises. These procedures support Army regulations (TCs 7-9 and 25-1), local range regulations, and established unit training SOPs. Conduct of the training exercise should clearly define and establish details and equipment needed to open and operate the range so that these actions do not adversely impact Soldiers' training time. The procedures outlined in this appendix should be followed in order to open the range and conduct effective training.

RANGE PROCEDURES

E-1. Before beginning an LFX, all personnel must receive an orientation on range operations. The orientation should outline the procedures for conducting the exercise, to include the duties of the nonfiring orders.

PERSONNEL AND DUTIES

D-2. To provide a safe and efficient range operation and effective instruction, certain duties may be required of personnel. The personnel may include—

- OIC.
- RSO.
- NCOIC.
- Ammunition detail.
- Unit armorer.
- Assistant instructor.
- Medical personnel.
- Control tower operators.
- Maintenance detail.

OFFICER IN CHARGE

D-3. The OIC is responsible for the overall operation of the range before, during, and after live-firing.

RANGE SAFETY OFFICER

D-4. The RSO—

- Is responsible for the safe operation of the range.
- Conducts a safety orientation before each scheduled LFX.
- Ensures that a brass and ammunition check is made before the unit leaves the range.
- Ensures that all personnel comply with the safety regulations and procedures prescribed for the conduct of an LFX.
- Ensures that all left-handed firers use left-handed firing devices.

NOTE: This officer should not be assigned duties other than those of the safety officer.

Noncommissioned Officer In Charge

D-5. The NCOIC assists the OIC and safety officer by performing duties as required; for example, he might supervise enlisted personnel who are supporting the LFX.

Ammunition Detail

D-6. This detail is composed of one or more ammunition handlers. The ammunition detail—

- Breaks down, issues, receives, accounts for, and safeguards live ammunition.
- Collects expended ammunition casings and other residue.

Unit Armorer

D-7. The unit armorer—

- Repairs rifles.
- Replaces parts.

Assistant Instructor

D-8. One assistant instructor is assigned for each one to ten firing points. Each assistant instructor—

- Ensures that all firers observe safety regulations and procedures.
- Assists firers having problems.

Medical Personnel

D-9. Medical personnel provide medical support, as required by regulations governing LFXs.

Control Tower Operators

D-10. Control tower operators—

- Raise and lower the targets.
- Time the exposures.
- Sound the audible signal.
- Give the fire commands.

NOTE: If possible, two men should be chosen to perform these functions.

Maintenance Detail

D-11. This detail should be composed of two segments: one to conduct small arms repair and one to perform minor maintenance on the target-holding mechanisms.

FIRING ORDER LINEUP

E-12. After the range cadre have given the safety and range briefings, they assemble the Soldiers in firing orders in correlation with the number of firing points on that range. After the firing order has been determined, firers have their weapons rodded, move to the firing line, and proceed to their assigned firing points, keeping their weapons pointed up and downrange at all times.

TOWER COMMANDS

E-13. Simple, standard fire commands are needed to avoid confusion and misunderstanding during LFXs.

GENERAL COMMANDS

E-14. The following general commands may be altered when necessary:

- "Firers, assume the ________ position."
- "Issue the firer ________ rounds of ammunition."
- "Coach, secure ________ rounds of ammunition."
- "Lock one round. Load."
- "Ready on the right?"
- "Ready on the left?"
- "Ready on the firing line?"
- "Commence firing when your targets appear."
- "Cease firing. Lock and clear all weapons."

Grouping Commands

E-15. Grouping commands include the following:

- "Firers, assume a good supported prone position."
- "Lock one of three single rounds. Load."
- "Ready on the right?"
- "Ready on the left?"
- "The firing line is ready."
- "Place your selector lever on SEMIAUTOMATIC."
- "Commence firing."
- "Cease fire. Lock and clear your weapons."
- "Clear on the right?"
- "Clear on the left?"
- "The firing line is clear."
- "Move down to your targets, and triangulate your shot group."
- "After all personnel have triangulated their targets, move back to the firing line."
- "At this time, make adjustments to your sights."
- "Repeat all firing commands until grouping standards are met."

ZERO COMMANDS

E-16. Tower commands are the same as grouping commands. Repeat all firing commands until zeroing standards are met.

Field Firing Exercises

E-17. Simple, standard fire commands are needed to avoid confusion during field firing exercises. Commands for exercises from stationary positions are as follows:

- "Firers, assume a good ________ position."
- "Lock one magazine of ________ rounds. Load."
- "Ready on the right?"
- "Ready on the left?"
- "The firing line is ready."
- "Place your selector lever on SEMIAUTOMATIC."
- "Scan your sector."
- "Cease fire. Lock and clear your weapon." (Place the selector lever in the SAFE position.)

E-18. The range officer relays his commands to the pit NCOIC by radio or telephone so he can keep abreast of the conduct of fire. Before each firing exercise, the range officer informs the pit NCOIC of the next exercise and any special instructions for target operation (for example, "The next firing will be for zero. Mark targets after each three-round shot group," or for slow fire, "The next firing will be ten rounds. Slow fire. Mark targets after each shot.").

E-19. Radio telephone operators (RTOs) relay commands to the pit and pass on special instructions to target operators as requested by assistant instructors.

NOTES: 1. Radio telephone operators never identify a firer on a particular firing point.

2. The command "MARK TARGET NUMBER ________" indicates that the target has been fired upon but has not withdrawn for marking.

PRACTICE RECORD FIRE AND RECORD FIRE

E-20. Simple, standard fire commands are needed to avoid confusion and misunderstanding during practice record fire and record fire.

Practice Record Fire

E-21. Practice record fire commands include the following:

- "Firers, assume a good supported (prone unsupported) position."
- "Scorers, point out the limits of your lane."
- "Firers, lock your first magazine. Load."
- "Scan your sector."
- "Cease fire."
- "Lock and clear all weapons."

Record Fire

E-22. Record fire commands include the following:

- "Firers, assume a good supported (prone unsupported) position."
- "Scorers, point out the limits of your sector."
- "Firers, lock your first 20-round magazine. Load."
- "Scan your sector."
- "Cease fire."
- "Lock and clear all weapons."

Rapid Fire Exercises

E-23. The following commands are used for rapid fire exercises:

- "Lock and clear all weapons."
- "Clear on the right?"
- "Clear on the left?"
- "The firing line is clear."
- "Firers, assume the ________ position."
- "Assistants, secure two magazines of five rounds each." ("Issue the firer one magazine of five rounds.")
- "Lock one magazine. Load."
- "Ready on the right?"
- "Ready on the left?"
- "Ready on the firing line?"
- "Watch your targets." ("Firers, assume the appropriate firing position and commence firing when the targets are presented.")

E-24. When all of the targets are withdrawn, the range officer checks for slow firers or malfunctions and then allows them to fire.

E-25. The pit NCOIC organizes, orients, and provides safety for the pit detail.

E-26. The success of KD firing depends on efficient operation of the targets and the close coordination between the pit NCOIC and range officer. All operators must know the proper procedure for operating and marking the target.

Marking Targets for Zeroing and Slow Fire

E-27. Targets are marked quickly after each shot or group of shots without command. During slow fire, the firer has a time limit of one minute for each shot. Twenty seconds is considered the maximum time limit for marking. A marker (spotter) is placed in the hit regardless of its location on the target. Each time the target is marked, the marker is removed from the previous hit, and the hole is pasted.

NOTE: 3-inch markers are used for 100, 200, and 300 meters; 5-inch markers are used for 500 meters.

Using Disk Markers

E-28. The target markers are painted black on one side and white on the opposite side. They are available in three dimensions:

- 1 1/2 inches (NSN 6920-00789-0864).
- 3 inches (NSN 6920-00-713-8255).
- 5 inches (NSN 6920-00-713-8254).

E-29. The disk spindle (NSN 6920-00-7138257) may also be procured through the supply channels.

RANGE OPERATIONS CHECKLIST

E-30. This checklist consists of nine sections, each covering a topic relating to range operations:

- Mission analysis.
- Double check.
- Become an expert.
- Determine requirements.
- Determine available resources.
- Foolproof.
- Occupy the range and conduct training.
- Close the range.
- KD range.

NOTE: The checklist should be modified to include local policy changes to regulations or SOPs.

E-31. The person responsible for the training must answer the questions in each section. To answer these questions, he should use the following procedures:

- Ask each question in order.
- Record a "Yes" answer by checking the GO column.
- Record a "No" or "Don't know" answer by checking the NO GO column.

- Refer to the checklist to find the GO and NO GO columns.
- After all of the questions in a section are answered, analyze the NO GOs.
  - Contact the people who reported them and ask if they have corrected each problem. If so, change the answer to GO.
  - If any NO GO remains, analyze it and implement a countermeasure for the shortfall. Afterward, check to ensure that the countermeasures work.
  - Before range operations start, be sure that a workable countermeasure is implemented for each safety hazard presented by a NO GO answer.

MISSION ANALYSIS

E-32. Mission analysis includes identifying the following:

- Who will be firing on the range? ________
- Number of personnel________ Units________
- What weapons and course will be used? Weapons________ Course________
- Where will the training be conducted? Range________
- When is the range scheduled for operations? Date________ Opens________ Closes________

DOUBLE CHECK

	GO	NO GO	REMARKS
1. Has sufficient ammunition been requested for the number of personnel?			
2. Are the range facilities adequate for the type of training to be conducted?			
3. Has enough time been scheduled to complete the training?			
4. Have conflicts that surfaced been resolved?			

BECOME AN EXPERT

	GO	NO GO	REMARKS
1. Review TMs and FMs on the weapons to be fired.			
2. Talk with the armorer and other personnel experienced with the weapons to be fired.			
3. Review AR 385-63.			
4. Visit range control and read the installation's range instructions.			
5. Reconnoiter the range (preferably while it is in use).			
6. Check ARTEPs to see if training tasks can be integrated into the range training plan.			

DETERMINE REQUIREMENTS

E-33. Determining the requirements involves identifying the equipment and personnel necessary to conduct the training.

Personnel

E-34. The following personnel are required for range operations:

- OIC.
- Safety officer.
- Assistant safety officer.
- NCOIC.
- Ammunition NCO.

- Ammunition personnel (determined by type of range).
- Target detail and target operators.
- Tower operator.
- Concurrent training instructors.
- Assistant instructors.
- Radio telephone operators.
- Guards (range requirements).
- Medic(s).
- Air guard.
- Armorer.
- Truck driver (range personnel and equipment).
- Mechanic for vehicles.

E-35. Further, the following must be checked:

- Have you overstaffed your range?

Equipment

E-36. The following equipment is required for range operations:

- Range packet and clearance form.
- Safety fan and diagram (if applicable).
- Other safety equipment (for example, aiming circle, compass).
- Publications pertaining to the training that will be conducted.
- Lesson plans, status reports, and reporting folder.
- Range flag and light (night firing).
- Radios.
- Field telephone and wire.
- 292 antenna (if necessary).
- PA set with backup bullhorn(s).
- Concurrent training markers.
- Training aids for concurrent training stations.
- Sandbags.
- Tentage (briefing tent, warm-up tent).
- Space heaters (if needed).
- Colored helmets for control personnel.
- Safety paddles and vehicle flag sets or lights.
- Ambulance or designated vehicle.
- Earplugs.
- Water for drinking and cleaning.
- Scorecards.
- Master score sheet.
- Armorer's tools and cleaning equipment for weapons.
- Brooms, shovels, and other cleaning supplies and equipment.
- Tables and chairs (if needed).
- Target accessories.
- Fire extinguishers.
- Tarp, stakes, and rope to cover the ammunition.
- Toilet paper.
- Spare weapons and repair parts (as needed).

- Tow bar and slave cables for vehicles.
- Fuel and oil for vehicles and target mechanisms.

DETERMINE AVAILABLE RESOURCES

E-37. Determining available resources involves the following:

- Fill personnel spaces.
- Keep unit integrity.
- Utilize NCOs.
- Coordinate with supporting organizations:
  - Ammunition.
  - Transportation.
  - Training aids.
  - Medics.
  - Weapons.
  - Other equipment.

FOOLPROOF

E-38. To foolproof range operations—

- Write an overall lesson plan for the range.
- Organize a plan for firing:
  - Determine the range organization.
  - Outline courses of fire to be used.
  - Have fire commands typed for use on the range.
  - Set the rotation of stations.
- Rehearse concurrent training instructors and assistants.
- Brief the RTO on unique range control radio procedures.
- Brief and rehearse the reporting NCO on range operation and all of his duties.
- Collect and concentrate equipment for use on the range in one location.
- Obtain training aids.
- Pick up targets from the range warehouse (if required).
- Report to range control for a safety briefing (if required) and sign for any special items.
- Publish the LOI:
  - Uniform of range and firing personnel (helmets and earplugs).
  - Mode of transportation, departure times, and places.
  - Methods of messing to be used.
  - Any special requirements being placed on units.

OCCUPY THE RANGE AND CONDUCT TRAINING

E-39. Certain actions must be performed to properly occupy the range and conduct training.

Occupy the Range

E-40. When occupying the range, perform the following actions:

- Request permission to occupy the range.
- Establish good communications.
- Have designated areas prepared:
  - Parking.
  - Ammunition point.

  - Medical station.
  - Water point.
  - Concurrent training.
  - Mess.
  - Helipad.
  - Armorer.
- Inspect the range for operational condition.
- Raise the flag when occupying or firing according to the local SOP.
- Check the ammunition to ensure that it is the correct type and quantity.
- Ensure that range personnel are in the proper uniform and the equipment is in position.
- Receive firing units.
- Conduct safety checks on weapons.
- Check for clean, fully operational weapons.
- Conduct a safety briefing (to include administrative personnel on the range).
- Organize personnel into firing orders (keep unit integrity, if possible).
- Request permission to commence firing from range control.

Conduct of Firing

E-41. When conducting firing, the following must be checked:

- Are communications to range control satisfactory?
- Are commands from the tower clear and concise?
- Are range areas policed?
- Is ammunition accountability maintained?
- Is the master score sheet updated?
- Is personnel accountability maintained?
- Are vehicles parked in the appropriate areas?
- Is the air guard on duty and alert?
- Are personnel in proper uniform?
- Are earplugs in use?
- Are troops responding properly to commands?
- Are on-the-spot corrections being made when troops use poor techniques or fail to hit the target?
- Is conservation of ammunition being enforced?
- Are weapons cleared before they are taken from the firing line?
- Are personnel checked for brass or ammunition before they leave the range?
- Is anyone standing around not involved in training or support?

Close the Range

E-42. To properly close the range, the following must be performed:

- Close downrange according to the local SOP.
- Remove all equipment and ammunition from the range.
- Police the range.
- Repaste and resurface targets as required by range instructions.
- Perform other maintenance tasks as required by the local SOP.
- Request a range inspector from range control when ready to be cleared.
- Submit an after-action report to headquarters.
- Report any noted safety hazards to the proper authorities.

KNOWN DISTANCE RANGE

E-43. Events held on the KD range require additional personnel and equipment.

Personnel

E-44. In addition to the personnel previously identified, the KD range requires—

- NCOIC of pit detail.
- Assistant safety officer for the pit area.

Equipment

E-45. In addition to the equipment previously identified, the KD range requires—

- A sound set for the pit area.
- Positive communication from the firing line to the pit area.
- Pasters.
- Glue and brushes for resurfacing targets.
- Lubricant for target frames.
- Proper targets mounted in target frames.
- Briefing on how to operate a KD range.
- Procedure for marking targets.
- Procedure for pit safety.

COMPUTER CONTROLLED RANGES

E-46. Modern computer-controlled ranges allow trainers to develop scenarios and control targets and battlefield simulation devices. This permits Soldiers and units to practice mission-essential tasks in a stressful environment. Computerized systems also provide performance feedback. Using data recorded during training, the computer generates AARs. This accurate feedback allows commander to assess the unit's performance and leaders to assess their unit's mission status and design training programs to overcome the identified shortcomings. The performance feedback highlights positive actions to reinforce correct procedures and to foster Soldiers' confidence, enabling Soldiers and leaders to recognize and correct their shortcomings.

RANGE SELECTION

E-47. This section identifies primary and alternate ranges used for training and qualification, with specific weapon systems and weapons based on applicable FMs. The following table lists primary and alternate ranges to satisfy weapon system training events.

Table E-1. Primary/alternate range selection.

WEAPON SYSTEM	TRAINING EVENT	RANGE	
		PRIMARY	ALTERNATE
M16-Series and M4	Zero	25-m range	AFF/ARF/MRF with 25m boots
	Sustainment/Record	AFF/ARF/MRF	25-m range
	Qualification	ARF/MRF/QTR	25-m range
	Night	AFF/ARF/MRF Night	25-m range
	CBRN Conditions	AFF/ARF/MRF	25-m range

AFF: Automated Field Fire Range
ARF: Automated Record Fire Range
MRF: Modified Record Fire Range
QTR: Qualification Training Range

Rifle/Machine Gun Zero Range (17801)

E-48. This range (Figure E-1) is used to train individual Soldiers on the skills necessary to align the sights and practice basic marksmanship techniques against stationary targets. The range is designed for conducting shot-grouping and zeroing exercises with the M16- and M4-series weapons, as well as crew-served machine guns. Its primary features include—

- 32 target frames at 25 meters.
- 16 target frames at 10 meters.
- 32 foxholes.

E-49. This range requires no automation. All targets are fixed at 25 meters from the firing line for M16- and M4-series weapons and 10 meters for machine gun.

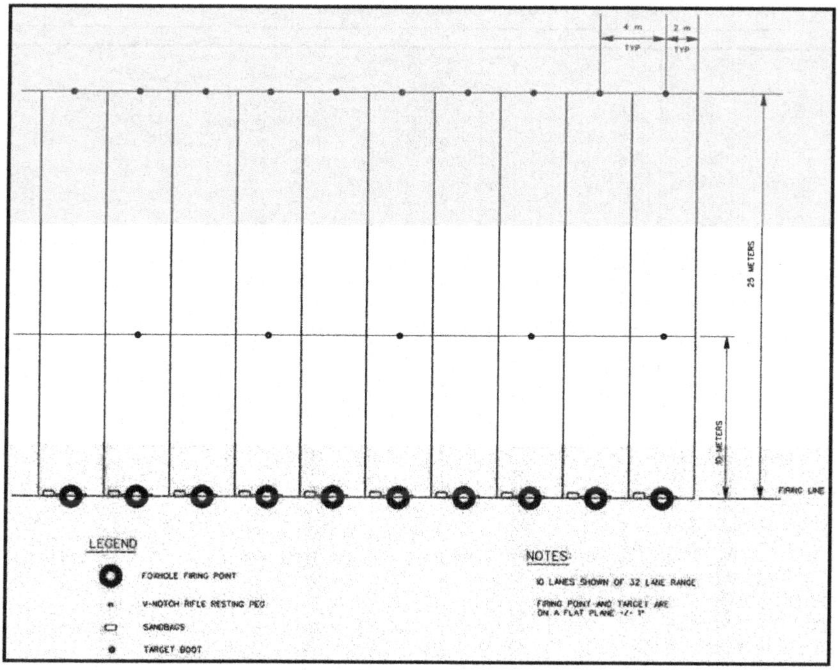

Figure E-1. Rifle/machine gun zero range (17801).

Automated Field Fire Range (17803)

E-50. The AFF range (Figure E-2) is used to train and familiarize Soldiers on the skills necessary to identify, engage, and hit stationary Infantry targets with M16- and M4-series weapons. Its primary features include—

- 96 stationary Infantry targets.
- 32 foxhole positions.

E-51. All targets are fully automated, and the event-specific target scenario is computer-driven and scored from the range operations center. The range operating system is fully capable of providing immediate performance feedback to the participants.

NOTE: This range can be used for automatic rifle practice.

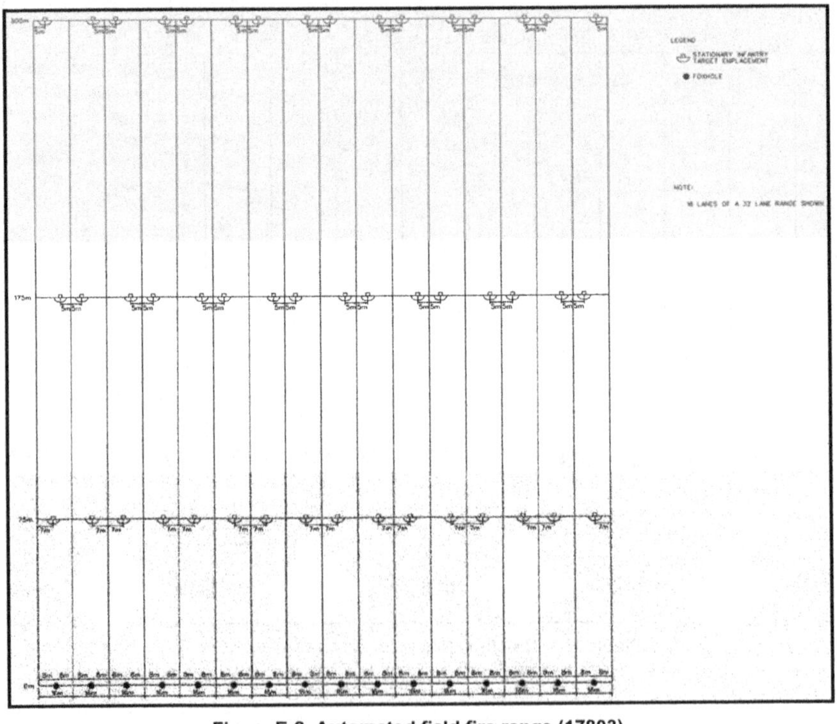

Figure E-2. Automated field fire range (17803).

Automated Record Fire Range (17805)

E-52. The ARF range (Figure E-3) is used to train and test individual Soldiers on the skills necessary to identify, engage, and defeat stationary Infantry targets for day/night qualification requirements with M16- and M4-series weapons. Its primary features include—

- 112 stationary Infantry targets.
- 16 foxholes.

E-53. All targets are fully automated, and the event-specific target scenario is computer-driven and scored from the range operations center. The range operating system is fully capable of providing immediate performance feedback to the participants.

NOTE: 1. To perform night fire, Soldiers fire at 50-meter targets from the night-fire line and baseline.

2. Range Set-up: Replace one of the 50m F-type silhouettes with an E- type silhouette. Low light illumination capability is required in both of the 50-meter target emplacements. Engage the F- type from the 25-meter night-firing line, and the E- type from the baseline.

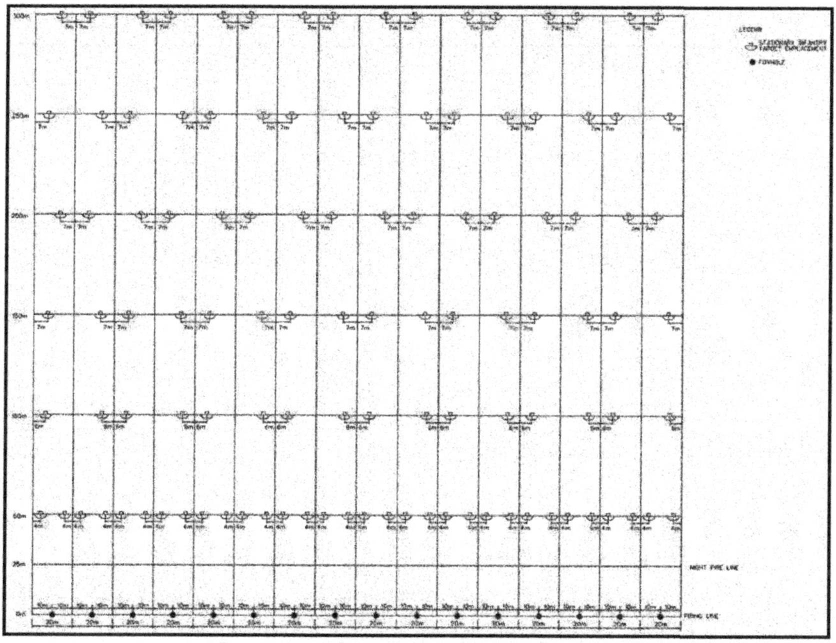

Figure E-3. Automated record fire range (17805).

Modified Record Fire Range (17806)

E-54. The modified record fire (MRF) range (Figure E-4) is used to train and test individual Soldiers on the skills necessary to identify, engage, and defeat stationary Infantry targets for day/night qualification requirements with M16- and M4-series weapons. This range combines the capabilities of the AFF (17803), ARF (17805), and automated night fire (17808) ranges to reduce land and maintenance requirements and increase efficiencies. Its primary features include—

- 144 stationary Infantry targets.
- 16 foxholes

E-55. All targets are fully automated, and the event-specific target scenario is computer-driven and scored from the range operations center. The range operating system is fully capable of providing immediate performance feedback to the participants.

NOTE: 1. To perform night fire, Soldiers fire at 50-meter targets from the night-fire line and baseline.

2. Range Set-up: Replace one of the 50m F-type silhouettes with an E- type silhouette. Low light illumination capability is required in both of the 50-meter target emplacements. Engage the F- type from the 25-meter night-firing line, and the E- type from the baseline.

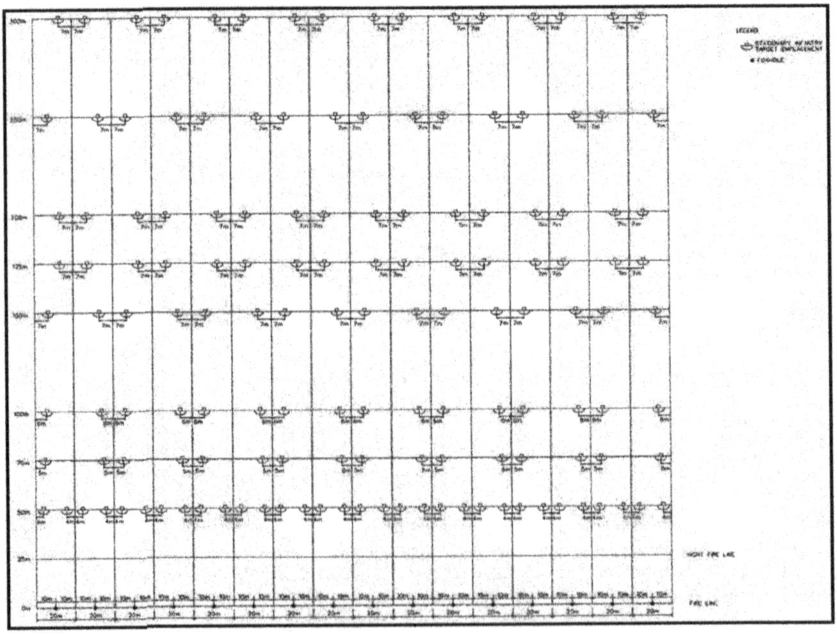

Figure E-4. Modified record fire range (17806).

Qualification Training Range (17809)

E-56. The qualification training (QTR) range (Figure E-5) is used to train and test Soldiers on the skills necessary to detect, identify, engage, and defeat stationary and moving Infantry targets in a tactical array with their prescribed weapons. This range combines the capabilities of the MRF range (17806), automated sniper field fire range (17812), combat pistol/military police firearms qualification course (17822), and multipurpose machine gun range (17833) to centralize training and reduce land, maintenance, and unit overhead requirements. Its primary features include—

- 429 stationary Infantry targets.
- 20 stationary armor targets.
- 20 moving Infantry targets.
- 10 stationary Infantry emplacements.
- 32 lanes rifle/machine gun zero.
- 15 lanes combat pistol qualification.
- 10 lanes sniper field fire.
- 16 lanes modified record fire.
- 10 lanes multipurpose machine gun.
- 2 target mechanisms each (400-meter and 700-meter).

E-57. All targets are fully automated, and the event-specific target scenario is computer-driven and scored from the range operations center. The range operating system is fully capable of providing immediate performance feedback to the participants.

NOTE: This range enhances throughput capability for units with multiple weapon densities by consolidating unit efforts to operate one training facility.

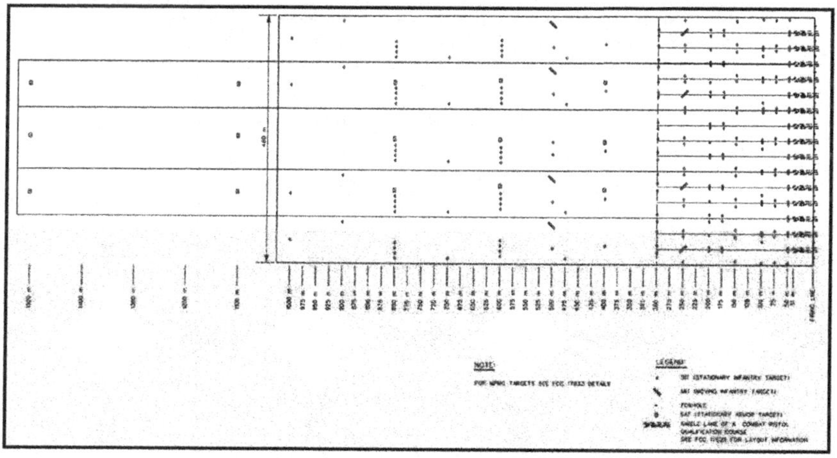

Figure E-5. Qualification training range (17809).

Appendix F

10-Meter Target Offsets and 25-Meter Zero Offsets

This section provides the 10-meter target offsets and the 25-meter zero offsets for M16- and M4-series weapons mounted with iron sights, optics, MILES, or aiming lasers. A blank, reproducible 10-meter target offset (Figure F-2) and an example of each weapon configuration (Figures F-3 through F-6) are provided. The M16A2 300-meter zeroing target is used for 25-meter zeroing with all weapon configurations.

MARKING 10-METER TARGET OFFSETS

F-1. To mark the proper 10-meter target offsets—

(1) Find the correct template for the weapon configuration.

(2) Starting from the center of the borelight circle on the offset, count the number of squares to the desired point of aim.

EXAMPLE

L2.0, U2.4

Starting from the center of the borelight circle (0.0, 0.0), move left 2 squares and up 2.4 squares.

NOTE: Each template also provides a number formula for the proper offset.

(3) Place the appropriate symbol or mark (Figure F-1).

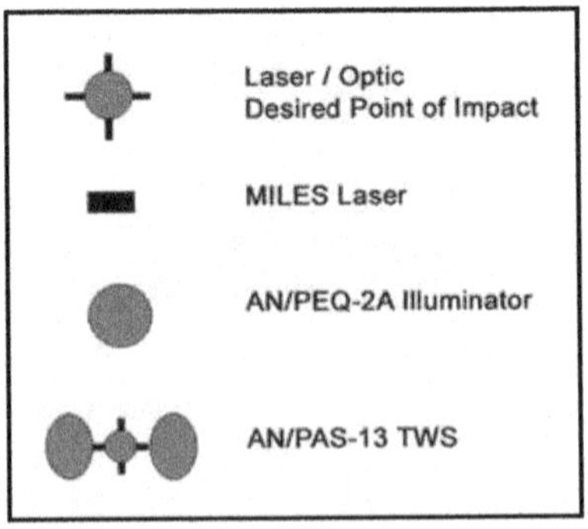

Figure F-1. 10-meter target offset symbols.

MARKING 25-METER ZERO OFFSETS

F-2. To mark the proper 25-meter zero offsets—

(1) Use only a M16A2 300-meter zeroing target.

(2) Find the correct target template for the weapon configuration.

(3) Count the number of squares, starting from the center of the 300-meter zeroing silhouette.

(4) Mark the designated strike point by drawing a small circle at the appropriate number of squares from the center of the 300-meter zeroing silhouette.

(5) Draw a 4-centimeter by 4-centimeter square, keeping the designated strike point at the center.

NOTES: 1. To reproduce the 10-meter target offset, copy the blank 10-meter target offset and place the example of the weapon being used on the back. This reproducible copy can be laminated and used repeatedly.

2. Table F-1 provides offset mounting information for various weapon configurations.

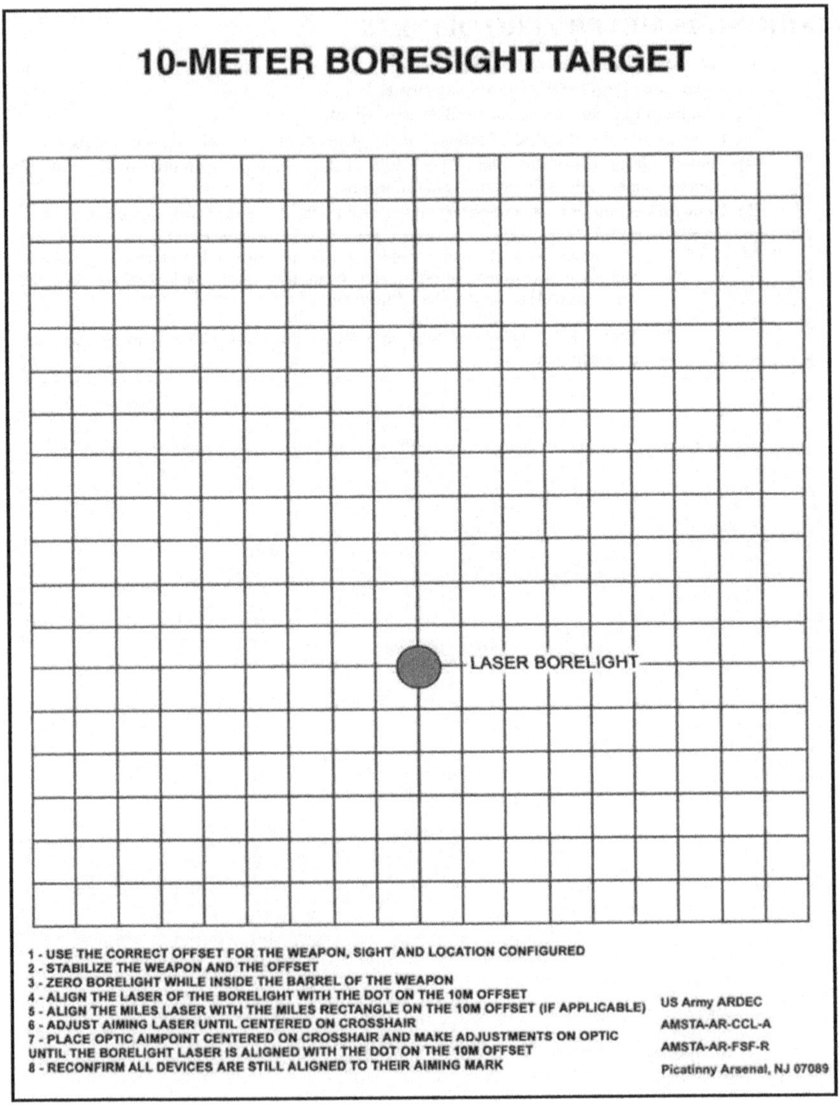

Figure F-2. Blank 10-meter target offset.

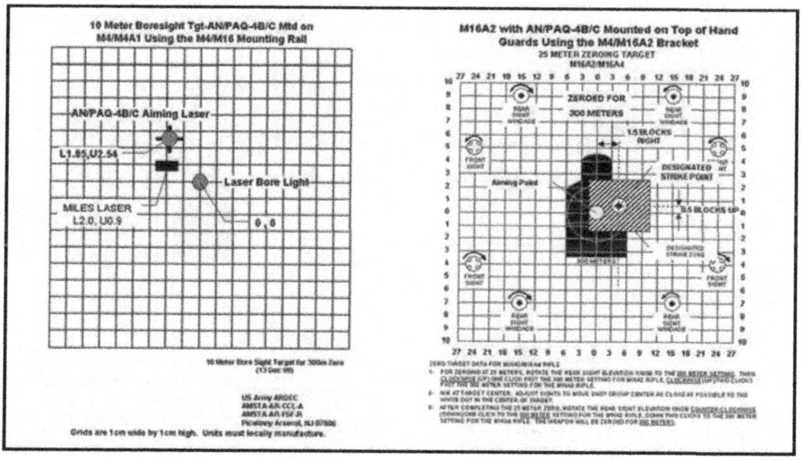

Figure F-3. M16A2 10-meter boresighting target/25-meter zeroing target offsets.

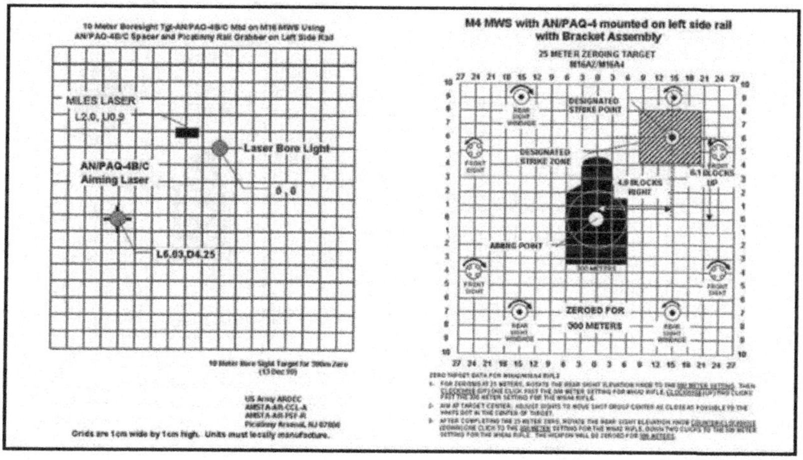

Figure F-4. M4 MWS 10-meter boresighting target/25-meter zeroing target offsets.

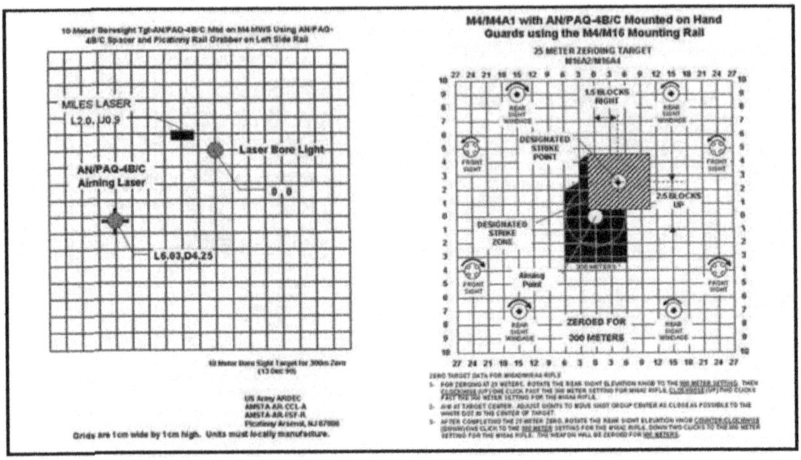

Figure F-5. M4/M4A1 10-meter boresighting target/25-meter zeroing target offsets.

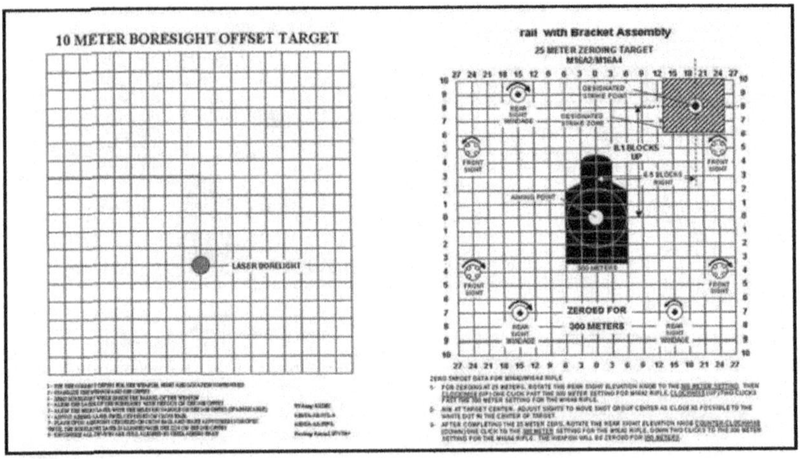

Figure F-6. M16A4 MWS 10-meter boresighting target/25-meter zeroing target offsets.

Table F-1. Offset mounting.

WEAPON	ACCESSORY	RAIL GRABBER	MOUNT	LOCATION	RANGE TO ZERO	ZERO OFFSET	BORESIGHT TARGET	MILES OFFSET
M16A2	Iron sight	N/A	N/A	N/A	300 m	0.0 0.0	0.0 4.2U	2.0L 0.9U
M16A2	M68	N/A	M68 gooseneck bracket	Carrying handle	300 m	0.0 1.4 cm DN	0.0 5.2U	2.0L 2.4U
M16A2	LTWS	TWS	TWS bracket assembly	Carrying handle	300 m	0.0 10D	0.0 13.4U	2.0L 2.4U
M16A2	TWS	N/A	TWS bracket assembly	Carrying handle	300 m	0.0 8.1D	0.0 11.5U	2.0L 2.4U
M16A2	AN/PVS-4	N/A	AN/PVS-4 mounting knob	Carrying handle	300 m	0.0 7.0D	0.0 9.4U	2.0L 0.9U
M16A2	AN/PAQ-4B/C	N/A	M4/M16 bracket	Hand guards	300 m	1.5R 0.5U	1.85L 2.54U	2.0L 0.9U
M16A2	AN/PEQ-2A/B	N/A	M4/M16 bracket	Hand guards	300 m	1.5L 0.5U	1.8R 2.4U	2.0L 0.9U
M16/M203	AN/PAQ-4B/C	N/A	Spacer and AN/PVS-4 mounting knob	Carrying handle	300 m	1.85R 2.6D	1.85L 8.6U	2.0L 3.9U
M16/M203	AN/PVS-4	N/A	AN/PVS-4 mounting bracket assembly	Carrying handle	300 m	4.2R 9.8D	TBD	2.0L 0.9U
M4/M4A1	BUIS	N/A	N/A	Upper receiver	300 m	0.0 0.0	0.0 4.01U	2.0L 0.9U
M4/M4A1	AN/PAQ-4B/C	N/A	M4/M16 bracket	Hand guards	300 m	1.5R 2.5U	1.85L 2.54U	2.0L 0.9U
M4/M4A1	LTWS	TWS	N/A	Upper receiver	300 m	0.0 4.5D	0.0 7.9U	TBD
M4/M4A1	TWS	Picatinny	TWS spacer and rail grabber	Upper receiver	300 m	0.0 5.7D	0.0 9.4U	2.0L 2.4U
M4/M4A1	AN/PEQ-2A/B	N/A	M4/M16 bracket	Hand guards	300 m	1.0L 0.3U	1.8R 2.4U	2.0L 0.9U
M4/M4A1	AN/PVS-4	Picatinny	Spacer and rail grabber	Upper receiver	300 m	0.0 3.4D	0.0 7.6U	2.0L 0.9U
M4/M4A1	M68	M68	Half-moon spacer	Upper receiver	300 m	0.0 1.4 cm DN	0.0 5.63U	2.0L 2.4U
M4/M203	BUIS	N/A	N/A	Upper receiver	300 m	0.0 0.0	0.0 6.01U	2.0L 0.9U
M4/M203	AN/PAQ-4B/C	N/A	Spacer and AN/PVS-4 mounting knob	Carrying handle	300 m	1.3R 1.9D	1.85L 8.6U	2.0L 0.9U
M4/M203	AN/PVS-4	Picatinny	Spacer and rail grabber	Upper receiver	300 m	0.0 3.4D	0.0 9.6U	2.0L 3.9U

NOTE: Target offsets not yet developed are indicated by TBD (to be developed).

Table F-1. Offset mounting (continued).

WEAPON	ACCESSORY	RAIL GRABBER	MOUNT	LOCATION	RANGE TO ZERO	ZERO OFFSET	BORESIGHT TARGET	MILES OFFSET
M4 MWS	BUIS	N/A	N/A	Upper receiver	300 m	0.0 0.0	0.0 4.01U	2.0L 0.9U
M4 MWS	AN/PVS-4	Picatinny	Spacer and rail grabber	Upper receiver	300 m	0.0 3.4D	0.0 7.6U	2.0L 0.9U
M4 MWS	M68	M68	Rail grabber	Upper receiver	300 m	0.0 1.4 cm DN	0.0 5.63U	2.0L 2.4U
M4 MWS	LTWS	TWS	N/A	Upper receiver	300 m	0.0 4.5D	0.0 7.9U	2.0L 2.4U
M4 MWS	TWS	TWS	Spacer	Upper receiver	300 m	0.0 5.7D	0.0 9.4U	2.0L 2.4U
M4 MWS	ANPEQ-2A	Insight	N/A	Left	300 m	TBD	4.5L 1.0D	2.0L 0.9U
M4 MWS	AN/PEQ-2A/B	Insight	N/A	Right	300 m	N/A	5.5R 5.4D	2.0L 0.9U
M4 MWS	AN/PEQ-2A/B	Insight	N/A	Top	300 m	1.5L 0.5D	2.9R 2.3U	2.0L 0.9U
M4 MWS	AN/PEQ-2A/B	Picatinny	Spacer	Top	300 m	N/A	1.95R 4.1U	2.0L 0.9U
M4 MWS	AN/PEQ-2A/B	Picatinny	Spacer	Right	300 m	N/A	6.35R 4.4D	2.0L 0.9U
M4 MWS	AN/PEQ-2A/B	Picatinny	Spacer	Left	300 m	6.9R 2.0U	6.2L 0.60D	2.0L 0.9U
M4MWS	AN/PEQ-2A/B	Insight	Training adapter	Top	300 m	2.0L 1.5D	N/A	2.0L 0.9U
M4 MWS	AN/PAQ-4B/C	Picatinny	AN/PAQ-4B/C bracket adapter	Top	300 m	4.9R 6.1U	1.75L 3.9U	2.0L 0.9U
M4 MWS	AN/PAQ-4B/C	Picatinny	AN/PAQ-4B/C bracket adapter (spacer)	Right	300 m	N/A	6.9R 0.9D	2.0L 0.9U
M4 MWS	AN/PAQ-4B/C	Insight	N/A	Top	300 m	N/A	1.75L 2.15U	2.0L 0.9U
M4MWS	AN/PAQ-4B/C	Insight	N/A	Right	300 m	N/A	4.35R 0.65D	2.0L 0.9U
M4MWS	AN/PAQ-4B/C	Insight	N/A	Left	300 m	N/A	4.30L 4.25D	2.0L 0.9U
M4 MWS M203	BUIS	N/A	N/A	Upper receiver	300 m	0.0 0.0	0.0 6.01U	2.0L 0.9U
M4 MWS M203	AN/PAQ-4B/C	Picatinny	Bracket adapter (spacer	Left	300 m	4.9R 6.1U	6.0L 4.0D	2.0L 3.9U
M4 MWS M203	AN/PVS-4	Picatinny	Spacer	Upper receiver	300 m	0.0 3.4D	0.0 9.6U	2.0L 3.9U
M16A4 MWS	BUIS	N/A	N/A	Upper receiver	300 m	0.0 0.0	0.0 4.01U	2.0L 0.9U
M16A4 MWS	AN/PAQ-4B/C	Picatinny	AN/PAQ-4B/C bracket adapter (spacer)	Left	300 m	6.5R 8.1U	6.03L 4.25D	2.0L 0.9U
M16A4 MWS	TWS	TWs	Spacer	Upper receiver	300 m	0.0 6.0D	0.0 9.4U	2.0L 2.4U
M16A4 MWS	M68	M68	N/A	Upper receiver	300 m	0.0 0.0	0.0 5.63U	2.0L 2.4U
M16A4 MWS	AN/PEQ-2A/B	Insight	N/A	Left	300 m	3.0R 3.0U	4.5L 1.0D	2.0L 0.9U
M16A4 MWS	AN/PVS-4	Picatinny	Spacer	Upper receiver	300 m	0.0 4.6D	0.0 7.6U	2.0L 0.9U

NOTE: Target offsets not yet developed are indicated by TBD (to be developed).

Table F-1. Offset mounting (continued).

WEAPON	ACCESSORY	RAIL GRABBER	MOUNT	LOCATION	RANGE TO ZERO	ZERO OFFSET	BORESIGHT TARGET	MILES OFFSET
M16A4 MWS M203	BUIS	N/A	N/A	Upper receiver	300 m	0.0 0.0	0.0 6.01U	2.0L 0.9U
M16A4 MWS M203	AN/PAQ-4B/C	Picatinny	AN/PAQ-4B/C bracket adapter (spacer	Left	300 m	6.5R 8.1U	6.0L 4.0D	2.0L 3.9U
M16A4 MWS M203	AN/PVS-4	Picatinny	Spacer	Upper receiver	300 m	0.0 4.6D	0.0 9.6U	2.0L 3.9U
M16A4 MWS	AN/PEQ-2A/B	Picatinny	Spacer	Left	300 m	6.0R 2.0U	6.2L 0.60D	2.0L 0.9U
M16A4 MWS	AN/PEQ-2A/B	Picatinny	Spacer	Right	300 m	TBD	6.35R 4.4D	2.0L 0.9U
M16A4 MWS	AN/PEQ-2A/B	Picatinny	Spacer	Top	300 m	TBD	1.95R 4.1U	2.0L 0.9U
M16A4 MWS	AN/PEQ-2A/B	Insight	N/A	Right	300 m	TBD	5.5R 5.4 D	2.0L 0.9U
M16A4 MWS	AN/PEQ-2A/B	Insight	N/A	Top	300 m	1.5L 0.5D	2.0R 2.3U	2.0L 0.9U
M16A4 MWS	AN/PEQ-2A/B	Insight	Training adapter	Top	300 m	2.0L 1.5D	TBD	2.0L 0.9D
M16A4 MWS	AN/PAQ-4B/C	Picatinny	AN/PAQ-4B/C bracket adapter	Top	300 m	4.9R 6.1U	1.75L 3.9U	2.0L 0.9U
M16A4 MWS	AN/PAQ-4B/C	Picatinny	AN/PAC-4B/C bracket adapter	Right	300 m	N/A	6.0R 0.9D	2.0L 0.9U
M16A4 MWS	AN/PAQ-4B/C	Insight	N/A	Top	300 m	N/A	1.75L 2.15U	2.0L 0.9U
M16A4 MWS	AN/PAQ-4B/C	Insight	N/A	Right	300 m	N/A	4.35R 0.65D	2.0L 0.9U
M16A4 MWS	AN/PAQ-4B/C	Insight	N/A	Left	300 m	N/A	4.30L 4.25D	2.0L 0.9U
NOTE: Target offsets not yet developed are indicated by TBD (to be developed).								

Glossary

SECTION I – ACRONYMS AND ABBREVIATIONS

AAR	after-action review
ACH	Army combat helmet
ACOG	advanced combat optical gunsight
AFF	automated field fire
AMU	Army Marksmanship Unit
AP	armor-piercing
AR	Army regulation
ARF	automated record fire
ARM	advanced rifle marksmanship
ARS	adapter rail system
ATC	Army training center
ATPIAL	advanced target pointer/illuminator aiming light
BDC	bullet drop compensated
BFA	blank firing attachment
BOLC	Basic Officer Leaders Course
BRM	basic rifle marksmanship
BUIS	backup iron sight
CBRN	chemical, biological, radiological, and nuclear
CCO	close combat optic
COA	course of action
cm	centimeter or centimeters
CRM	composite risk management
DA Pam	Department of the Army pamphlet
EST	Engagement Skills Trainer
FM	field manual
FOV	field of view
fps	feet per second
ft	foot or feet
HWTS	heavy weapon thermal sight
I2	image-intensifier
IBA	interceptor body armor
IET	initial entry training
in	inch or inches
IR	infrared
KD	known distance
lb	pound or pounds
LBE	load-bearing equipment
LFX	live-fire exercise
LMTS	Laser Marksmanship Training System
LOGCAP	Logistics Civil Augmentation Program
LOMAH	location of misses and hits

LWTS	light weapon thermal sight
m	meter or meters
MDMP	military decision-making process
METL	mission-essential task list
METT-TC	mission, enemy, terrain and weather, troops and support available, time available, civil considerations
min	minute or minutes
mm	millimeter or millimeters
MOA	minute of angle
MOPP	mission oriented protective posture
mph	miles per hour
MRF	modified record fire
MWS	modular weapon system
NCOES	Noncommissioned Officers Education System
NCOIC	noncommissioned officer in charge
NFOV	narrow field of view
NGO	nongovernmental organizations
NOD	night observation devices
NSN	national stock number
NVD	night vision device
OIC	officer in charge
OSUT	one-station unit training
OTV	outer tactical vest
PASGT	personnel armor system for ground troops
PH	probability of hit
PMCS	preventive maintenance checks and services
PMI	preliminary marksmanship instruction
POI	program of instruction
QTR	qualification training
RETS	Remote Electronic Target System
RFA	rimfire adapter
RSO	range safety officer
RTO	radio telephone operator
SAPI	small arms protective insert
SAW	squad automatic weapon
SDM	squad designated marksman
SDZ	surface danger zone
sec	second or seconds
SRC	short-range combat
SRM	short-range marksmanship
SRT	special reaction team
SRTA	short-range training ammunition
STRAC	Standards in Training Commission (DA Pam 350-38)
STX	situation training exercise
TADSS	training aids, devices, simulators, and simulations
TC	training circular
TSC	training support centers
TM	technical manual

TPIAL	target pointer illuminator/aiming light
TWS	thermal weapon sight
UO	urban operations
USACRC	United States Army Combat Readiness Center
WFF	warfighting functions
WFOV	wide field of view

SECTION II – TERMS

Active Army: All Regular Army (RA) forces in the Active Army.

adjusted aiming point: An aiming point that allows for gravity, wind, target movement, zero changes, and MOPP firing.

advanced marksmanship: Normally refers to marksmanship skills taught during ARM.

advanced rifle marksmanship: Normally refers to the formal marksmanship instruction received by infantrymen upon completion of BRM during OSUT.

aiming: A marksmanship fundamental; refers to the precise alignment of the rifle sights with the target.

aiming card: The M15A1 aiming card is a cardboard sleeve with a moveable insert. The rear sight aperture, front sight post, and target are pictured. This training device is used in conjunction with aiming instructions.

aiming point: A place on a target in which the rifle sights are aligned normally the target center of mass.

alibi target: A target or additional target a soldier is allowed to engage during qualification firing when unable to complete a record fire scenario due to circumstances beyond his control; for example, a target mechanism, weapon, or ammunition malfunction.

alternate course: Alternatives to standard qualification courses.

ammunition lot: A quantity of cartridges, each of which is made by one manufacturer under uniform conditions and is expected to work in a uniform manner.

ammunition lot number: Code number that identifies a particular quantity of ammunition from one manufacturer.

aperture: The hole in the rear sight.

armorer: One who services and makes repairs on small arms and performs similar duties to keep small arms ready for use.

Army Training and Evaluation Program: A guide for the training and evaluation of critical unit combat missions – crew/squad through battalion/task force echelon.

Army Training Center: Conducts OSUT and BRM. Locations are Fort Benning, GA; Fort Jackson, SC; Fort Knox, KY.

artificial illumination: Any light from a man-made source.

assault course: An area of ground used for training soldiers in attacking an enemy in close combat.

automatic fire: A firing mode that causes the weapon to continue firing as long as the trigger is held or until all ammunition has been expended.

ball: The projectile; the bullet.

ball ammunition: General-purpose standard service ammunition with a solid core bullet.

ball and dummy: An exercise that substitutes a dummy round for a live round without the firer knowing it. An excellent exercise for identifying and correcting trigger jerks.

ballistics: A science that deals with the motion and flight characteristics of projectiles.

barrel erosion: Wearing away of the surface of the bore due to the combined effects of gas washing, coring, and mechanical abrasion.

basic marksmanship: Fundamental marksmanship skills taught in BRM during IET and OSUT.

basic rifle marksmanship: The formal course of marksmanship instruction received by all soldiers.

battlesight zero: A sight setting that soldiers keep on their weapons. It provides the highest probability of hitting most high-priority combat targets with minimum adjustment to the aiming point, a 250 meter sight setting as on the M16A1 rifle, and a 300 meter sight setting as on the M16A2 rifle.

blank ammunition: A complete cartridge without the bullet used to simulate weapon firing.

blank firing adapter: A device that fits in the muzzle of the rifle; used only with blank ammunition.

brass: An alloy of copper and zinc used to make cartridge cases and bullet jackets. Also, a common name for expended cases.

breath control: The third marksmanship fundamental; refers to the control of breathing to help keep the rifle steady during firing.

bullet: The projectile or ball; the part that goes downrange. It may also be used to refer to the complete cartridge.

bull's-eye target: Any target with a round black circle and scoring rings. Normally used in competitive marksmanship training.

buttplate: Metal or rubber covering of the end of the stock on the rifle.

cadre coach: A trainer with expertise and knowledge exceeding that of the firer.

caliber: Diameter of the bore; for example, the M16-series rifle bore is 5.56mm (.223 inch).

cartridge: A complete round of ammunition.

center of mass: A point that is horizontally (left and right) and vertically (up and down) at the center of the target.

chambering: The step in the cycle of operation that refers too fully seating the round in the chamber of the rifle.

chamber plug: A range safety device that is a small plastic plug designed to fit into the chamber of the M16. A handle extends out the ejection port so safety personnel can see at a glance that the rifle is clear of ammunition.

clock method: Method of calling shots by referring to the figures on an ordinary clock dial assumed to have the target at its center. Also a method of determining the strength and direction of wind.

coach: Any individual who assists firers on the firing line.

coach-and-pupil method: Method of training in which pairs of pupils take turns practicing a procedure explained by the instructor/trainer.

cocking: The step in the cycle of operation that refers to the rearward movement of the bolt riding over the hammer, resetting the weapon for subsequent firing.

collective firing proficiency: Units delivering effective fire in a tactical setting. It requires individual skill plus command and control to engage all targets within an assigned sector.

concurrent training: Training that occurs at the same time that other unit members are using the primary training facilities.

cookoff: A round that fires as a result of a hot chamber without the trigger being pulled. It can occur any time until the weapon is cooled.

crack and thump: A method to determine the general direction and distance to an enemy firer who is shooting at you.

cradle: A vise-like mechanism that holds a weapon in a secure position for test firing.

cross dominance: A soldier with a dominant hand and a dominant eye that are not the same; for example, a right-hander firer with a dominant left eye.

cycle of operation: The eight steps involved in firing a round of ammunition: feeding, chambering, locking, firing, unlocking, extracting, ejecting, and cocking.

cyclic rate of fire: The maximum rate at which a weapon will fire in the automatic mode.

dime-washer exercise: A dry-fire exercise used to practice trigger squeeze.

downrange feedback: Used to describe any training technique that provides precise knowledge of bullet strike (whether hit or miss).

dry fire: A technique used to simulate the firing of a live round with an empty weapon. Any application of the fundamentals of marksmanship without live ammunition may be referred to as dry fire.

dry-fire moving target trainer: A small-motorized scaled target device used to teach the engagement of moving personnel targets.

dummy ammunition: A cartridge without a primer or powder. Primarily used for ball-and-dummy exercises on the live-fire line.

effective wind: The average of all the varying winds encountered.

ejection: The step in the cycle of operation that removes the expended cartridge from the weapon out of the ejection port.

elevation adjustment: Rotating the front sight post to cause the bullet to strike higher or lower on the target.

expert: The highest qualification rating.

external ballistics: What happens to the bullet between the time it leaves the rifle and the time it arrives at the target.

extraction: The step in the cycle of operation that pulls the round from the chamber.

eye relief: The distance from the firing eye to the rear sight. Eye relief is a function of stock weld.

feedback: Obtaining knowledge of performance.

feedback target: Targets designed for use at 75, 175, or 300 meters; includes an overprinted grid similar to a zero target.

feeding: The step in the cycle of operation that is the forward movement of the bolt, stripping the top round from the magazine and moving it toward the chamber.

field firing: Training on the standard field firing range with target banks at 75, 175, and 300 meters.

firing: The step in the cycle of operation that refers to pulling the trigger, releasing the hammer to strike the firing pin, which strikes the primer. The primer ignites and, in turn, ignites the powder charge within the cartridge case.

firing hand: The right hand of a right-handed firer. The left hand of a left-handed firer.

firing pin: Plunger in the bolt of a rifle that strikes the primer.

fleeting target: A moving target remains within observing or firing distance for such a short period that it affords little time for deliberate adjustment and fire against it.

functioning: (See cycle of operation.)

fundamentals of rifle marksmanship: The four essential elements needed to hit targets: steady position, aiming, breath control, and trigger squeeze.

gravity: The natural pull of all objects to the center of the earth.

grouping: A live-fire exercise with the objective of shooting tight shoot groups.

gun bore line: A reference line established by the linear extension of the bore axis of a gun.

headspace: The distance between the face of the bolt(fully closed) and the face of a fully chambered cartridge.

hold-off: (See adjusted aiming point.)

horizontal dispersion: The left-to-right displacement of bullets on a target.

immediate action: procedures applied to rapidly reduce any rifle stoppage without determining its cause.

individual firing proficiency: Individual firing skills; for example, an individual's performance on the record fire course.

Infantry Remoted Target System (IRETS): (See RETS.)

infrared aiming light: A unique night sighting system that uses infrared light to assist in the aiming process.

initial entry training: Indicates the first training received by a new soldier, includes the MOS-producing portion of his training such a one-station unit training (OSUT).

initial pressure: The applications of about half of the total trigger pressure it takes to fire a rifle.

instructor-trainer ratio: The number of soldiers for which each instructor/trainer is responsible.

internal ballistics: What happens to the bullet before it leaves the muzzle of the rifle.

known distance: Describes the older range complexes with large target frames behind a large berm and firing lines at 100 yards or 100-meter increments. (See FM 25-7.)

laser: Light amplification by simulated emission of radiation.

lead: Distance ahead of a moving target that a rifle must be aimed to hit the target.

lead rule: Provides the soldier guidance on how to adjust his aiming point to hit moving targets.

line of sight: A line between the rifle and the aiming point, extending from the firing eye through the center of the rear aperture, across the tip of the front sight post, and onto the target.

location of misses and hits : A projectile location system that provides immediate and precise information to the firer concerning bullet strike (hit or miss).

locking: The step in the cycle of operation that is a counterclockwise rotation of the bolt, securing it into the barrel locking lugs.

long-range sight: The aperture marked L on the M16A1 rifle equipped with standard sights; provides for a zero at 375 meters. The M16A1 rifle equipped with LLLSS has an aperture marked L, but it is a regular sight.

Low-Light Level Sight System (LLLSS): A sighting system for low visibility firing that replaces the standard front and rear sights on the M16A1 rifle.

marksman: The designation given to the lowest qualification rating.

maximum effective range: The greatest distance at which a soldier may be expected to deliver a target hit.

maximum effective rate of fire: The highest rates of fire that can be maintained and still achieve target hits.

maximum range: The longest distance a projectile will travel when fired from a weapon held at the optimum angle.

minute of angle: A angle that would cover 1 inch at a distance of 100 yards, 2 inches at 200 yards, and so on. Each click of sight adjustment on the M16A1 rifle with standard sights is equal to one minute of angle.

Multiple Integrated Laser Engagement System (MILES): A tactile shooting device that uses a low-powered laser to activate detectors placed on people and vehicles.

muzzle velocity: The speed of a projectile as it leaves the muzzle of the weapon.

natural point of aim: The direction of the body/rifle combination is oriented while in a stable, relaxed firing position.

natural respiratory pause: The temporary cessation of breathing between an exhale and inhale.

night firing: Firing performed under all conditions of limited visibility.

nonfiring hand: The opposite of the firing hand.

optical sight: Sight with lenses, prisms, or mirrors used in lieu of iron sights.

Paige sighting device: A device with a small-scaled target that fits into the muzzle of the weapon, allowing the soldier to practice aiming.

pasters: Small white or black gum-backed paper used for covering bullet holes.

peep sight: The rear sight; a sight with a small aperture (hole).

peer coach: A soldier with shooting experience and knowledge equal to that of the firer he is coaching.

pit: The target area behind the large berm of a KD range.

plastic practice ammunition: Ammunition with a plastic projectile, high-muzzle velocity (the light weight causes it to lose velocity rapidly with a maximum range of 250 meters or less) designed for use in close-in training areas; frangible bullet.

point of aim: The exact spot on a target the rifle sights are aligned with.

point of impact: The point that a bullet strikes; usually considered in relation to point of aim.

pop, no kick: A firing condition when the primer ignites and the powder charge does not. This normally results in lodging the bullet inside the barrel.

pop-up target: A silhouette target that is activated remotely so it can suddenly appear and fall when struck by a bullet.

practice record: Firing conducted on a qualification course for practice.

predetermined fire: A technique of aligning the rifle during good visibility so the rifle can be aligned and fired on designated areas when they cannot be seen due to darkness, smoke, or fog.

preparatory marksmanship training: All marksmanship training that takes place before live fire.

primer: A small explosive device in the center base of the cartridge case that is struck by the firing pin to fire the round.

probability of hit: Ranging from 0 to 1.0, it refers to the odds of a given round hitting the target at a given range.

qualification firing: Firing on any authorized course that results in meeting qualification requirements; may also be called record fire. (See record fire.)

quick fire: A technique of fire used to engage surprise targets at close range.

range card: Small chart on which ranges and directions to various targets and other important points in the area under fire are recorded.

rapid semiautomatic fire: A firing procedure that results in an accurate shot being fired every one or two seconds.

receiver: That portion of a firearm that holds the barrel and houses the bolt and firing mechanism.

recoil: The rearward motion or kick of a gun upon firing.

record fire: Any course of fire used to determine if qualification standards are met. The standard record fire course consists of 40 target exposures at ranges between 50 and 300 meters. The standard course requires 23 hits to qualify as marksman, 30 for sharpshooter, and 36 for expert.

reduced range ammunition: Ammunition that is designed to be a ballistic match with service ammunition to an appropriate range for training (may be less than maximum effective range) and a reduced maximum range.

regular rear sight: The M16A1 rifle rear sight that is zeroed for 250 meters (the unmarked aperture on rifles with standard sights and the aperture marked L on rifles equipped with LLLSS).

reinforcement training: Training conducted that is over and above scheduled training.

remedial action: A procedure applied after immediate action has failed to correct a malfunction, which determines the cause of the malfunction.

remedial training: Additional training presented to soldiers who have demonstrated special shooting problems.

Remote Electronic Target System: Range complexes. Some ranges include moving targets.

Reserve Components: Includes Army National Guard and Army Reserve forces.

ricochet fire: Fire in which the projectile glances from a surface after impact.

Riddle sighting device: A small magnetic device with a scaled target that attaches to the front sight assembly, allowing the soldier to practice aiming.

rifle cant: Any leaning of the rifle to the left or right from a vertical position during firing.

rim-fire adapter: The caliber .22-rim fire adapter (M261) consists of a bolt and a magazine insert, which allows standard .22 caliber ammunition to be fired in the M16 rifle.

round: May refer to a complete cartridge or to a bullet.

scaled-silhouette target: Any target that is reduced in size. When it is observed from 25 meters, it looks the same size as though at a greater range.

sector of fire: An area assigned to an individual, weapon, or unit to be covered by fire.

semiautomatic fire: A mode of fire that allows one round to be fired each time the trigger is pulled.

serviceability checks: A technical inspection of the rifle to determine if it is safe to fire and in working condition. (May not ensure accuracy.)

service ammunition: Standard ammunition used by the military. Ammunition designed for combat.

service rifle: The primary rifle of a military force.

service school: Branch schools such as the US Army Infantry School at Fort Benning, Ga. and the Armor School at Fort Knox, Ky.

sharpshooter: The middle rating of qualification.

shot group: A number of shots fired using the same aiming point, which accounts for rifle, ammunition, and firer variability. Three shots are enough, but any number of rounds may be fired in a group.

shot group analysis: A procedure for analyzing the size of shot groups on a target to determine firer error.

sight alignment: Placing the center tip of the front sight post in the exact center of the rear aperture.

sighter rounds: Rounds fired that allow the bullet strike to be observed in relation to the aiming point.

sight picture: Placing correct sight alignment on a selected aiming point on a target.

sight radius: The distance from the front sight post to the rear sight aperture of a rifle.

sighting device (M16): A small metal device with a tinted square of glass that is placed on the carrying handle, allowing a coach to see what the firer sees through the sights.

silhouette target: A target that represents the outline of a man.

spotters: A round cardboard disk placed in bullet holes with a small wooden peg so the bullet strike can be observed from the firing line.

squad automatic weapon: A lightweight, one-man, 5.56mm machine gun.

starlight scope: A weapon scope that amplifies ambient light so targets can be seen and effectively engaged during darkness. The AN/PVS-2 and AN/PVS-4 are used on the M16 rifle.

steady position: The first marksmanship fundamental, which refers to the establishment of a position that allows the weapon to be held still while it is being fired.

stock weld: The contact of the cheek with the stock of the weapon.

supported position: Any position that uses something other than the body to steady the weapon (artificial support).

suppressive fire: Any engagement that does not have a definite or visible target. Firing in the general direction of known or suspected enemy location.

sustained rate of fire: Rate of fire that a weapon can continue to deliver for an indefinite period without overheating.

terminal ballistics: What happens to the bullet when it comes in contact with the target.

tight shot group: A shot group with all bullet holes close together.

tracer ammunition: Ammunition with a substance at the rear of the bullet that ignites soon after firing. It burns brightly so the trajectory of the bullet can be seen.

tracking: Engaging moving targets where the lead is established and maintained; moving with the target as the trigger is squeezed.

train the trainer: Describes any training that is designed to train marksmanship instructors or coaches.

trainfire: A marksmanship program using pop-up targets in a realistic environment.

trajectory: The flight path the bullet takes from the rifle to the target.

trapping: A technique for engaging moving targets. The aiming point is established forward of the target. The rifle is held stationary and fired as the target approaches the aiming point.

trigger squeeze: The fourth fundamental; squeezing the trigger so that the movement of firing is a surprise, the lay of the weapon is not disturbed, and a large target hit can be expected.

unit marksmanship: All marksmanship training that is conducted by units.

unlocking: The step in the cycle of operation that refers to the clockwise rotation of the bolt after firing, freeing the bolt from the barrel locking lugs.

unsupported position: Any position that requires the firer to hold the weapon steady using only his body (bone support).

vertical dispersion: The up-and-down displacement of bullets on a target.

Weaponeer: A training device that simulates the firing of the M16 rifle to provide performance feedback.

windage adjustment: Moving the rear sight aperture to cause the bullet to strike left or right on the target.

wind value: The effect the wind will have on the trajectory of the bullet.

wobble area: The natural movement of the weapon/sight on and around an aiming point when the weapon is being held in a steady position.

zero criterion: The standard or requirement for zeroing; 4cm or smaller group at 25 meters.

zeroing: Adjusting the rifle sights so bullets hit the aiming point at a given range.

zero target: A scaled-silhouette target with a superimposed grid for use at 25 meters.

References

SOURCES USED

These sources were used to prepare this publication.

AR 385-10, The Army Safety Program. 23 August 2007.

AR 385-63, Range Safety. 19 May 2003.

DA Pam 350-38, Standards in Weapons Training. 16 August 2004.

DA Pam 385-63, Range Safety. 10 April 2003.

FM 5-19, Composite Risk Management. 21 August 2006.

TM 9-1005-249-10, , Operator's Manual for Rifle, 5.56-mm, M16 (NSN 1005-00-856-6885), Rifle 5.56-mm, M16A1 (1005-00-073-9421). 11 May 1990.

TM 9-1005-319-10, Operator's Manual for Rifle, 5.56 MM, M16A2 W/E (NSN 1005-01-128-9936) (EIC: 4GM); Rifle, 5.56 MM, M16A3 (1005-01-357-5112); Rifle, 5.56 MM, M16A4 (1005-01-383-2872) (EIC: 4F9); Carbine, 5.56 MM, M4 W/E (1005-01-231-0973) (EIC: 4FJ); Carbine, 5.56 MM, M4A1 (1005-01-382-0953) (EIC: 4GC), 1 October 1998. 01 October 1998.

DOCUMENTS NEEDED

These documents must be available to the intended users of this publication.

ARMY PUBLICATIONS

AR 350-1, Army Training and Leader Development. 03 August 2007.

FM 7-0, Training the Force. 22 October 2002.

FM 7-1, Battle Focused Training. 15 September 2003.

TC 7-9, Infantry Live-Fire Training. 30 September 1993.

TC 25-1, Training Land. 15 March 2004.

TC 25-8, Training Ranges. 5 April 2004.

TM 9-1005-319-23&P, Unit and Direct Support Maintenance Manual (Including Repair Parts and Special Tools List) FOR RIFLE, 5.56MM, M16A2 W/E (NSN 1005-01-128-9936) (EIC: 4GM) Rifle, 5.56MM, M16A3, (1005-01-357-5112) Rifle, 5.56MM, M16A4, W/E (1005-01-383-2872) (EIC: 4F9) Carbine, 5.56MM, M4 (1005-01-231-0973) (EIC: 4FJ) and Carbine, 5.56MM, M4A1 (1005-01-382-0953) (EIC: 4GC), 1 May 1991.

TM 9-1240-413-13&P, Operator and Field Maintenance Manual Including Repair Parts and Special Tools List for M68 Sight, Reflex, w/Quick Release and Sight Mount (COMP M2: NSN 1240-01-411-1265) (COMP M4: NSN 1240-01-540-3690). 19 March 2008.

TM 9-5855-1914-13&P, Operator and Field Maintenance Manual Including Repair Parts and Special Tools List for the Advanced Target Pointer Illuminator Aiming Light (ATPIAL) AN/PEQ-15 (NSN 5855-01-534-5931). 16 February 2007.

TM 9-6920-746-12&P, Operator's and Organizational Maintenance Manual (Including Repair Parts and Special Tools List) M2 Practice Bolt Plastic Ammunition for Rifle 5.56mm, M16 Series (NSN 1005-01-184-4041). 09 September 1986.

TM 11-5855-213-10, Operator's Manual for Night Vision Sight, Individual Served Weapon, AN/PVS-4 (NSN 5855-00-629-5334) (EIC: IPJ). 01 February 1993.

TM 11-5855-261-10, Operator's Manual for Infrared Aiming Light, AN/PAQ-4 (NSN 5855-01-107-5925). 28 May 1981.

TM 11-5855-301-12&P, Operator's and Unit Maintenance Manual (Including Repair Parts and Special Tools List) Light, Aiming, Infrared, AN/PAQ-4B (NSN 5855-01-361-1362), AN/PAQ-4C (5855-01-398-4315).15 May 2000.

TM 11-5855-306-10, Operator's Manual for Monocular Night Vision Device, AN/PVS-14 (NSN 5855-01-432-0524) (EIC: IPX). 15 May 2007.

TM 11-5855-308-12&P, Operator's and Unit Maintenance Manual (Including Repair Parts and Special Tools List), Target Pointer Illuminator/Aiming Light AN/PEQ-2A (NSN 5855-01-447-8992). 15 May 2000.

TM 11-5855-312-10, Operator's Manual Sight, Thermal AN/PAS-13B(V)2 (NSN 5855-01-464-3152); AN/PAS-13B(V)3 (5855-01-464-3151). 15 February 2005.

TM 11-5855-316-10, Operator's Manual AN/PAS-13C(V)1 Sight, Thermal (NSN 5855-01-523-7707); AN/PAS-13C(V)2 Sight, Thermal (NSN 5855-01-523-7713) (MWTS); AN/PAS-13C(V)3 Sight, Thermal (NSN 5855-01-523-7715). 15 June 2006.

TM 11-5855-317-10, Operator's Manual For Sight, Thermal AN/PAS-13D(V)2 (NSN 5855-01-524-4313); AN/PAS-13D(V)3 NSN 5855-01-524-4314) (HWTS). 15 February 2007.

Army Forms

DA Form 2028, Recommended Changes to Publications and Blank Forms.

DA Form 3595-R, Record Fire Scorecard.

DA Form 3601-R, Single Target—Field Fire I Scorecard.

DA Form 5239-R, 100-, 200-, and 300-Meter Downrange Feedback Scorecard.

DA Form 5241-R, Single and Multiple Targets—Field Fire II Scorecard.

DA Form 5789-R, Record Fire Scorecard—Known-Distance Course (LRA).

DA Form 5790-R, Record Fire Scorecard—Scaled Target Alternate Course.

DA Form 7489-R, Record Night Fire Scorecard (LRA).

DA Form 7566, Composite Risk Management Worksheet.

DA Form 7649-R, Squad Designated Marksman—Record Fire I and II.

DA Form 7650-R, Squad Designated Marksman—Position Evaluation.

READINGS RECOMMENDED

These sources contain relevant supplementary information.

AR 190-11, Physical Security of Arms, Ammunition, and Explosives. 15 November 2006.

AR 350-38, Training Device Polices and Management. 15 October 1993.

DA Pam 350-9, Index and Description of Army Training Devices. 03 September 2002.

DA Pam 385-64, Ammunition and Explosive Safety Standards. 15 December 1999.

FM 1-02, Operational Terms and Graphics. 21 September 2004.

FM 3-06, Urban Operations. 26 October 2006.

FM 3-06.11, Combined Arms Operations in Urban Terrain. 28 February 2002.

FM 3-11.4, Multiservice Tactics, Techniques, and Procedures for Nuclear, Biological, and Chemical (NBC) Protection. 02 June 2003.

FM 4-25.11, First Aid. 23 December 2002.

FM 20-3, Camouflage, Concealment, and Decoys. 30 August 1999.

FM 21-10, Field Hygiene and Sanitation. 21 June 2000

FM 3-21.75, The Warrior Ethos and Soldier Combat Skills. 28 January 2008.

FM 22-6, Guard Duty. 17 September 1971.

FM 25-4, How to Conduct Training Exercises. 10 September 1984.

STP 21-1-SMCT, Soldier's Manual of Common Tasks, Skill Level 1. 14 December 2007.

STP 21-24-SMCT, Soldier's Manual of Common Tasks (SMCT), Skill Levels 2-4. 2 October 2006.

TC 25-20, A Leader's Guide to After-Action Reviews. 30 September 1993.

TC 90-1, Training for Urban Operations. 19 May 2008.

TM 9-1005-249-23&P, Unit and Direct Support Maintenance Manual (Including Repair Parts and Special Tools List) for Rifle, 5.56-mm, M16 (NSN 1005-00-856-6885) (EIC: 4F7); Rifle, 5.56-mm, M16A1 (1005-00-073-9421) (EIC: 4FC). 19 June 1991.

TM 9-6920-363-12&P, Operator's and Organizational Maintenance Manual (Including Repair Parts and Special Tools List) for Conversion Kit (Cal .22 Rimfire Adapter) M261 (NSN 1005-01-010-1561) for Rifle: 5.56-mm, M16 and M16A1. Reprinted with Changes 1 and 2. 21 August 1984.

INTERNET WEBSITES

U.S. Army Publishing Directorate, http://www.apd.army.mil/

Reimer Doctrine and Training Digital Library, http://www.train.army.mil

Index

B

M

N

P

SINGLE TARGET
FIELD FIRE I SCORECARD

For use of this form see FM 3-22.9; the proponent agency is TRADOC.

ID CODE	UNIT	DATE *(YYYYMMDD)*	EVALUATOR'S ID CODE

TABLE 1
SUPPORTED FIRING POSITION

RD	RANGE (m)	TIME (sec)	HIT	MISS
1	75	6		
2	175	8		
3	300	10		
4	175	8		
5	75	6		
6	300	10		
7	300	10		
8	75	75		
9	175	175		
10	175	8		
11	300	10		
12	175	8		
13	75	6		
14	300	10		
15	175	8		
16	75	6		
17	300	10		
18	75	6		
TOTAL				

TABLE 2
SUPPORTED FIRING POSITION

RD	RANGE (m)	TIME (sec)	HIT	MISS
1	75	6		
2	175	8		
3	300	10		
4	175	8		
5	75	6		
6	300	10		
7	300	10		
8	75	6		
9	175	8		
10	175	8		
11	300	10		
12	175	8		
13	75	6		
14	300	10		
15	175	8		
16	75	6		
17	300	10		
18	75	8		
TOTAL				

TABLE 3
PRONE FIRING POSITION

RD	RANGE (m)	TIME (sec)	HIT	MISS
1	75	6		
2	175	8		
3	300	10		
4	175	8		
5	75	6		
6	300	10		
7	300	10		
8	75	6		
9	175	8		
TOTAL				

TABLE 4
KNEELING FIRING POSITION

RD	RANGE (m)	TIME (sec)	HIT	MISS
1	75	6		
2	175	8		
3	75	6		
4	175	8		
5	75	6		
6	175	8		
7	75	6		
8	75	6		
9	175	8		
TOTAL				

SCORE

TABLE	HIT	MISS	NO FIRE
1			
2			
3			
4			

REMARKS

DATE INITIALED *(YYYYMMDD)*	SCORER'S INITIALS

100-, 200- AND 300-METER DOWNRANGE FEEDBACK SCORECARD

For use of this form see FM 3-22.9; the proponent agency is TRADOC.

ID CODE	UNIT	DATE *(YYYYMMDD)*	EVALUATOR'S ID CODE

ZERO CONFIRMATION

RANGE (m)	HIT	MISS
300		

ZERO CONFIRMATION REFIRE

RANGE (m)	HIT	MISS
300		

TABLE 1 -- DOWNRANGE SILHOUETTES, SUPPORTED FIRING POSITION

RDS	RANGE (m)	HIT	MISS
5	100		
10	200		
5	300		
TOTAL			

TABLE 2 -- DOWNRANGE SILHOUETTES, PRONE FIRING POSITION

RDS	RANGE (m)	HIT	MISS
5	100		
10	200		
5	300		
TOTAL			

ZERO-SHOT LOCATIONS

300M

WIND SPEED (mph):

WIND DIRECTION (use arrow):

SHOT LOCATIONS

SUPPORTED FIRING POSITION

TABLE 1

PRONE UNSUPPORTED FIRING POSITION

TABLE 2

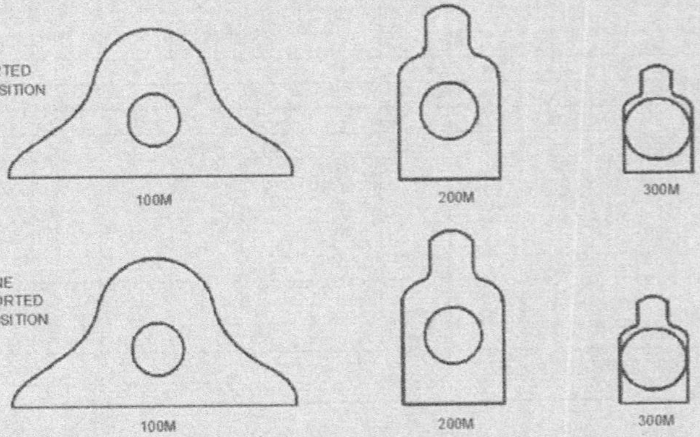

REMARKS

DATE INITIALED *(YYYYMMDD)*	SCORER'S INITIALS

SINGLE AND MULTIPLE TARGETS
FIELD FIRE II SCORECARD

For use of this form see FM 3-22.9; the proponent agency is TRADOC.

ID CODE	UNIT	DATE *(YYYYMMDD)*	EVALUATOR'S ID CODE

TABLE 1 — SUPPORTED FIRING POSITION

RD	RANGE (m)	TIME (sec)	HIT	MISS
1	75	5	☐	☐
2	175	7	☐	☐
3	75	11	☐	☐
4	300		☐	☐
5	75	9	☐	☐
6	175		☐	☐
7	75	10	☐	☐
8	300		☐	☐
9	175	11	☐	☐
10	300		☐	☐
		TOTAL		

TABLE 2 — SUPPORTED FIRING POSITION

RD	RANGE (m)	TIME (sec)	HIT	MISS
1	175	7	☐	☐
2	75	10	☐	☐
3	300		☐	☐
4	75	9	☐	☐
5	175		☐	☐
6	300	9	☐	☐
7	75	9	☐	☐
8	175		☐	☐
9	175	11	☐	☐
10	300		☐	☐
11	75	9	☐	☐
12	175		☐	☐
13	175	11	☐	☐
14	300		☐	☐
15	75	5	☐	☐
16	175	11	☐	☐
17	300		☐	☐
18	75	9	☐	☐
19	175		☐	☐
20	75	10	☐	☐
21	300		☐	☐
22	175	7	☐	☐
		TOTAL		

TABLE 3 — PRONE FIRING POSITION

RD	RANGE (m)	TIME (sec)	HIT	MISS
1	75	6	☐	☐
2	175	8	☐	☐
3	75	13	☐	☐
4	300		☐	☐
5	75	11	☐	☐
6	175		☐	☐
7	75	12	☐	☐
8	300		☐	☐
9	175	13	☐	☐
10	300		☐	☐
11	75	11	☐	☐
12	175		☐	☐
13	175	8	☐	☐
14	75	6	☐	☐
15	75	11	☐	☐
16	175		☐	☐
17	75	12	☐	☐
18	300		☐	☐
19	75	11	☐	☐
20	175		☐	☐
21	175	13	☐	☐
22	300		☐	☐
		TOTAL		

SCORE

TABLE	HIT	MISS
2		
3		
TOTAL		

REMARKS

DATE INITIALED *(YYYYMMDD)*	SCORER'S INITIALS

RECORD FIRING SCORECARD -- KNOWN DISTANCE COURSE

For use of this form see FM 3-22.9; the proponent agency is TRADOC.

ID CODE	UNIT	DATE *(YYYYMMDD)*	EVALUATOR'S ID CODE

TABLE 1 PRONE SUPPORTED OR FOXHOLE SUPPORTED FIRING POSITION (TIME: 120 SECONDS)								TABLE 2 PRONE UNSUPPORTED FIRING POSITION (TIME: 60 SECONDS)				TABLE 3 KNEELING FIRING POSITION (TIME: 60 SECONDS)			
RD	RANGE (m)	HIT	MISS	RD	RANGE (m)	HIT	MISS	RD	RANGE (m)	HIT	MISS	RD	RANGE (m)	HIT	MISS
1		☐	☐	11		☐	☐	1		☐	☐	1		☐	☐
2		☐	☐	12		☐	☐	2		☐	☐	2		☐	☐
3		☐	☐	13		☐	☐	3		☐	☐	3		☐	☐
4	300	☐	☐	14	300	☐	☐	4	300	☐	☐	4	300	☐	☐
5		☐	☐	15		☐	☐	5		☐	☐	5		☐	☐
6	E-SIL	☐	☐	16	E-SIL	☐	☐	6	E-SIL	☐	☐	6	F-SIL	☐	☐
7		☐	☐	17		☐	☐	7		☐	☐	7		☐	☐
8		☐	☐	18		☐	☐	8		☐	☐	8		☐	☐
9		☐	☐	19		☐	☐	9		☐	☐	9		☐	☐
10		☐	☐	20		☐	☐	10		☐	☐	10		☐	☐
TOTAL				TOTAL				TOTAL				TOTAL			

SCORE

TABLE	HIT	MISS	NO FIRE
1			
2			
3			

FIRER'S QUALIFICATION SCORE

QUALIFICATION SCORE RATINGS *(Check One)*

☐ 36-40 -- EXPERT

☐ 26-32 -- MARKSMAN

☐ 33-37 -- SHARPSHOOTER

☐ 25 AND BELOW -- UNQUALIFIED

LIGHT

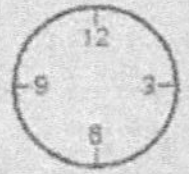

WIND

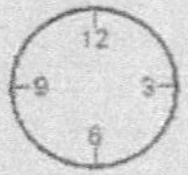

NIGHT FIRE EXERCISE

DATE *(YYYYMMDD)*	HIT	MISS	GO	NO GO
			☐	☐

CBRN FIRE EXERCISE

DATE *(YYYYMMDD)*	HIT	MISS	GO	NO GO
			☐	☐

THE FIRER WILL BE ISSUED 40 ROUNDS. THE ROUNDS WILL BE PRELOADED IN FOUR 10-ROUND MAGAZINES - TWO FOR TABLE 1, ONE FOR EACH REMAINING TABLE.

ZERO			
ELEV			
WIND			

DATE INITIALED *(YYYYMMDD)*	SCORER'S INITIALS
DATE INITIALED *(YYYYMMDD)*	OFFICER'S INITIALS

This scorecard is used to score known distance course record fire qualification when the known distance range is used. This course is used only when the standard record fire course is not available.

CONDUCT OF FIRE

The uniform for qualification ia a helmet, LBE/LBV, and interceptor body armor with front and back SAPI plates (if available). No other armor is required.

(1) Table 1 -- Supported Firing Position. The firer is given two 10-round magazines to engage an E-silhouette at 300 yards within 120 seconds.

(2) Table 2 -- Unsupported Firing Position. The firer is given a 10-round magazine to engage an E-silhouette at 200 yards within 60 seconds.

(3) Table 3 -- Unsupported Firing Position. The firer is given a 10-round magazine to engage an F-silhouette at 100 yards within 60 seconds.

SCORING

Scoring is conducted in the pits, with the results provided after each table. A hit is scored for any bullet hole that is within or touches some part of the silhouette facing.

RECORD FIRING SCORECARD -- SCALED TARGET ALTERNATE COURSE

For use of this form see FM 3-22.9; the proponent agency is TRADOC.

ID CODE	UNIT	DATE *(YYYYMMDD)*	EVALUATOR'S ID CODE

TABLE 1 — PRONE SUPPORTED OR FOXHOLE SUPPORTED FIRING POSITION (TIME: 120 SECONDS)

RD	RANGE (m)	HIT	MISS	RD	RANGE (m)	HIT	MISS
1	300	☐	☐	11	150	☐	☐
2	300	☐	☐	12	150	☐	☐
3	250	☐	☐	13	100	☐	☐
4	250	☐	☐	14	100	☐	☐
5	200	☐	☐	15	100	☐	☐
6	200	☐	☐	16	100	☐	☐
7	200	☐	☐	17	100	☐	☐
8	200	☐	☐	18	100	☐	☐
9	150	☐	☐	19	50	☐	☐
10	150	☐	☐	20	50	☐	☐
TOTAL				TOTAL			

TABLE 2 — PRONE UNSUPPORTED FIRING POSITION (TIME: 60 SECONDS)

RD	RANGE (m)	HIT	MISS
1	300	☐	☐
2	250	☐	☐
3	200	☐	☐
4	200	☐	☐
5	150	☐	☐
6	150	☐	☐
7	100	☐	☐
8	100	☐	☐
9	100	☐	☐
10	50	☐	☐
TOTAL			

TABLE 3 — KNEELING FIRING POSITION (TIME: 60 SECONDS)

RD	RANGE (m)	HIT	MISS
1	150	☐	☐
2	150	☐	☐
3	100	☐	☐
4	100	☐	☐
5	100	☐	☐
6	100	☐	☐
7	100	☐	☐
8	100	☐	☐
9	50	☐	☐
10	50	☐	☐
TOTAL			

SCORE

TABLE	HIT	MISS	NO FIRE
1			
2			
3			

FIRER'S QUALIFICATION SCORE

QUALIFICATION SCORE RATINGS *(Check One)*

- ☐ 36-40 -- EXPERT
- ☐ 23-29 -- MARKSMAN
- ☐ 30-35 -- SHARPSHOOTER
- ☐ 22 AND BELOW -- UNQUALIFIED

FIRER ISSUED 40 ROUNDS TO ENGAGE 10 TARGETS. NO MORE THAN 4 ROUNDS PER TARGET. THE ROUNDS WILL BE PRELOADED IN ONE 20-ROUND MAGAZINE FOR TABLE 1, ONE 10-ROUND MAGAZINE FOR TABLE 2, AND ONE 10-ROUND MAGAZINE FOR TABLE 3.

NIGHT FIRE EXERCISE

DATE *(YYYYMMDD)*	HIT	MISS	GO	NO GO
			☐	☐

CBRN FIRE EXERCISE

DATE *(YYYYMMDD)*	HIT	MISS	GO	NO GO
			☐	☐

REMARKS

DATE INITIALED *(YYYYMMDD)*	SCORER'S INITIALS
DATE INITIALED *(YYYYMMDD)*	OFFICER'S INITIALS

This scorecard is used to score alternate course record fire qualification when the 25-meter scaled silhouette target (NSN 6920-01-167-1396) is used. The alternate course is used only when standard record fire and known distance ranges are unavailable.

NOTE: If zeroing/grouping exercises are not performed on the day of record fire, six rounds of training/sustainment ammunition will be fired for 25-meter zero confirmation prior to conducting the qualification course.

CONDUCT OF FIRE

Alternate course qualification firers will have one 20-round magazine and two 10-round magazines. To ensure that firers do not forget which targets they engaged and shoot a given target more than the prescribed number of times, firers should adhere to the following guideline: Engage targets on the sheet from left to right and nearest to farthest. Engagement should follow this order: 50m, 100m left, 100m center, 100m right, 150m left, 150m right, 200m left, 200m right, 250m, and 300m.

Though the time between each firing position is not specified, enough time should be allotted to allow the firer to clear his weapon, quickly change firing positions, and reload before beginning the next firing table. The range RSO ensures that enough time is given between each change in firing positions to facilitate the timely flow of the record fire qualification table.

(1) Table 1-- Prone Supported Firing Position or (at the unit commander's discretion) Foxhole Supported Firing Position. The firer is given one 20-round magazine to engage 10 silhouettes on the target sheet. Table 1 includes 2 rounds for each silhouette. Firing must be completed in 120 seconds. No more than 2 hits are scored for each silhouette.

(2) Table 2 -- Prone Unsupported Firing Position. The firer is given one 10-round magazine to engage 10 silhouettes on the same target sheet. Table 2 includes 1 round for each silhouette. Firing must be completed in 60 seconds. No more than 1 hit is scored for each silhouette.

(3) Table 3 -- Kneeling Firing Position. The firer is given one 10-round magazine to engage 10 silhouettes on the same target sheet. Table 3 includes 2 rounds for each silhouette positioned 50 to 150 meters away. Firing must be completed in 60 seconds. No more than 2 hits are scored for each silhouette.

SCORING

The same target sheet is used for every 40-round qualification table that a firer completes. One hit is awarded for each round that strikes within or touches some part of the silhouette. A maximum of four hits for each silhouette on the same target sheet are scored.

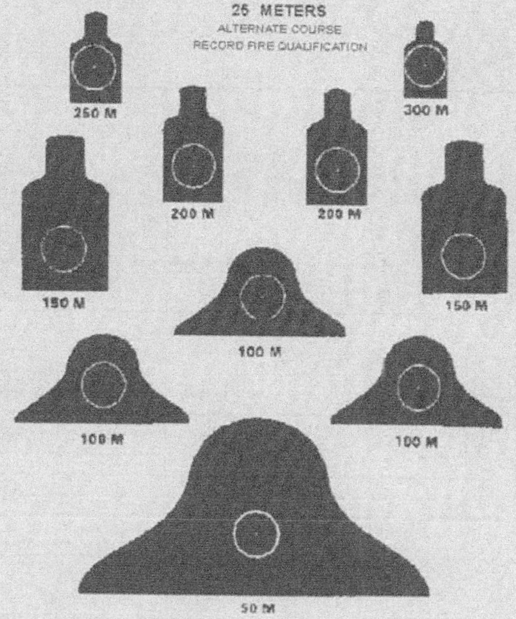

THE WHITE DOT ON EACH TARGET SHOWS THE CENTER OF MASS AIMING POINT. BULLETS SHOULD HIT WITHIN THE CIRCLE, BUT ARE SCORED AS HITS IF THEY HIT ANY PART OF THE SILHOUETTE.

RECORD NIGHT FIRE SCORECARD

For use of this form see FM 3-22.9; the proponent agency is TRADOC.

ID CODE	UNIT	DATE *(YYYYMMDD)*	EVALUATOR'S ID CODE

TABLE 1
FOXHOLE SUPPORTED FIRING POSITION

RD	RANGE (m)	TIME (sec)	HIT	MISS	NO FIRE	RD	RANGE (m)	TIME (sec)	HIT	MISS	NO FIRE
1	50L	3	☐	☐	☐	11	100	8	☐	☐	☐
2	200	6	☐	☐	☐	12	200		☐	☐	☐
3	100	4	☐	☐	☐	13	150	8	☐	☐	☐
4	150	5	☐	☐	☐	14	50R		☐	☐	☐
5	100	8	☐	☐	☐	15	100	8	☐	☐	☐
6	150	5	☐	☐	☐	16	150		☐	☐	☐
7	50R	3	☐	☐	☐	17	200	6	☐	☐	☐
8	200	6	☐	☐	☐	18	150	5	☐	☐	☐
9	150	5	☐	☐	☐	19	50L	9	☐	☐	☐
10	250	7	☐	☐	☐	20	250		☐	☐	☐
TOTAL						TOTAL					

TABLE 2
PRONE UNSUPPORTED FIRING POSITION

RD	RANGE (m)	TIME (sec)	HIT	MISS	NO FIRE	RD	RANGE (m)	TIME (sec)	HIT	MISS	NO FIRE
1	100	5	☐	☐	☐	11	150	8	☐	☐	☐
2	200	8	☐	☐	☐	12	50R	5	☐	☐	☐
3	150	6	☐	☐	☐	13	100	12	☐	☐	☐
4	50L	12	☐	☐	☐	14	200		☐	☐	☐
5	200		☐	☐	☐	15	150	12	☐	☐	☐
6	150	12	☐	☐	☐	16	50L		☐	☐	☐
7	100		☐	☐	☐	17	100	12	☐	☐	☐
8	50R	12	☐	☐	☐	18	150		☐	☐	☐
9	150		☐	☐	☐	19	200	12	☐	☐	☐
10	100	8	☐	☐	☐	20	100		☐	☐	☐
TOTAL						TOTAL					

SCORE

TABLE	HIT	MISS	NO FIRE
1			
2			
TOTAL			

FIRER'S QUALIFICATION SCORE

QUALIFICATION SCORE RATINGS *(Check One)*

- ☐ 35-40 -- EXPERT
- ☐ 17-23 -- MARKSMAN
- ☐ 24-34 -- SHARPSHOOTER
- ☐ 16 AND BELOW -- UNQUALIFIED

AIMING DEVICE USED *(Check One)*

- ☐ AN/PEQ-2A/B
- ☐ AN/PAQ-4B/C
- ☐ AN/PVS-4

REMARKS

DATE INITIALED *(YYYYMMDD)*	SCORER'S INITIALS
DATE INITIALED *(YYYYMMDD)*	OFFICER'S INITIALS

SQUAD DESIGNATED MARKSMAN
RECORD FIRE I AND II

For use of this form, see FM 3-22.9; the proponent agency is TRADOC.

ID CODE	UNIT	DATE *(YYYYMMDD)*	EVALUATOR'S ID CODE

RECORD FIRE I - MECHANICAL SIGHT ADJUSTMENT

TARGET RANGE/POSITION	HIT	MISS
100m unsupported	☐	☐
100m unsupported	☐	☐
100m unsupported	☐	☐
100m unsupported	☐	☐
200m supported	☐	☐
200m supported	☐	☐
200m unsupported	☐	☐
200m unsupported	☐	☐
300m supported	☐	☐
300m supported	☐	☐
300m unsupported	☐	☐
300m unsupported	☐	☐
400m supported	☐	☐
400m supported	☐	☐
400m supported	☐	☐
400m supported	☐	☐
500m supported	☐	☐
500m supported	☐	☐
500m supported	☐	☐
500m supported	☐	☐
TOTAL		

☐ **PASS** (14-20) ☐ **FAIL** (13 and below)

DATE INITIALED *(YYYYMMDD)*

DATE INITIALED *(YYYYMMDD)*

RECORD FIRE II - ADJUSTED AIMING POINTS/OPTIC

TARGET RANGE/POSITION	HIT	MISS
100m unsupported	☐	☐
100m unsupported	☐	☐
100m unsupported	☐	☐
100m unsupported	☐	☐
200m supported	☐	☐
200m supported	☐	☐
200m unsupported	☐	☐
200m unsupported	☐	☐
300m supported	☐	☐
300m supported	☐	☐
300m unsupported	☐	☐
300m unsupported	☐	☐
400m supported	☐	☐
400m supported	☐	☐
400m supported	☐	☐
400m supported	☐	☐
500m supported	☐	☐
500m supported	☐	☐
500m supported	☐	☐
500m supported	☐	☐
TOTAL		

☐ **PASS** (14-20) ☐ **FAIL** (13 and below)

SCORER'S INITIALS

OFFICER'S INITIALS

SQUAD DESIGNATED MARKSMAN POSITION EVALUATION

For use of this form, see FM 3-22.9; the proponent agency is TRADOC.

☐ FOXHOLE SUPPORTED ☐ PRONE UNSUPPORTED	ID CODE	UNIT	DATE *(YYYYMMDD)*	EVALUATOR'S ID CODE

PHASE	EYE RELIEF	TRIGGER FINGER	ELBOWS	NON-FIRING HAND	LEGS
EXAMPLE	How far is nose from charging handle?	What part of 1st joint (tip, middle, joint)?	Is platform stable?	Where is it located (Soldier's choice, stable)?	Where are they located (stability for upper body)?
Phase I -- Steady Position					
Phase II -- Borelight Dry Fire					
Phase V -- Practice Qualification					
100m					
200m					
300m					
400m					
500m					
Phase VI -- Record Qualification					
100m					
200m					
300m					
400m					
500m					

DATE INITIALED *(YYYYMMDD)*	SCORER'S INITIALS

RANGER CREED

Recognizing that I volunteered as a Ranger, fully knowing the hazards of my chosen profession, I will always endeavor to uphold the prestige, honor, and high esprit do corps of the Rangers.

Acknowledging the fact that a Ranger is a more elite Soldier who arrives at the cutting edge of battle by land, sea, or air, I accept the fact that as a Ranger my country expects me to move further, faster, and fight harder than any other soldier.

Never shall I fail my comrades I will always keep myself mentally alert, physically strong, and morally straight and I will shoulder more than my share of the task whatever it may be, one hundred percent and then some.

Gallantly will I show the world that I am a specially selected and well trained Soldier. My courtesy to superior officers, neatness of dress, and care of equipment shall set the example for others to follow.

Energetically will I meet the enemies of my country. I shall defeat them on the field of battle for I am better trained and will fight with all my might. Surrender is not a Ranger word. I will never leave a fallen comrade to fall into the hands of the enemy and under no circumstances will I ever embarrass my country.

Readily will I display the intestinal fortitude required to fight on to the Ranger objective and complete the mission, though I be the lone survivor.

STANDING ORDERS, ROGERS' RANGERS

1. Don't forget nothing.
2. Have your musket clean as a whistle, hatchet scoured, sixty rounds powder and ball, and be ready to march at a minute's warning.
3. When you're on the march, act the way you would if you was sneaking up on a deer. See the enemy first.
4. Tell the truth about what you see and what you do. There is an army depending on us for correct information. You can lie all you please when you tell other folks about the Rangers, but don't never lie to a Ranger or officer.
5. Don't never take a chance you don't have to.
6. When we're on the march we march single file, far enough apart so one shot can't go through two men.
7. If we strike swamps, or soft ground, we spread out abreast, so it's hard to track us.
8. When we march, we keep moving till dark, so as to give the enemy the least possible chance at us.
9. When we camp, half the party stays awake while the other half sleeps.
10. If we take prisoners, we keep' em separate till we have had time to examine them, so they can't cook up a story between' em.
11. Don't ever march home the same way. Take a different route so you won't be ambushed.
12. No matter whether we travel in big parties or little ones, each party has to keep a scout 20 yards ahead, 20 yards on each flank, and 20 yards in the rear so the main body can't be surprised and wiped out.
13. Every night you'll be told where to meet if surrounded by a superior force.
14. Don't sit down to eat without posting sentries.
15. Don't sleep beyond dawn. Dawn's when the French and Indians attack.
16. Don't cross a river by a regular ford.
17. If somebody's trailing you, make a circle, come back onto your own tracks, and ambush the folks that aim to ambush you.
18. Don't stand up when the enemy's coming against you. Kneel down, lie down, hide behind a tree.
19. Let the enemy come till he's almost close enough to touch, then let him have it and jump out and finish him up with your hatchet.

--MAJOR ROBERT ROGERS, 1759

RANGER HISTORY

The history of the American Ranger is a long and colorful saga of courage, daring and outstanding leadership. It is a story of men whose skills in the art of fighting have seldom been surpassed. Only the highlights of their numerous exploits are told here.

Rangers primarily performed defensive missions until Benjamin Church's Company of Independent Rangers from Plymouth Colony proved successful in raiding hostile Indians during King Phillip's War in 1675. In 1756, Major Robert Rogers, a native of New Hampshire, recruited nine companies of American colonists to fight for the British during the French and Indian War. Ranger techniques and methods of operation were an inherent characteristic of the American frontiersmen; however, Major Rogers was the first to capitalize on them and incorporate them into the fighting doctrine of a permanently organized fighting force.

The method of fighting used by the first Rangers was further developed during the Revolutionary War by Colonel Daniel Morgan, who organized a unit known as "Morgan's Riflemen". According to General Burgoyne, Morgan's men were "....the most famous corps of the Continental Army, all of them crack shots."

Francis Marion, the "Swamp Fox," organized another famous Revolutionary War Ranger element known as "Marion's Partisans." Marion's Partisans, numbering anywhere from a handful to several hundred, operated both with and independent of other elements of General Washington's Army. Operating out of the Carolina swamps, they disrupted British communications and prevented the organization of loyalists to support the British cause, substantially contributing to the American victory.

The American Civil War was again the occasion for the creation of special units such as Rangers. John S. Mosby, a master of the prompt and skillful use of cavalry, was one of the most outstanding Confederate Rangers. He believed that by resorting to aggressive action he could compel his enemies to guard a hundred points. He would then attack one of the weakest points and be assured numerical superiority.

With America's entry into the Second World War, Rangers came forth to add to the pages of history. Major William O. Darby organized and activated the 1st Ranger Battalion on June19, 1942 at Carrickfergus, North Ireland. The members were all hand-picked volunteers; 50 participated in the gallant Dieppe Raid on the northern coast of France with British and Canadian commandos. The 1st, 3rd, and 4th Ranger Battalions participated with distinction in the North African, Sicilian and Italian campaigns. Darby's Ranger Battalions spearheaded the Seventh Army landing at Gela and Licata during the Sicilian invasion and played a key role in the subsequent campaign which culminated in the capture of Messina. They infiltrated German lines and mounted an attack against Cisterna, where they virtually annihilated an entire German parachute regiment during close in, night, bayonet, and hand-to-hand fighting.

The 2nd and 5th Ranger Battalions participated in the D-Day landings at Omaha Beach, Normandy. It was during the bitter fighting along the beach that the Rangers gained their official motto. As the situation became critical on Omaha Beach, the division commander of the 29th Infantry Division stated that the entire force must clear the beach and advance inland. He then turned to Lieutenant Colonel Max Schneider, Commander of the 5th Ranger Battalion, and said, "Rangers, lead the way." The 5th Ranger Battalion spearheaded the breakthrough and thus enabled the allies to drive inland away from the invasion beaches.

The 6th Ranger Battalion, operating in the Pacific, conducted Ranger type missions behind enemy lines which involved reconnaissance and hard-hitting, long-range raids. They were the first American contingent to return to the Philippines, destroying key coastal installations prior to the invasion. A reinforced company from the 6th Ranger Battalion formed the rescue force which liberated American and allied prisoners of war from the Japanese prison camp at Cabanatuan.

Another Ranger-type unit was the 5307th Composite Unit (Provisional), organized and trained as a long-range penetration unit for employment behind enemy lines in Japanese occupied Burma. The unit commander was Brigadier General (later Major General) Frank D. Merrill. Its 2,997 officers and men became popularly known as "Merrill's Marauders."

The men composing Merrill's Marauders were volunteers from the 5th, 154th, and 33rd Infantry Regiments and from other Infantry regiments engaged in combat in the southwest and South Pacific. These men responded to a call from then Chief of Staff, General George C. Marshall, for volunteers for a hazardous mission. These volunteers were to have a high state of physical ruggedness and stamina and were to come from jungle-trained and jungle-tested units.

Prior to their entry into the Northern Burma Campaign, Merrill's Marauders trained in India under the overall supervision of Major General Orde C. Wingate, British Army. There, they were trained from February to June 1943 in long-range penetration tactics and techniques of the type developed and first employed by General Wingate. The operations of the Marauders were closely coordinated with those of the Chinese 22nd and 38th Divisions in a drive to recover northern Burma and clear the way for the construction of Ledo Road, which was to link the Indian railhead at Ledo with the old Burma Road to China. The Marauders marched and fought through jungle and over mountains from Hukwang Valley in northwest Burma to Myitkyina and the Irrawaddy River. In 5 major and 30 minor engagements, they met and defeated the veteran soldiers of the Japanese 18th Division. Operating in the rear of the main force of the Japanese, they prepared the way for the southward advances of the Chinese by disorganizing supply lines and communications. The climax of the Marauder's operations was the capture of Myitkyina Airfield, the only all-weather strip in northern Burma. This was the final victory of "Merrill's Marauders," which disbanded in August 1944. Remaining personnel were consolidated into the 475th Infantry Regiment, which fought its last battle on February 3, and 4,1945, at Loi-Kang Ridge, China. This Infantry Regiment would serve as the forefather of today's 75th Ranger Regiment.

Shortly after the outbreak of the Korean War in June 1950, the 8th Army Ranger Company was formed of volunteers from American units in Japan. The Company was trained in Korea and distinguished itself in combat during the drive to the Yalu River, performing task force and spearhead operations. In November 1950 during the massive Chinese intervention, this small unit, though vastly outnumbered, withstood five enemy assaults on its position.

In September 1950, a Department of the Army message called for volunteers to be trained as Airborne Rangers. In the 82nd Airborne Division, five thousand regular Army paratroopers volunteered, and from that number nine hundred men were selected to form the initial eight Airborne Ranger Companies. An additional nine companies were formed from volunteers of regular Army and National Guard Infantry Divisions. These seventeen Airborne Ranger companies were activated and trained at Fort Benning, Georgia, with most receiving additional training in the mountains of Colorado.

In 1950 and 1951, some 700 men of the 1st, 2nd, 3rd, 4th, 5th, and 8th Airborne Ranger companies fought to the front of every American Infantry Division in Korea. Attacking by land, water, and air, these six Ranger companies conducted raids, deep penetrations, and ambush operations against North Korean and Chinese forces. They were the first Rangers in history to make a combat jump. After the Chinese intervention, these Rangers were the first Americans to re-cross the 38th parallel. The 2nd Airborne Ranger Company was the only African American Ranger unit in the history of the American Army. The men of the six Ranger companies who fought in Korea paid the bloody price of freedom. One in nine of this gallant brotherhood died on the battlefields of Korea.

Other Airborne Ranger companies led the way while serving with infantry divisions in the United States, Germany, and Japan. Men of these companies volunteered and fought as members of line infantry units in Korea. One Ranger, Donn Porter, would be posthumously awarded the Medal of Honor. Fourteen Korean War Rangers became general officers and dozens became colonels, senior noncommissioned officers, and leaders in civilian life. They volunteered for the Army, the Airborne, the Rangers, and for combat. The first men to earn and wear the coveted Ranger Tab, these men are the original Airborne Rangers.

In October 1951, the Army Chief of Staff, General J. Lawton Collins directed, "Ranger training be extended to all combat units in the Army." The Commandant of the Infantry School was directed to establish a Ranger Department. This new department would develop and conduct a Ranger course of instruction. The objective was to raise the standard of training in all combat units. This program was built upon what had been learned from the Ranger Battalions of World War II and the Airborne Ranger companies of the Korean conflict.

During the Vietnam Conflict, fourteen Ranger companies consisting of highly motivated volunteers served with distinction from the Mekong Delta to the DMZ. Assigned to separate brigade, division, and field force units, they conducted long-range reconnaissance and exploitation operations into enemy-held areas providing valuable combat intelligence. Initially designated at LRRP, then LRP companies, these units were later designated as C, D,E,F,G,H,I,K,L,M,N,O and P (Ranger) 75th Infantry.

Following Vietnam, recognizing the need for a highly trained and highly mobile reaction force, the Army Chief of Staff, General Abrams directed the activation of the first battalion-sized Ranger units since World War II, the 1st and 2nd Battalions (Ranger), 75th Infantry. The 1st Battalion was trained at Fort Benning, Georgia and was activated February 8, 1974 at Fort Stewart, Georgia with the 2nd Battalion being activated on October 3, 1974. The 1st Battalion is now located at Hunter Army Airfield, Georgia and the 2nd Battalion at Fort Lewis, Washington.

The farsightedness of General Abrams' decision, as well as the combat effectiveness of the Ranger battalions, was proven during the United States' invasion of the island of Grenada in October 1983 to protect American citizens there, and to restore democracy. As expected, Rangers led the way! During this operation, code named "Urgent Fury," the Ranger battalions conducted a daring, low level airborne assault (from 500 feet) to seize the airfield at Point Salines, and then continued operations for several days to eliminate pockets of resistance, and rescue American medical students.

As a result of the demonstrated effectiveness of the Ranger battalions, the Department of the Army announced in 1984 that it was increasing the strength of Ranger units to its highest level in 40 years. To do this, it activated another Ranger battalion as well as a Ranger Regimental Headquarters. These new units, the 3rd Battalion (Ranger), 75th Infantry, and Headquarters Company (Ranger), 75th Infantry, have increased the Ranger strength of the Army to over 2,000 soldiers actually assigned to Ranger units. On February 3, 1986, the 75th Infantry was renamed the 75th Ranger Regiment.

On December 20, 1989, the 75th Ranger Regiment was once again called upon to demonstrate its effectiveness in combat. For the first time since its reorganization in 1984, the Regimental Headquarters and all three Ranger battalions were deployed on Operation "Just Cause" in Panama. During this operation, the 75th Ranger Regiment spearheaded the assault into Panama by conducting airborne assaults onto Torrijos/Tocumen Airport and Rio Hato Airfield to facilitate the restoration of democracy in Panama, and protect the lives of American citizens. Between December 20, 1989 and January 7, 1990, numerous follow-on missions were performed in Panama by the Regiment.

Early in 1991, elements of the 75th Ranger Regiment deployed to Saudi Arabia in support of Operation Desert Storm.

In August 1993, elements of the 75th Ranger Regiment deployed to Somalia in support of Operation Restore Hope, and returned November 1993.

In 1994, elements of the 75th Ranger Regiment deployed to Haiti in support of Operation Uphold Democracy.

In 2000 – 2001, elements of the 75th Ranger Regiment deployed to Kosovo in support of Operation Joint Guardian.

Since September 11, 2001, the 75th Ranger Regiment has led the way in the Global War on Terrorism. In October 2001, elements of the 75th Ranger Regiment deployed to Afghanistan in support of Operation Enduring Freedom. In March 2003, elements of the Regiment deployed in support of Operation Iraqi Freedom.

The performance of these Rangers significantly contributed to the overall success of these operations and upheld the Ranger tradition of the past. As in the past, the Regiment stands ready to execute its mission to conduct special operations in support of the United States' policies and objectives.

RANGER MEDAL OF HONOR WINNERS

Millett, Lewis L. Sr	Captain	Feb 7 1951	Co. E 2/27th Infantry
Porter, Donn F.*	Sergeant	Sept 7 1952	Co. G 2/14th Infantry
Mize, Ola L.	Sergeant	June 10-11 1953	Co. K 3/15th Infantry
Dolby, David C.	Staff Sergeant	May 21 1966	Co. B 1/8th (ABN) Calvary
Foley, Robert F.	Captain	Nov 5 1966	Co. A 2/27th Infantry
Zabitosky, Fred M.	Staff Sergeant	Feb 19 1968	5th Special Forces
Bucha, Paul W.	Captain	May 16-19 1968	Co. D 3/187 Infantry
Rabel, Laszlo*	Staff Sergeant	Nov 13 1968	74th Infantry (LRRP)
Howard, Robert L.	Sergeant First Class	Dec 30 1968	5th Special Forces
Law, Robert D. *	Specialist 4	Feb 22 1969	Co. I 75th Infantry (Ranger)
Kerrey, J. Robert	Lieutenant	Mar 14 1969	Seal Team 1
Doane, Stephen H.*	1st Lieutenant	Mar 25 1969	Co. B 1/5th Infantry
Pruden, Robert J.*	Staff Sergeant	Nov 22 1969	Co. G 75th Infantry (Ranger)
Littrell, Gary L.	Sergeant First Class	April 4-8 1970	Advisory Team 21 (Ranger)
Lucas, Andre C.*	Lt Colonel	Jul 1-23 1970	HHC 2/506 Infantry
Gordon, Gary I. *	Master Sergeant	Oct 3 1993	Task Force Ranger
Shughart, Randall D. *	Sergeant First Class	Oct 3 1993	Task Force Ranger

*Awarded posthumously

SUMMARY OF CHANGES

THROUGHOUT	UPDATED doctrine, technology, terms; clarified verbiage
CHAPTER 1	UPDATED duties and responsibilities ADDED FO
CHAPTER 2	UPDATED WARNO ADDED sample WARNO UPDATED OPORD ADDED terrain model
CHAPTER 3	ADDED PLOT-CR ADDED risk estimate distances
CHAPTER 4	IDENTIFIED TL positions during movement
CHAPTER 5	REORGANIZED combat operations
CHAPTER 7	ADDED section on antennas
CHAPTER 8	UPDATED aircraft
CHAPTER 9	UPDATED rope bridge equipment UPDATED sequence and responsibilities
CHAPTER 10	UPDATED equipment and descriptions UPDATED climbing commands
CHAPTER 11	NA
CHAPTER 12	ADDED first aid techniques
CHAPTER 13	ADDED figure 'Lighting a time fuse with a match' ADDED expedient explosives ADDED table of US demolitions ADDED breaching and timber cutting charges
CHAPTER 15	NEW chapter covering vehicle convoy operations
APPENDIX A	NEW - Resources
APPENDIX B	NEW - Quick Reference Cards

TABLE OF CONTENTS

Chapter 1
LEADERSHIP

Leadership, the most essential element of combat power, gives purpose, direction, and motivation in combat. The leader balances and maximizes maneuver, firepower, and protection against the enemy. This chapter discusses how he does this by exploring the principles of leadership (BE, KNOW, DO); the duties, responsibilities, and actions of an effective leader; and the leader's assumption of command.

1-1. **PRINCIPLES**. The principles of leadership are BE, KNOW, and DO.

BE

- Technically and tactically proficient
- Able to accomplish to standard all tasks required for the wartime mission.
- Courageous, committed, and candid.
- A leader with integrity.

KNOW

- The four major factors of leadership and how they affect each other are--
  ---Led
  ---Leader
  ---Situation
  ---Communications
- Yourself, and the strengths and weaknesses in your character, knowledge, and skills. Seek continual self-improvement, that is, develop your strengths and work to overcome your weaknesses.
- Your Rangers, and look out for their well-being by training them for the rigors of combat, taking care of their physical and safety needs, and disciplining and rewarding them.

DO--

- Seek responsibility and take responsibility for your actions; exercise initiative; demonstrate resourcefulness; and take advantage of opportunities on the battlefield that will lead to you to victory; accept fair criticism, and take corrective actions for your mistakes.
- Assess situations rapidly, make sound and timely decisions, gather essential information, announce decisions in time for Rangers to react, and consider the short- and long-term effects of your decision.
- Set the example by serving as a role model for your Rangers. Set high but attainable standards; be willing do what you require of your Rangers; and share dangers and hardships with them.
- Keep your subordinates informed to help them make decisions and execute plans within your intent, encourage initiative, improve teamwork, and enhance morale.
- Develop a sense of responsibility in subordinates by teaching, challenging, and developing them. Delegate to show you trust them. This makes them want more responsibility.
- Ensure the Rangers understand the task; supervise them, and ensure they accomplish it. Rangers need to know what you expect: when and what you want them to do, and to what standard.
- Build the team by training and cross-training your Rangers until they are confident in their technical and tactical abilities. Develop a team spirit that motivates them to go willingly and confidently into combat.
- Know your unit's capabilities and limitations, and employ them accordingly.

1-2. **DUTIES, RESPONSIBILITIES, AND ACTIONS**. To complete all assigned tasks, every Ranger in the patrol must do his job. Each must accomplish his specific duties and responsibilities and be a part of the team.

PLATOON LEADER (PL)

- Is responsible for what the patrol does or fails to do. This includes tactical employment, training, administration, personnel management, and logistics. He does this by planning, making timely decisions, issuing orders, assigning tasks, and supervising patrol activities. He must know his Rangers and how to employ the patrol's weapons. He is responsible for positioning and employing all assigned or attached crew-served weapons and employment of supporting weapons.
- Establishes time schedule using backwards planning. Considers time for execution, movement to the objective, and the planning and preparation phase of the operation.
- Takes the initiative to accomplish the mission in the absence of orders. Keeps higher informed by using periodic situation reports (SITREP).
- Plans with the help of the platoon sergeant (PSG), squad leaders, and other key personnel (team leaders, FO, attachment leaders).
- Stays abreast of the situation through coordination with adjacent patrols and higher HQ; supervises, issues FRAGOs, and accomplishes the mission.
- If needed to perform the mission, requests more support for his patrol from higher headquarters.
- Directs and assists the platoon sergeant in planning and coordinating the patrol's sustainment effort and casualty evacuation (CASEVAC) plan.
- During planning, receives on-hand status reports from the platoon sergeant and squad leaders.
- Reviews patrol requirements based on the tactical plan.
- Ensures that all-round security is maintained at all times.
- Supervises and spot-checks all assigned tasks, and corrects unsatisfactory actions.
- During execution, positions himself where he can influence the most critical task for mission accomplishment; usually with the main effort, to ensure that his platoon achieves its decisive point
- Is responsible for positioning and employing all assigned and attached crew-served weapons.
- Commands through his squad leaders IAW the intent of the two levels higher commanders.
- Conducts rehearsals.

PLATOON SERGEANT (PSG)

The PSG is the senior NCO in the patrol and second in succession of command. He helps and advises the patrol leader, and leads the patrol in the leader's absence. He supervises the patrol's administration, logistics, and maintenance, and he prepares and issues paragraph 4 of the patrol OPORD.

DUTIES

- Organizes and controls the patrol CP IAW the unit SOP, patrol leader's guidance, and METT-TC factors.
- Receives squad leader's requests for rations, water, and ammunition. Work with the company first sergeant or XO to request resupply. Directs the routing of supplies and mail.
- Supervises and directs the patrol medic and patrol aid-litter teams in moving casualties to the rear.
- Maintains patrol status of personnel, weapons, and equipment; consolidates and forwards the patrol's casualty reports (DA Forms 1155 and 1156); and receives and orients replacements.
- Monitors the morale, discipline, and health of patrol members.
- Supervises task-organized elements of patrol:
  - — Quartering parties.
  - — Security forces during withdrawals.
  - — Support elements during raids or attacks.
  - — Security patrols during night attacks.
- Coordinates and supervises patrol resupply operations.
- Ensures that supplies are distributed IAW the patrol leader's guidance and direction.
- Ensures that ammunition, supplies, and loads are properly and evenly distributed (a critical task during consolidation and reorganization).
- Ensures the casualty evacuation plan is complete and executed properly.

- Ensures that the patrol adheres to the platoon leader's time schedule.
- Assists the platoon leader in supervising and spot-checking all assigned tasks, and corrects unsatisfactory actions.

ACTIONS DURING MOVEMENT AND HALTS
- Takes actions necessary to facilitate movement.
- Supervises rear security during movement.
- Establishes, supervises, and maintains security during halts.
- Knows unit location.
- Performs additional tasks as required by the patrol leader and assists in every way possible. Focuses on security and control of patrol.

ACTIONS AT DANGER AREAS
- Directs positioning of near-side security (usually conducted by the trail squad or team).
- Maintains accountability of personnel.

ACTIONS ON THE OBJECTIVE AREA
- Assists with ORP occupation.
- Supervises, establishes, and maintains security at the ORP.
- Supervises the final preparation of men, weapons, and equipment in the ORP IAW the patrol leader's guidance.
- Assists the patrol leader in control and security.
- Supervises the consolidation and reorganization of ammunition and equipment.
- Establishes, marks, and supervises the planned CCP, and ensures that the personnel status (to include WIA/KIA) is accurately reported to higher.
- Performs additional tasks assigned by the patrol leader and reports status to platoon leader.

ACTIONS IN THE PATROL BASE
- Assists in patrol base occupation.
- Assist in establishing and adjusting perimeter.
- Enforces security in the patrol base.
- Keeps movement and noise to a minimum.
- Supervises and enforces camouflage.
- Assigns sectors of fire.
- Ensures designated personnel remain alert and equipment is maintained to a high state of readiness.
- Requisitions supplies, water, and ammunition, and supervises their distribution.
- Supervises the priority of work and ensures its accomplishment.
  - Security plan.
    - Ensures crew-served weapons have interlocking sectors of fire.
    - Ensures Claymores are emplaced to cover dead space.
    - Ensures range cards and sector sketch are complete.
  - Alert plan.
  - Evacuation plan.
  - Withdrawal plan.
  - Alternate patrol base.
  - Maintenance plan.
  - Hygiene plan.
  - Messing plan.
  - Water plan.
  - Rest plan.
- Performs additional tasks assigned by the patrol leader and assists him in every way possible.

SQUAD LEADER (SL)

Is responsible for what the squad does or fails to do. He is a tactical leader who leads by example.

DUTIES
- Completes casualty feeder reports and reviews the casualty reports completed by squad members.
- Directs the maintenance of the squad's weapons and equipment.

- Inspects the condition of Rangers' weapons, clothing, and equipment.
- Keeps the PL and PSG informed on status of squad.
- Submits ACE report to PSG.

ACTIONS THROUGHOUT THE MISSION

- Obtains status report from team leaders and submits reports to the PL and PSG.
- Makes a recommendation to the PL/PSG when problems are observed.
- Delegates priority task to team leaders, and supervises their accomplishment IAW squad leader's guidance.
- Uses initiative in the absence of orders.
- Follows the PL's plan and makes recommendations.

ACTIONS DURING MOVEMENT AND HALTS

- Ensures heavy equipment is rotated among members and difficult duties are shared.
- Notifies PL of the status of the squad.
- Maintains proper movement techniques while monitoring route, pace, and azimuth.
- Ensures the squad maintains security throughout the movement and at halts.
- Prevents breaks in contact.
- Ensures subordinate leaders are disseminating information, assigning sectors of fire, and checking personnel.

ACTION IN THE OBJECTIVE AREA

- Ensures special equipment has been prepared for actions at the objective.
- Maintains positive control of squad during the execution of the mission.
- Positions key weapons systems during and after assault on the objective.
- Obtains status reports from team leaders and ensures ammunition is redistributed and reports status to PL.

ACTIONS IN THE PATROL BASE

- Ensures patrol base is occupied according to the plan.
- Ensures that his sector of the patrol base is covered by interlocking fires; makes final adjustments, if necessary.
- Sends out LP or OPs in front of assigned sector. (METT-TC dependent).
- Ensures priorities of work are being accomplished, and reports accomplished priorities to the PL and PSG.
- Adheres to time schedule.
- Ensures personnel know the alert and evacuation plans, and the locations of key leaders, OPs, and the alternate patrol base.

WEAPONS SQUAD LEADER

Is responsible for all that the weapons squad does or fails to do. His duties are the same as those of the squad leader. Also, he controls the machine guns in support of the patrol's mission. He advises the PL on employment of his squad.

DUTIES

- Supervises machine gun teams to ensure they follow priorities of work.
- Inspects machine gun teams for correct range cards, fighting positions, and understanding of fire plan.
- Supervises maintenance of machine guns, that is, ensures that maintenance is performed correctly, that deficiencies are corrected and reported, and that the performance of maintenance does not violate the security plan.
- Assists PL in planning.
- Positions at halts and danger areas and according to the patrol SOP any machine guns not attached to squads.
- Rotates loads. Machine gunners normally get tired first.
- Submits ACE report to PSG.
- Designates sectors of fire, principal direction of fire (PDF), and secondary sectors of fire for all guns.
- Gives fire commands to achieve maximum effectiveness of firepower:
  - — Shifts fires.
  - — Corrects windage or elevation to increase accuracy.
  - — Alternates firing guns.
  - — Controls rates of fire and fire distribution.
- Knows locations of assault and security elements, and prevents fratricide.
- Reports to the PL.

TEAM LEADER (TL)

Controls the movement of his fire team and the rate and placement of fire. To do this, leads from the front and uses the proper commands and signals. Maintains accountability of his Rangers, weapons, and equipment. Ensures his Rangers maintain unit standards in all areas, and are knowledgeable of their tasks and the operation. The following checklist outlines specific duties and responsibilities of team leaders during mission planning and execution. The TL leads by example:

ACTIONS DURING PLANNING AND PREPARATION

- Warning Order.
  - Assists in control of the squad.
  - Monitors squad during issuance of the order.
- OPORD Preparation.
  - Posts changes to schedule.
  - Posts and updates team duties on warning order board.
  - Submits ammunition and supply requests.
  - Picks up ammunition and supplies.
  - Distributes ammunition and special equipment.
  - Performs all tasks given in the SL's special instructions paragraph.
- OPORD Issuance and Rehearsal.
  - Monitors squad during issuance of the order.
  - Assists SL during rehearsals.
- Takes actions necessary to facilitate movement.
- Enforces rear security.
- Establishes, supervises, and maintains security at all times.
- Performs additional tasks as required by the SL, and assists him in every way possible, particularly in control and security.

ACTION IN THE ORP

- Assists in the occupation of the ORP.
- Helps supervise, establish, and maintain security.
- Supervises the final preparation of Rangers, weapons, and equipment in the ORP IAW the SL's guidance.
- Assists in control of personnel departing and entering the ORP.
- Reorganizes perimeter after the reconnaissance party departs.
- Maintains communication with higher headquarters.
- Upon return of reconnaissance party, helps reorganize personnel and redistribute ammunition and equipment; ensures accountability of all personnel and equipment is maintained.
- Disseminates PIR to his team.
- Performs additional tasks assigned by the SL.

ACTIONS IN THE PATROL BASE

- Inspects the perimeter to ensure team has interlocking sectors of fire; prepares team sector sketch.
- Enforces the priority of work and ensures it is properly accomplished.
- Performs additional tasks assigned by the SL and assists him in every way possible.

MEDIC

Assists the PSG in directing aid and litter teams; monitors the health and hygiene of the platoon.

DUTIES

- Treats casualties, conducts triage, and assists in CASEVAC under the control of the PSG.
- Aids the PL or PSG in field hygiene matters. Personally checks the health and physical condition of platoon members.
- Requests Class VIII (medical) supplies through the PSG.
- Provides technical expertise to and supervision of combat lifesavers.
- Ensures casualty feeder reports are correct and attached to each evacuated casualty.
- Carries out other tasks assigned by the PL or PSG.

RADIO OPERATOR Is responsible for establishing and maintaining communications with higher headquarters and within the patrol. *DUTIES DURING PLANNING* • Enters the net at the specified time. • Ensures that all frequencies, COMSEC fills, and net IDs, are preset in squad/platoon radios. • Informs SL and PL of changes to call signs, frequencies, challenge and password, and number combination based on the appropriate time in the ANCD. • Ensures the proper function of all radios and troubleshoots and reports deficiencies to higher. • Weatherproofs all communications equipment. *DUTIES DURING EXECUTION* • Serves as en route recorder during all phases of the mission. • Records all enemy contact and reports it to higher in a SALUTE format. • Reports all OPSKEDs to higher. • Consolidates and records all PIR.
FORWARD OBSERVER (FO) Works for the PL. Serves as the eyes of the FA and mortars. Is mainly responsible for locating targets, and for calling for and adjusting indirect fire support. Knows the terrain where the platoon is operating; knows the tactical situation. Knows the mission, the concept, and the unit's scheme of maneuver and priority of fires. *DUTIES DURING PLANNING* • Selects targets to support the platoon's mission based on the company OPORD, platoon leader's guidance, and analysis of METT-TC factors. • Prepares and uses situation maps, overlays, and terrain sketches. *DUTIES DURING EXECUTION* • Informs the FIST headquarters of platoon activities and of the fire support situation • Selects new targets to support the platoon's mission based on the company OPORD, the platoon leader's guidance, and an analysis of METT-TC factors. • Calls for and adjusts fire support. • Operates as a team with the radio operator. • Selects OPs. • Maintains communications as prescribed by the FSO. • Maintains the current 8-digit coordinate of his location at all times.

1-3. **ASSUMPTION OF COMMAND**. Any platoon/squad member might have to take command of his element in an emergency, so every Ranger must be prepared to do so. During an assumption of command, situation permitting, the Ranger assuming command accomplishes the following tasks (not necessarily in this order) based on METT-TC:

- *INFORMS* the unit's subordinate leaders of the command and notifies higher.
- *CHECKS* security.
- *CHECKS* crew-served weapons.
- *PINPOINTS* location.
- *COORDINATES* and *CHECKS* equipment.
- *CHECKS* personnel status.
- *ISSUES* FRAGO (if required).
- *REORGANIZES* as needed, maintaining unit integrity when possible.
- *MAINTAINS* noise and light discipline.
- *CONTINUES* patrol base activities, especially security, if assuming command in a patrol base.
- *RECONNOITERS* or, at the least, conducts a map reconnaissance.
- *FINALIZES* plan.
- *EXECUTES* the mission.

NOTES

Chapter 2
OPERATIONS

This chapter provides techniques and procedures used by Infantry platoons and squads throughout the planning and execution phases of tactical operations. Specifically, it discusses the troop-leading procedures, combat intelligence, combat orders, and planning techniques and tools needed to prepare a platoon to fight. These topics are time sensitive and apply to all combat operations. When they have time, leaders can plan and prepare in depth. If they have less time, they must rely on previously rehearsed actions, battle drills, and standing operating procedures (SOPs).

2-1. TROOP-LEADING PROCEDURES

Troop leading procedures comprise the following steps. They are what a leader does to prepare his unit to accomplish a tactical mission. The TLP starts when the leader is alerted for a mission or receives a change or new mission. He can perform Steps 3 through 8 in any order, or at the same time. (He can also use the tools of the tactician shown in Figure 2-1):

1. Receive the mission. 2. Issue a warning order. 3. Make a tentative plan. 4. Initiate movement.	5. Reconnoiter. 6. Complete the plan. 7. Issue the complete order. 8. Supervise.

Figure 2-1. TOOLS OF THE TACTICIAN RELATIONSHIP

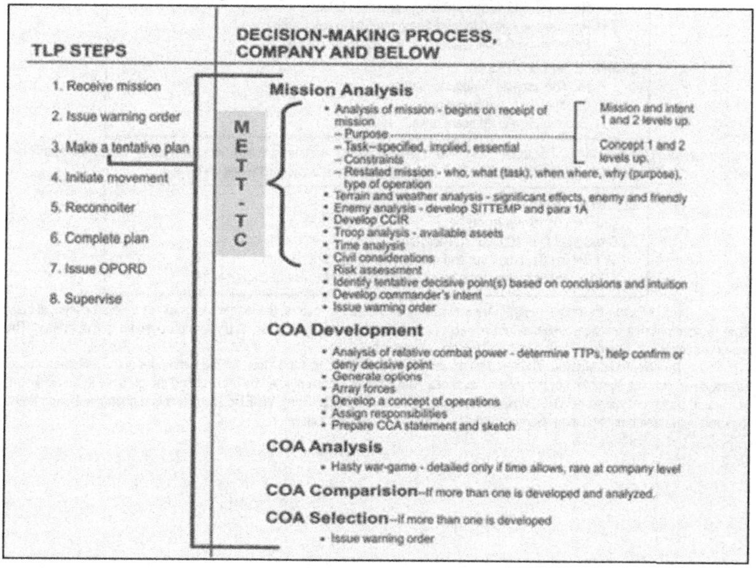

a. **STEP 1--RECEIVE THE MISSION.** The leader may receive the mission in a warning order, an operation order (OPORD), or a fragmentary order (FRAGO). He should use no more than one third of the available time for his own planning and for issuing his OPORD. The remaining two thirds is for subordinates to plan and prepare for the operation. Leaders should also consider other factors such as available daylight and travel time to and from orders and rehearsals.

b. **STEP 2--ISSUE A WARNING ORDER.** The leader provides initial instructions in a warning order. The warning order contains enough information to begin preparation as soon as possible. The warning order mirrors the five-paragraph OPORD format. A warning order may include--

- The mission or nature of the operation (mission statement).
- Time and place for issuance of the operation (coordinating instructions).
- Who is participating in the operation (coordinating instructions).
- Time of the operation (timeline).

c. **STEP 3--MAKE A TENTATIVE PLAN.** The leader uses METT-TC to develop an estimate of the situation, which he will use as the basis for his tentative plan. This set of actions is referred to as the leader's mission analysis:

(1) **Conduct a detailed mission analysis.**

(a) *Concept and Intent.* Higher commanders' concept and intent two levels up. This information is found in paragraph 1b for two levels up and in paragraphs 2 and 3 for one higher.

(b) *Unit Tasks.* Tasks that are clearly stated in the order (Specified Tasks) or tasks that become apparent as the OPORD is analyzed (Implied Tasks).

EXAMPLE SPECIFIED TASKS

- Retain hill 545 to prevent envelopment of B Co.
- Provide one squad to the 81-mm platoon to carry ammo.
- Establish an OP vic GA124325 NLT 301500 NOV 89.

EXAMPLE IMPLIED TASKS

- Provide security during movement.
- Conduct resupply operations.
- Coordinate with adjacent units.

(c) *Unit Limitations.* The leader next determines all control measures or instructions in the OPORD that restrict his freedom of action; these are called limitations. In every operation, there are some limitations on the unit.

COMMON LIMITATIONS

Graphic control measures.
Cross the LD at 100030 OCT 94 (time).
MOPP4 in effect (uniform and environmental considerations).
ADA weapons status, tight; warning status, yellow (rules of engagement).

(d) *Mission-Essential Task(s).* After reviewing all the above factors, the leader identifies his mission-essential task(s). Failure to accomplish a mission-essential task results in the unit's failure to accomplish its primary purpose for that operation. The mission essential task should be found in the maneuver paragraph.

(e) *Restated Mission.* The restated mission statement becomes the focus for the remainder of the estimate process. This is a clear, concise statement of the mission essential task(s) to be accomplished by the unit and the purpose to be achieved. The mission statement will state WHO, WHAT (the task), WHEN (the critical time), WHERE (usually a grid coordinate), and WHY (the purpose the unit must achieve). Some examples of restated missions follow:

EXAMPLE RESTATED MISSIONS

> (WHO) 1st Platoon attacks (WHAT) to seize (WHERE) HILL 482 vic NB 457371 (OBJ BLUE) (WHEN) NLT 090500Z Dec 92 L 482 (WHY) to enable the company's main effort to destroy enemy command bunker.
>
> (WHO) 1st Platoon, C Company defends (WHAT) to destroy from (WHERE) AB163456 to AB163486 to AB123486 to AB123456 (WHEN) NLT 281530Z Oct 97 (WHY) to prevent enemy forces from enveloping B Company, 1-66 Infantry (L) from the south.

(2) Analyze the situation and develop a course of action. Each COA must be--suitable, acceptable, feasible, distinguishable, and complete.

(3) Upon developing a COA, the unit leader will assign C2 headquarters, complete generic task organization assigning all organic and attached elements, and prepare COA statement and sketch.

(4) With the restated mission from Step 1 to provide focus, the leader continues the estimate process using the remaining factors of METT-TC.

(a) What is known about the **ENEMY**?

Composition	This is an analysis of the forces and weapons that the enemy can bring to bear. Determine what weapons systems they have available, and what additional weapons and units are supporting him.
Disposition	The enemy's disposition is how he is arrayed on the terrain, such as in defensive positions, in an assembly area, or moving in march formation.
Strength	Percentage strength.
Recent Activities	Identify recent and significant enemy activities that may indicate future intentions.
Reinforcement Capabilities	Determine positions for reserves and estimated time to counterattack or reinforce.
Possible Courses of Action	Determine the enemy's possible COAs. This will help the leader determine how to best array his own forces against the enemy and fight the battle.

(b) How will **TERRAIN** and **WEATHER** affect the operation? Analyze terrain using OAKOC.

- **Observation and Fields of Fire**. Determine locations that provide the best observation and fields of fire along friendly and enemy avenues of approach, near the objective, and on key terrain. The analysis of fields of fire is mainly concerned with the ability to cover the terrain with direct fire.
- **Avenues of Approach**. Analyze and identify friendly and enemy avenues of approach. Identify avenues of approach en route to the objective, on and around the objective, and for possible enemy counterattack. Also, consider aerial and subterranean avenues.

Offensive Considerations (Friendly)	• How can these avenues support my movement? • What are the advantages/disadvantages of each? (Consider enemy, speed, cover, and concealment.) • What are the likely enemy counterattack routes?

Offensive Considerations (Enemy)	• How can the enemy use these approaches? • Which avenue is most dangerous? Least? (Prioritize each approach.) • Which avenues would support a counterattack?

- **Cover and Concealment**. The analysis of cover and concealment is often inseparable from the fields of fires and observation. Weapon positions must have both to be effective and to be survivable. Infantry units are capable of improving poor cover and concealment by digging in and camouflaging their positions. When moving, the terrain is used to provide cover and concealment.
- **Obstacles**. Identify the existing and reinforcing obstacles and hindering terrain that will affect mobility.
- **Key Terrain**. Key terrain is any location or area that the seizure, retention, or control of affords a marked advantage to either combatant. Using the map, aerial photos, and information already gathered, look for key terrain that dominates avenues of approach or the objective area. Next, look for decisive terrain that if held or controlled will have an extraordinary impact on the mission.

(5) Analyze courses of action (war-game). This analysis is conducted by war-gaming friendly courses of action against the enemy's most probable courses of action. The leader can--and should--war-game with his subordinates.

(6) Compare courses of action. The leader compares the COAs and selects the one that is most likely to accomplish the assigned mission. He considers the advantages and disadvantages for each COA. He also considers how the critical events impact on COAs.

(7) Make a decision. The leader selects the COA that he believes has the best chance of accomplishing the mission.

d. **STEP 4--START NECESSARY MOVEMENT**. The unit may need to begin movement while the leader is still planning or forward reconnoitering. This step may occur anytime during the TLP.

e. **STEP 5--RECONNOITER**. If time allows, the leader makes a personal reconnaissance. When time does not allow, the leader must make a map or aerial photo reconnaissance. Sometimes the leader must rely on others, such as scouts, to conduct the reconnaissance.

f. ***STEP 6--COMPLETE THE PLAN***. The leader completes his plan based on the reconnaissance and any changes in the situation.

g. **STEP 7--ISSUE THE COMPLETE ORDER**. Platoon and squad leaders normally issue oral operations orders to aid subordinates in understanding the concept of the mission. Leaders may require subordinates to repeat part of the order, demonstrate it on a terrain model, or sketch their understanding of the operation. Leaders should also quiz their Rangers to ensure that all Rangers understand the mission.

h. **STEP 8--SUPERVISE AND REFINE**. The leader supervises the unit's preparation for combat by conducting rehearsals and inspections.

(1) ***Rehearsals***. Rehearsals include the practice of having squad leaders brief their planned actions in execution sequence to the platoon leader. The leader should conduct rehearsals on terrain that resembles the actual ground, and in similar light conditions. At a minimum, the leader should conduct a backbrief (least preferred), full force rehearsal (most preferred), or reduced force rehearsal (key leaders), and a rock drill.

(a) *Purpose*. The leader uses rehearsals to--

- Practice essential tasks (improve performance).
- Reveal weaknesses or problems in the plan.
- Coordinate the actions of subordinate elements.
- Improve Ranger understanding of the concept of the operation (foster confidence in Rangers).

(b) *Times and Tasks*. The platoon may begin rehearsals of battle drills and other SOP items before the receipt of the operation order. Once the order has been issued, it can rehearse mission specific tasks. Some important tasks to rehearse include--

- Actions on the objective.
- Assaulting a trench, bunker, or building.
- Actions at the assault position.
- Breaching obstacles (mine and wire).
- Using special weapons or demolitions.
- Actions on unexpected enemy contact.

(c) *Types.*

- Backbrief.
  - -- Key leaders sequentially brief the actions required during operation.
  - -- Patrol leader controls.
  - -- Conducted twice: right after FRAGO (confirmation brief), and again after subordinates develop own plan.
- Reduced force.
  - -- Conducted when time is key constraint.
  - -- Conducted when security must be maintained.
  - -- Key leaders normally attend.
  - -- Mock-ups, sand tables, and small scale replicas used.
- Full force.
  - -- Most effective type.
  - -- First executed in daylight and open terrain.
  - -- Secondly conduct in same conditions as operation.
  - -- All Rangers participate.
  - -- May use force on force.
- Techniques.
  - -- Force on force.
  - -- Map (limited value and limited number of attendees).
  - -- Radio (cannot mass leaders; confirms communications).
  - -- Sand table or terrain model (key leaders; includes all control measures).
  - -- Rock drill (similar to sand table/terrain model; subordinates actually move themselves). Very effective for coordinating timing and sequence of events--especially when air assets are used.

(d) *Inspections.* Squad leaders should conduct initial inspections shortly after receipt of the warning order. The platoon sergeant spot-checks throughout the unit's preparation for combat. The platoon leader and platoon sergeant make a final inspection of--

- Weapons and ammunition.
- Uniforms and equipment.
- Mission-essential equipment.
- Soldier's understanding of the mission and their specific responsibilities.
- Communications.
- Rations and water.
- Camouflage.
- Deficiencies noted during earlier inspections.

2-2. **COMBAT INTELLIGENCE**. Gathering information is one of the most important aspects of conducting patrolling operations. The following details what information to collect and how to report it:

a. **Reports**. All information must be quickly, completely, and accurately reported. Use the SALUTE report format for reporting and recording information.

SIZE	– Seven enemy personnel
ACTIVITY	– Traveling SW
LOCATION	– GA123456
UNIT/UNIFORM	– OD uniforms with red six-point star on left shoulder
TIME	– 210200JAN99
EQUIPMENT	– Carry one machine gun and one rocket launcher

b. **Field Sketches**. Try to include a sketch with each report. Include only any aspects of military importance such as targets, objectives, obstacles, sector limits, or troop dispositions and locations (use symbols from FM 1-02). Use notes to explain the drawing, but they should not clutter the sketch.

c. **Captured Documents**. The leader collects documents and turns them in with his reports. He marks each document with the time and place of capture.

d. **Prisoners**. If prisoners are captured during a patrolling operation, they should be treated IAW the Geneva Convention and handled by the 5-S rule:

(1) Search
(2) Silence
(3) Segregate
(4) Safeguard
(5) Speed to rear

e. **Debriefs**. Immediately upon return from a mission, the unit is debriefed. The intelligence officer will generally have a unit-specific format for debriefing a patrol.

2-3. **WARNING ORDER**. Warning orders give subordinates advance notice of upcoming operations. This gives them time to prepare. A warning order must be brief, but complete. *A warning order only authorizes execution when it clearly says so.* (Figure 2-2 shows an example format; Figure 2-3 shows an example warning order.)

Figure 2-2. WARNING ORDER FORMAT

WARNING ORDER ________

*Roll call, pencil/pen/paper, RHB, map, protractor, leader's monitor, hold all questions till the end
References: Refer to higher headquarters' OPORD, and identify map sheet for operation.
Time Zone Used throughout the Order: (Optional)
Task Organization: (Optional; see paragraph 1c.)

1. **SITUATION** (Higher's OPORD para 1a[1]--[3])

a. **Enemy Forces**. Include significant changes in enemy composition, dispositions, and courses of action. Information not available for inclusion in the initial WARNO can be included in subsequent warning orders. (who, what, where)

b. **Friendly Forces**. (Optional) only address if essential to the WARNO.

(1) Higher commander's mission (who, what, where, why).
(2) Higher commander's intent. (Higher's [go to mapboard] OPORD para 1b[2], give task and purpose)

c. **Attachments and Detachments**. Initial task organization, only address major unit changes.

(1) Orient relative to each point on the compass.
(2) Box in the entire AO with grid lines.
(3) Trace each Zone using boundaries.
(4) Familiarize by identifying 3 natural (terrain) and 3 man-made features in each zone.
(5) Point out the enemy and friendly locations on the map board.

2. **MISSION**. State mission twice (who, what, when, where, and why) and concisely state task and purpose.

3. **EXECUTION**.

a. **Concept of Operation**. Provide as much information as available. The concept should describe the form of maneuver or defensive technique, critical events, decisive point of the operation and why it is decisive, task and purpose for the main and supporting efforts, purposes of the warfighting systems, such as engineers, fire support, intelligence etc), acceptable risk, and the desired endstate. After the desired endstate is stated, leaders should talk through the concept of the operation like a story. Use a simple sketch to assist in relaying the information.

b. **Tasks to Maneuver Units**. Provide specified tasks to subordinate units. These are tactical instructions on how to execute the mission for each element in task organization. Planning guidance consists of tasks assigned to subordinate units, special teams, and key individuals.

c. **Tasks to Combat Support Units**. See paragraph 3b.

d. **Coordinating Instructions**. Include any information available at that time. If you know it, then at least cover--

- Uniform and equipment common to all. Consider the factors of METT-TC and tailor the load for each Ranger.
- Time line. (State 4 W's and specified and implied times. Reverse plan. Use 1/3-2/3 rule.)
- CCIR.
- Risk guidance.
- Deception guidance.
- Specific priorities, in order of completion.
- Guidance on orders and rehearsals.
- Orders group meeting (attendees, location, and time).
- Earliest movement time and degree of notice.

4. **SERVICE SUPPORT**. Include any known logistics preparation for the operation.

a. Special equipment. Identify requirements and any coordination measures the unit needs to take to transfer the special equipment. (State the equipment you will use, need, or want for the mission.)

b. Transportation. Identify requirements and any coordination needed to pre-position assets. (State unit's means of infil/exfil.)

5. **COMMAND AND SIGNAL**.

a. **Command**. State the chain of command if different from unit SOP.

b. **Signal**. Identify current SOI edition, and pre-position signal assets to support operation.

- Give subordinates guidance on tasks to complete for preparation of the OPORD and the mission.
- Give time, place, and uniform of the OPORD.
- Give a time hack and ask for questions.

NOTES

Figure 2-3. EXAMPLE WARNING ORDER

WARNING ORDER (SQUAD)

ROLL CALL, CAMP DARBY SPECIAL (CDS) 1:50,000 MAP, PEN, PAPER, PENCIL, PROTRACTOR, RANGER HANDBOOK (RHB), HOLD ALL QUESTIONS, AND TLs MONITOR AND TASK-ORGANIZE

1. **SITUATION**.
   a. **Enemy**.
      WHO: The Aragon Liberation Front (ALF).
      WHAT: Has been conducting aggressive offensive operations.
      WHERE: In Zones A, B, C.
   b. **Friendly**:
      (1) **Higher's Mission**.
      WHO: 2nd PLT, C CO.
      WHAT: Conducts area ambushes to destroy enemy personnel and equipment.
      WHERE: Zone D.
      WHY: To prevent enemy logistical resupply.
      (2) **Higher Commander's Intent**.
      - Rapid infiltration into Zone D.
      - Find, fix, and finish enemy forces in Zone D.
      - Enemy personnel and equipment located and targeted.
      - Squads ready to conduct future combat operations.
   c. **Attachments/Detachments**. MG Tm #1/2 PLT 160630-190630.
2. **MISSION**.
   WHO: 1st SQD, 2nd PLT
   WHAT: Conducts a point ambush to destroy enemy personnel and equipment.
   WHERE: GA 1805 7878.
   WHEN: NLT 25 1830 DEC 2007.
   WHY: To prevent enemy logistical resupply and gather PIR.
3. **EXECUTION**.

a. **Concept of Operation**. (Orient Rangers to sketch or terrain model). We are currently located at Camp Darby, GA 1962 7902, which is on a ridge. Our ground tactical plan will begin when we will move generally southwest to our ORP at Grid GA 187778. Terrain feature there is a draw for 1,000 meters, by foot and it will take 1 hour. Here will we finalize the preparing of M, W, and E. We will then move generally southwest to our objective at Grid GA 184774. Terrain feature there is a ridge for 200 meters, by foot, and it will take us 30 minutes to an hour. The reason for this is stealth on the objective while we occupy our positions.

b. **Tasks to Maneuver Units**.

(1) HQs: HQ is responsible for command and control and communications. Commo/HQ is 2nd in the OOM. M240 will be placed in a position to provide supporting fires in the kill zone. RTO will be the recorder en route and during actions on the objective. SL will be located where he can best control his squad and observe the kill zone. SL is responsible for dissemination of information.

(2) ATM: ATM will be 1st OOM/responsible for navigation, and for clearing and securing the far side at all linear danger areas. ATM is flank security for AOO, will provide early warning and seal off objective/1-2 man EPW, and search TM/1-2 man aid and litter TM/1-2 man demo TM/1 man (P), 1 man (A) ORP clearing TM/ 2-2 man flank security. TM for AOO/ 1 compassman, 1 pace man, or ATL will be the sec. TM LDR along the most likely avenue of approach.

(3) BTM: BTM 3RD OOM/responsible for securing the near side of all linear danger areas/BTM. Is assault for AOO. Must provide suppressive fire and assault thru kill zone/1-2 man EPW and search team/1-2 man aid and litter team/1-2 man demolitions team; 1 man (P); 1 man (A) ORP clearing team; 1-2 man S/O team; 1-2 man security team for reentry; 1 compassman; 1 pace man; and BTL. You will be the assault element leader for AOO.

c. **Tasks to Combat Support Units**. None.

d. **Coordinating Instructions**.

(1) Uniform and equipment common to all. At a minimum, each Ranger wears or carries--

(a) A ballistic helmet w/camouflaged band and luminous tape; clean and serviceable ACUs, camelback, black belt w/buckle, blackened combat boots, serviceable OD or black socks, serviceable OD/brown T-shirt, ID tags w/breakaway lanyard around neck (silenced). Right cargo pocket; (RHB, CDS [map], protractor, pencil, pen, and paper) waterproofed, left cargo pocket; patrol cap w/ luminous tape, ID card in right breast pocket, SL will carry sterile fire support overlay in left breast pocket.

(b) LCE, consisting of pistol belt, suspenders, 2 ammo pouches, seven 30-round magazines, one first aid dressing with pouch, one compass and pouch, two 1-quart canteens w/covers and cups, one whistle, and four M67 fragmentation grenades. All are tied down IAW 4th Bn SOP, except that the pace cord is tied to suspenders.

(c) Rucksack, complete; one set serviceable ACUs, two 2-quart canteens w straps; E-tool with cover; two pair socks; two brown T-shirts; weapons cleaning kit; one waterproof bag; a sewing kit; a patrol cap; MREs; poncho and liner; wet weather top; personal hygiene kit; one brown towel; and camouflage stick. RTO also packs and ties off the ANCD.

(d) Optional bug juice, 550 cord, foot powder, black tape, knife, all other items will be approved by SL 15 minutes prior to completion time of OPORD.

(2) Time schedule.

WHEN	WHAT	WHERE	WHO	REF
*0630	WARNING ORDER	BAY AREA	All	3D1
0700	INITIAL INSPECTION	BAY AREA	All	
*0730	Required Ammo/Supplies	CO TOC	BTL	4B1
*0745	P/U-Ammo/Supplies	CO TOC	BTL/Detail	4B1
0800	Test Fire	T/F AREA	All	
*0830	S-2/S-3 Coord	CO BAY AO	SL/RTO	3D1
*0830	Fire Spt Coord	BN TOC	SL/ RTO	3A2
0850	Adj Unit Coord	BN TOC	ATL/compass man	3D1
*0900	Enter Net	BAY AREA	RTO	5B1
0930	Complete Plan	BAY AREA	All	
*1000	Issue OPORD	BAY AREA	All	3D1
1300	Rehearsal	BAY AREA	All	
1400	Final Inspection	BAY AREA	All	
1430	Conduct Move	BAY AREA	All	
1800	In ORP	GA	All	
2000	In Position	GA 184774	All	
*2300	Mission Complete	GA 186874	All	3D1
*0200	Reenter FFU	GA	All	
0500	S-2 Debrief	BN TOC	All	
* Specified time				

4. **SERVICE AND SUPPORT**.

a. **Special Equipment**.

(1) HQs: One hand-held radio and extra batteries; one AN/PRC-119 radio and extra batteries, one ANCD, one AN/PSN-11 and extra batteries, one M-22 binos, one AN/PVS-14 and extra batteries; one sheets overlay material, three sheets carbon paper; two HE smoke, two red smoke, one M240-B MG complete w/tripod, flex mount, T&E, and Pintle, one AN/PVS-4 and extra batteries, one AN/PVS-7D, and extra batteries.

(2) ATM: One hand-held radio and extra batteries, one M-24 binos, two AN/PVS-14 and extra batteries, two HE smoke, two yellow smoke, one thermite grenade, two gags, two flex cuffs, two sandbags, one CLS bag, one pole-less litter, two M18A1 claymore mines, one AT-4, one LAW, one M-203 vest, two M-249 SAW pouches, one small metal flashlight, and extra batteries secured to rifle.

(3) BTM: One hand-held radio and extra batteries, one M-24 binos, two AN/PVS 14 and extra batteries, two HE smoke, two yellow smoke, one thermite grenade; two each gags, flex cuffs, and sandbags; one CLS bag, one pole-less Litter, two M18A1 claymore mines, one AT-4, one LAW, one M-203 vest, two M249 SAW pouches, one small metal flashlight, and extra batteries secured to rifle.

b. **Ammo**. (per weapon system).

M16/M4	**M249**	**M203**	**M240**	**SL AND TLs**
210 Rounds 5.56-MM Ball	625 Rounds 5.56-MM (4:1 Mix)	16 Rounds 40-MM HEDP 5 Rounds 40-MM Shot 4 Rounds 40-MM Illumination 2 Rounds 40-MM TP-T	825 Rounds 7.62-MM (4:1 Mix)	42 Rounds 5.56-MM Tracer

c. **Transportation**. NA

5. **COMMAND AND SIGNAL**.

a. **Command**.

PERSONNEL	**POS**	**SOC**
HQ/SUPPORT		
BARNETT	SL	1
MARENGO	RTO	4
CASELY	MG	11
WADE	AG	12
ATM/SECURITY		
TRUITT	ATL	3
SILCOX	GRN	5
WISE	R	6
EDMUNDS	R	7
BTM/ASSAULT		
SPENCER	BTL	2
CHRISTENSEN	GRN	8
RUFUS	R	9
ROBINSON	R	10

SUBORDINATE GUIDANCE:	
ATL:	Write paragraph 1 of squad OPORD, and for drawing all sketches, terrain models, routes, and nonsterile fire support overlays.
BTL:	You are second in the chain of command and in charge at all times during my absence. Write paragraph 4 of squad OPORD Write the supply, DX, and ammo lists, draw and issue all items, ensure that everyone conducts a test fire, and all equipment is tied down IAW 4th RTBn SOP, and update the squad status card and hand receipt.
RTO:	Write paragraph 5 of squad OPORD, ensure all radios are operational with proper freqs loaded, and ensure we enter the net on time.
TL:	TLs: You will update the Warning Order board with all the correct information. As a task is accomplished, you will line it out. Also, post your succession of command (SOC), duty positions, and job descriptions (specialty teams and key individuals). Come and see me for further guidance at the conclusion of this WARNO.

b. **Signal.**

TACTICAL FREQUENCY	62,000
MEDEVAC FREQUENCY	35,000
CHALLENGE	BAYONET
PASSWORD	MILLETT
RUNNING PASSWORD	RANGER
NUMBER COMBO	7
TIME, PLACE, AND UNIFORM FOR OPORD. TIME HACK. ANY QUESTIONS?	

NOTES

2-4. **OPERATIONS ORDER**. An operations order (OPORD) is a directive issued by a leader to his subordinates in order to effect the coordinated execution of a specific operation. A five-paragraph format (shown below) is used to organize the briefing, to ensure completeness, and to help subordinate leaders understand and follow the order. Use a terrain model or sketch along with a map to explain the order. When possible, such as in the defense, give the order while observing the objective. The platoon/squad leader briefs his OPORD orally off notes that follow the five-paragraph format. Before the issuance of the OPORD, the leader ensures that the following resources are in place: pencil, pen, paper; RHB; map; protractor; leader's monitor. Then he calls roll and says "*Please hold all questions till the end.*" Figure 2-4 shows an example OPORD format. Figure 2-5 shows an OPORD shell.

Figure 2-4. EXAMPLE OPORD FORMAT

OPERATIONS ORDER

[Plans and orders normally contain a code name and are numbered consecutively within a calendar year.]

References: [The heading of the plan or order includes a list of maps, charts, datum, or other related documents the unit will need to understand the plan or order. The user need not reference the SOP, but may refer to it in the body of the plan or order. He refers to a map by map series number (and country or geographic area, if required), sheet number and name, edition, and scale, if required. "*Datum*" refers to the mathematical model of the earth that applies to the coordinates on a particular map. It is used to determine coordinates. Different nations use different datum for printing coordinates on their maps. The datum is usually referenced in the marginal information of each map.]

Time zone used throughout the order: The time zone used throughout the order (including annexes and appendixes) is the time zone applicable to the operation. Operations across several time zones use Zulu time.

Task organization: Describe the allocation of forces to support the commander's concept. Task organization may be shown in one of two places: preceding paragraph one, or in an annex, if the task organization is long and complicated.

**Weather and light data, and general forecast (Only discuss what will affect your patrol. Note effects on friendly and enemy personnel and equipment.):*

High
Moonrise
Sunrise
Low
Moonset
Sunset
Wind Speed
Moon Phase
BMNT
Wind Direction
Percent Illumination
EENT

**This is the information the leader received from higher's OPORD.*

Terrain: Make a statement concerning the effects terrain will have on both friendly and enemy forces in the area of operation using the OAKOC format.

1. **SITUATION**.

a. **Enemy Forces**. The enemy situation in higher headquarters' OPORD (paragraph 1a) is the basis for this, but the leader refines this to provide the detail required by his subordinates.

(1) Enemy's composition, disposition, strength.

(2) Recent activities.

(3) Known/suspected locations and capabilities.

(4) Describe the enemy's most likely and most dangerous course of action on a map.

b. **Friendly Forces**. This information is in paragraphs 1b, 2, and 3 of higher headquarters' OPORD.

(1) State the mission of the Higher Unit (2 levels up).

(2) State the mission of the Higher Unit (1 level up).

(3) State intent 2 levels up.

(4) State locations of units to the left, right, front, and rear. State those units' tasks and purposes; and say how those units will influence your unit's patrol.

(a) Show other units locations on map board.

(b) Include statements about the influence each of the above patrols will have on your mission, if any.

- Obtain this information from higher's OPORD. It gives each leader an idea of what other units are doing and where they are going. This information is in paragraph 3a (1) (execution, concept of the operation, maneuver).
- Also include any information obtained when the leader conducts adjacent unit coordination.

c. **Attachments and Detachments**. Do not repeat information already listed under Task Organization. Try to put all information in the Task Organization. However, when not in the Task Organization, list units that are attached or detached to the headquarters that issues the order. State when attachment or detachment is to be effective if different from when the OPORD is effective, such as on order, on commitment of the reserve). Use the term "*remains attached*" when units will be or have been attached for some time.

2. **MISSION**. State the mission derived during the planning process. There are no subparagraphs in a mission statement. Include the 5 W's: Who, What (task), Where, When, and Why (purpose).

- Make it a clear and concise statement and read it twice.
- Go to map and point out the exact location of the OBJ and point out the unit's present location.

3. **EXECUTION**. State the commander's intent.

a. **Concept of Operation**. The concept statement should be concise and clear. It describes how the unit will accomplish its mission from start to finish. The number of subparagraphs, if any, is based on what the leader considers appropriate for the level of leadership and complexity of the operation. Generally, the concept should describe the form of maneuver or defensive technique and critical events; identify what the decisive point of the operation is and why; describe the task and purpose for the main effort and supporting efforts; state the purpose of the warfighting functions, such as engineers, fire support, or intelligence; define acceptable risk; and identify the desired endstate. After stating the desired endstate, leaders should talk through the concept of the operation like a story, using a simple sketch to help them convey the information. Figure 1 shows the six warfighting functions (WFF):

Figure 1. Warfighting functions.

- Fire support
- Movement and Maneuver
- Protection
- Command and Control
- Intelligence
- Sustainment (formerly called "CSS")

(1) ***Maneuver***. The maneuver paragraph details the mechanics of the operations. This subparagraph specifically addresses all subordinate units and attachments by name, giving each its mission in the form of a task and purpose. The main effort's mission must be supported by the subordinate unit's missions. Actions on the objective will comprise the majority of this paragraph and should include a detailed plan for engagement/disengagement criteria, an alternate plan in the event of compromise or unplanned movement of enemy forces, and a withdrawal plan.

(2) ***Fires***. State scheme of fires to support the overall concept. This paragraph should state which maneuver unit has priority of fires. Fires should be planned using the PLOT-CR format (purpose, location, observer, trigger, communication method, resources). A target list worksheet and overlay are referenced here, if applicable. Specific targets are discussed and pointed out on the terrain model (see Chapter 3, Fire Support).

(3) ***Casualty Evacuation***.

(a) *Plan*. State a detailed CASEVAC plan during each phase of the operation. Include CCP locations, tentative extraction points, and methods of extraction.

b. **Tasks to Maneuver Units**. Clearly state the missions or tasks for each maneuver unit that reports directly to the headquarters issuing the order. List the units in the task organization, including reserves. Use a separate subparagraph for each maneuver unit. Only state the tasks that are necessary for comprehension, clarity, and emphasis. Place tactical tasks that affect two or more units in Coordinating Instructions (subparagraph 3d). Platoon leaders task their subordinate squads. Those squads may be tasked to provide any of the following special teams: reconnaissance and security, assault, support, aid and litter, EPW and search, clearing, and demolitions. Detailed instructions may also be given to the platoon sergeant, RTO, compassman, and paceman.

(1) ***Tasks to Combat Support Units***. Add subparagraphs here only as necessary. List CS units in subparagraphs from the task organization. Use CS subparagraphs to list only those specific tasks that CS units must accomplish and that are not specified or implied elsewhere. Include organization for combat if not clear from task organization.

c. **Coordinating Instructions**. This is always the last subparagraph under paragraph 3. List only the instructions that apply to two or more units, and which are seldom covered in unit SOPs. Refer the user to an annex for more complex instructions. The following information is required:

1. Time schedule (state time, place, uniform, and priority of rehearsals, backbriefs, inspections, and movement).

2. Give the commander's critical information requirements (CCIR), which includes priority intelligence requirements (PIR), essential elements of friendly information (EEFI), and friendly force information requirements (FFIR):

a. PIR is intelligence that the commander must have for planning and decision making.

b. EEFI are critical aspects of friendly operations that, if known by the enemy, would compromise, lead to failure, or limit success of the operation.

c. FFIR include what the commander needs to know about friendly forces available for the operation. It can include personnel status, ammunition status, and leadership capabilities.

3. Risk reduction control measures. These are measures unique to the operation. They supplement the unit SOP and can include mission-oriented protective posture, operational exposure guidance, vehicle recognition signals, and fratricide prevention measures.

4. Rules of engagement (ROE).

5. Environmental considerations.

6. Force protection

7. Movement plan. Use terrain model, sketch, or both. State azimuths, directions, and grid coordinates. Use names of real Rangers on the sketches.

a. Sketch out the order of movement, the formation, and the movement technique.

b. Sketch actions at halts (long and short).

c. Routes (use terrain model):

1) Orient the terrain model (North, South, East, and West).

2) Identify the grid lines, both North-South and East-West.

3) Brief the legend.

4) Box in the AO with distinguishable natural or man-made terrain features such as roads, ridgelines, rivers, and streams.

5) Brief all the routes depicted on the terrain model from start to finish using specific azimuths and distance.

a) The primary route from the start point to the ORP, from the ORP to the objective, and from the objective to the patrol base while describing the terrain over which the squad/platoon will travel. (Include near/far side rally points for danger areas.)

b) The alternate route from the start point to the ORP, from the ORP to the objective, and from the objective to the patrol base while describing the terrain over which the squad/platoon will travel. (Include near/far side rally points for danger areas.)

6) Brief the fire support plan (if not given in fires paragraph). When covering fires the leader should cover the following (PLOT-CR):

- *Purpose.*
- *Location and terrain feature to the targets.*
- *Observer.*
- *Trigger.*
- *Communication method.*
- *Resources.*

7) State where the sterile fire support overlay is being carried.

a) Rally points and actions at all rally points (include grid location and terrain reference, and use sketch).

b) Actions at danger areas. This is a general plan for unknown linear, small open areas and large open areas; specific plan for all known danger areas that unit will encounter along the route (use sketch).

c) Actions on enemy contact; battle drills (use sketch).

4. **SERVICE SUPPORT**. Address service support in the areas shown below as needed to clarify the service support concept. Subparagraphs can include:

a. **General**. Refer to the SOPs that govern the sustainment operations of the unit. Provide current and proposed company trains locations, casualty, and damaged equipment collection points and routes.

b. **Materiel and Services**.

(1) ***Supply***.

Class I--Rations plan.
Class III--Petroleum.
Class V--Ammunition.
Class VII--Major end items.
Class VIII--Medical.
Class IX--Repair parts.
Distribution Methods.

(2) ***Transportation***.

(3) ***Services***. (Laundry and showers.)

(4) ***Maintenance***. (Weapons and equipment.)

(a) *Medical Evacuation and Hospitalization*. Method of evacuating dead and wounded, friendly and enemy personnel. Include priorities and location of CCP.

(b) *Personnel Support*. Method of handling EPWs and designation of the EPW collection point.

5. **COMMAND AND SIGNAL**.

This paragraph states where command and control facilities and key leaders are located during the operation. Use the OPORD shell to save time and organize your thoughts while preparing an OPORD.

a. **Command**.

- Location of the higher unit commander and CP.
- Location of key personnel and CP during each phase of the operation.

- Succession of command.
- Adjustments to the patrol SOP.

b. **Signal**.
- SOI index in effect.
- Methods of communication in priority.
- Pyrotechnics and signals, to include arm and hand signals (demonstrate).
- Code words such as OPSKEDs.
- Challenge and password (use behind friendly lines).
- Number combination (use forward of friendly lines).
- Running password.
- Recognition signals (near/far and day/night).

c. **Special Instructions to RTOs**.

d. **Actions after Issuance of OPORD--**
(1) Issue annexes.
(2) Give time hack.
(3) ASK for questions.

Figure 2-5. OPORD SHELL

OPORD

DTG:

REFERENCE:

TASK ORGANIZATION:

1ST PLT	2ND PLT	TM CONTROL:

1. **SITUATION**.

BMNT: SR: SS: EENT: MR: MS: HI: LO: ILLUM:

a. **Enemy**.

Terrain

(1) Higher mission:

Cdr's intent:

(2) Units to N, S, E, W.

End state

Terrain

End state

2. **MISSION**:

3. **EXECUTION**:

a. **Concept of Operations**.

Cdr's intent:

(ME): T: P:	(SE1): T: P:	(SE2): T: P:	(SE3): T: P:	(SE4): T: P:	(SE5): T: P:

b. **Maneuver**

PHASE 1:
ME:
SE1:
SE2:
SE3:
SE4:
SE5:
FIRES:
MTRS:
C2:

PHASE 2:
ME:
SE1:
SE2:
SE3:
SE4:
SE5:
FIRES:
MTRS:
C2:

PHASE 3:
ME:
SE1:
SE2:
SE3:
SE4:
SE5:
FIRES:
MTRS:
C2:

PHASE 4:
ME:
SE1:
SE2:
SE3:
SE4:
SE5:
FIRES:
MTRS:
C2:

c. **Tasks to Maneuver Units**.

(1) Risk-reduction control measures.
Rules of engagement.
(2) Environmental considerations.
(3) Non-Airborne personnel.

(ME): (SE1): (SE2):

(SE3): (SE4): (SE5):

d. **Tasks to Subordinate Units**

Mortars

Coordinating instructions:

PIR:

EEFI:

FFIR:

Timeline.

Coordinating Instructions.

WHEN	WHO	WHAT	WHERE

4. **SERVICE SUPPORT**.

a. **General**.

b. **Material and Services**.

(1) Supply:

Class I.

Class III:

Class V:

Class VIII:

Class IX:

(2) LOGSTAT.

(3) The chain of command is responsible to ensure that ammo lasts between resupplies.

c. **Medical Evacuation**.

(1) Air MEDEVAC is on call through [this] frequency. JP capable; use standard 9-line, 24-hr coverage.

(2) Establish friendly CCPs in AAs. Adjust CCPs as mission progresses.

(3) Troop medical clinic located vic [this location].

(4) Precedence for evacuation of WIA is US military, host nation military, host nation civilian, and enemy military.

(5) Precedence for evacuation of KIA is same. Evacuation of KIA will be as per mission constraints.

d. **Personnel**.

(1) Method of handling EPWs:

ANNEXES:

5. **COMMAND AND SIGNAL**.

a. **Command**.

(1) TF CP:

(2) Succession of Command:

b. **Signal**.

(1) Challenge and Password:

(2) Number Combination:

(3) Running Password:

(4) Recognition:

(a) Far:

(b) Near:

(7) Code word:

(8) OPSKED:

(9) Reports. SALUTE to higher on enemy contact.

2-5. **FRAGMENTARY ORDER**. A FRAGO is an abbreviated form of an operations order, usually issued on a day-to-day basis that eliminates the need for restating information contained in a basic operations order. It is issued after an OPORD to change or modify that order or to execute a branch or sequel to that order. Figure 2-6 shows an annotated FRAGO format:

Figure 2-6. ANNOTATED FRAGO FORMAT

FRAGMENTARY ORDER____________

Time Zone referenced throughout order:

Task Organization:

Weather and Light Data:

High:	BMNT:	Moonrise:
Low:	Sunrise:	Moonset:
Wind Speed:	Sunset:	% Illum:
Wind Direction:	EENT:	
Forecast:		

Terrain (changes that will effect operation in new area of operations):

OAKOC: Observation and fields of fire, avenues of approach, key terrain, observation, and cover and concealment.

1. **SITUATION**. (Brief changes from base OPORD specific to this day's operation.)

a. **Enemy Situation**.

(1) Composition, disposition and strength.

(2) Capabilities.

(3) Recent activities.
(4) Most likely COA.

b. **Friendly Situation.**

(1) Higher mission.
(2) Adjacent patrols task/purpose.
(3) Adjacent patrol objective/route (if known).

2. **MISSION** (Who, what [task], when, where, why [purpose]—from higher HQ's maneuver paragraph).

3. **EXECUTION**

a. **Concept of Operation**. (Explain how unit will accomplish mission in general terms. Identify mission essential task, designate the main effort and how the supporting efforts support the main effort.)

(1) ***Maneuver***. (Assign task/purpose and discuss actions on the objective in detail. Use a sketch or terrain model to brief.)

(2) ***Fires***. (In support of today's mission, portions that do not change are not briefed.) PLOT-CR:
(a) Purpose.
(b) Location.
(c) Observer.
(d) Trigger.
(e) Communication method.
(f) Resource.

(3) ***CASEVAC***.

b. **Tasks to Maneuver Units**. (List unique tasks that apply to each squad.)

c. **Coordinating Instructions**.

(1) ***Timeline***.
(a) Hit time:
(b) ORP time:
(c) Movement time from PB:
(d) Final inspection:
(e) Rehearsals:
(f) FRAGO complete:

(2) ***Movement Plan***. (Use a sketch to brief.)
(a) Routes (primary and alternate):
(b) OOM:
(c) Formations:
(d) Movement technique:

(3) ***PIR***. (Specific to this mission):

(4) ***Rehearsal Plan***.

(5) ***Patrol Base Plan***. (If not IAW SOP):
(a) Teams:
(b) Occupation plan:
(c) Operations plan (security plan, alert plan, Black and Gold):
(d) Priorities of work:

(6) ***Air Assault Plan***. (Provided with higher's FRAGO, if applicable):
(a) Number, type, and ACL aircraft:
(b) PZ grid, DOL, and PZ posture time:
(c) Load time/lift time/flight time:
(d) Number of lifts and composition:
(e) Air checkpoints en route:
(f) LZ grid/DOL/LZ time:
(g) Actions after getting off aircraft:
(h) Actions on contact on LZ:

(7) ***Linkup Plan***. (If applicable):
(a) Time of linkup:
(b) Location of linkup site:
(c) Stationary element:
(d) Moving element:
(e) Rally points:
(f) Actions at linkup point:
(g) Near/far recognition signals (day and night):

4. **SUPPORT**. (Only cover changes from base order that apply for today's mission).
a. **Material and Services**.
(1) Any changes in classes of supply:
(2) Resupply plan:
(3) Water resupply plan:
(4) Aerial resupply plan (if applicable) :
(5) Truck plan:
(6) Maintenance issues specific to plan:
b. **Medical Evacuation Plan**. (Specific to mission):
(1) CCP point and markings:
(2) Aid and litter duties (if not SOP):

5. **COMMAND AND SIGNAL**.
a. **Command**.
(1) Location of Company CP:
(2) Location of PL:
(3) Location of key leaders:
(4) Succession of command:
b. **Signal**.
(1) Location of radios:
(a) During movement:
(b) During actions on the objective:
(2) SOI in effect:

(a) Bn command freq:	Co CP call sign:
(b) Bn MEDEVAC freq:	PL call sign:
(c) Retrans freq:	PSG call sign:
(d) Company freq:	1 SL call sign:
(e) Platoon freq:	2 SL call sign:
	3 SL call sign:
	WSL call sign:

(3) Pyro signals in use:
(4) Running password/number combination:
(5) OPSKEDs in effect:
(a) The field FRAGO should take no more than 40 minutes to issue, with 30 minutes for the target. The proposed planning guide is as follows:

Paragraphs 1 and 2:	5 minutes
Paragraph 3:	20 to 30 minutes
Paragraphs 4 and 5:	5 minutes

(b) The FRAGO should focus on actions on the objective. The PL may use subordinates to prepare para 1, 4, 5 and routes and fires for the FRAGO. It is acceptable for subordinates to brief the portions of the FRAGO they prepare.
(c) Use of sketches and a terrain model are critical to allow rapid understanding of the operation/FRAGO.
(d) Rehearsals are critical as elements of the constrained planning model. The FRAGO used in conjunction with effective rehearsals reduces preparation time and allows the PL more time for movement and recon.
(e) Planning in a field environment will necessarily reduce the amount of time leaders have for in-depth mission planning. The TLP gives leaders a framework to plan missions and produce orders when time is short.

2-6. **ANNEXES**. Operation order annexes are necessary to complete the plan and to provide greater clarity and understanding during complex or critical aspects of the operation. Information issued in annex form include the: aerial re-supply, truck movement, air assault, patrol base, small boat, linkup, and stream crossing annexes. Annexes are prepared only if the subject is not addressed thoroughly enough in the OPORD; brevity remains the standard. Annexes are always issued after the operation order.

AIR MOVEMENT ANNEX

1. **SITUATION**.
   a. **Enemy Situation**.
      (1) Enemy air capability.
      (2) Enemy ADA capability.
      (3) Include in weather: % Illum, illum angle, NVG window, ceiling, and visibility.
   b. **Friendly Situation**.
      (1) Unit(s) supporting operation.
      (2) Friendly ADA status.
2. **MISSION**.
3. **EXECUTION**.
   a. **Concept of Operation**.
   b. **Subunit Missions**.
   c. **Coordinating instructions**.
      (1) ***Pickup Zone***.
         (a) Name/Number/
         (b) Coordinates/
         (c) Load Time/
         (d) Takeoff Time/
         (e) Markings/
         (f) Control/
         (g) Landing Formation/
         (h) Approach/Departure Direction.
         (i) Alternate PZ Name/Number.
         (j) Penetration Points.
         (k) Extraction Points.
      (2) ***Landing Zone***.
         (a) Name/Number.
         (b) Coordinates.
         (c) H-Hour.
         (d) Markings.
         (e) Control.
         (f) Landing Formation/Direction.
         (g) Alt LZ Name/Number.
         (h) Deception Plan.
         (i) Extraction LZ.
      (3) ***Laager Site***.
         (a) Communications.
         (b) Security Force.
      (4) ***Flight Routes and Alternates***.
      (5) ***Abort Criteria***.
      (6) ***Down Aircraft/Crew*** (designated area of recovery [DAR]).
      (7) ***Special Instructions***.
      (8) ***Cross-FLOT Considerations***.
      (9) ***Aircraft Speed***.

(10) ***Aircraft Altitude***.
(11) ***Aircraft Crank Time***.
(12) ***Rehearsal Schedule/Plan***.
(13) ***Actions on Enemy Contact*** (en route and on the ground).

4. **SERVICE SUPPORT**.
   a. **Forward Area Refuel/Rearm Points**.
   b. **Classes I, III, and V** (specific).

5. **COMMAND AND SIGNAL**.
   a. **Command**.
   b. **Signal**.
      (1) Air/ground call signs and frequencies.
      (2) Air/ground emergency code.
      (3) Passwords/number combinations.
      (4) Fire net/Quickfire net.
      (5) Time zone.
      (6) Time hack.

AERIAL RESUPPLY ANNEX

1. **SITUATION**.
   a. **Enemy Forces** (include weather).
   b. **Friendly Forces**.
   c. **Attachments and Detachments**.

2. **MISSION**.

3. **EXECUTION**.
   a. **Concept of Operation**.
      (1) ***Maneuver***.
      (2) ***Fires***.
   b. **Tasks to Combat Units**.
      (1) ***Command and Control***.
      (2) ***Security***.
      (3) ***Marking***.
      (4) ***Recovery/Transport***.
   c. **Tasks to Combat Support Units**.
   d. **Coordinating Instructions**.
      (1) Flight Route.
         (a) General.
         (b) Checkpoints.
         (c) Communication checkpoint (CCP).
            - Marking of CCP.
            - Report time.
         (d) Heading from CCP.
      (2) Landing/Drop Zone.
         (a) Location.
            - Primary.
            - Alternate.
         (b) Marking.
            - Near.
            - Far.

(3) Drop Information.
(a) Date/time of resupply (and alternates).
(b) Code letter at DZ/LZ.
(c) Length of DZ in seconds or dimensions of LZ.
(d) Procedures for turning off DZ/LZ.
(e) Formation, altitude, and air speed.
- En route.
- At DZ/LZ.

(4) Actions on enemy contact during resupply.
(5) Abort Criteria: En route and at DZ/LZ.
(6) Actions at DZ/LZ.
- Rehearsals.

4. **SERVICE SUPPORT.**

5. **COMMAND AND SIGNAL.**
a. **Command.**
(1) Location of patrol leader.
(2) Location of assistant patrol leader.
(3) Location of members not involved in resupply.
b. **Signal.**
(1) Air to ground call-signs and frequencies (primary and alternate).
(2) Long range visual signals.
(3) Short range visual signals.
(4) Emergency procedures and signals.
(5) Air drop communication procedures.
(6) Code words.

PATROL BASE ANNEX

1. **SITUATION**
a. Enemy Forces
b. Friendly Forces
c. Attachments and Detachments

2. **MISSION.**

3. **EXECUTION.**
a. **Concept of Operation.**
(1) ***Maneuver.***
(2) ***Fires.***
b. **Tasks to Combat Units.**
(1) ***Teams.***
- Security
- Recon
- Surveillance
- LP/OPs

(2) ***Individuals.***
c. **Tasks to Combat Support Units.**
d. **Coordinating instructions.**
(1) Occupation plan.
(2) Operations plan.
- Security Plan.
- Alert Plan.

- Priority of work.
- Evacuation plan.
- Alternate patrol base (used when primary is unsuitable or compromised).

4. **SERVICE SUPPORT.**
   a. Water plan.
   b. Maintenance plan.
   c. Hygiene plan.
   d. Messing plan.
   e. Rest plan.

5. **Command and Signal.**
   a. **Command.**
      (1) Location of patrol leader.
      (2) Location of assistant patrol leader.
      (3) Location of patrol CP.
   b. **Signal.**
      (1) Call signs and frequencies.
      (2) Code words.
      (3) Emergency signals.

SMALL BOAT ANNEX

1. **SITUATION.**
   a. **Enemy Forces.**
      (1) **Weather.**
         (a) Tide.
         (b) Surf.
         (c) Wind.
      (2) **Terrain.**
         (a) River width.
         (b) River depth and water temperature.
         (c) Current.
         (d) Vegetation.
      (3) **Identification, Location, Activity, and Strength.**
   b. **Friendly Forces** (unit furnishing support).
   c. **Attachments and Detachments.**
   d. **Organization for Movement.**

2. **MISSION.**

3. **EXECUTION.**
   a. **Concept of Operation.**
      (1) ***Maneuver.***
      (2) ***Fires.***
   b. **Tasks to Combat Units.**
      (1) Security.
      (2) Tie-down teams.
         (a) Load equipment.
         (b) Secure equipment.
      (3) Designation of coxswains and boat commanders.
      (4) Selection of navigator(s) and observer(s).

c. **Coordinating Instructions**.
   (1) Formations and order of movement.
   (2) Route and alternate route.
   (3) Method of navigation.
   (4) Actions on enemy contact.
   (5) Rally points.
   (6) Embarkation plan.
   (7) Debarkation plan.
   (8) Rehearsals.
   (9) Time schedule.

4. **SERVICE SUPPORT**.
   a. **Ration Plan**.
   b. **Weapons and Ammunition**.
   c. **Uniform and Equipment**.
      (1) Method of distribution of paddles and life jackets.
      (2) Disposition of boat, paddles and life jackets upon debarkation.

5. **COMMAND AND SIGNAL**.
   a. **Command**.
      (1) Location of patrol leader.
      (2) Location of assistant patrol leader.
   b. **Signal**.
      (1) Signals used between boats and in boats.
      (2) Code words.

STREAM CROSSING ANNEX

1. **SITUATION**.
   a. **Enemy Forces**.
      (1) ***Weather***.
      (2) ***Terrain***.
         (a) River width.
         (b) River depth and water temperature.
         (c) Current.
         (d) Vegetation.
         (e) Obstacles.
      (3) ***Enemy*** (location, identification, activity).
   b. **Friendly Forces**.
   c. **Attachments and Detachments**.

2. **MISSION**.

3. **EXECUTION**.
   a. **Concept of Operation**.
      (1) ***Maneuver***.
      (2) ***Fires***.
   b. **Tasks to Combat Units**.
      (1) ***Elements***.
      (2) ***Teams***.
      (3) ***Individuals***.

c. **Tasks to Combat Support Units.**
d. **Coordinating Instructions.**
(1) Crossing procedure/techniques.
(2) Security.
(3) Order of crossing.
(4) Actions on enemy contact.
(5) Alternate plan.
(6) Rallying points.
(7) Rehearsal plan.
(8) Time schedule.

4. **SERVICE SUPPORT.**

5. **COMMAND AND SIGNAL.**
a. **Command.**
(1) Location of patrol leader.
(2) Location of assistant patrol leader.
(3) Location of CP.
b. **Signal.**
(1) Emergency signals.
(2) Signals.

TRUCK ANNEX

1. **SITUATION.**
a. **Enemy Forces.**
b. **Friendly Forces.**
c. **Attachments and Detachments.**

2. **MISSION.**

3. **EXECUTION.**
a. **Concept of Operation.**
(1) ***Maneuver.***
(2) ***Fires.***
b. **Tasks to Combat Units.**
c. **Tasks to Combat Support Units.**
d. **Coordinating Instructions.**
(1) Time of departure and return.
(2) Loading plan and order of movement.
(3) Route (primary and alternate).
(4) Air Guards.
(5) Actions on enemy contact (vehicle ambush) during movement, loading, and downloading.
(6) Actions at the detrucking point.
(7) Rehearsals.
(8) Vehicle speed, separation, and recovery plan.
(9) Broken vehicle instructions.

4. **SERVICE SUPPORT**.

5. **COMMAND AND SIGNAL**.
   a. **Command**. Location of PL and PSG
   b. **Signal**.
      (1) Radio call signs and frequencies
      (2) Code words

2-7. **COORDINATION CHECKLISTS**. The following checklists are items that a platoon/squad leader must check when planning for a combat operation. In some cases, he will coordinate directly with the appropriate staff section, in most cases this information will be provided by the company commander or platoon leader. The platoon/squad leader, to keep him from overlooking anything that may be vital to his mission, may carry copies of these checklists:

Figure 2-7. Intelligence coordination checklist
Figure 2-8. Operations coordination checklist
Figure 2-9. Fire support coordination checklist
Figure 2-10. Coordination with forward unit checklist
Figure 2-11. Adjacent unit coordination checklist
Figure 2-12. Rehearsal area coordination checklist
Figure 2-13. Army aviation coordination checklist
Figure 2-14. Vehicular movement coordination checklist

Figure 2-7. INTELLIGENCE COORDINATION CHECKLIST

INTELLIGENCE COORDINATION CHECKLIST

In this coordination, the leader is informed of any changes in the situation as given in the operation order of mission briefing. He must keep himself constantly updated to ensure the plan is sound:

1. Identification of enemy unit.
2. Weather and light data.
3. Terrain update.
   a. Aerial photos.
   b. Trails and obstacles not on map.
4. Known or suspected enemy locations.
5. Weapons.
6. Probable course of action.
7. Recent enemy activities.
8. Reaction time of reaction forces.
9. Civilians on the battlefield.
10. Update to CCIR.

Figure 2-8. OPERATIONS COORDINATION CHECKLIST

OPERATIONS COORDINATION CHECKLIST

This coordination occurs with the platoon leader/company commander so that the platoon/squad leader can confirm his mission and operational plan, receive any last-minute changes to his mission or plan, and to update his subordinates or issue a FRAGO, if required:

1. Mission backbrief.
2. Identification of friendly units.
3. Changes in the friendly situation.
4. Route selection, LZ/PZ/DZ selection.
5. Linkup procedures.
6. Transportation/movement plan.

7. Resupply (in conjunction with S4).
8. Signal plan.
9. Departure and re-entry of forward units.
10. Special equipment requirements.
11. Adjacent units operating in the area of operations.
12. Rehearsal areas.
13. Method of insertion/extraction.

Figure 2-9. FIRE SUPPORT COORDINATION CHECKLIST

FIRE SUPPORT COORDINATION CHECKLIST

The platoon/squad leader will normally coordinate the following with the platoon forward observer (FO):

1. Mission backbrief.
2. Identification of supporting unit.
3. Mission and objective.
4. Route to and from the objective (include alternate routes).
5. Time of departure and expected time of return.
6. Unit target list (from fire plan).
7. Type of available support (artillery, mortar, naval gunfire, and aerial support to include Army, Navy, and Air Force) and their locations.
8. Ammunition available (to include different fuses).
9. Priority of fires.
10. Control measures.
    a. Checkpoints.
    b. Boundaries.
    c. Phase lines.
    d. Fire support coordination measures.
    e. Priority targets (target list).
    f. RFA (restrictive fire area).
    g. RFL (restrictive fire line).
    h. NFA (no-fire areas).
    i. Precoordinated authentication.
11. Communication (include primary and alternate means, emergency signals and code words).

Figure 2-10. COORDINATION WITH FORWARD UNIT CHECKLIST

COORDINATION WITH FORWARD UNIT CHECKLIST

A platoon/squad that requires foot movement through a friendly forward unit must coordinate with that unit's commander for a safe and orderly passage. If no time and place has been designated for coordination with the forward unit, the platoon/squad leader should set a time and place when he coordinates with the S-3. He must talk with someone at the forward unit who has the authority to commit that unit in assisting the platoon/squad during departure. Coordination entails a two-way exchange of information:

1. Identification (yourself and your unit).
2. Size of platoon/squad.
3. Time(s) and place(s) of departure and return, location(s) of departure point(s), ERRP, and detrucking points.
4. General area of operations.
5. Information on terrain and vegetation.
6. Know or suspected enemy positions or obstacles.
7. Possible enemy ambush sites.
8. Latest enemy activity.
9. Detailed information on friendly positions such as crew-served weapons, FPF.

10. Fire and barrier plan.
   a. Support the unit can furnish. How long and what can they do?
      (1) Fire support.
      (2) Litter teams.
      (3) Navigational signals and aids.
      (4) Guides.
      (5) Communications.
      (6) Reaction units.
      (7) Other.
   b. Call signs and frequencies.
   c. Pyrotechnic plan.
   d. Challenge and password, running password, number combination.
   e. Emergency signals and code words.
   f. If the unit is relieved, pass the information to the relieving unit.
   g. Recognition signals.

Figure 2-11. ADJACENT UNIT COORDINATION CHECKLIST

ADJACENT UNIT COORDINATION CHECKLIST

Immediately after the operation order of mission briefing, the platoon/squad leader should check with other platoon/squad leaders who will be operating in the same areas. If the leader is not aware of any other units operating is his area, he should check with the S3 during the operations coordination. The S3 can help arrange this coordination if necessary. The platoon/squad leaders should exchange the following information with other units operating in the same area:

1. Identification of the unit.
2. Mission and size of unit.
3. Planned times and points of departure and reentry.
4. Route(s).
5. Fire support and control measures.
6. Frequencies and call signs.
7. Challenge and password, running password, number combination.
8. Pyrotechnic plan.
9. Any information that the unit may have about the enemy.
10. Recognition signals.

Figure 2-12. REHEARSAL AREA COORDINATION CHECKLIST

REHEARSAL AREA COORDINATION CHECKLIST

This coordination is conducted with the platoon leader/company commander to facilitate the unit's safe, efficient and effective use of rehearsal area prior to its mission:

1. Identification of your unit.
2. Mission.
3. Terrain similar to objective site.
4. Security of the area.
5. Availability of aggressors.
6. Use of blanks, pyrotechnics, and ammunition.
7. Mock-ups available.
8. Time the area is available (preferably when light conditions approximate light conditions of patrol).
9. Transportation.
10. Coordination with other units using area.

Figure 2-13. ARMY AVIATION COORDINATION CHECKLIST

ARMY AVIATION COORDINATION CHECKLIST

This coordination is conducted with the platoon leader/company commander and/or S3 Air to facilitate the time, detailed and effective use of aviation assets as they apply to your tactical mission:

1. **SITUATION**.
   a. **Enemy Situation**.
      (1) Enemy air capability.
      (2) Enemy ADA capability.
      (3) Include in Weather: Percent illum, illum angle, NVG window, ceiling, and visibility.
   b. **Friendly Situation**.
      (1) Unit(s) supporting operation, Axis of movement/corridor/routes.
      (2) Friendly ADA status.

2. **MISSION**.

3. **EXECUTION**.
   a. **Concept of the Operation**. Overview of what requesting unit wants to accomplish with the air assault/air movement.
   b. **Tasks to Combat Units**.
      (1) Infantry.
      (2) Attack aviation.
   c. **Tasks to Combat Support Units**.
      (1) Artillery.
      (2) Aviation (lift).
   d. **Coordinating Instructions**.
      (1) ***Pickup Zone***.
      - Direction of landing.
      - Time of landing/flight direction.
      - Location of PZ/alternate PZ.
      - Loading procedures.
      - Marking of PZ (panel, smoke, SM, lights).
      - Flight route planned (SP, ACP, RP).
      - Formations: PZ, en route, LZ.
      - Code words:
        - PZ secure (prior to landing), PZ clear (lead bird and last bird).
        - Alternate PZ (at PZ, en route, LZ), names of PZ/alt PZ.
      - TAC air/artillery.
      - Number of pax per bird and for entire lift.
      - Equipment carried by individuals.
      - Marking of key leaders.
      - Abort criteria (PZ, en route, LZ).

      (2) ***Landing Zone***.
      - Direction of landing.
      - False insertion plans.
      - Time of landing (LZ time).
      - Location of LZ and Alternate LZ.
      - Marking of LZ (panel, smoke, SM, lights).
      - Formation of landing.
      - Code words, LZ name, alternate LZ name.
      - TAC air/artillery preparation, fire support coordination.
      - Secure LZ or not?

4. **SERVICE AND SUPPORT**.
   a. Number of aircraft per lift and number of lifts.
   b. Refuel/rearm during mission or not?
   c. Special equipment/aircraft configuration for weapons carried by unit personnel.
   d. Bump plan.
5. **COMMAND AND SIGNAL**.
   a. Frequencies, call signs and code words.
   b. Locations of air missions commander ground tactical commander, and air assault task force commander.

Figure 2-14. VEHICULAR MOVEMENT COORDINATION CHECKLIST

VEHICULAR MOVEMENT COORDINATION CHECKLIST

This is coordinated with the supporting unit through the platoon sergeant/first sergeant to facilitate the effective, detailed, and efficient use of vehicular support and/or assets:

1. Identification of the unit.
2. Supporting unit identification.
3. Number and type of vehicles and tactical preparation.
4. Entrucking point.
5. Departure/loading time.
6. Preparation of vehicles for movement.
   a. Driver responsibilities.
   b. Platoon/squad responsibilities.
   c. Special supplies/equipment required.
7. Availability of vehicles for preparation/rehearsals/inspection (time and location).
8. Routes.
   a. Primary.
   b. Alternate.
   c. Checkpoints.
9. Detrucking points.
   a. Primary.
   b. Alternate.
10. March internal/speed.
11. Communications (frequencies, call signs, codes).
12. Emergency procedures and signals.

2-8. **DOCTRINAL TERMS**

TACTICAL TASK--A clearly defined, measurable activity accomplished by individuals and organizations. Tasks are specific activities that contribute to the accomplishment of encompassing missions or other requirements. A task should be definable, measurable, and decisive (achieve the purpose).

Actions by Friendly Force		**Effects on Enemy Force**
Assault	Follow and Assume	Block
Attack-by-Fire	Follow and Support	Canalize
Breach	Linkup	Contain
Bypass	Occupy	Defeat
Clear	Reconstitution	Destroy
Combat Search and Rescue	Reduce	Disrupt
Consolidation and Reorganization	Retain	Fix
Control	Secure	Interdict
Counterreconnaissance	Seize	Isolate
Disengagement	Support-by-Fire	Neutralize
Exfiltrate	Suppress	Penetrate
		Turn

PURPOSE (in order to)--The desired or intended result of the tactical operation stated in terms related to the enemy or the desired situation. The why of the mission statement. The most important component of the mission statement.

Allow	Divert	Prevent
Cause	Enable	Protect
Create	Envelop	Support
Deceive	Influence	Surprise
Deny	Open	

OPERATION--A military action or the carrying out of a military action to gain the objectives of any battle or campaign.

Types of operations include:	Forms of Offensive Maneuver:	Types of Defensive Techniques:	Other Operations:
Movement to Contact Search and Attack Attack Ambush Demonstration Feint Raid Spoiling Attack Exploitation Pursuit	Envelopment Frontal Attack Infiltration Penetration Turning Movement	Area Defense Mobile Defense Retrograde Ops Delay Withdrawal Retirement	Reconnaissance Ops Security Ops Information Ops Combined Arms Breach Passage of Lines Relief in Place River Crossing Ops Troop Movement Admin Approach March Road March

2-9. **TERRAIN MODEL**. The terrain model (Figure 2-15) is an outstanding visual means used during the planning process to communicate the patrol routes and also detailed actions on the objective. At a minimum, the model is required to during planning to display routes to the objective and to highlight prominent terrain features the patrol will encounter during movement. A second terrain model is usually constructed of the objective area, enlarged to a sufficient size and detail to brief the patrol's actions on the objective.

a. **Checklist**.

(1) North-seeking arrow.

(2) Scale.

(3) Grid lines.

(4) Objective location.

(5) Exaggerated terrain relief, water obstacles.

(6) Friendly patrol locations.

(7) Targets (indirect fires, including grid and type of round).

(8) Routes, primary and alternate.

(9) Planned RPs (ORP, L/URP, RP).

(10) Danger areas (roads, trails, open areas).

(11) Legend.

(12) Blowup of objective area.

b. **Construction**. The following are some field-expedient techniques to aid in terrain model construction:

(1) Use a 3 x 5 card, MRE box, or piece of paper to label the objective or key sites.

(2) Grid lines can be made using string from the guts of 550 cord or colored tape. (Grid lines are identified by writing numbers on small pieces of paper.)

(3) Trees and vegetation are replicated by using moss, green or brown spray paint, pine needles, crushed leaves, or cut up grass.

(4) Water is designated by blue chalk, blue spray paint, blue yarn, tin foil, or MRE creamer.

(5) North seeking arrows are made from sharpened twigs, pencils, or colored yarn.

(6) Enemy positions are designated using red yarn, M-16 rounds, toy Rangers, or poker chips.

(7) Friendly positions such as security elements, support by fire, and assault elements are made using M-16 rounds, toy Rangers, poker chips, small MRE packets of sugar and coffee, or pre-printed acetate cards.

(8) Small pieces of cardboard or paper can identify target reference points and indirect fire targets. Ensure grids are shown for each point.

(9) Breach, support by fire, and assault positions are made using colored yarn or string so that these positions can be easily identified.

(10) Bunkers and buildings are constructed using MRE boxes or tongue depressors/sticks.

(11) Perimeter wire is constructed from a spiral notebook.

(12) Key phase lines are constructed with colored string or yarn.

(13) Trench lines are replicated by colored tape or yarn, by digging a furrow and coloring it with colored chalk or spray paint.

Note: All symbols used on the terrain model must be clearly identified in the legend.

Figure 2-15. TERRAIN MODEL

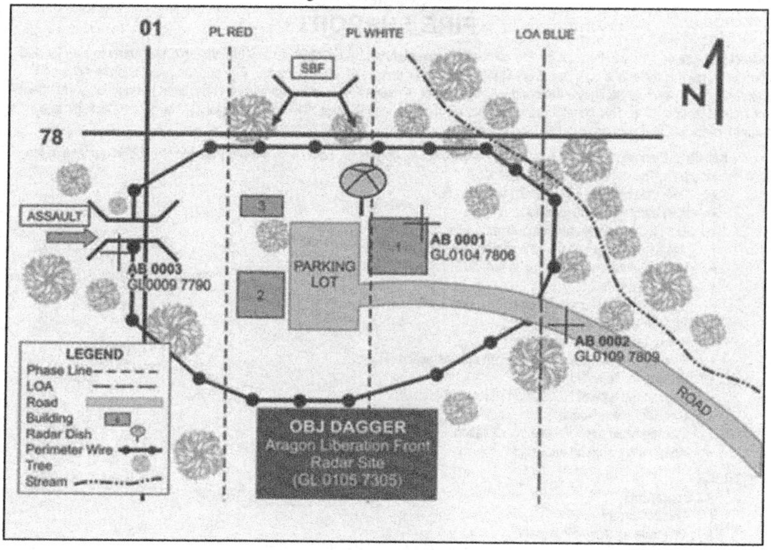

NOTES

Chapter 3
FIRE SUPPORT

Indirect fire support can greatly increase the combat effectiveness and survivability of any Infantry unit. The ability to plan for and effectively use this asset is a task that every Ranger and small unit leader should master. Fire support assets can assist a unit suppress, fix, destroy, or neutralize the enemy. Employment of indirect fire support should be considered throughout every offensive and defensive operation. This chapter discusses planning, tasks, capabilities, risk estimate distances, target overlays, close air support, elements and sequence of call for fire, and example call-for-fire transmissions.

3-1. **PLANNING**. Planning is the continual process of selecting targets on which fires are prearranged to support a phase of the commander's plan. The principles of planning follow:

- Consider what the commander wants to do.
- Plan early and continuously.
- Exploit all available targeting assets.
- Use all available lethal and nonlethal fire support means.
- Use the lowest echelon able to furnish effective support.
- Observe all fires.
- Use the most effective fire support asset available.
- Provide adequate fire support.
- Avoid unnecessary duplication.
- Provide for safety of friendly forces and installations.
- Provide for flexibility.
- Furnish the type of fire support requested.
- Consider the airspace.
- Provide rapid and effective coordination.
- Keep all fire support informed.

3-2. **TASKS**.

a. **All Operations**.

(1) Locate targets.
(2) Integrate all available assets.
(3) Destroy, neutralize, or suppress all enemy direct and indirect fire systems.
(4) Provide illumination and smoke.
(5) Provide fires in support of JAAT and SEAD missions.
(6) Deliver scatterable mines.
(7) Prepare for future operations.
(8) Provide positive clearance of fires.

b. **Offensive Operations**.

(1) Support the movement to contact, chance contact.
(2) Soften enemy defenses before the attack by arranging short, violent preparations, where required.
(3) Provide support during the attack by attacking high payoff targets.
(4) Plan for deep and flanking fires.
(5) Plan fires during consolidation.
(6) Provide counterfires.

c. **Defensive Operations**.

(1) Execute all plans as the commander intends.
(2) Disorganize, delay, and weaken the enemy before the attack.
(3) Provide counterfire.
(4) Provide fires in support of planned engagement areas (EAs).
(5) Attack high payoff targets.
(6) Plan fires in support of barriers and obstacle plans.
(7) Plan for deep, flanking, and rear fires.

(8) Provide fires in support of counterattacks.
(9) Plan final protective fire (FPFs).

3-3. **CAPABILITIES**. Tables 3-1 and 3-2 show capabilities of field artillery and mortars.

Table 3-1. CAPABILITIES OF FIELD ARTILLERY

WEAPON	MAX RANGE (meters)	MIN RANGE (meters)	MAX RATE Rds per Min	Burst Radius	SUSTAINED RATE Rds per Min
105-mm Howitzer M119, Towed	14,000 m	0	6 for 2 min	35 m	3 Rds for 30 min, then 1 rd per min
155-mm Howitzer M198, Towed	18 m,100 m 30,000 m (RAP)	0	4 for 3 min 2 for 30 min	50 m	1 rd per min Temp Dependent
155-mm Howitzer M109A6 SP	18 m, 100 m 30,000 m (RAP)	0	4 for 3 min	50 m	1 for 60 min 0.5

Table 3-2. CAPABILITIES OF MORTARS

WPN	MUNITIONS AVAILABLE	MAX HE RANGE	MIN RANGE (meters)	MAX RATE (Rds per Min)	BURST RADIUS	SUSTAINED RATE (Rds per Min)
60mm	HE, WP, Illum	3,500 m	70 m (HE)	30 for 4 min	30 m	20
81mm	HE, WP, Illum	5,600 m	70 m (HE)	25 for 2 min	38 m	8
120mm	HE, Smoke, Illum	7,200 m	180 m (HE)	15 for 1 min	60 m	5

DANGER

DANGER CLOSE

1. **When the target is within 600 meters of any friendly troops (for mortars and field artillery), announce DANGER CLOSE in the Method of Engagement portion of the call for fire.**
2. **When adjusting 5-inch or smaller naval guns on targets within 750 meters, announce DANGER CLOSE. For larger naval guns, announce DANGER CLOSE for targets within 1,000 meters. Failure to adhere to this guidance can result in fratricide.**
3. **Avoid making corrections using the bracketing method of adjustment, because doing so can cause serious injury or death. Use only the creeping method of adjustment during DANGER CLOSE missions. Make corrections of no more than 100 meters by creeping the rounds to the target.**

3-4. **RISK ESTIMATE DISTANCES**. RED applies to combat only. Minimum safe distances (Table 3-3) apply to training IAW AR 350-1. RED takes into account the bursting radius of particular munitions and the characteristics of the delivery system and associates this combination with a percentage representing the likelihood of becoming a casualty, that is, the percentage of risk. RED is defined as the minimum distance friendly troops can approach the effects of friendly fires without suffering appreciable casualties of 0.1% PI or higher.

> **WARNING**
>
> **[Commanders] use risk-estimate distance formulas to determine acceptable risk levels in combat only. Specifically, use them to identify the risk to your Rangers at various distances from their targets. Risk estimate distances apply only in combat. In training, use minimum safe distances (MSD).**

a. **Casualty Criterion**. The casualty criterion is the 5-minute assault criterion for a prone soldier in winter clothing and helmet. Physical incapacitation means that a soldier is physically unable to function in an assault within a 5-minute period after an attack. A PI value of less than 0.1 percent can be interpreted as being less than or equal to one chance in one thousand.

Table 3-3. RISK ESTIMATE DISTANCES FOR MORTARS AND CANNON ARTILLERY

	Risk Estimate Distances (Meters)					
	10 % PI			0.1 % PI		
System	1/3 range	2/3 range	Max range	1/3 range	2/3 range	Max range
60-mm mortar	60	65	65	100	150	175
81-mm mortar	75	80	80	165	185	230
120-mm mortar	100	100	100	150	300	400
105-mm howitzer	85	85	90	175	200	275
155-mm howitzer	100	100	125	200	280	450
155-mm DPICM	150	180	200	280	300	475

b. **Risk**. Using echelonment of fires within the specified RED for a delivery system requires the unit to assume some risks. The maneuver commander determines by delivery system how close he will allow fires to fall in proximity to his forces. The maneuver commander makes the decision for this risk level, but he relies heavily on the FSO's expertise.

3-5. **TARGET OVERLAYS**.

a. Figure 3-1 shows contents of fire support overlay.

Figure 3-1. CONTENTS OF FIRE SUPPORT OVERLAY

- Unit and official capacity of person making overlay.
- Date the overlay was prepared.
- Map Sheet Number.
- Effective Period of Overlay (DTG).
- Priority target.
- ORP Location.
- Call signs and frequencies. (PRI/ALT)
- Routes--Primary/Alternate.
- Phase Lines/Checkpoints used by the patrol.
- Spares.
- Index marks to position overlay on map.
- Objective.
- Target Symbols.
- Description, location and remarks column, complete.

b. Using a PLOT-CR checklist helps ensure the leader's fire support plan is complete:
   (1) Purpose--Purpose of the planned indirect fires.
   (2) Location--Planned targets with an 8-digit grid (minimum).
   (3) Observer--Planned observer to see the impact of the rounds and adjust.
   (4) Trigger--Method to initiate fires.
   (5) Communication--Method to communicate between the observer and the supporting unit.
   (6) Resources--Planned allocated resource for each target.

c. Sterile overlay includes---
   (1) Index marks to position overlay on map.
   (2) Target symbols.

d. Target overlay symbols (Figure 3-2) include--
   (1) Point target:
   (2) Linear target:
   (3) Circular target:

Figure 3-2. TARGET OVERLAY SYMBOLS--POINT (TOP), LINEAR (CENTER), AND CIRCULAR (BOTTOM)

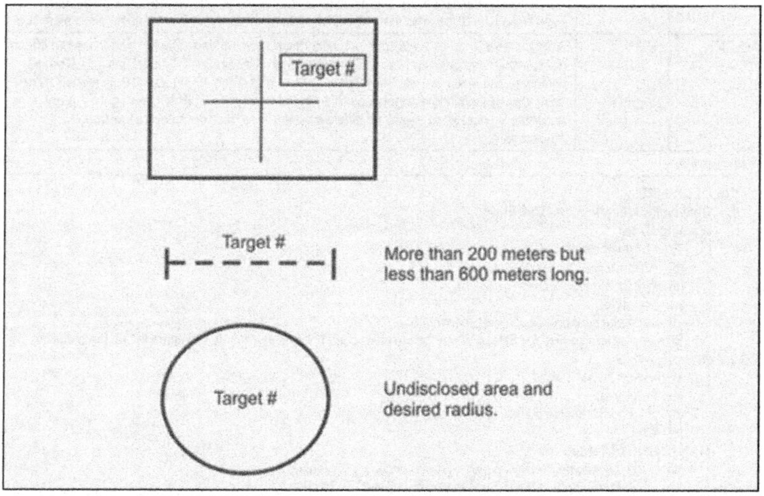

3-6. **CLOSE AIR SUPPORT**. The two types of close air support requests are planned and immediate. Planned requests are processed by the Army chain to corps for approval. Immediate requests are initiated at any level and processed by the battalion S-3 FSO, and air liaison officer.

a. Format for requesting immediate CAS:
   (1) Observer identification.
   (2) Warning order (request close air).
   (3) Target location (grid).

(4) Target description. (Description must include, as a minimum: type and number of targets, activity or movement; point or area targets, include desired results on target and time on target.

(5) Location of friendly forces.

(6) NAV details (elevation).

(7) Threats--ADA, small arms, and so on.

(8) Hazards--friendly aircraft in area.

(9) Wind direction.

b. Close air support capabilities are shown in Table 3-4.

Table 3-4. CLOSE AIR SUPPORT

AIRCRAFT	SERVICE	CHARACTERISTICS
A-10 *	AF	Specialized CAS aircraft; subsonic; typical load 6,000 lbs, max load 16,000 lbs; 30-mm gun.
F-16 *	AF	A multirole aircraft; complements the F-4 and F-15 in an air-to-air role. Most accurate ground delivery system in the inventory; supersonic; typical load 6,000 lbs, max load 10,600 lbs.
F-18 *	N/MC	A multirole fighter scheduled to replace the F-4. A wide variety of air-to-surface weapons. Typical load 7,000 lbs, max load 17,000 lbs, 20-mm gun mounted in the nose; air-to-air missiles.
AC-130	AF/R	A specialized CAS/RACO aircraft, propeller driven. Two models: The "A" model has two 40-mm guns, two 20-mm guns and two 7.62-mm miniguns. The similar "H" model has no 7.62-mm miniguns, and has a 105-mm howitzer in place of one of the 40-mm guns. Both models have advanced sensors and a target-acquisition system, including FLIR and low light TV. Very accurate. Vulnerable to enemy air defense systems, so must operate in a low-threat environment.
*FM capability.		

3-7. **CALL FOR FIRE**.

a. **Observer's Identification - Call Signs**.

b. **Warning Order**.

(1) ***Type of mission***.

(a) Adjust fire.

(b) Fire for effect.

(c) Suppress.

(d) Immediate suppression/immediate smoke.

(2) ***Size of element to Fire for Effect***. When the observer does not specify what size element to fire, the battalion FDC will decide.

(3) ***Method of Target Location***.

(a) Polar plot.

(b) Shift from a known point.

(c) Grid.

(4) **Location of Target**.

(a) *Grid Coordinate - 6 Digit*. 8 digit if greater accuracy is required.

(b) *Shift from a Known Point*. Send observer-target (OT) direction:

- Mils (nearest 10).
- Degrees.
- Cardinal direction.
- Send lateral shift (right/left, nearest 10 mils).
- Send range shift (add/drop, nearest 100 mils).
- Send vertical shift (up/down). Use only if it exceeds 35 mils (nearest 5 mils).

(c) *Polar Plot*. Send direction. (Nearest 10 mils)

- Send distance (nearest 100 mils).
- Send vertical shift (nearest 5 mils).

(5) ***Description of Target***.
(a) Type.
(b) Activity.
(c) Number.
(d) Degree of protection.
(e) Size and shape (length/width or radius).

(6) ***Method of Engagement, Type of Adjustment***. When the observer does not request a specific type of fire control adjustment, area fire is issued.
(a) Area fire--moving target.
(b) Precision fire--point target.

(7) ***Danger Close***. Use when friendly troops are within--
(a) 600 mils for mortars.
(b) 600 mils for artillery.
(c) 750 mils for naval guns 5 inches or smaller.

(8) ***Mark***. Use to orient observer or to indicate targets.

(9) ***Trajectory***.
(a) Low angle (standard).
(b) High angle (mortars or if requested).

(10) ***Ammunition***. Use HE quick used unless observer specifies otherwise.
(a) Projectile (HE, illum, ICM, smoke, and so on).
(b) Fuse (quick, time, and so on).
(c) Volume of fire (observer may request the number of rounds to be fired).

(11) ***Distribution***.
(a) 100 mil sheaf (standard).
(b) Converged sheaf (used for small hard targets).
(c) Special sheaf (any length, width, and attitude).
(d) Open sheaf (separate bursts).
(e) Parallel sheaf (linear target).

(12) ***Method of Fire and Control***.
(a) *Method of Fire*. Specific guns and a specific interval between rounds. Normally adjust fire, one gun is used with a 5-second interval between rounds.
(b) *Method of Control*.

- AT MY COMMAND - FIRE--Remains in effect until observer announces, "*Cancel at my command.*"
- CANNOT OBSERVE--Observer can't see the target.
- TIME ON TARGET--Observer tells FDC when he wants the rounds to impact.
- CONTINUOUS ILLUMINATION. Calculated by the FDC, otherwise observer indicates interval between rounds in seconds.
- COORDINATED ILLUMINATION. Observer may order the interval between illum and HE shells.
- CEASE LOADING--Stop loading rounds.
- CHECK FIRING--Stop firing at once.
- CONTINUOUS FIRE--Load and fire as fast as possible.
- REPEAT--Fire another round(s) with or without adjustments.

(13) ***Authentication***. Challenge and reply.

(14) ***Message to Observer***.
(a) Battery(ies) to fire for effect.
(b) Adjusting battery.
(c) Changes to the initial call for fire.
(d) Number of rounds (per tube) to be fired for effect.
(e) Target numbers.

(f) Additional information:
- Time of flight. Moving target mission.
- Probable error in range. 38 meters or greater (normal mission).
- Angle T. 500 mils or greater.

c. **Correction of Errors**. When FDC has made an error when reading back he fire support data, the observer announces "*CORRECTION*" and transmits the correct data in its entirety.

d. **Call-for-Fire Transmissions**. Example call-for-fire transmissions are shown in Table 3-5.

Table 3-5. EXAMPLE CALL-FOR-FIRE TRANSMISSIONS.

GRID MISSION	
OBSERVER	**FIRING UNIT**
F24, this is J42, ADJUST FIRE, OVER.	J42, this is F24, AJUST FIRE, OUT.
GRID WM180513, DIRECTION 0530, OVER.	GRID WM180513, DIRECTION 0530, OUT.
Infantry platoon dug in, OVER.	Infantry platoon dug in, OUT.
	SHOT OVER.
SHOT OUT.	
	SPLASH, OVER.
SPLASH OUT.	
End of mission, 15 casualties, platoon dispersed, OVER.	End of mission, 15 casualties, platoon dispersed, OUT.
SHIFT FROM KNOWN POINT	
OBSERVER	**FIRING UNIT**
J42, THIS IS F24, ADJUST FIRE, SHIFT AB1001, OVER.	F24, THIS IS J42, ADJUST FIRE, SHIFT AB1001, OUT.
DIRECTION 2420, RIGHT 400, ADD 400, OVER.	DIRECTION 2420, RIGHT 400, ADD 400, OUT.
FIVE T-72 TANKS AT POL SITE, OVER.	FIVE T-72 TANKS AT POL SITE, AUTHENTICATE JULIET NOVEMBER, OVER.
I AUTHENTICATE TANGO, OVER.	
	SHOT, OVER.
SHOT OUT.	SPLASH, OVER.
SPLASH OUT.	
END OF MISSION, 2 TANKS DESTROYED, 3 IN WOODLINE, OVER	END OF MISSION, 2 TANKS DESTROYED, 3 IN WOODLINE, OUT

POLAR	
OBSERVER	**FIRING UNIT**
J42, this is F24, adjust fire, polar OVER.	F24, this is J42, adjust fire, polar, OUT
Direction 2300, distance 4,000, OVER.	Direction 2300, distance 4,000, OUT.
Infantry platoon dug in, OVER	Infantry platoon dug in, OUT
	Shot OVER.
Shot OUT.	Splash, OVER.
Splash OUT.	
End of mission, 15 casualties, platoon dispersed, OVER.	End of mission, 15 casualties, platoon dispersed, OUT.

NOTES

Chapter 4
MOVEMENT

To survive on the battlefield, stealth, dispersion, and security must be enforced in all tactical movements. The leader must be skilled in all movement techniques.

4-1. **FORMATIONS**. Movement formations are comprised of elements and Soldiers in relation to each other. Fire teams, squads and platoons use several movement formations. Formations provide the leader control based on a METT-TC analysis. Leaders position themselves where they can best command and control formations. The formations below allow the fire team leader to lead by example, *"Follow me, and do as I do."* All Soldiers in the team must be able to see their leader. The formations in Figure 4-1 reflect fire team formations. Squad formations are very similar with more Soldiers. Squads can operate in lines and files similar to fire teams. When squads operate in wedges or in echelon, the fire teams use those formations and simply arrange themselves in column or one team behind the other. Squads may also use the vee, with one team forming the lines of the vee, and the SL at the apex for command and control. Platoons work on the same basic formations as the Squads. When operating as a platoon, the platoon leader must carefully select the location for his machine guns in the movement formation.

Figure 4-1. FORMATIONS

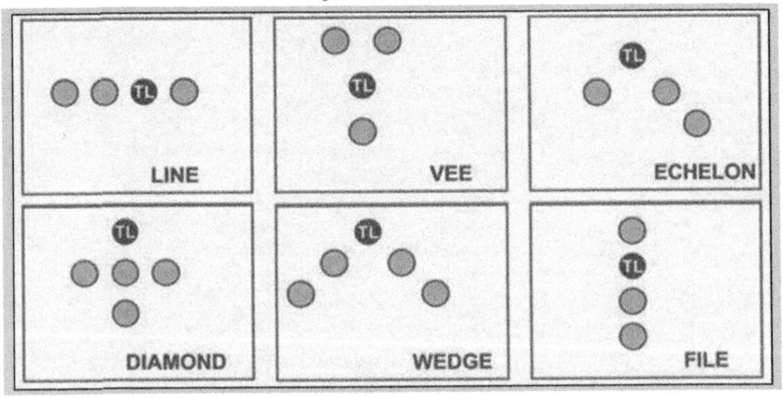

a. **Techniques**. A movement technique is the method a unit uses to traverse terrain. There are three movement techniques: traveling, traveling overwatch, and bounding overwatch. The selection of a movement technique is based on the likelihood of enemy contact and the need for speed. Factors to consider for each technique are control, dispersion, speed, and security. Movement techniques are not fixed formations. They refer to the distances between Soldiers, teams, and squads that vary based on METT-TC. Soldiers must be able to see their fire team leaders. The platoon leader should be able to see his lead squad leader. Leaders control movement with hand-and-arm signals and use radios only when needed.

b. **Standards**.

(1) Unit moves on designated route or arrives at specified location IAW OPORD maintaining accountability of all assigned/attached personnel.

(2) Unit uses movement formation and technique ordered by the leader based on METT-TC.

(3) Leaders remain oriented (within 200 meters) and follow planned route unless METT-TC dictates otherwise.

(4) Unit maintains 360 degree security and remains 100% alert during movement.

(5) Unit maintains 360 degree security and a minimum of 75% security during halts.

(6) If contact with the enemy is made, it is made with the smallest element possible.

(7) Control measures are used during movement such as head counts, rally points, or phase lines.

c. **Fundamentals.**

(1) ***Land Navigation***. Mission accomplishment depends on successful land navigation. The patrol should use stealth and vigilance to avoid chance contact. Designate a primary and alternate compass and pace man per patrol.

Note: The point man will not be tasked to perform compass or pace duties. The point man's sole responsibility is forward security for the element.

(2) ***Avoidance of Detection***. Patrols must use stealth, and use the cover and concealment of the terrain to its maximum advantage. Whenever possible, move during limited visibility in order to maximize technological advantages gained by night vision devices and to hinder the enemy's ability to detect the patrol. Exploit the enemy's weaknesses, and attempt to time movements to coincide with other operations that distract the enemy. The enemy threat and terrain determines which of the three movement techniques will be used:

(a) Fire teams maintain visual contact, but the distance between them is such that the entire patrol does not become engaged if contact is made. Fire teams can spread their formations as necessary to gain better observation to the flanks. Although widely spaced, men retain their relative position in their wedge and follow their team leader. Only in extreme situations should the file be used.

(b) The lead squad must secure the front along with assuming responsibility for navigation. For a long movement, the PL may rotate the lead squad responsibilities. The fire team/squad in the rear is charged with rear security.

(c) Vary movement techniques to meet the changing situation.

(d) With the exception of fire team leaders, leaders move inside their formations where they can maintain the best control.

(3) ***Security***. The patrol must use both active and passive security measures constantly. Assign subunits responsibility for security at danger areas, patrol bases, and most importantly in the objective area.

(4) ***Fire Support***. Plan fire support (mortars, artillery, tactical air, attack helicopter, naval gunfire).

(5) ***Three-Dimensional Battlefield***. 360 Degree Security is achieved through high and low security. Within a fire team, squad, and so on, the leader must assign appropriate sectors of fire to their subordinate in order to ensure all aspects of the battlefield are covered. This includes trees, multiple storied structures, tunnels, sewers, ditches.

d. **Movement Techniques.**

(1) The traveling technique is used when enemy contact is not likely but speed is necessary.

(2) The traveling overwatch technique is used when enemy contact is possible.

(3) The bounding overwatch technique is used when enemy contact is likely, or when crossing a danger area.

e. **Traveling**. In the traveling technique, the distance between individuals is about 10 meters with 20 meters between squads. It has the following characteristics:

(1) More control than traveling overwatch but less than bounding overwatch.

(2) Minimum dispersion.

(3) Maximum speed.

(4) Minimum security.

f. **Traveling Overwatch**. The traveling overwatch technique is the basic movement technique. The distance between individuals is about 20 meters, between teams about 50 meters.

(1) In platoon traveling overwatch, the lead squad must be far enough ahead of the rest of the platoon to detect or engage any enemy before the enemy observes or fires on the main body. However, it must be close enough to be supported by the platoon's small arm's fires. This is normally between 50 to 100 meters, depending on terrain, vegetation, and light and weather conditions.

(2) In a column formation, only the lead squad should use the traveling overwatch; however, if greater dispersion is desired, all squads may use it.

(3) In other formations, all squads use traveling overwatch unless the platoon leader specifies otherwise.

(4) Traveling overwatch has the following characteristics:

- Good control.
- Good dispersion.
- Good speed.
- Good security forward.

g. **Bounding Overwatch**. In the bounding overwatch technique (Figure 4-2), the distance between men remains about 20 meters. The distance between teams and squads varies.

(1) The squad or platoon has a bounding element and an overwatch element. The bounding element moves while the overwatch element occupies an overwatch position that can cover the route of the bounding element by fire. Each bound is within supporting range of the overwatch element.

(2) There are two types of bounding, successive and alternating (Figure 4-2). Successive is nothing more than one squad moving to a position, then the overwatching squad moving to a position generally online with the first squad. Alternating bounding is when one squad moves into position, then the overwatching squad moves to a position in front of the first squad.

Figure 4-2. SQUAD BOUNDING OVERWATCH

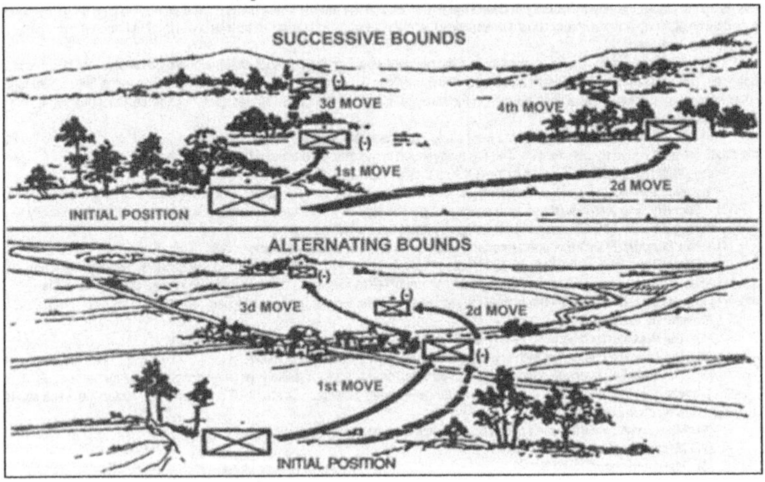

(3) The length of a bound depends on the terrain, visibility, and control.

(4) Before a bound, the leader gives the following instructions to his subordinates:

- Direction of the enemy if known
- Position of overwatch elements
- Next overwatch position
- Route of the bounding element
- What to do after the bounding element reaches the next position
- How the elements receive follow-on orders

(5) The characteristics of bounding overwatch are:

- Maximum control
- Maximum dispersion
- Minimum speed
- Maximum security

h. **Platoon Bounding Overwatch** (Figure 4-3).

(1) ***Method One.*** When platoons use bounding overwatch, one squad bounds and one squad overwatches; the third squad awaits orders. Forward observers stay with the overwatching squad to call for fire. Platoon leaders normally stay with the overwatching squad who use machine guns and attached weapons to support the bounding squad.

(2) ***Method Two.*** Another way is to have one squad use bounding overwatch and have the other two squads use traveling or traveling overwatch technique.

Figure 4-3. PLATOON BOUNDING OVERWATCH

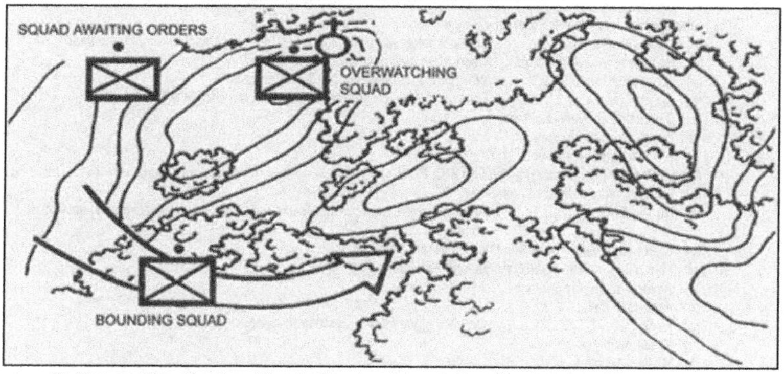

(3) ***Movement Considerations.*** When deciding where to move the bounding element, consider--

- Where the enemy is likely to be.
- The mission.
- The routes to the next overwatch position.
- The weapons ranges of the overwatching unit.
- The responsiveness of the rest of the unit.
- The fields of fire at the next overwatch position.

4-2. **TACTICAL MARCHES**. Platoons conduct two types of marches with the company: foot marches and motor (road) marches.

a. **Purpose/General**. A successful foot march is when troops arrive at their destination at the prescribed time, physically able to execute their tactical mission.

b. **Standard**.

(1) The unit crosses the start point and release point at the time specified in the order.

(2) The unit follows the prescribed route, rate of march, and interval without deviation unless required otherwise by enemy action or higher headquarters action.

c. **Fundamentals**.

(1) Effective control.

(2) Detailed planning.

(3) Rehearsals.

d. **Considerations**.

(1) ***METT-TC.***

- Mission ....Task and purpose.
- Enemy ....Intentions, capabilities, and course of action.
- Terrain and weather.................Road condition/trafficability, and visibility.

- Troops/equipment....................Condition of Soldiers and their loads, numbers and types of weapons and radios.
- Time..Start time, release time, rate of march, time available.
- Civilians ....Movement through populated areas, refugees, OPSEC.

(2) ***Task Organization***:
- Security...................................Advance and trail teams.
- Main body ....Two remaining line squads and weapons squad.
- Headquarters...........................Command and control.
- Control measures.

(3) ***Start Point and Release Point*** (given by higher):
  (a) Checkpoints.............................At checkpoints report to higher and use checkpoint to remain oriented.
  (b) Rally or rendezvous points ......Used when elements become separated.
  (c) Location of leaders ....Where they can best control their elements.
  (d) Commo plan ....Locations of radios, frequencies, call signs, and OPSKEDs.
  (e) Dispersion between Soldiers--
  - 3 to 5 meters / day.
  - 1 to 3 meters / night.

(4) ***March Order***. May be issued as an OPORD, FRAGO, or annex to either (must use operational overlay or strip map)"
  (a) Formations and order of movement
  (b) Route of march........................Assembly area, start point, release point, rally points, checkpoints, break/halt points.
  (c) Start point time, release point time, and rate of march.
  (d) March interval for squads, teams, and individuals.
  (e) Actions on enemy contact........Air and ground.
  (f) Actions at halts.
  (g) Fires ....Detailed plan of fire support for the march.
  (h) Water supply plan.
  (i) MEDEVAC plan.

e. **Duties and Responsibilities**.

(1) ***Platoon Leader***:
- Before ....Issues warning order and FRAGO; inspects and supervises.
- During ....Ensures unit makes movement time; ensures interval is maintained and that unit remains oriented; maintains security; checks condition of men; and enforces water discipline and field sanitation.
- After ....Ensures men are prepared to accomplish their mission, supervises SLs, and ensures medical coverage is provided to men as needed.

(2) ***Platoon Sergeant***:
- Before ....Assists PL, makes recommendations, and enforces uniform and packing lists.
- During ....Controls stragglers, helps platoon leader maintain proper interval and security.
- At Halts....................................Enforces security, ensures welfare of men, enforces field sanitation and litter discipline, and enforces use of preventive medicine.
- After...Coordinates for water, rations, and medical supplies. Recovers casualties.

(3) ***Squad Leaders***:
- Before ....Provides detailed instruction to TLs, inspects boots and socks for serviceability and proper fit, inspects adjustment of equipment, inspects to ensure canteens are full, and ensures equal distribution of loads.
- During ....Controls squad, maintains proper interval between men and equipment, enforces security, remains oriented.

- At Halts....................................Ensures security is maintained, provides men for water resupply as detailed. Physically checks the men in his squad, ensures they drink water, and that they change socks as necessary. Rotates heavy equipment. (Units should plan this in detail to avoid confusion before, during, and after halts.)
- After..Occupies squad sector of assembly area, conducts foot inspection, reports condition of men to PL, and prepares men to accomplish the mission.

(4) ***Security Squad***:
- Lead Team.
  - -- Serves as point element for platoon.
  - -- Recons route to SP.
  - -- Calls in checkpoints.
  - -- Provides early warning.
  - -- Maintains rate of march.
  - -- Moves 10 to 20 meters in front of main body.
- Trail Team.
  - -- Provides rear security.
  - -- Moves 10 to 20 meters behind main body.

(5) ***Medic***:
- Assesses and treats march casualties.
- Advise chain of command on evacuation and transportation requirements of casualties.

(6) ***Individual***:
- Maintains interval.
- Follows TL's examples.
- Relays hand and arm signals; remains alert during movement and at halts.
- Remains alert during movement and at halts.

4-3. **MOVEMENT IN LIMITED VISIBILITY CONDITIONS**. During hours of limited visibility, a platoon will use surveillance, target acquisition, and night observation (STANO) devices to enhance their effectiveness. Leaders must be able to control, navigate, maintain security, and move during limited visibility.

a. **Control**. When visibility is poor, the following methods aid in control.

(1) Use of night vision devices.
(2) Leaders move closer to the front.
(3) Platoon reduces speed.
(4) Use of luminescent tape on equipment.
(5) Reduce intervals between men and elements.
(6) Conducts headcount regularly.

b. **Navigation**. While navigating during limited visibility the same techniques are used as during the day; however, leaders exercise more care to keep the patrol oriented.

c. **Security**.

(1) Enforce strict noise and light discipline.
(2) Use radio-listening silence.
(3) Use camouflage.
(4) Use terrain to avoid detection by enemy surveillance or night vision devices.
(5) Make frequent listening halts; Conduct SLLS (Stop, Look, Listen, Smell).
(6) Mask the sounds of movement when possible. (Rain, wind, and flowing water will mask the sounds of movement).

d. **Rally Points**. Plan actions to be taken at rally points in detail. All elements must maintain communications at all time. There are two techniques for actions at rally points:

(1) ***Minimum Force***: Patrol members assemble at the rally point, and the senior leader assumes command. When the minimum force (designated in the OPORD) is assembled and organized the patrol will continue the mission.

(2) ***Time Available***: The senior leader determines if the patrol has enough time remaining to accomplish the mission.

e. **Actions at Halts**. During halts, post security and cover all approaches into the sector with key weapons.

(1) ***Short Halt***. Typically 1-2 minutes long. Soldiers seek immediate cover and concealment and take a knee. Leaders assign sectors of fire.

(2) ***Long Halt***. More than 2 minutes. Soldiers assume the prone position behind cover and concealment. (Ensure soldiers have clear fields of fire.) Leaders assign sectors of fire.

4-4. **DANGER AREAS**. A danger area is any place on a unit's route where the leader determines his unit may be exposed to enemy observation or fire. Some examples of danger areas are open areas, roads and trails, built-up areas, enemy positions, and natural and man-made obstacles. Bypass danger areas whenever possible.

a. **Standards**.

(1) The unit prevents the enemy from surprising the main body.
(2) The unit moves all personnel and equipment across the danger area.
(3) The unit prevents decisive engagement by the enemy.

b. **Fundamentals**.

(1) Designate near and far side rally points.
(2) Secure near side, left and right flank, and rear security.
(3) Recon and secure the far side.
(4) Cross the danger area.
(5) Plan for fires on all known danger areas.

c. **Technique for Crossing Danger Areas**.

(1) ***Linear Danger Area*** (LDA, Figure 4-4) ***Actions for a Squad***:

STEP 1: The alpha team leader (ATL) observes the linear danger area and sends the hand and arm signal to the SL who determines to bound across.

STEP 2: SL directs the ATL to move his team across the LDA far enough to fit the remainder of the squad on the far side of the LDA. Bravo team moves to the LDA to the right or left to provide an overwatch position prior to A team crossing.

STEP 3: SL receives the hand and arm signal that it is safe to move the rest of the squad across (B team is still providing overwatch).

STEP 4: SL moves himself, radio operator, and B team across the LDA. (A team provides overwatch for squad missions.)

STEP 5: A team assumes original azimuth at SL's command or hand and arm signal.

Figure 4-4. LINEAR DANGER AREA

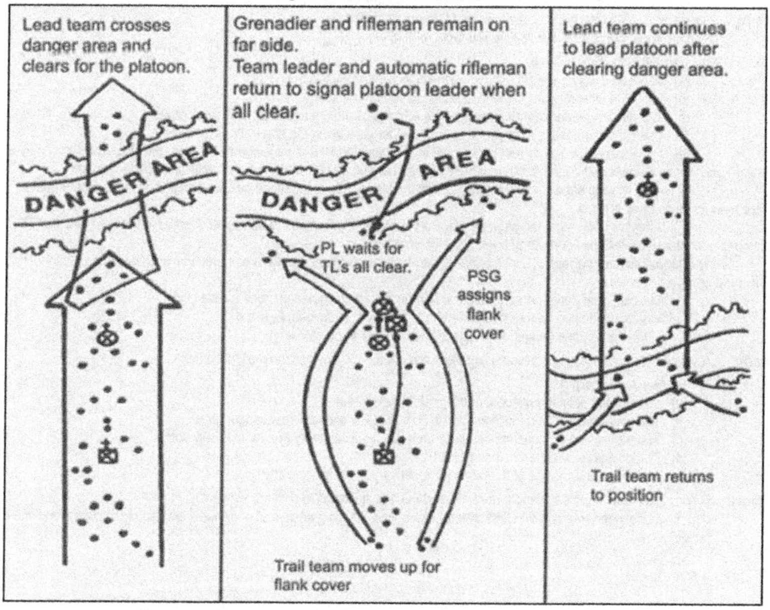

(2) ***LDA Crossing for a Platoon***:

(a) The lead squad halts the platoon, and signals danger area.

(b) The platoon leader moves forward to the lead squad to confirm the danger area and decides if current location is a suitable crossing site.

(c) The platoon leader confirms danger area/crossing site and establishes near and far side rally points.

(d) On the platoon leader's signal, the A team of the lead squad establishes an overwatch position to the left of the crossing site. Prior to crossing, the compassman with the lead two squads confirm azimuth and pace data.

(e) B team of the lead squad establishes an overwatch position to the right of the crossing site.

(f) Once overwatch positions are established, the platoon leader gives the second squad in movement the signal to bound across by fire team.

(g) Once across, the squad is now lead in movement and continues on azimuth

(h) Once Stop, Look, Listen, and Smell (SLLS) is conducted, squad leader signals platoon leader all clear.

- Day time—hand and arm signal such as thumbs up.
- Night time—Clandestine signal such as IR, red lens.

(i) Platoon leader receives all clear and crosses with radio operator, FO, WSL, and 2 gun teams.

(j) Once across, PL signals the 3rd squad in movement to cross.

(k) PSG with medic and one gun team crosses after 2nd squad is across (sterilizing central crossing site).

(l) PSG signals security squad to cross at their location.

Notes: Platoon leader will plan for fires at all known LDA crossing sites.
Nearside security in overwatch will sterilize signs of the patrol .

(3) ***Danger Area (Small/Open)***

(a) The lead squad halts the platoon and signals danger area.
(b) The PL moves forward to the lead squad to confirm the danger area.
(c) The platoon leader confirms danger area and establishes near and far side rally points.
(d) The PL designates lead squad to bypass danger area using the detour-bypass method.
(e) Paceman suspends current pace count and initiates an interim pace count. Alternate pace/compass man moves forward and offsets compass 90 degrees left or right as designated and moves in that direction until clear of danger area.
(f) After moving set distance (x-meters as instructed by PL). Lead squad assumes original azimuth, and primary pace man resumes original pace.
(g) After the open area, the alternate pace/compass man offsets his compass 90 degrees left or right and leads the platoon/squad the same distance (x-meters) back to the original azimuth.

(4) ***Danger Areas (Series)***: A series of danger areas is two or more danger areas within an area that can be either observed or covered by fire.

- Double linear danger area (use linear danger area technique and cross as one LDA).
- Linear/small open danger area (use by-pass/contour technique. Figure 4-5).
- Linear/large open danger area (use platoon wedge when crossing).

Note: A series of danger areas is crossed using the technique which provides the most security.

(5) ***Danger Area (Large)***:

(a) Lead squad halts the platoon, and signals danger area.
(b) The platoon leader moves forward with RTO and FO and confirms danger area.
(c) The platoon leader confirms danger area and establishes near and far side rally points.
(d) PL designates direction of movement.
(e) PL designates change of formation as necessary to ensure security.

Notes: Before point man steps into danger area, PL and FO adjust targets to cover movement.
If far side of danger area is within 250 meters, PL establishes overwatch, and designates lead squad to clear wood line on far side.

Figure 4-5. SMALL OPEN AREA

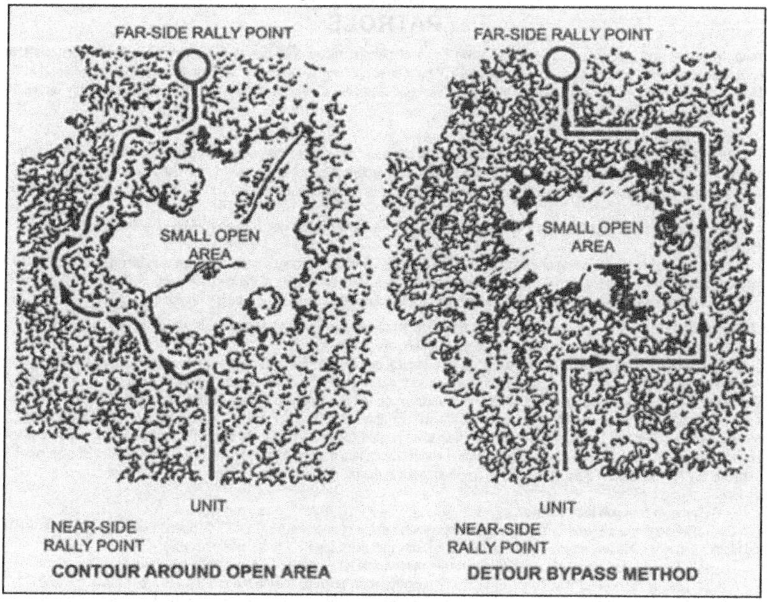

NOTES

Chapter 5
PATROLS

Infantry platoons and squads primarily conduct two types of patrols: reconnaissance and combat. This chapter describes the principles of patrolling, planning considerations, types of patrols, supporting tasks, patrol base, and movement to contact (FM 7-8, FM 3-0, and FM 101-5-1). In this chapter, the terms "*element*" and "*team*" refer to the squads, fire teams, or buddy teams that perform the tasks as described.

5-1. **PRINCIPLES.** All patrols are governed by five principles:

a. **Planning.** Quickly make a simple plan and effectively communicate it to the lowest level. A great plan that takes forever to complete and is poorly disseminated isn't a great plan. Plan and prepare to a realistic standard and rehearse everything.

b. **Reconnaissance.** Your responsibility as a Ranger leader is to confirm what you think you know, and to learn that which you do not already know.

c. **Security.** Preserve your force as a whole. Every Ranger and every rifle counts; anyone could be the difference between victory and defeat.

d. **Control.** Clear understanding of the concept of the operation and commander's intent, coupled with disciplined communications, to bring every man and weapon available to overwhelm the enemy at the decisive point.

e. **Common Sense.** Use all available information and good judgment to make sound, timely decisions.

5-2. **PLANNING.** This paragraph provides the planning considerations common to most patrols. It discusses task organization, initial planning and coordination, completion of the plan, and contingency planning.

a. **Task Organization.** A patrol is a mission, not an organization. To accomplish the patrolling mission, a platoon or squad must perform specific tasks, for example, secure itself, cross danger areas, recon the patrol objective, breach, support, or assault. As with other missions, the leader tasks elements of his unit in accordance with his estimate of the situation, identifying those tasks his unit must perform and designating which elements of his unit will perform which tasks. Where possible, in assigning tasks, the leader should maintain squad and fire team integrity. The chain of command continues to lead its elements during a patrol. Squads and fire teams may perform more than one task in an assigned sequence; others may perform only one task. The leader must plan carefully to ensure that he has identified and assigned all required tasks in the most efficient way. Elements and teams for platoons conducting patrols include--

(1) Elements common to all patrols:

(a) *Headquarters Element.* The headquarters consists of the platoon leader (PL), RTO, platoon sergeant (PSG), FO, RTO, and medic. It may include any attachments that the PL decides that he or the PSG must control directly.

(b) *Aid and Litter Team.* Aid and litter teams are responsible for buddy aid and evacuation of casualties.

(c) *Enemy Prisoner of War Team.* EPW teams control enemy prisoners using the five S's and the leader's guidance.

(d) *Surveillance Team.* The surveillance team keeps watch on the objective from the time that the leader's reconnaissance ends until the unit deploys for actions on the objective. They then rejoin their parent element.

(e) *En Route Recorder.* Part of the HQ element, maintains communications with higher and acts as the recorder for all CCIR collected during the mission.

(f) *Compass Man.* The compass man assists in navigation by ensuring the patrol remains on course at all times. Instructions to the compass man must include initial and subsequent azimuths. As a technique, the compass man should preset his compass on the initial azimuth before the unit moves out, especially if the move will be during limited visibility conditions. The platoon or squad leader should also designate an alternate compass man.

(g) *Point/Pace Man.* As required, the PL designates a primary and alternate point man and a pace man for the patrol. The pace man aids in navigation by keeping an accurate count of distance traveled. The point man selects the actual route through the terrain, guided by the compass man or team leader. In addition, the point man also provides frontal security.

(2) Elements common to all combat patrols:

(a) *Assault Element.* The assault element seizes and secures the objective and protects special teams as they complete their assigned actions on the objective.

(b) *Security Element.* The security element provides security at danger areas, secures the ORP, isolates the objective, and supports the withdrawal of the rest of the patrol once actions on the objective are complete. The security element may have separate security teams, each with an assigned task or sequence of tasks.

(c) *Support Element*. The support element provides direct and indirect fire support for the unit. Direct fires include machine guns, medium and light antiarmor weapons, small recoilless rifles. Indirect fires available may include mortars, artillery, CAS, and organic M203 weapon systems.

(d) *Demolition Team*. Demolition teams are responsible for preparing and detonating the charges to destroy designated equipment, vehicles, or facilities on the objective.

(e) *EPW and Search Teams*. The assault element may provide two-man (buddy teams) or four-man (fire team) search teams to search bunkers, buildings, or tunnels on the objective. These teams will search the objective or kill zone for any PIR that may give the PL an idea of the enemy concept for future operations. Primary and alternate teams may be assigned to ensure enough prepared personnel are available on the objective.

(f) *Breach Element*. The breach team conducts initial penetration of enemy obstacles to seize a foothold and allow the patrol to enter an objective. This is typically done IAW METT-TC and the steps outlined in the "*Conduct an Initial Breach of a Mined Wire Obstacle*" battle drill in Chapter 6 of this Handbook.

(3) Elements common to all reconnaissance patrols:

(a) *Reconnaissance Team*. Reconnaissance teams reconnoiter the objective area once the security teams are in position. Normally these are two-man teams (buddy teams) to reduce the possibility of detection.

(b) *Reconnaissance and Security Teams*. R&S teams are normally used in a zone reconnaissance, but may be useful in any situation when it is impractical to separate the responsibilities for reconnaissance and security.

(c) *Security Element*. When the responsibilities of reconnaissance and security are separate, the security element provides security at danger areas, secures the ORP, isolates the objective, and supports the withdrawal of the rest of the platoon once the recon is complete. The security element may have separate security teams, each with an assigned task or sequence of tasks.

b. **Initial Planning and Coordination**. Leaders plan and prepare for patrols using the troop-leading procedures and the estimate of the situation, as described in Chapter 2. Through an estimate of the situation, leaders identify required actions on the objective (mission analysis) and plan backward to departure from friendly lines and forward to reentry of friendly lines. Because patrolling units act independently, move beyond the direct-fire support of the parent unit, and operate forward of friendly units, coordination must be thorough and detailed. Coordination is continuous throughout planning and preparation. PLs use checklists to preclude omitting any items vital to the accomplishment of the mission.

(1) ***Coordination with Higher Headquarters***. This coordination includes Intelligence, Operations, and Fire Support IAW Chapter 2, *Coordination Checklists* (**page 2-34**). This initial coordination is an integral part of Step 3 of Troop-Leading Procedures, *Make a Tentative Plan*.

(2) ***Coordination with Adjacent Units***. The leader also coordinates his unit's patrol activities with the leaders of other units that will be patrolling in adjacent areas at the same time, IAW Chapter 2-7, *Coordination Checklists* (**page 2-34**).

c. **Completion of Plan**. As the PL completes his plan, he considers--

(1) ***Specified and Implied Tasks***. The PL ensures that he has assigned all specified tasks to be performed on the objective, at rally points, at danger areas, at security or surveillance locations, along the route(s), and at passage lanes. These make up the maneuver and tasks to maneuver units subparagraphs of the Execution paragraph.

(2) ***Key travel and Execution Times***. The leader estimates time requirements for movement to the objective, leader's reconnaissance of the objective, establishment of security and surveillance, completion of all assigned tasks on the objective, and passage through friendly lines. Some planning factors are--

- Movement: Average of 1 Kmph during daylight hours in Woodland Terrain; Average limited visibility ½ Kmph. Add additional time for restrictive, or severely restrictive terrain such as mountains, swamps, or thick vegetation.
- Leader's recon: NLT 1.5 hr.
- Establishment of security and surveillance: 0.5 hr.

(3) ***Primary and Alternate Routes***. The leader selects primary and alternate routes to and from the objective. The return routes should differ from the routes to the objective. The PL may delegate route selection to a subordinate, but is ultimately responsible for the routes selected.

(4) ***Signals***. The leader should consider the use of special signals. These include hand-and-arm signals, flares, voice, whistles, radios, and infrared equipment. Primary and alternate signals must be identified and rehearsed so that all Soldiers know their meaning.

(5) ***Challenge and Password Forward of Friendly Lines***. The challenge and password from the unit's ANCD must not be used beyond the FLOT.

(a) *Odd-Number System.* The leader specifies an odd number. The challenge can be any number less than the specified number. The password will be the number that must be added to it to equal the specified number, for example, the number is 7, the challenge is 3, and the password is 4.

(b) *Running Password.* ANCDs may also designate a running password. This code word alerts a unit that friendly Soldiers are approaching in a less than organized manner and possibly under pressure. The number of Soldiers approaching follows the running password. For example, if the running password is "*Ranger,*" and five friendly Soldiers are approaching, they would say "*Ranger five.*"

(6) ***Location of Leaders***. The PL considers where he and the PSG and other key leaders are located during each phase of the mission. The PL positions himself where he can best control the actions of the patrol. The PSG is normally located with the assault element during a raid or attack to help the PL control the use of additional assaulting squads, and will assist with securing the OBJ. The PSG will locate himself at the CCP to facilitate casualty treatment and evacuation. During a reconnaissance mission, the PSG will stay behind in the ORP to facilitate the transfer of Intel to the higher headquarters, and control the recon elements movement into and out of the ORP.

(7) ***Actions on Enemy Contact***. Unless required by the mission, the unit avoids enemy contact. The leader's plan must address actions on chance contact at each phase of the patrol mission. The unit's ability to continue will depend on how early contact is made, whether the platoon is able to break contact successfully (so that its subsequent direction of movement is undetected), and whether the unit receives any casualties because of the contact. The plan must address the handling of seriously wounded Soldiers and KIAs. The plan must also address the handling of prisoners who are captured because of chance contact and are not part of the planned mission.

(8) ***Contingency Plans.*** The leader leaves his unit for many reasons throughout the planning, coordination, preparation, and execution of his patrol mission. Each time the leader departs the patrol main body, he must issue a five-point contingency plan to the leader left in charge of the unit. The contingency plan is described by the acronym GOTWA, as follows. The patrol leader will additionally issue specific guidance stating what tasks are to be accomplished in the ORP in his absence:

- G: Where the leader is GOING.
- O: OTHERS he is taking with him.
- T: TIME he plans to be gone.
- W: WHAT to do if the leader does not return in time.
- A: The unit's and the leader's ACTIONS on chance contact while the leader is gone.

(9) ***Rally Points***. The leader considers the use and location of rally points. A rally point is a place designated by the leader where the unit moves to reassemble and reorganize if it becomes dispersed. Soldiers must know which rally point to move to at each phase of the patrol mission should they become separated from the unit. They must also know what actions are required there and how long they are to wait at each rally point before moving to another.

(a) *Criteria*. Rally points must be--

- Easily identifiable in daylight and limited visibility.
- Show no signs of recent enemy activity.
- Covered and concealed.
- Away from natural lines of drift and high-speed avenues of approach.
- Defendable for short periods of time.

(b) *Types*. The most common types of rally points are initial, en route, objective, and near-and-far-side rally points.

(10) ***Objective Rally Point***. The ORP is typically 200 to 400m from the objective, or at a minimum, one major terrain feature away. Actions at the ORP include--

- Conduct SLLS and pinpoint location.
- Leaders Recon of the Objective.
- Issuing a FRAGO, if needed.
- Making final preparations before continuing operations, for example, recamouflaging, preparing demolitions, lining up rucksacks for quick recovery. Preparing EPW bindings, first aid kits, litters, and inspecting weapons.
- Accounting for Soldiers and equipment after actions at the objective are complete.
- Reestablishing the chain of command after actions at the objective are complete.
- Disseminating information from reconnaissance, if contact was not made.

(11) ***Leader's reconnaissance of the objective***. The plan must include a leader's reconnaissance of the objective once the platoon or squad establishes the ORP. Before departing, the leader must issue a 5-point contingency plan. During his

reconnaissance, the leader pinpoints the objective, selects reconnaissance, security, support, and assault positions for his elements, and adjusts his plan based on his observation of the objective. Each type of patrol requires different tasks during the leader's reconnaissance. The platoon leader will bring different elements with him. (These are discussed separately under each type of patrol) The leader must plan time to return to the ORP, complete his plan, disseminate information, issue orders and instructions, and allow his squads to make any additional preparations. During the Leader's Reconnaissance for a Raid or Ambush, the PL will leave surveillance on the OBJ.

(12) ***Actions on the objective***. Each type of patrol requires different actions on the objective. Actions on the objective are discussed under each type of patrol.

5-3. **RECONNAISSANCE PATROLS**. Recon patrols are one of the two types of patrols. provide timely and accurate information on the enemy and terrain. They confirm the leader's plan before it is executed. Units on reconnaissance operations collect specific information (*priority intelligence requirements* [PIR]) or general information (*information requirements* [IR]) based on the instructions from their higher commander. The two types of recon patrols discussed here are area and zone. This section discusses the fundamentals of reconnaissance, task standards for the two most common types of recon, and actions on the objective for those types of recon.

a. **Fundamentals of Reconnaissance**. In order to have a successful area reconnaissance, the platoon leader must apply the fundamentals of the reconnaissance to his plan during the conduct of the operation.

(1) ***Gain all required information***. The parent unit tells the patrol leader (PL) what information is required. This is in the form of the IR and PIR. The platoon's mission is then tailored to what information is required. During the entire patrol, members must continuously gain and exchange all information gathered, but cannot consider the mission accomplished unless all PIR has been gathered.

(2) ***Avoid detection by the enemy***. A patrol must not let the enemy know that it is in the objective area. If the enemy knows he is being observed, he may move, change his plans, or increase his security measures. Methods of avoiding detection are--

(a) Minimize movement in the objective area (area reconnaissance).

(b) Move no closer to the enemy than necessary.

(c) If possible, use long-range surveillance or night vision devices.

(d) Camouflage, stealth, noise, and light discipline.

(e) Minimize radio traffic.

(3) ***Employ security measures***. A patrol must be able to break contact and return to the friendly unit with what information is gathered. If necessary, break contact and continue the mission. Security elements are emplaced so that they can overwatch the reconnaissance elements and suppress the enemy so the reconnaissance element can break contact.

(4) ***Task organization***. When the platoon leader receives the order, he analyzes his mission to ensure he understands what must be done. Then he task organizes his platoon to best accomplish the mission IAW METT-TC. Recons are typically squad-sized missions.

b. **Task Standards**.

(1) ***Area Reconnaissance***. The area recon patrol collects all available information on PIR and other intelligence not specified in the order for the area. The patrol completes the recon and reports all information by the time specified in the order. The patrol is not compromised.

(2) ***Zone Reconnaissance***. The zone recon patrol determines all PIR and other intelligence not specified in the order for its assigned zone. The patrol reconnoiters without detection by the enemy. The patrol completes the recon and reports all information by the time specified in the order.

c. **Actions on the Objective, Area Reconnaissance** (Figure 5-1).

(1) The element occupies the ORP as discussed in the section on occupation of the ORP. The RTO calls in spare for occupation of ORP. The leader confirms his location on map while subordinate leaders make necessary perimeter adjustments.

(2) The PL organizes the platoon in one of two ways: separate recon and security elements, or combined recon and security elements.

(3) The PL takes subordinates leaders and key personnel on a leader's recon to confirm the objective and plan.

(a) Issues a 5-point contingency plan before departure.

(b) Establishes a suitable release point that is beyond sight and sound of the objective if possible, but that is definitely out of sight. The RP should also have good rally point characteristics.

(c) Allow all personnel to become familiar with the release point and surrounding area.

(d) Identifies the objective and emplaces surveillance. Designates a surveillance team to keep the objective under surveillance. Issues a contingency plan to the senior man remaining with the surveillance team. The surveillance team is positioned with one man facing the objective, and one facing back in the direction of the release point.

(e) Takes subordinate leaders forward to pinpoint the objective, emplace surveillance, establish a limit of advance, and choose vantage points.

(f) Maintains commo with the platoon throughout the leader's recon.

(4) The PSG maintains security and supervises priorities of work in the ORP.

(a) Reestablishes security at the ORP.

(b) Disseminates the PLs contingency plan.

(c) Oversees preparation of recon personnel (personnel recamouflaged, NVDs and binos prepared, weapons on safe with a round in the chamber).

(5) The PL and his recon party return to the ORP.

(a) Confirms the plan or issues a FRAGO.

(b) Allows subordinate leaders time to disseminate the plan.

(6) The patrol conducts the recon by long-range observation and surveillance if possible.

(a) R&S elements move to observation points that offer cover and concealment and that are outside of small-arms range.

(b) Establishes a series of observation posts (OP) if information cannot be gathered from one location.

(c) Gathers all PIR using the SALUTE format.

(7) If necessary, the patrol conducts its recon by short-range observation and surveillance.

(a) Moves to an OP near the objective.

(b) Passes close enough to the objective to gain information.

(c) Gathers all PIR using the SALUTE format.

(8) R&S teams move using a technique such as the cloverleaf method to move to successive OP's. In this method, R&S teams avoid paralleling the objective site, maintain extreme stealth, do not cross the limit of advance, and maximize the use of available cover and concealment.

(9) During the conduct of the recon, each R&S team will return to the release point when any of the following occurs:

- They have gathered all their PIR.
- They have reached the limit of advance.
- The allocated time to conduct the recon has elapsed.
- Contact has been made.

(10) At the release point, the leader will analyze what information has been gathered and determine if he has met the PIR requirements.

(11) If the leader determines that he has not gathered sufficient information to meet the PIR requirements, or if the information he and the subordinate leader gathered differs drastically, he may have to send R&S teams back to the objective site. In this case, R&S teams will alternate areas of responsibilities. For example, if one team reconnoitered from the 6 – 3 – 12, then that team will now recon from the 6 – 9 – 12.

(12) The R&S element returns undetected to the ORP by the specified time.

(a) Disseminates information to all patrol members through key leaders at the ORP, or moves to a position at least one terrain feature or one kilometer away to disseminate. To disseminate, the leader has the RTO prepare three sketches of the objective site based on the leader's sketch and provides the copies to the subordinate leaders to assist in dissemination.

(b) Reports any information requirements and/or any information requiring immediate attention to higher headquarters, and departs for the designated area.

(13) If contact is made, move to the release point. The recon element tries to break contact and return to the ORP, secure rucksacks, and quickly move out of the area. Once they have moved a safe distance away, the leader will inform higher HQ of the situation and take further instructions from them.

(a) While emplacing surveillance, the recon element withdraws through the release point to the ORP, and follows the same procedures as above.

(b) While conducting the reconnaissance, the compromised element returns a sufficient volume of fire to allow them to break contact. Surveillance can fire an AT-4 at the largest weapon on the objective. All elements will pull off the objective and move to the release point. The senior man will quickly account for all personnel and return to the ORP. Once in the ORP, follow the procedures previously described.

Figure 5-1. ACTIONS ON THE OBJECTIVE, AREA RECONNAISSANCE

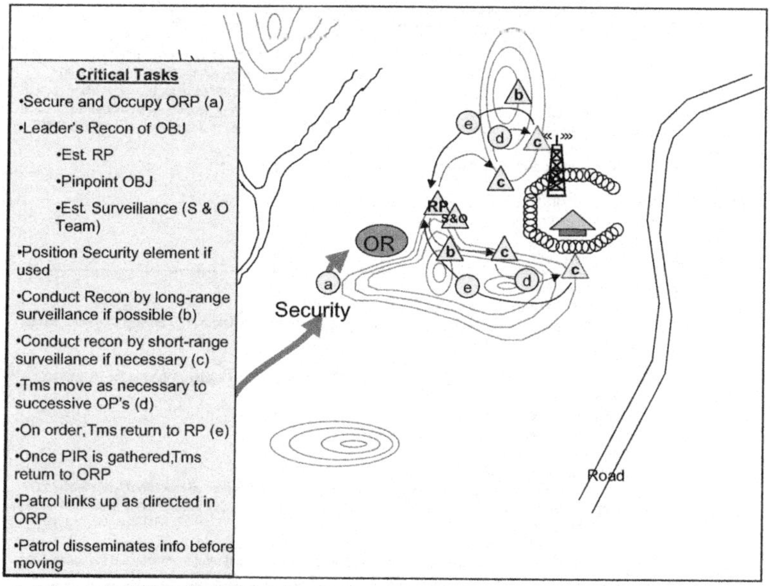

d. **Actions on the Objective, Zone Reconnaissance**.

(1) The element occupies the initial ORP as discussed in the section occupation of the ORP. The radio operator calls in spare for occupation of ORP. The leader confirms his location on map while subordinate leaders make necessary perimeter adjustments.

(2) The recon team leaders organize their recon elements.

(a) Designate security and recon elements.

(b) Assign responsibilities (point man, pace man, en route recorder, and rear security), if not already assigned.

(c) Designates easily recognizable rally points.

(d) Ensure local security at all halts.

(3) The patrol recons the zone.

(a) Moves tactically to the ORPs.

(b) Occupies designated ORPs.

(c) Follows the method designated by the PL (fan, converging routes, or box method, Figure 5-2).

Figure 5-2. COMPARISON OF ZONE RECONNAISSANCE METHODS

FAN METHOD	*CONVERGING ROUTES METHOD*	*BOX METHOD*
• Uses a series (fan) of ORPs. • Patrol establishes security at first ORP. • Each recon element moves from ORP along a different fan-shaped route. Route overlaps with that of other recon elements. This ensures recon of entire area. • Leader maintains reserve at ORP. • When all recon elements return to ORP, PL collects and disseminates all info before moving to next ORP.	• PL selects routes from ORP thru zone to a linkup point at the far side of the zone from the ORP. Each recon element moves and recons along a specified route. They converge (link up) at one time and place.	• PL sends recon elements from the first ORP along routes that form a box. He sends other elements along routes throughout the box. All teams link up at the far side of the box from the ORP.

(d) The recon teams reconnoiter.

- During movement, the squad will gather all PIR specified by the order.
- Recon team leaders will ensure sketches are drawn or digital photos are taken of all enemy hard sites, roads, and trails.
- Return to the ORP, or link up at the rendezvous point on time.
- When the squad arrives at new rendezvous point or ORP, the recon team leaders report to the PL with all information gathered.

(e) The PL continues to control the recon elements.

- PL moves with the recon element that establishes the linkup point.
- PL changes recon methods as required.
- PL designates times for the elements to return to the ORP or to linkup.
- PL collects all information and disseminates it to the entire patrol. PL will brief all key subordinate leaders on information gathered by other squads, establishing one consolidated sketch if possible, and allow team leaders time to brief their teams.
- PL and PSG account for all personnel.

(f) The patrol continues the recon until all designated areas have been reconned, and returns undetected to friendly lines.

5-4. **COMBAT PATROLS**. Combat patrols are the second type of patrol. Combat patrols are further divided into raids and ambushes Units conduct combat patrols to destroy or capture enemy Soldiers or equipment; destroy installations, facilities, or key points; or harass enemy forces. Combat patrols also provide security for larger units. This section describes overall combat patrol planning considerations, task considerations for each type of combat patrol, and finally actions on the objective for each type.

a. **Planning Considerations (General)**. In planning a combat patrol, the PL considers the following:

(1) ***Tasks to Maneuver Units***. Normally the platoon headquarters element controls the patrol on a combat patrol mission. The PL makes every try to maintain squad and fire team integrity as he assigns tasks to subordinates units.

(a) The PL must consider the requirements for assaulting the objective, supporting the assault by fire, and security of the entire unit throughout the mission.

- For the assault on the objective, the PL considers the required actions on the objective, the size of the objective, and the known or presumed strength and disposition of the enemy on and near the objective.
- The PL considers the weapons available, and the type and volume of fires required to provide fire support for the assault on the objective.
- The PL considers the requirement to secure the platoon at points along the route, at danger areas, at the ORP, along enemy avenues of approach into the objective, and elsewhere during the mission.
- The PL will also designate engagement/disengagement criteria.

(b) The PL assigns additional tasks to his squads for demolition, search of EPWs, guarding of EPWs, treatment and evacuation (litter teams) of friendly casualties, and other tasks required for successful completion of patrol mission (if not already in the SOP).

(c) The PL determines who will control any attachments of skilled personnel or special equipment.

(2) ***Leader's Reconnaissance of the Objective***. In a combat patrol, the PL has additional considerations for the conduct of his reconnaissance of the objective from the ORP.

(a) *Composition of the leader's reconnaissance party.* The platoon leader will normally bring the following personnel.

- Squad leaders to include the weapons squad leader.
- Surveillance team.
- Forward observer.
- Security element (dependent on time available).

(b) *Conduct of the leader's reconnaissance.* In a combat patrol, the PL considers the following additional actions in the conduct of the leader's reconnaissance of the objective.

- The PL designates a release point approximately half way between the ORP and this objective. The PL posts the surveillance team. Squads and fire teams separate at the release point, and then they move to their assigned positions.
- The PL confirms the location of the objective or kill zone. He notes the terrain and identifies where he can emplace claymores to cover dead space. Any change to his plan is issued to the squad leaders (while overlooking the objective if possible).
- If the objective is the kill zone for an ambush, the leader's reconnaissance party should not cross the objective; to do so will leave tracks that may compromise the mission.
- The PL confirms the suitability of the assault and support positions and routes from them back to the ORP.
- The PL issues a five-point contingency plan before returning to the ORP.

b. **Ambush**.

(1) ***Planning Considerations***. An ambush is a surprise attack from a concealed position on a moving or temporarily halted target. Ambushes are classified by category--hasty or deliberate; type--point or area; and formation--linear or L-shaped. The leader uses a combination of category, type, and formation in developing his ambush plan. The key planning considerations include--

(a) Coverage of entire kill zone by fire.

(b) METT-TC.

(c) Use of existing or reinforcing obstacles, including Claymores, to keep the enemy in the kill zone.

(d) Security teams (typically equipped with hand-held antitank weapons such as AT-4 or LAW; Claymores; and various means of communication.

(e) Protect the assault and support elements with claymores or explosives.

(f) Use security elements or teams to isolate the kill zone.

(g) Assault through the kill zone to the limit of advance (LOA). (The assault element must be able to move quickly through its own protective obstacles.)

(h) Time the actions of all elements of the platoon to preclude loss of surprise. In the event any member of the ambush is compromised, he may immediately initiate the ambush.

(i) When the ambush must be manned for a long time, use only one squad to conduct the entire ambush and determining movement time of rotating squads from the ORP to the ambush site.

(2) ***Categories***.

(a) *Hasty Ambush*. A unit conducts a hasty ambush when it makes visual contact with an enemy force and has time to establish an ambush without being detected. The actions for a hasty ambush must be well rehearsed so that Soldiers know what to do on the leader's signal. They must also know what action to take if the unit is detected before it is ready to initiate the ambush.

(b) *Deliberate Ambush*. A deliberate ambush is conducted at a predetermined location against any enemy element that meets the commander's engagement criteria. The leader requires the following detailed information in planning a deliberate ambush: size and composition of the targeted enemy, and weapons and equipment available to the enemy.

(3) ***Types***.

(a) Point ambush. In a point ambush, Soldiers deploy to attack an enemy in a single kill zone.

(b) Area ambush. In an area, Soldiers deploy in two or more related point ambushes.

(4) ***Formations***. (Figure 5-3).

(a) *Linear Ambush*. In an ambush using a linear formation, the assault and support elements deploy parallel to the enemy's route. This positions both elements on the long axis of the kill zone and subjects the enemy to flanking fire. This formation can be used in close terrain that restricts the enemy's ability to maneuver against the platoon, or in open terrain provided a means of keeping the enemy in the kill zone can be effected.

(b) *L-Shaped Ambush*. In an L-shaped ambush, the assault element forms the long leg parallel to the enemy's direction of movement along the kill zone. The support element forms the short leg at one end of and at right angles to the assault element. This provides both flanking (long leg) and enfilading fires (short leg) against the enemy. The L-shaped ambush can be used at a sharp bend in a trail, road, or stream. It should not be used where the short leg would have to cross a straight road or trail.

Figure 5-3. AMBUSH FORMATIONS

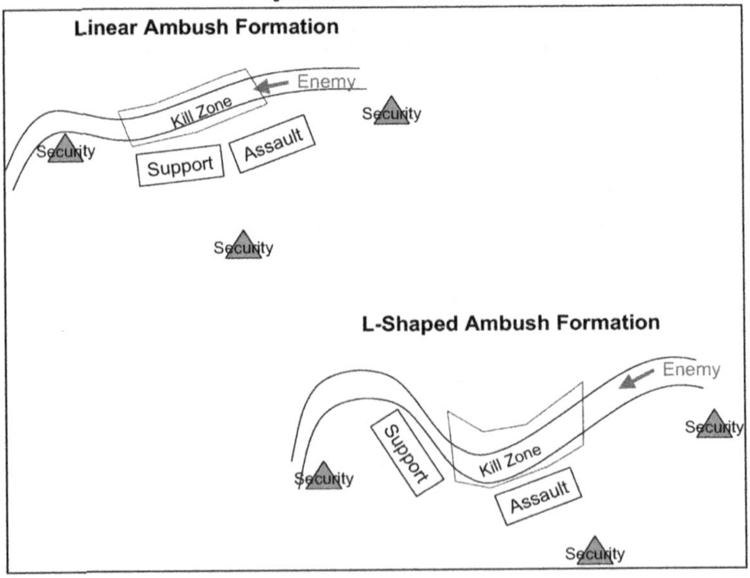

c. **Hasty Ambush.**

(1) ***Task Standards.*** The platoon moves quickly to concealed positions. The ambush is not initiated until the majority of the enemy is in the kill zone. The unit does not become decisively engaged. The platoon surprises the enemy. The patrol captures, kills, or forces the withdrawal of all of the enemy within the kill zone. On order, the patrol withdraws all personnel and equipment in the kill zone from observation and direct fire. The unit does not become decisively engaged by follow-on elements. The platoon continues follow-on operations.

(2) ***Actions on the Objective (Hasty Ambush).*** See Figure 5-4.

(a) Using visual signals, any Soldier alerts the unit that an enemy force is in sight. The Soldier continues to monitor the location and activities of the enemy force until his team or squad leader relieves him, and gives the enemy location and direction of movement.

(b) The platoon or squad halts and remains motionless.

- The PL gives the signal to conduct a hasty ambush, taking care not to alert the enemy of the patrol's presence.
- The leader determines the best nearby location for a hasty ambush. He uses arm-and-hand signals to direct the unit members to covered and concealed positions.

(c) The leader designates the location and extent of the kill zone.

d) Teams and squads move silently to covered and concealed positions, ensuring positions are undetected and have good observation and fields of fire into the kill zone.

(e) Security elements move out to cover each flank and the rear of the unit. The leader directs the security elements to move a given distance, set up, and then rejoin the unit on order or, after the ambush (the sound of firing ceases). At squad level, the two outside buddy teams normally provide flank security as well as fires into the kill zone. At platoon level, fire teams make up the security elements.

(f) The PL assigns sectors of fire and issues any other commands necessary such as control measures.

(g) The PL initiates the ambush, using the greatest casualty-producing weapon available, when the largest percentage of enemy is in the kill zone. The PL--

- Controls the rate and distribution of fire.
- Employs indirect fire to support the ambush.
- Orders cease fire.
- (If the situation dictates) Orders the patrol to assault through the kill zone.

(h) The PL designates personnel to conduct a hasty search of enemy personnel and process enemy prisoners and equipment.

(i) The PL orders the platoon to withdraw from the ambush site along a covered and concealed route.

(j) The PL gains accountability, reorganizes as necessary, disseminates information, reports the situation, and continues the mission as directed.

d. **Deliberate (Point/Area) Ambush.**

(1) ***Task Standards.*** The ambush is emplaced NLT the time specified in the order. The patrol surprises the enemy and engages the enemy main body. The patrol kills or captures all enemy in the kill zone and destroys equipment based on the commander's intent. The patrol withdraws all personnel and equipment from the objective, on order, within the time specified in the order. The patrol obtains all available PIR from the ambush and continues follow-on operations.

(2) ***Actions on the Objective (Deliberate Ambush).*** See Figure 5-5.

(a) The PL prepares the patrol for the ambush in the ORP.

(b) The PL prepares to conduct a leader's reconnaissance. He--

- Designates the members of the leader's recon party (typically includes squad leaders, surveillance team, FO, and possibly the security element.
- Issues a contingency plan to the PSG.

Figure 5-4. ACTIONS ON THE OBJECTIVE – HASTY AMBUSH

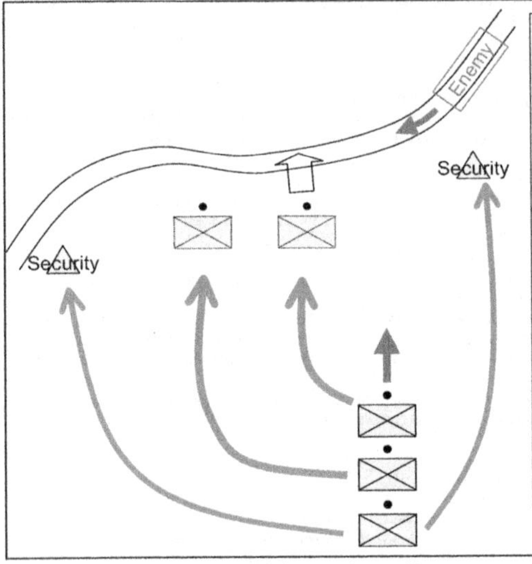

Critical Tasks

- Patrol detects an enemy unit; PL is notified
- Patrol halts and remains motionless
- PL gives signal for Hasty Ambush
- PL directs elements to covered and concealed positions
- Security Elements move to flanks of patrol
- PL establishes control measures
- PL initiates and controls ambush
- PL directs a hasty search
- Patrol consolidates, reorganizes, withdraws, reports, and continues mission

(c) The PL conducts his leader's reconnaissance. He--

- Ensures the leader's recon party moves undetected.
- Confirms the objective location and suitability for the ambush.
- Selects a kill zone.
- Posts the surveillance team at the site and issues a contingency plan.
- Confirms suitability of assault and support positions, and routes from them to the ORP.
- Selects the position of each weapon system in the support-by-fire position, and then designates sectors of fire.
- Identifies all offensive control measures to be used. Identifies the PLD, the assault position, LOA, any boundaries or other control measures. If available, the PL can use infrared aiming devices to identify these positions on the ground.

(d) The PL adjusts his plan based on info from the reconnaissance. He--

- Assigns positions.
- Designates withdrawal routes.

(e) The PL confirms the ambush formation.

(f) The security team(s) occupy first, securing the flanks of the ambush site, and providing early warning. The security element must be in position before the support and assault elements move forward of the release point. A security team remains in the ORP if the patrol plans to return to the ORP after actions on the objective. If the ORP is abandoned, a rear security team should be emplaced.

(g) Support element leader assigns sectors of fire. He--

- Emplaces claymores and obstacles as designated.
- Identifies sectors of fire and emplaces limiting stakes to prevent friendly fires from hitting other elements.
- Overwatches the movement of the assault element into position.

(h) Once the support element is in position, or on the PLs order, the assault element--

- Departs the ORP and moves into position.
- Upon reaching the PLD, the assault element transitions from the movement formation to the battle formation.
- Identifies individual sectors of fire as assigned by the PL. Emplaces aiming stakes or uses metal-to-metal contact with the machine gun tripods to prevent fratricide on the objective.
- Emplaces claymores to help destroy the enemy in the kill zone.
- Camouflages positions.

(i) The security element spots the enemy and notifies the PL, and reports the direction of movement, size of the target, and any special weapons or equipment carried. The security element also keeps the platoon leader informed if any enemy forces are following the lead force.

(j) The PL alerts other elements, and determines if the enemy force is too large, or if the ambush can engage the enemy successfully.

(k) The PL initiates the ambush using the highest casualty-producing device. He may use a command-detonated claymore. He must also plan a backup method for initiating the ambush, in case his primary means fails. This should also be a casualty-producing device such as his individual weapon. He passes this information to all Rangers, and practices it during rehearsals.

(l) The PL ensures that the assault and support elements deliver fire with the heaviest, most accurate volume possible on the enemy in the kill zone. In limited visibility, the PL may use infrared lasers to further define specific targets in the kill zone.

(m) Before assaulting the target, the PL gives the signal to lift or shift fires.

(n) The assault element--

- Assaults before the remaining enemy can react.
- Kills or captures enemy in the kill zone.
- Uses individual movement techniques or bounds by fire teams to move.
- Upon reaching the limit of advance, halts and establishes security. If needed, it reestablishes the chain of command and remains key weapon systems. All Soldiers will load a fresh magazine or drum of ammunition using the buddy system. ACE reports will be submitted through the chain of command. The PL will submit an initial contact report to higher.

(o) The PL directs special teams (EPW search, aid and litter, demo) to accomplish their assigned task once the assault element has established its LOA.

- Once the kill zone had been cleared, collect and secure all EPWs and move them out of the kill zone before searching bodies. Coordinate for an EPW exchange point to link up with higher to extract all EPWs and treat them IAW the five S's.
- Search from one side to the other and mark bodies that have been searched to ensure the area is thoroughly covered. Units should use the "*clear out, search in*" technique, clear from the center of the objective out ensuring the area is clear of all enemy combatants; then search all enemy personnel towards the center of the objective. Search all dead enemy personnel using two-man search techniques.
  - -- As the search team approaches a dead enemy Soldier, one-man guards while the other man searches. First, he kicks the enemy weapon away.
  - -- Second, he rolls the body over (if on the stomach) by lying on top and when given the go ahead by the guard (who is positioned at the enemy's head), the searcher rolls the body over on him. This is done for protection in case the enemy Soldier has a grenade with the pin pulled underneath him.
  - -- The searchers then conduct a systematic search of the dead Soldier from head to toe removing all papers and anything new (different type rank, shoulder boards, different unit patch, pistol, weapon, or NVD). They note if the enemy has a fresh or shabby haircut and the condition of his uniform and boots. They note the radio frequency, and then they secure the SOI, maps, documents, and overlays.
  - -- Once the body has been thoroughly searched, the search team will continue in this manner until all enemy personnel in and near the kill zone have been searched.
- Identify, collect, and prepare all equipment to be carried back or destroyed.
- Evacuate and treat friendly wounded first, then enemy wounded, time permitting.

- The demolition team prepares dual-primed explosives or incendiary grenades and awaits the signal to initiate. This is normally the last action performed before the unit departs the objective and may signal the security elements to return to the ORP.
- Actions on the objective with stationary assault line; all actions are the same with the exception of the search teams. To provide security within the teams to the far side of the kill zone during the search, they work in three-Ranger teams. Before the search begins, the Rangers move all KIAs to the near side of the kill zone.

(p) If enemy reinforcements try to penetrate the kill zone, the flank security will engage to prevent the assault element from being compromised.

(q) The platoon leader directs the unit's withdrawal from the ambush site:

- Elements normally withdraw in the reverse order that they established their positions.
- The elements may return to the RP or directly to the ORP, depending on the distance between elements.
- The security element of the ORP must be alert to assist the platoon's return to the ORP. It maintains security for the ORP while the rest of the platoon prepares to leave.
- If possible, all elements should return to the location at which they separated from the main body. This location should usually be the RP.

(r) The PL and PSG direct actions at the ORP, to include accountability of personnel and equipment and recovery of rucksacks and other equipment left at the ORP during the ambush.

(s) The platoon leader disseminates information, or moves the platoon to a safe location (no less than one kilometer or one terrain feature away from the objective) and disseminates information.

(t) As required, the PL and FO execute indirect fires to cover the platoon's withdrawal.

Figure 5-5. ACTIONS ON THE OBJECTIVE -- DELIBERATE AMBUSH

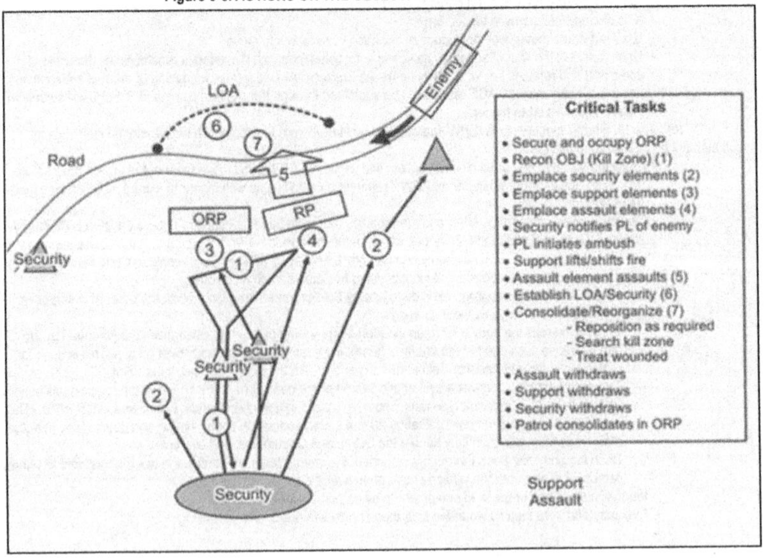

e. **Perform Raid**. The patrol initiates the raid NLT the time specified in the order, surprises the enemy, assaults the objective, and accomplishes its assigned mission within the commander's intent. The patrol does not become decisively engaged en route to the objective. The patrol obtains all available PIR from the raid objective and continues follow-on operations.

(1) ***Planning Considerations***. A Raid is a form of attack, usually small scale, involving a swift entry into hostile territory to secure information, confuse the enemy, or destroy installations followed by a planned withdrawal. Squads do not conduct raids. The sequence of platoon actions for a raid is similar to those for an ambush. Additionally, the assault element of the platoon may have to conduct a breach of an obstacle. It may have additional tasks to perform on the objective such as demolition of fixed facilities. Fundamentals of the raid include--

- Surprise and speed. Infiltrate and surprise the enemy without being detected.
- Coordinated fires. Seal off the objective with well-synchronized direct and indirect fires.
- Violence of action. Overwhelm the enemy with fire and maneuver.
- Planned withdrawal. Withdraw from the objective in an organized manner, maintaining security.

(2) ***Actions on the Objective (Raid)***. See Figure 5-6.

(a) The patrol moves to and occupies the ORP IAW the patrol SOP. The patrol prepares for the leader's recon.

(b) The PL, squad leaders, and selected personnel conduct a leader's recon.

- PL leaves a five point contingency plan with the PSG.
- PL establishes the RP, pinpoints the objective, contacts the PSG to prep men, weapons, and equipment, emplaces the surveillance team to observe the objective, and verifies and updates intelligence information.
- Leader's recon verifies location of and routes to security, support, and assault positions.
- Security teams are brought forward on the leader's reconnaissance and emplaced before the leader's recon leaves the RP
- Leaders conduct the recon without compromising the patrol.
- Leaders normally recon support by fire position first, then the assault position.

(c) The PL confirms, denies, or modifies his plan and issues instructions to his squad leaders.

- Assigns positions and withdrawal routes to all elements.
- Designates control measures on the objective (element objectives, lanes, limits of advance, target reference points, and assault line).
- Allows SLs time to disseminate information, and confirm that their elements are ready.

(d) Security elements occupy designated positions, moving undetected into positions that provide early warning and can seal off the objective from outside support or reinforcement.

(e) The support element leader moves the support element to designated positions. The support element leader ensures his element can place well-aimed fire on the objective.

(f) The PL moves with the assault element into the assault position. The assault position is normally the last covered and concealed position before reaching the objective. As it passes through the assault position the platoon deploys into its assault formation; that is, its squads and fire teams deploy to place the bulk of their firepower to the front as they assault the objective.

- Makes contact with the surveillance team to confirm any enemy activity on the objective.
- Ensures that the assault position is close enough for immediate assault if the assault element is detected early.
- Moves into position undetected, and establish local security and fire control measures.

(g) Element leaders inform the PL when their elements are in position and ready.

(h) The PL directs the support element to fire.

(i) Upon gaining fire superiority, the PL directs the assault element to move towards the objective.

- Assault element holds fire until engaged, or until ready to penetrate the objective.
- PL signals the support element to lift or shift fires. The support element lifts or shifts fires as directed, shifting fire to the flanks of targets or areas as directed in the FRAGO.

(j) The assault element attacks and secures the objective. The assault element may be required to breech a wire obstacle. As the platoon, or its assault element, moves onto the objective, it must increase the volume and accuracy of fires. Squad leaders assign specific targets or objectives for their fire teams. Only when these direct fires keep the enemy suppressed can the rest of the unit maneuver. As the assault element gets closer to the enemy, there is more emphasis on suppression and less on maneuver. Ultimately, all but one fire team may be suppressing to allow that one fire team to break into the enemy position. Throughout the assault, Soldiers use proper individual movement techniques, and fire teams retain their basic shallow wedge formation. The platoon does not get "*on-line*" to sweep across the objective.

- Assault element assaults through the objective to the designated LOA.
- Assault element leaders establish local security along the LOA, and consolidate and reorganize as necessary. They provide ACE reports to the PL and PSG. The platoon establishes security, operates key weapons, provides first aid, and prepares wounded Soldiers for MEDEVAC. They redistribute ammunition and supplies, and they relocate selected weapons to alternate positions if leaders believe that the enemy may have pinpointed them during the attack. They adjust other positions for mutual support. The squad and team leader provide ammunition, casualty, and equipment (ACE) reports to the platoon leader. The PL/PSG reorganizes the patrol based on the contact.
  -- On order, special teams accomplish all assigned tasks under the supervision of the PL, who positions himself where he can control the patrol.
  -- Special team leaders report to PL when assigned tasks are complete.

(k) On order or signal of the PL, the assault element withdraws from the objective. Using prearranged signals, the assault line begins an organized withdrawal from the objective site, maintaining control and security throughout the withdrawal. The assault element bounds back near the original assault line, and begins a single file withdrawal through the APL's choke point. All Rangers must move through the choke point for an accurate count. Once the assault element is a safe distance from the objective and the headcount is confirmed, the platoon can withdraw the support element. If the support elements were a part of the assault line, they withdraw together, and security is signaled to withdraw. Once the support is a safe distance off the objective, they notify the platoon leader, who contacts the security element and signals them to withdraw. All security teams link up at the release point and notify the platoon leader before moving to the ORP. Personnel returning to the ORP immediately secure their equipment and establish all-round security. Once the security element returns, the platoon moves out of the objective area as soon as possible, normally in two to three minutes.

- Before withdrawing, the demo team activates demo devices and charges.
- Support element or designated personnel in the assault element maintain local security during the withdrawal.
- Leaders report updated accountability and status (ACE report) to the PL and PSG.

(l) Squads withdraw from the objective in the order designated in the FRAGO to the ORP.

- Account for personnel and equipment.
- Disseminate information.
- Redistribute ammunition and equipment as required.

(m) The PL reports mission accomplishment to higher and continues the mission.

- Reports raid assessment to higher.
- Informs higher of any IR/PIR gathered.

NOTES

Figure 5-6. ACTIONS ON THE OBJECTIVE -- RAID

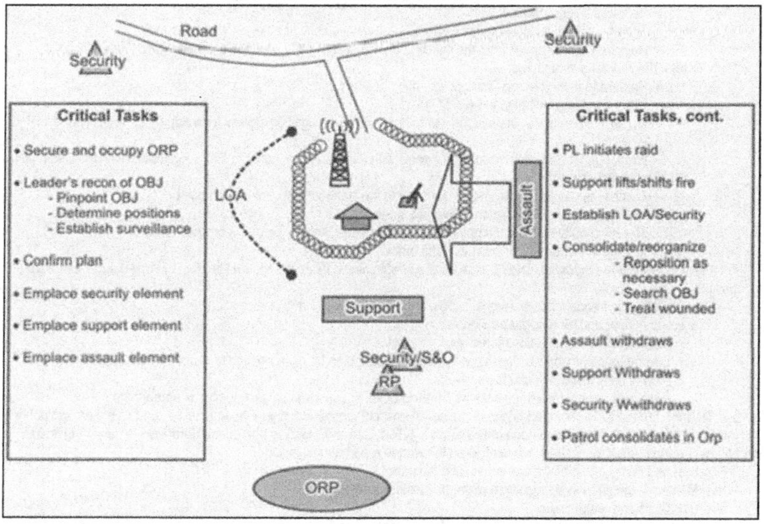

5-5. **SUPPORTING TASKS**. This section covers Linkup, Patrol Debriefing, and Occupation of an ORP.

a. **Linkup**. A linkup is a meeting of friendly ground forces. Linkups depend on control, detailed planning, communications, and stealth.

(1) ***Task Standard***. The units link up at the time and place specified in the order. The enemy does not surprise the main bodies. The linkup units establish a consolidated chain of command.

(2) ***Linkup Site Selection***. The leader identifies a tentative linkup site by map reconnaissance, other imagery, or higher headquarters designates a linkup site. The linkup site should have the following characteristics:

- Ease of recognition.
- Cover and concealment.
- No tactical value to the enemy.
- Location away from natural lines of drift.
- Defendable for a short period of time.
- N=Multiple access and escape routes.

(3) ***Execution***. Linkup procedure begins as the unit moves to the linkup point. The steps of this procedure are:

(a) The stationary unit performs linkup actions.

- Occupies the linkup rally point (LRP) NLT the time specified in the order.
- Establishes all-around security, establishes commo, and prepares to accept the moving unit.
- The security team clears the immediate area around the linkup point. It then marks the linkup point with the coordinated recognition signal. The security team moves to a covered and concealed position and observes the linkup point and immediate area around it.

(b) The moving unit--
- Performs linkup actions.
- The unit reports its location using phase lines, checkpoints, or other control measures.
- Halts at a safe distance from the linkup point in a covered and concealed position (the linkup rally point).

(c) The PL and a contact team--
- Prepare to make physical contact with the stationary unit.
- Issue a contingency plan to the PSG.
- Maintain commo with the platoon; verify near and far recognition signals for linkup (good visibility and limited visibility).
- Exchange far and near recognition signals with the linkup unit; conduct final coordination with the linkup unit.

(d) The stationary unit--
- Guides the patrol from its linkup rally point to the stationary unit linkup rally point.
- Linkup is complete by the time specified in the order.
- The main body of the stationary unit is alerted before the moving unit is brought forward.

(e) The patrol continues its mission IAW the order.

(4) ***Coordination Checklist***. The PL coordinates or obtains the following information from the unit that his patrol will link up with:
- Exchange frequencies, call signs, codes, and other communication information.
- Verify near and far recognition signals.
- Exchange fire coordination measures.
- Determine command relationship with the linkup unit; plan for consolidation of chain of command.
- Plan actions following linkup.
- Exchange control measures such as contact points, phase lines, contact points, as appropriate.

b. **Debrief**. Immediately after the platoon or squad returns, personnel from higher headquarters conduct a thorough debrief. This may include all members of the platoon or the leaders, RTOs, and any attached personnel. Normally the debriefing is oral. Sometimes a written report is required. Information on the written report should include--
- Size and composition of the unit conducting the patrol.
- Mission of the platoon such as type of patrol, location, and purpose.
- Departure and return times.
- Routes. Use checkpoints, grid coordinates for each leg or include an overlay.
- Detailed description of terrain and enemy positions that were identified.
- Results of any contact with the enemy.
- Unit status at the conclusion of the patrol mission, including the disposition of dead or wounded Soldiers.
- Conclusions or recommendations.

c. **Objective Rally Point**. The ORP is a point out of sight, sound, and small arms range of the objective area. It is normally located in the direction that the platoon plans to move after completion of actions on the objective. The ORP is tentative until the objective is pinpointed.

(1) ***Occupation of the ORP***. See Figure 5-7.

(a) The patrol halts beyond sight and sound of the tentative ORP (200-400m in good visibility, 100-200m in limited visibility).

(b) The patrol establishes a security halt IAW the unit SOP.

(c) After issuing a five-point contingency plan to the PSG, the PL moves forward with a recon element to conduct a leader's recon of the ORP.

(d) For a squad-sized patrol, the PL moves forward with a compass man and one member of each fire team to confirm the ORP.
- After physically clearing the ORP location, the PL leaves two Rangers at the 6 o'clock position facing in opposite directions.
- The PL issues a contingency plan and returns with the compass man to guide the patrol forward.
- The PL guides the patrol forward into the ORP, with one team occupying from 3 o'clock through 12 o'clock to 9 o'clock, and the other occupying from 9 o'clock through 6 o'clock to 3 o'clock.

(e) For a platoon-size patrol, the PL, RTO, WSL, three ammo bearers, a team leader, a SAW gunner, and riflemen go on the leaders recon for the ORP and position themselves at 10, 2, and 6 o'clock.

- The first squad in the order of march is the base squad, occupying from 10 to 2 o'clock
- The trail squads occupy from 2 to 6 o'clock and 6 to 10 o'clock, respectively.
- The patrol headquarters element occupies the center of the triangle.

(2) ***Actions in the ORP***. The unit prepares for the mission in the ORP. Once the leader's recon pinpoints the objective, the PSG generally lines up rucks IAW unit SOP in the center of the ORP.

Figure 5-7. OCCUPATION OF THE ORP

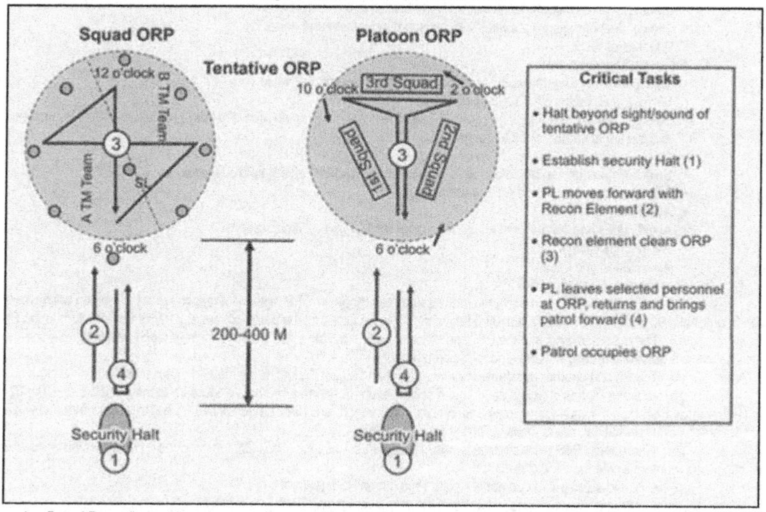

d. **Patrol Base**. A patrol base is a security perimeter that is set up when a squad or platoon conducting a patrol halts for an extended period. Patrol bases should not be occupied for more than a 24-hour period (except in emergency). A patrol never uses the same patrol base twice.

(1) ***Use***. Patrol bases are typically used–

- To avoid detection by eliminating movement.
- To hide a unit during a long detailed reconnaissance.
- To perform maintenance on weapons, equipment, eat and rest.
- To plan and issue orders.
- To reorganize after infiltrating on an enemy area.
- To establish a base from which to execute several consecutive or concurrent operations.

(2) ***Site Selection***. The leader selects the tentative site from a map or by aerial reconnaissance. The site's suitability must be confirmed and secured before the unit moves into it. Plans to establish a patrol base must include selecting an alternate patrol base site. The alternate site is used if the first site is unsuitable or if the patrol must unexpectedly evacuate the first patrol base.

(3) ***Planning Considerations***. Leaders planning for a patrol base must consider the mission and passive and active security measures. A patrol base (PB) must be located so it allows the unit to accomplish its mission.

- Observation posts and communication with observation posts.
- Patrol or platoon fire plan.
- Alert plan.
- Withdrawal plan from the patrol base to include withdrawal routes and a rally point, rendezvous point, or alternate patrol base.
- A security system to make sure that specific Soldiers are awake at all times.
- Enforcement of camouflage, noise, and light discipline.
- The conduct of required activities with minimum movement and noise.
- Priorities of Work.

(4) ***Security Measures***.

- Select terrain the enemy would probably consider of little tactical value.
- Select terrain that is off main lines of drift.
- Select difficult terrain that would impede foot movement, such as an area of dense vegetation, preferably bushes and trees that spread close to the ground.
- Select terrain near a source of water.
- Select terrain that can be defended for a short period and that offers good cover and concealment.
- Avoid known or suspected enemy positions.
- Avoid built-up areas.
- Avoid ridges and hilltops, except as needed for maintaining communications.
- Avoid small valleys.
- Avoid roads and trails.

(5) ***Occupation***. See Figure 5-8.

(a) A PB is reconned and occupied in the same manner as an ORP, with the exception that the platoon will typically plan to enter at a 90 degree turn. The PL leaves a two-man OP at the turn, and the patrol covers any tracks from the turn to the PB.

(b) The platoon moves into the PB. Squad-sized patrols will generally occupy a cigar-shaped perimeter; platoon-sized patrols will generally occupy a triangle-shaped perimeter.

(c) The PL and another designated leader inspect and adjust the entire perimeter as necessary.

(d) After the PL has checked each squad sector, each SL sends a two-man R&S team to the PL at the CP. The PL issues the three R&S teams a contingency plan, reconnaissance method, and detailed guidance on what to look for (enemy, water, built-up areas or human habitat, roads, trails, or possible rally points).

(e) Where each R&S team departs is based on the PLs guidance. The R&S team moves a prescribed distance and direction, and reenters where the PL dictates.

- Squad-sized patrols do not normally send out an R&S team at night.
- R&S teams will prepare a sketch of the area to the squad front if possible.
- The patrol remains at 100 % alert during this recon.
- If the PL feels the patrol was tracked or followed, he may elect to wait in silence at 100 % alert before sending out R&S teams.
- The R&S teams may use methods such as the "*I*," the "*Box*," or the "*T*." Regardless of the method chosen, the R&S team must be able to provide the PL with the same information.
- Upon completion of R&S, the PL confirms or denies the patrol base location, and either moves the patrol or begins priorities of work.

(6) ***Passive (Clandestine) Patrol Base (Squad)***.

- The purpose of a passive patrol base is for the rest of a squad or smaller size element.
- Unit moves as a whole and occupies in force.
- Squad leader ensures that the unit moves in at a 90-degree angle to the order of movement.
- A claymore mine is emplaced on route entering patrol base.
- Alpha and Bravo teams sit back-to-back facing outward, ensuring that at least one individual per team is alert and providing security.

(7) ***Priorities of Work (Platoon and Squad).*** Once the PL is briefed by the R&S teams and determines the area is suitable for a patrol base, the leader establishes or modifies defensive work priorities in order to establish the defense for the patrol base. Priorities of work are not a laundry list of tasks to be completed; to be effective, priorities of work must consist of a task, a given time, and a measurable performance standard. For each priority of work, a clear standard must be issued to guide the element in the successful accomplishment of each task. It must also be designated whether the work will be controlled in a centralized or decentralized manner. Priorities of work are determined IAW METT-TC. Priorities of Work may include, but are not limited to the following tasks:

(a) *Security (Continuous).*

- Prepare to use all passive and active measures to cover all of the perimeter all of the time, regardless of the percentage of weapons used to cover that all of the terrain.
- Readjust after R&S teams return, or based on current priority of work (such as weapons maintenance).
- Employ all elements, weapons, and personnel to meet conditions of the terrain, enemy, or situation.
- Assign sectors of fire to all personnel and weapons. Develop squad sector sketches and platoon fire plan.
- Confirm location of fighting positions for cover, concealment, and observation and fields of fire. SLs supervise placement of aiming stakes and claymores.
- Only use one point of entry and exit, and count personnel in and out. Everyone is challenged IAW the unit SOP.
- Hasty fighting positions are prepared at least 18 inches deep (at the front), and sloping gently from front to rear, with a grenade sump if possible.

(b) *Withdrawal Plan.* The PL designates the signal for withdrawal, order of withdrawal, and the platoon rendezvous point and/or alternate patrol base.

(c) *Communication (Continuous).* Commo must be maintained with higher headquarters, OP's, and within the unit. May be rotated between the patrol's RTOs to allow accomplishment of continuous radio monitoring, radio maintenance, act as runners for PL, or conduct other priorities of work.

(d) *Mission Preparation and Planning.* The PL uses the patrol base to plan, issue orders, rehearse, inspect, and prepare for future missions.

(e) *Weapons and Equipment Maintenance.* The PL ensures that machine guns, weapon systems, commo equipment, and night vision devices (as well as other equipment) are maintained. These items are not disassembled at the same time for maintenance (no more than 33 percent at a time), and weapons are not disassembled at night. If one machine gun is down, then security for all remaining systems is raised.

(f) *Water Resupply.* The PSG organizes watering parties as necessary. The watering party carries canteens in an empty rucksack or duffel bag, and must have commo and a contingency plan prior to departure.

(g) *Mess Plan.* At a minimum, security and weapons maintenance are performed prior to mess. Normally no more than half the platoon eats at one time. Rangers typically eat 1 to 3 meters behind their fighting positions.

- Rest/sleep plan management. The patrol conducts rest as necessary to prepare for future operations.
- Alert Plan and Stand-to. The PL states the alert posture and the stand-to time. He develops the plan to ensure all positions are checked periodically, OP's are relieved periodically, and at least one leader is always alert. The patrol typically conducts stand-to at a time specified by unit SOP (such as 30 minutes before and after BMNT or EENT).
- Resupply. Distribute or cross-load ammunition, meals, equipment, etc.
- Sanitation and personal hygiene. The PSG and medic ensure a slit trench is prepared and marked. All Soldiers will brush teeth, wash face, shave, wash hands, armpits, groin, feet, and darken (brush shine) boots daily. The patrol will not leave trash behind.

Figure 5-8. PATROL BASE

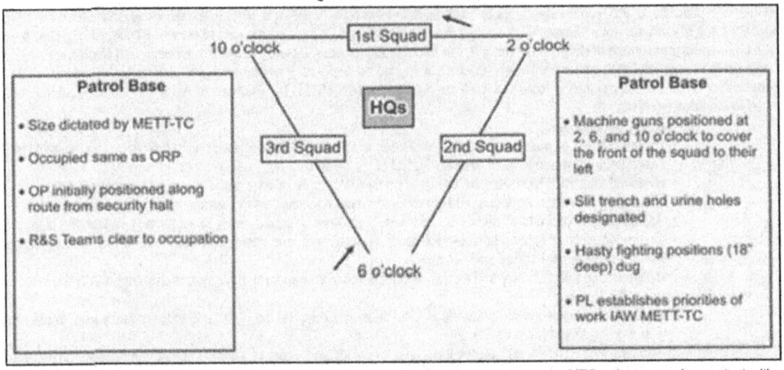

5-6. **MOVEMENT TO CONTACT**. The MTC is one of the five types of offensive operations. An MTC gains or regains contact with the enemy. Once contact is made, the unit develops the situation. Normally a platoon conducts an MTC as part of a larger force.

a. **Techniques**. The two techniques of conducting a movement to contact are search and attack and approach march.

(1) ***Search and Attack (S&A)***. This technique is used when the enemy is dispersed, is expected to avoid contact, disengage or withdraw, or you have to deny his movement in an area. The search and attack technique involves the use of multiple platoons, squads, and fire teams coordinating their actions to make contact with the enemy. Platoons typically try to find the enemy and then fix and finish him. They combine patrolling techniques with the requirement to conduct hasty or deliberate attacks once the enemy has been found.

(a) *Planning Considerations.*

- The factors of METT-TC.
- The requirement for decentralized execution.
- The requirement for mutual support.
- The length of operations.
- Minimize Soldier's load to improve stealth and speed.
- Resupply and MEDEVAC.
- Positioning key leaders and equipment.
- Employment of key weapons.
- Requirement for patrol bases.
- Concept for entering the zone of action.
- The concept for linkups while in contact.

(b) *Critical Performance Measures.*

- The platoon locates the enemy without being detected.
- Once engaged, fixes the enemy in position and maneuvers against the enemy.
- Maintains security throughout actions to avoid being flanked.

(2) ***Approach March.*** The concept of the approach march is to make contact with the smallest element, allowing the commander the flexibility of destroying or bypassing the enemy. A platoon uses the approach march method as part of a larger unit. It can be tasked as the advance guard, move as part of the main body, or provide flank or rear security for the company or battalion. They may also receive on-order missions as part of the main body.

(a) *Fundamentals*. These basics are common to all movements to contact.

- Make enemy contact with smallest element possible.
- Rapidly develop combat power upon enemy contact.

- Provide all-round security for the unit.
- Support higher unit's concept.
- Reports all information rapidly and accurately and strives to gain and maintain contact with the enemy.
- Requires decentralized execution.

(b) *Planning Considerations.* The following issues should be considered heavily for MTC operations:

- Factors of METT-TC.
- Reduced Soldier's load.

(c) *Critical Performance Measures.*

- PL selects the appropriate movement formation based on likelihood of enemy contact.
- Maintains contact, once contact is made, until ordered to do otherwise.

b. **Task Standards**. The platoon moves NLT the time specified in the order, the platoon makes contact with the smallest element possible, and the main body is not surprised by the enemy. Once the platoon makes contact, it maintains contact. The platoon destroys squad and smaller-sized elements, and fixes elements larger than a squad. The platoon maintains sufficient fighting force capable of conducting further combat operations. Reports of enemy locations and contact are forwarded. If not detected by the enemy, the PL initiates a hasty attack. The platoon sustains no casualties from friendly fire. The platoon is prepared to initiate further movement within 25 minutes of contact, and all personnel and equipment are accounted for.

NOTES

Chapter 6
BATTLE DRILLS

Infantry battle drills describe how platoons and squads apply fire and maneuver to commonly encountered situations. They require leaders to make decisions rapidly and to issue brief oral orders quickly.

Section I. INTRODUCTION

This section defines and describes the format for battle drills.

6-1. **DEFINITION.** FM 7-1 defines a battle drill as "*a collective action rapidly executed without applying a deliberate decision-making process.*"

a. Characteristics of a battle drill are—

- They require minimal leader orders to accomplish and are standard throughout the Army.
- Sequential actions are vital to success in combat or critical to preserving life.
- They apply to platoon or smaller units.
- They are trained responses to enemy actions or leader's orders.
- They represent mental steps followed for offensive and defensive actions in training and combat.

b. A unit's ability to accomplish its mission often depends on Soldiers and leaders to execute key actions quickly. All Soldiers and their leaders must know their immediate reaction to enemy contact, as well as follow-up actions. Drills are limited to situations requiring instantaneous response; therefore, Soldiers must execute drills instinctively. This results from continual practice. Drills provide small units with standard procedures essential for building strength and aggressiveness.

(1) They identify key actions that leaders and Soldiers must perform quickly.

(2) They provide for a smooth transition from one activity to another; for example, from movement to offensive action to defensive action.

(3) They provide standardized actions that link Soldier and collective tasks at platoon level and below. (Soldiers perform individual tasks to CTT or SDT standard.)

(4) They require the full understanding of each individual and leader, and continual practice.

6-2. **FORMAT.** Drills include a ***title***, a ***situation*** to cue the unit or leader to start the drill, the ***required actions*** in sequence, and supporting illustrations. Where applicable, drills are cross-referenced with material in other chapters, or other drills, or both. Training standards for battle drills are in the mission training plan (MTP).

Section II. DRILLS

This section provides the battle drills for small units.

BATTLE DRILL 1
REACT TO CONTACT

SITUATION: A squad or platoon receives fires from enemy individual or crew-served weapons.

REQUIRED ACTIONS: See Figure 6-1.

1. Soldiers immediately assume the nearest covered positions.
2. Soldiers return fire immediately on reaching the covered positions.
3. Squad/team leaders locate and engage known or suspected enemy positions with well-aimed fire, and they pass information to the platoon/squad leader.
4. Fire team leaders control the fire of their Soldiers by using standard fire commands (initial and supplemental) containing the following elements:
    - Alert
    - Direction
    - Description of target
    - Range
    - Method of fire (manipulation and rate of fire)
    - Command to commence firing

5. Soldiers maintain contact (visual or oral) with the Soldiers on their left and right.

6. Soldiers maintain contact with their team leaders and indicate the location of enemy positions.

7. The leaders (visual or oral) check the status of their personnel.

8. The squad/team leaders maintain visual contact with the platoon/squad leader.

9. The platoon/squad leader moves up to the squad/fire team in contact and links up with its leader.
   a. The platoon leader brings his RTO, platoon FO, the nearest squad's squad leader, and one machine gun team.
   b. The squad leader of the trail squad moves to the front of his lead fire team.
   c. The PSG also moves forward with the second machine gun team and links up with the platoon leader, ready to assume control of the base-of-fire element.

10. The platoon/squad leader determines whether or not his unit must move out of an EA.

11. The platoon/squad leader determines whether his unit can gain and maintain suppressive fires with the element already in contact (based on the volume and accuracy of enemy fires against the element in contact). The platoon/squad leader assesses the situation. He identifies—
   - The location of the enemy position and obstacles.
   - The size of the enemy force engaging the unit in contact. The number of enemy automatic weapons, the presence of any vehicles, and the employment of indirect fires are indicators of the enemy strength.
   - Vulnerable flanks.
   - Covered and concealed flanking routes to the enemy position.

12. The platoon/squad leader determines the next COA, such as fire and movement, assault, breach, knock out bunker, enter and clear a building or trench.

13. The platoon/squad leader reports the situation to the company commander/platoon leader and begins to maneuver his unit.

14. Platoon leader directs platoon FO to call for and adjust indirect fires (mortars or artillery). Squad leaders relay requests through the platoon leader. The platoon/squad leader in conjunction with the platoon FO maintains accurate battle tracking of all friendly elements to facilitate quick clearance of fires.

15. Leaders relay all commands and signals from the platoon chain of command.

16. The PSG positions the base of fire element to observe and to provide supporting fires.

Note: Once the platoon has executed the React to Contact drill, the platoon leader quickly assesses the situation, for example, enemy size and location. He picks a COA. The platoon leader reports the situation to the company commander.

Figure 6-1. REACT TO CONTACT

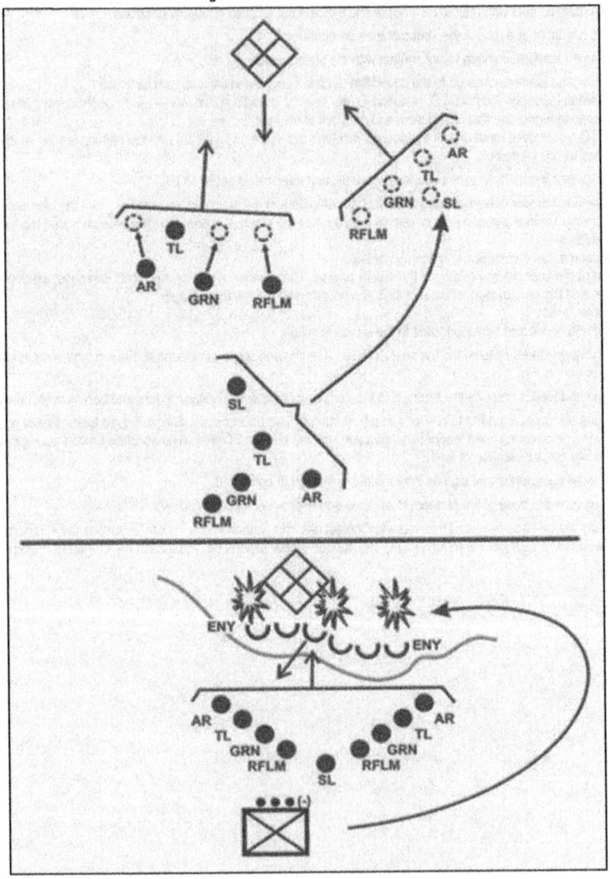

BATTLE DRILL 2
BREAK CONTACT

SITUATION: The squad or platoon is under enemy fire and must break contact.

REQUIRED ACTIONS: See Figure 6-2.

1. The platoon/squad leader directs fire support for the disengagement. He accomplishes this by—
   - Directing one squad/fire team in contact to support the disengagement of the remainder of the unit.
   - Considering the use of indirect fires for breaking contact.
   - Clearing the location, task, purpose, and method of conducting the fire mission with the platoon FO.

2. The platoon/squad leader orders a distance and direction, a terrain feature, or last ORP for the movement of the first squad/fire team. In conjunction with the platoon FO, the leader maintains accurate battle tracking of all friendly elements to facilitate quick clearance of fires.

3. The base of fire squad/team continues to suppress the enemy. The platoon/squad leader directs the platoon FO to execute the fire mission, if needed.

4. The moving squad/team assumes the overwatch position. The squad/team should use M203 grenade launchers, throw fragmentation and concussion grenades, and use smoke grenades to mask movement. The platoon/squad leader directs the platoon FO to execute smoke mission to screen friendly elements movement, if needed.

5. The moving squad/team takes up the designated position and engages the enemy position.

6. The platoon leader directs the base-of-fire element to move to its next location. Based on the terrain and the volume and accuracy of the enemy's fire, the moving fire team or squad may need to use fire and movement techniques.

7. The platoon/squad continues to bound away from the enemy until—
   - It breaks contact. It must continue to suppress the enemy as it does this.
   - It passes through a higher-level SBF position.
   - Its squad/fire teams are in the assigned position to conduct the next mission.

8. The leader should consider changing his unit's direction of movement once contact is broken. This will reduce the ability of the enemy to place effective indirect fires on the unit.

9. If the platoon/squad becomes disrupted, Soldiers stay together and move to the last designated rally point.

10. The platoon/squad leaders account for Soldiers, report, reorganize as necessary, and continue the mission. With the platoon FO, the platoon/squad develops a quick fire plan to support the hasty defense or new route of march.

Figure 6-2. BREAK CONTACT

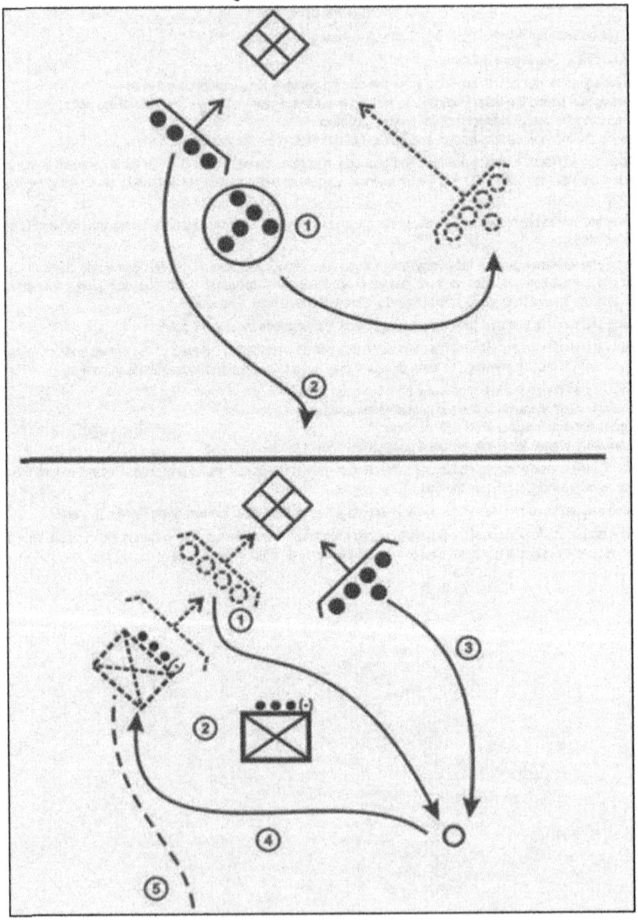

BATTLE DRILL 3
REACT TO AMBUSH

SITUATION: The squad or platoon enters a kill zone and the enemy initiates an ambush with a casualty-producing device and a high volume of fire.

REQUIRED ACTIONS: See Figure 6-3.

1. Near ambush (within hand-grenade range).
   a. Soldiers in the kill zone return fire immediately. How they go about this depends on the terrain. If—
      - Cover is not available, Soldiers immediately, without order or signal, assume prone position and throw concussion or fragmentation and smoke grenades.
      - Cover is available, without order or signal, Soldiers seek the nearest covered position, assume the prone position, and throw fragmentation or concussion and smoke grenades.

   b. Immediately after the explosion of the concussion or fragmentation grenades, Soldiers in the kill zone return fire and assault through the ambush position using fire and movement.
   c. Soldiers not in the kill zone identify the enemy's position. Fire is shifted as the personnel in the kill zone begin to assault.
   d. Soldiers in the kill zone continue the assault to eliminate the ambush or until contact is broken.
   e. The platoon/squad conducts consolidation and reorganization.

2. Far ambush (out of hand-grenade range). Soldiers receiving fire immediately return fire, take up covered positions, and suppress the enemy by—
   - Destroying or suppressing enemy crew-served weapons.
   - Sustaining suppressive fires.

3. Soldiers (squad/teams) not receiving fires move by a covered and concealed route to a vulnerable flank of the enemy position and assault using fire and movement techniques.

4. Soldiers in the kill zone continue suppressive fires and shift fires as the assaulting squad/team fights through the enemy position.

5. The platoon FO calls for and adjusts indirect fires as directed by the platoon leader. On order, he shifts or ceases fires to isolate the enemy position or to attack them with indirect fires as they retreat.

6. The platoon/squad leader reports, reorganizes as necessary, and continues the mission.

Figure 6-3. REACT TO AMBUSH

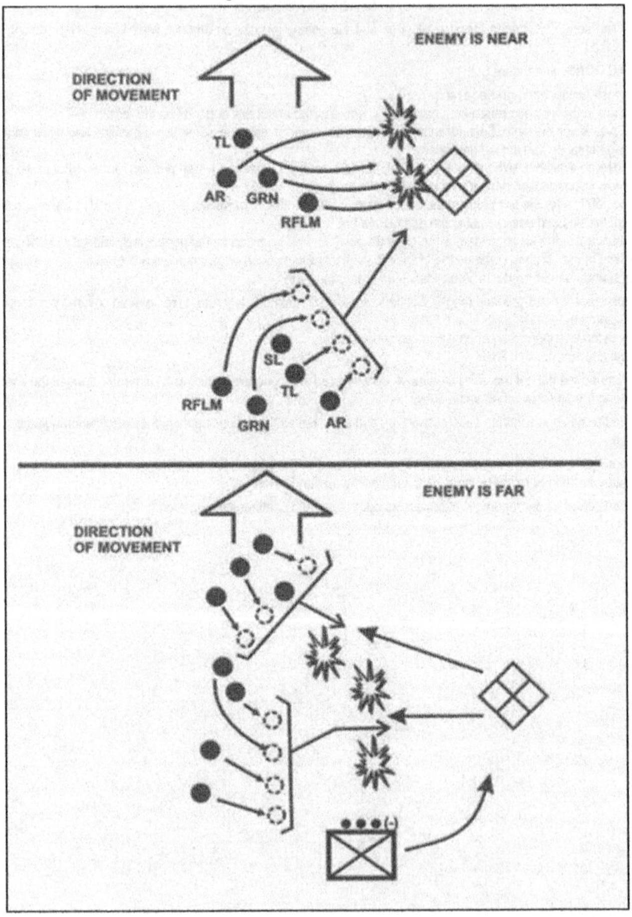

BATTLE DRILL 4
KNOCK OUT BUNKERS

SITUATION: While moving as part of a larger force, the platoon identifies the enemy in bunkers.

REQUIRED ACTIONS: See Figures 6-4 and 6-5.

1. The platoon initiates contact.
   a. The squad in contact establishes a base of fire.
   b. The platoon leader, his RTO, platoon FO, and one machine gun team move forward to link up with the squad leader of the squad in contact.
   c. The PSG moves forward with the second machine gun team and assumes control of the base-of-fire squad.
   d. The base-of-fire squad—
      - Destroys or suppresses enemy crew-served weapons first.
      - Obscures the enemy position with smoke (M203).
      - Sustains suppressive fires at the lowest possible level.
   e. The platoon leader directs platoon FO to call for and adjust indirect fires. The platoon leader in conjunction with the platoon FO maintains accurate battle tracking of all friendly elements to facilitate quick clearance of fires.
2. The platoon leader determines that he can maneuver by identifying—
   - Enemy bunkers, other supporting positions, and any obstacles.
   - Size of the enemy force engaging the platoon. The number of enemy automatic weapons, the presence of any vehicles, and the employment of indirect fires are indicators of enemy strength.
   - A vulnerable flank of at least one bunker.
   - A covered and concealed flanking route to the flank of the bunker.
3. The platoon leader determines which bunker is to be assaulted first and directs one squad (not in contact) to knock it out.
4. If necessary, the PSG repositions a squad, fire team, or machine gun team to isolate the bunker as well as to continue suppressive fires.
5. The assaulting squad, with the platoon leader and his RTO, move along the covered and concealed route and take action to knock out the bunker. The following occurs.
   a. On the platoon leader's signal, the support squad ceases or shifts fires to the opposite side of the bunker from which the squad is assaulting.
   b. At the same time, the platoon FO shifts indirect fires to isolate enemy positions.
6. The assaulting squad leader reports to the platoon leader and reorganizes his squad.
7. The platoon leader—
   - Directs the supporting squad to move up and knock out the next bunker.
   - Directs the assaulting squad to continue and knock out the next bunker.
   - Rotates squads as necessary.
8. The platoon leader reports, reorganizes as necessary, and continues the mission. The company follows up the success of the platoon attack and continues to assault enemy positions.

Figure 6-4. KNOCK OUT A BUNKER (SQUAD)

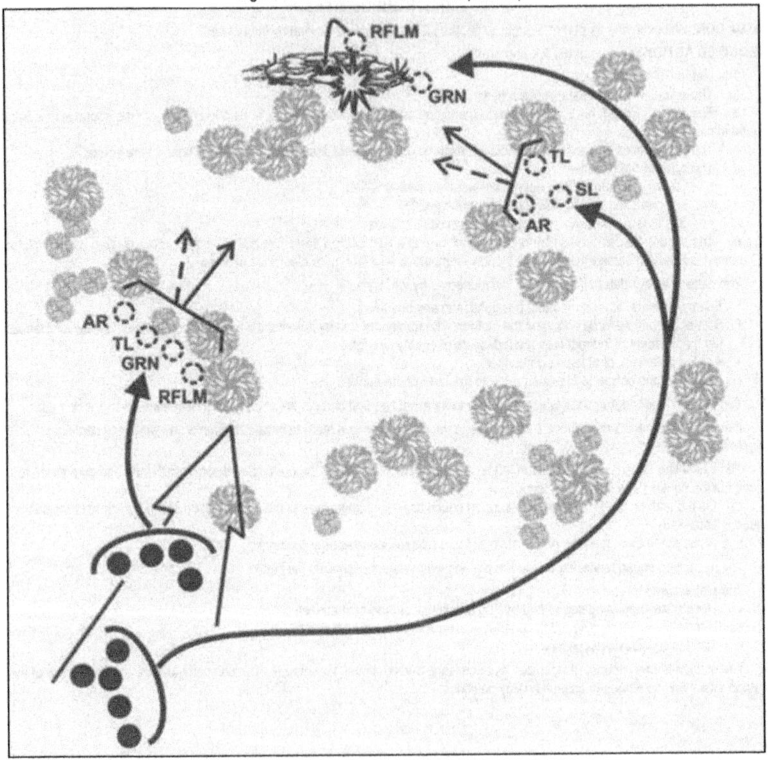

Figure 6-5. KNOCK OUT BUNKERS (PLATOON)

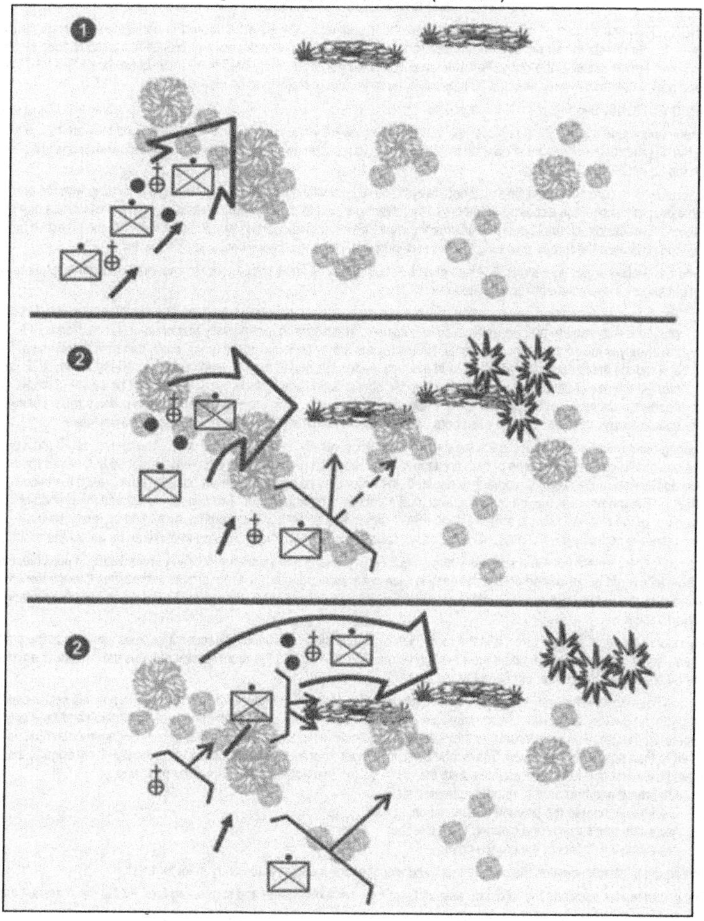

BATTLE DRILL 5
ENTER BUILDING/CLEAR ROOM

SITUATION: Operating as part of a larger force (in daylight or darkness), the squad is tasked to participate in clearing a building. The platoon leader directs the squad to enter the building or to clear a room. An entry point or breach has already been identified or will be created before initiating the entry. For a detailed discussion of urban entry breaching techniques, see FM 3.06-11. Enemy forces and noncombatants may or may not be present in the room and or building to be cleared.

REQUIRED ACTIONS: See Figures 6-6 through 6-14.

1. Platoon and squad leaders must consider the task and purpose they have been given and the method they are to use to achieve the desired results. They must operate IAW the ROE and must be aware of the affects that platoon weapons will have on the type and composition of the buildings.

2. To seize or gain control of a building may not always require committing troops into the structure or closing with the enemy. Before initiating this action and exposing members of the clearing squad to direct enemy contact and risking casualties, the platoon leader should consider/direct employment of all organic, crew-served, and supporting weapon systems onto the objective area to suppress and neutralize the threat, providing the mission, purpose, building composition, and ROE permit.

3. When conducting urban operations, Soldiers must be equipped at all times with a properly-mounted and immediately-useable NVD or light source to illuminate the immediate area.

Note: The following discussion assumes that only the platoon's organic weapons are supporting the infantry squad. Urban situations may require precise application of firepower. This situation is especially true of an urban environment in which the enemy is mixed with noncombatants. Noncombatants may be found in the room, which can restrict the use of fires and reduce the combat power available to a squad leader. His squad may have to operate in no-fire areas. ROE can prohibit the use of certain weapons until a specific hostile action takes place. All Soldiers must be aware of the ROE. Leaders must include the specific use of weapons in their planning for precision operations in urban terrain. Leaders should always consider the use of snipers or designated marksmen to apply precise fires to the objective.

4. Clearing team members must approach the entry point quickly, quietly, and in standard order. The squad leader must ensure he is in a position to control the actions of both fire teams. This approach preserves the element of surprise and allows for quick entry and domination of the room. If a breach is required, the order may be slightly modified based on the breach technique (FM 3-06.11). The members of the fire team are assigned numbers one through four. The rifleman is # 1 and the grenadier is # 3. If one member of the clearing team is armed with the SAW rather than an M16 rifle or carbine, he should be designated # 4. The team leader is normally the # 2 man, because he will have the most immediate decision to make as he enters the room.

5. The entire team enters the room as quickly and smoothly as possible and clears the doorway immediately. If possible, the team moves from a covered or concealed position already in their entry order. Ideally, the team arrives and passes through the entry point without having to stop. If the team must stop to effectively "*stack*" outside the entry point, it must do so only momentarily, and it must provide cover.

6. The door is the focal point of anyone in the room. It is known as the "*fatal funnel,*" because it focuses attention at the precise point where the individual team members are the most vulnerable. Moving into the room quickly reduces the chance of anyone being hit by enemy fire directed at the doorway.

7. For this battle to be effectively employed, each member of the team must know his sector of fire and how his sector overlaps and links with the sectors of the other team members. Team members do not move to the points of domination and then engage their targets. Rather, they engage targets as they calmly and quickly move to their designated points. Engagements must not slow movement to their points of domination. Team members may shoot from as short a range as 1 to 2 inches. They engage the most immediate threat first and then the less immediate threats in sector. Immediate threats are personnel who—

- Are armed and prepared to return fire immediately.
- Block movement to the position of domination.
- Are within arm's reach of a clearing team member.
- Are within 3 to 5 feet of the breach point.

8. The squad leader designates the assault team and identifies the location of the entry point for them.

9. The squad leader positions the follow-on assault team to provide overwatch and supporting fires for the initial assault team.

10. Assault team members use available cover and concealment, and move as close to the entry point as possible.

a. If an explosive breach or a ballistic breach is to be performed by a supporting element, the assault team remains in a covered position until the breach is made. They may provide overwatch and fire support for the breaching element if necessary.

b. All team members must signal one another that they are ready before the team moves to the entry point.

c If stealth is a consideration, team members avoid the use of verbal signals, which may alert the enemy and remove the element of surprise.

d. Assault team members must move quickly from the covered position to the entry point, minimizing the time they are exposed to enemy fire.

11. The assault team enters through the entry point or breach. Unless a grenade will be thrown prior to entry, the team should avoid stopping outside the point of entry.

a. If required, the # 2 man throws a grenade of some type (fragmentary, concussion, stun) into the room before entry.

b. The use of grenades should be consistent with the ROE and building structure. The grenade should be cooked off before being thrown, if applicable to the type of grenade used.

c. If stealth is not a factor, the thrower should sound off with a verbal indication ("*Frag out*," "*Concussion out*," "*Stun out*") that a grenade of some type is being thrown. If stealth is a factor, only visual signals are given as the grenade is thrown.

CAUTION

If walls and floors are thin, fragments from fragmentation grenades and debris created by concussion grenades can injure Soldiers outside the room. If the structure has been stressed by previous explosive engagements, the use of these grenades could cause it to collapse. Leaders must determine the effectiveness of these types of grenades compared to possibilities of harm to friendly troops.

12. On the signal to go, or after the grenade detonates, the assault team moves through the entry point (Figure 6-6 through Figure 6-9) and quickly takes up positions inside the room that allow it to completely dominate the room and eliminate the threat. Unless restricted or impeded, team members stop movement only after they have cleared the door and reached their designated point of domination. In addition to dominating the room, all team members are responsible for identifying possible loopholes and mouseholes in the ceiling, walls, and floor.

Note: Where enemy forces may be concentrated and the presence of noncombatants is highly unlikely, the assault team can precede their entry by throwing a fragmentation or concussion grenade (structure dependent) into the room, followed by bursts of automatic small-arms fire by the # 1 man as he enters. Carefully consider the ROE and building composition before employing this method.

13. The # 1 and # 2 men are initially concerned with the area directly to their front, then along the wall on either side of the door or entry point. This area is in their path of movement, and it is their primary sector of fire. Their alternate sector of fire is from the wall they are moving toward, back to the opposite far corner.

14. The # 3 and # 4 men start at the center of the wall opposite their point of entry and clear to the left if moving toward the left, or to the right if moving toward the right. They stop short of their respective team member (either the # 1 man or the # 2 man).

15. The team members move toward their points of domination, engaging all threat or hostile targets in sequence in their sector. Team members must exercise fire control and discriminate between hostile and noncombatant room occupants. (The most practical way to do this is to identify whether or not the target has a weapon in his or her hands.) Shooting is done without stopping, using reflexive shooting techniques. Because the Soldiers are moving and shooting at the same time, they must move using the careful hurry. Figure 6-10, shows all four team members at their points of domination for a room with a center door and their overlapping sectors of fire.

16. The first man (rifleman) enters the room and eliminates the immediate threat. He has the option of going left or right, normally moving along the path of least resistance to one of two corners. When using a doorway as the point of entry, the team uses the path of least resistance, which they determine initially based on the way the door opens. If the door opens inward, the first man plans to move away from the hinges. If the door opens outward, he plans to move toward the hinged side. Upon entering, his direction is influenced by the size of the room, the enemy situation, and furniture or other obstacles that hinder or channel his movement.

17. The direction each man moves in should not be preplanned unless the exact room layout is known. Each man should go in a direction opposite the man in front of him (Figure 6-6). Every team member must know the sectors and duties of each position.

18. As the first man goes through the entry point, he can usually see into the far corner of the room. He eliminates any immediate threat and continues to move along the wall if possible and to the first corner, where he assumes a position of domination facing into the room.

Figure 6-6. FIRST MAN ENTERS A ROOM, FOLLOWED BY TEAM LEADER

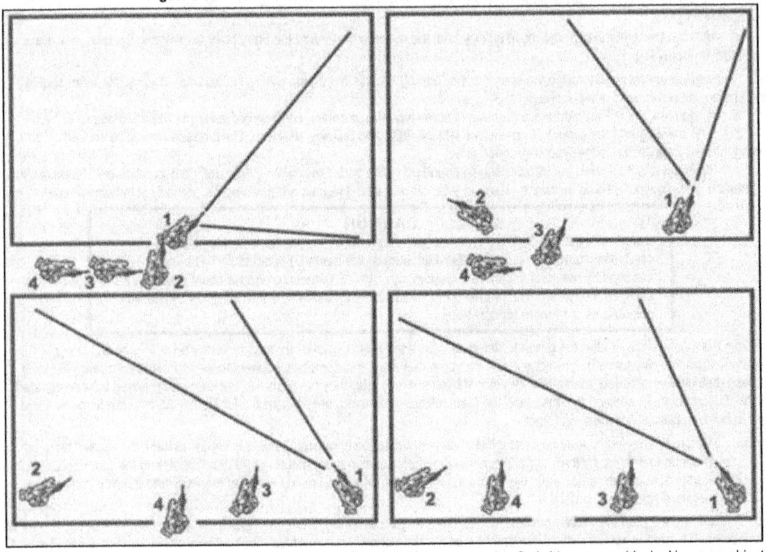

Note: Team members must always stay within 1 meter of the wall. If a team member finds his progress blocked by some object that will force him more than 1 meter from the wall, he must either step over it (if able) or stop where he is and clear the rest of his sector from where he is. If this action creates dead space in the room, the team leader directs which clearing actions to take once other members of the team have reached their points of domination.

19. The second man (normally the team leader), entering almost simultaneously with the first, moves in the opposite direction, following the wall (Figure 6-7). The second man must clear the entry point, clear the immediate threat area, clear his corner, and move to a dominating position on his side of the room. The second man must also immediately determine if he is entering a center door or corner door and act accordingly (Figure 6-7 and Figure 6-10).

Figure 6-7. SECOND MAN (TEAM LEADER) ENTERS A ROOM

20. The third man (normally the grenadier) simply goes opposite of the second man inside the room, moves at least 1 meter from the entry point, and takes a position that dominates his sector (Figure 6-8).

Figure 6-8. THIRD MAN ENTERS A ROOM

21. The fourth man (normally the SAW gunner) moves opposite of the third man, clears the doorway by at least 1 meter, and moves to a position that dominates his sector (Figure 6-9).

Figure 6-9. FOURTH MAN IN A ROOM

Note: If the path of least resistance takes the first man to the left, then all points of domination are the mirror image of those shown in the diagrams.

22. Points of domination should not be in front of doors or windows so team members are not silhouetted to the outside of the room (Figure 6-10). No movement should mask the fire of any of the other team members.

Figure 6-10. POINTS OF DOMINATION AND SECTORS OF FIRE, CENTER DOOR VERSUS CORNER DOOR

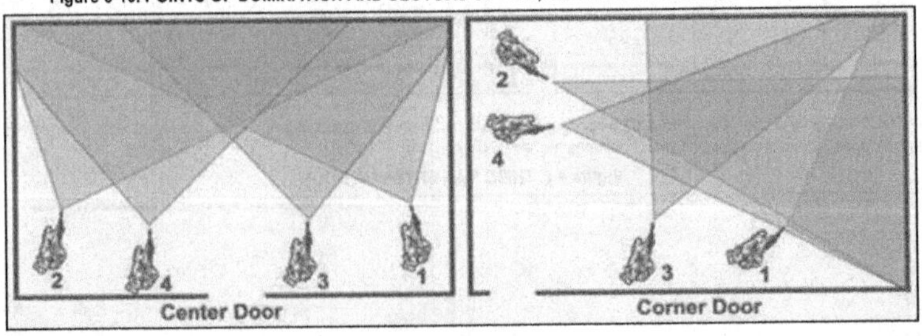

23. On order, any member of the assault team may move deeper into the room overwatched by the other team members. The team leader must control this action.

24. Once the room is cleared, the team leader signals to the squad leader that the room has been cleared. The squad leader marks the room IAW unit SOP (see FM 3.06-11). The squad leader determines whether or not his squad can continue to clear through the building.

25. The squad reorganizes as necessary. Leaders redistribute the ammunition. The squad leader reports to the platoon leader when the room is clear.

Note: If at any point a team member experiences a weapon malfunction in the presence of enemy combatants, he has to make an immediate decision. If the enemy is outside immediate danger distance from the Soldier, then the Soldier should drop to one knee, which indicates that he has experienced a weapon malfunction and prevents fratricide by ensuring that the next Soldier's fields of fire are clear. Once on a knee, the Soldier must remain there until the team leader directs him to stand up. If the kneeling Soldier corrects his weapon malfunction, he can continue to engage targets from his kneeling position. If, however, the enemy Soldier is so close that he presents an immediate threat, and if dropping to a knee would expose the US Soldier to immediate harm, then the US Soldier must try to subdue or disable the enemy Soldier. He can strike the enemy with the muzzle of his weapon or a bayonet to the face, throat, or chest. Or, he could grapple with the enemy soldier and take him to the ground immediately to clear the other team members' fields of fire. Once they have cleared their sectors and eliminated any other threats in the room, the other members of the clearing team assist.

26. Although this battle drill is effective, leaders might have to modify it for the situation. Some reasons and methods for modifying the technique are shown in Table 6-1.

Table 6-1. REASONS FOR AND METHODS OF MODIFYING ENTRY TECHNIQUES

REASON	METHOD
Objective rooms are consistently small	Clear in teams of 2 or 3
Shortage of personnel	Clear in teams of 2 or 3
Enemy poses no immediate threat	2 or 3 men search rooms to ensure no enemy or noncombatants are present
No immediate threat, speed is critical	1 to 3 men visually search each room

27. When full four-man teams are not available for room clearing, two- and three-man teams can be used. If the # 1 or # 2 man discovers that the room is very small, he can yell, "*Short room*" or "*Short,*" which tells the # 3 or # 4 man (whoever following the # 1 or # 2 man) to stay outside the room. Figures 6-11 and 6-12 show the points of domination and sectors of fire for a two-man clearing team. Figures 6-13 and 6-14 show the actions for a three-man team.

DANGER

RICOCHETS POSE A HAZARD. ALL SOLDIERS MUST BE AWARE OF THE TYPE OF WALL CONSTRUCTION OF THE ROOM BEING CLEARED. THE WALLS OF AN ENCLOSED ROOM PRESENT MANY RIGHT ANGLES. COMBINED WITH HARD SURFACES SUCH AS CONCRETE, A BULLET MAY CONTINUE TO RICOCHET AROUND A ROOM UNTIL ITS ENERGY IS SPENT. AFTER HITTING THREAT PERSONNEL, BALL AMMUNITION MAY PASS THROUGH THE BODY AND RICOCHET. BODY ARMOR AND THE KEVLAR HELMET PROVIDE SOME PROTECTION FROM THIS HAZARD.

Figure 6-11. POINTS OF DOMINATION AND SECTORS OF FIRE (TWO-MAN TEAM, CENTER DOOR)

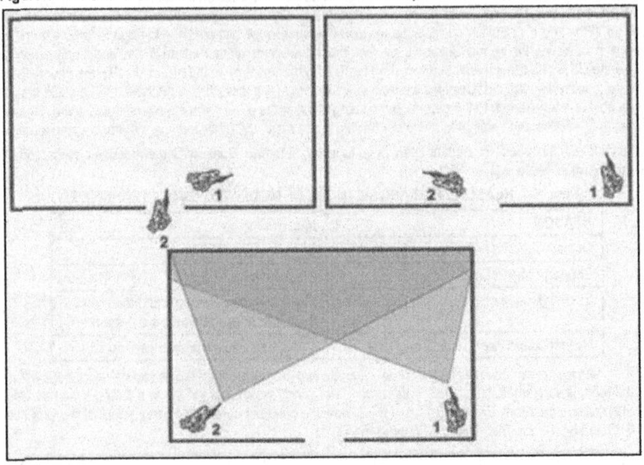

Figure 6-12. POINTS OF DOMINATION AND SECTORS OF FIRE (TWO-MAN TEAM, CORNER DOOR)

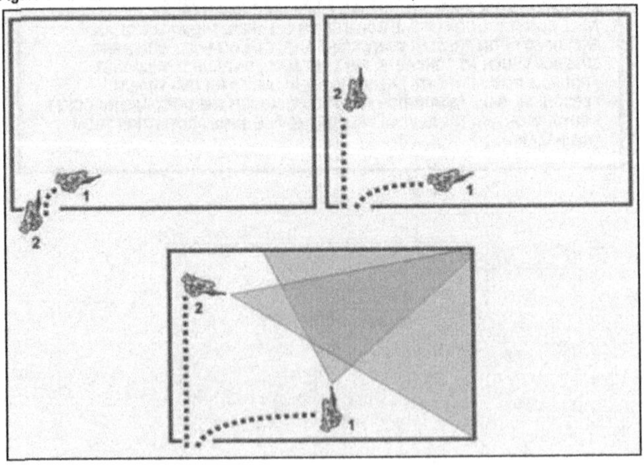

Figure 6-13. POINTS OF DOMINATION AND SECTORS OF FIRE (THREE-MAN TEAM, CENTER DOOR)

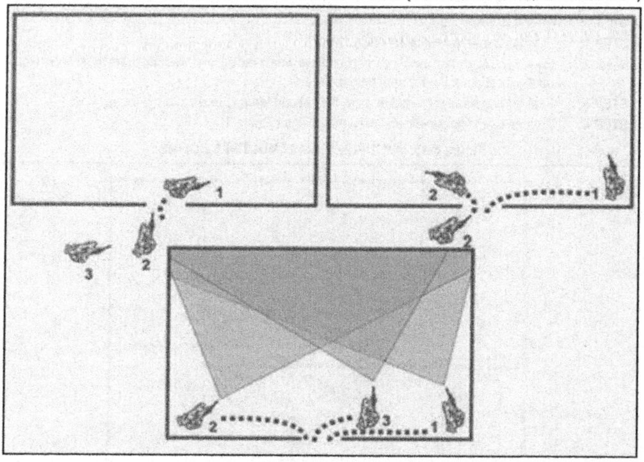

Figure 6-14. POINTS OF DOMINATION AND SECTORS OF FIRE (THREE-MAN TEAM, CORNER DOOR)

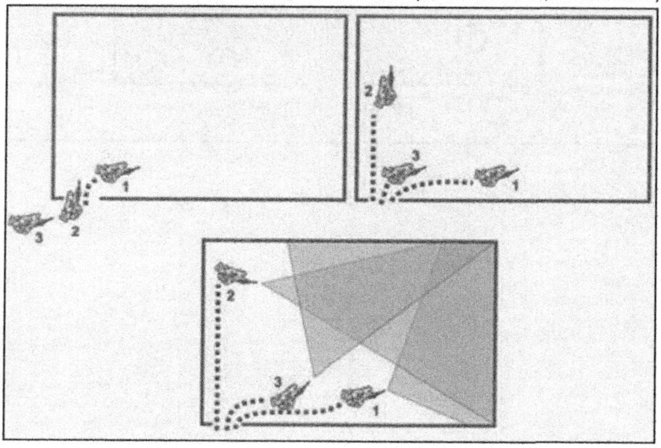

28. While moving through a building Soldiers may encounter the following architectural building features.

a. **Multiple Team/Multiple Rooms**. Figure 6-15 shows this.

STEP 1: First team enters and clears Room 1.

STEP 2: Squad leader determines the direction from which the second clearing team must enter Room 1, based on the location of the Room 2 entry point.

STEP 3: First team collapses inward to allow the second team to move into the room.

STEP 4: Second team 'stacks left' and prepares to enter Room 2.

Figure 6-15. MULTIPLE TEAMS / MULTIPLE ROOMS

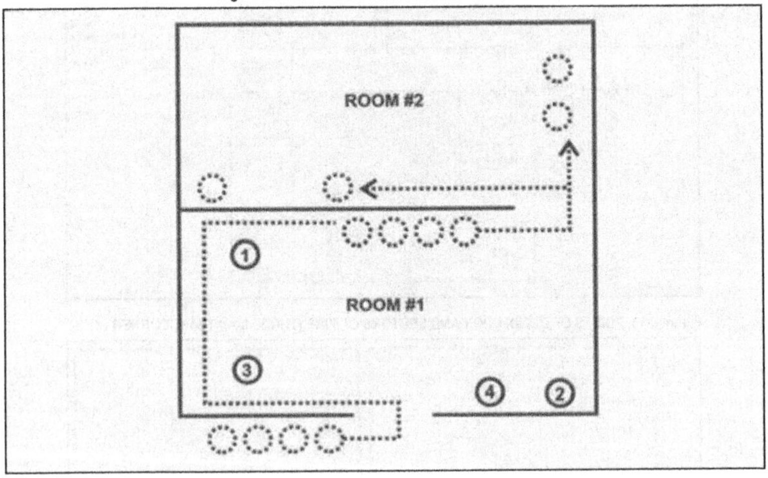

b. **Open Stairwell**. This is a gap between flights of stairs that allows a person to look up and down between the flights. Figure 6-16 shows an open stairwell.

STEP 1: The # 1 man pulls security on the highest point he can see / engage

STEP 2: The # 2 man moves up the stairs on the inside with the # 3 man to a point that he can see / engage the next landing, where he turns around and continues to move up to the next landing.

STEP 3: The # 3 man moves up the stairs with the # 2 man on the outside and engages the threat on the immediate landing.

STEP 4: The # 4 man moves up the stairs with the # 1 man, on the squeeze, the # 2 man turns around to engage the next landing.

STEP 5: The flow continues with the # 2 man picking up the sector of the # 1 man had. The # 3 man picks up where the # 2 man was. The # 4 man picks up where the # 3 man was. The # 1 man picks up where the # 4 man was.

Note: Most stairwells require a second team.

Figure 6-16. OPEN STAIRWELL

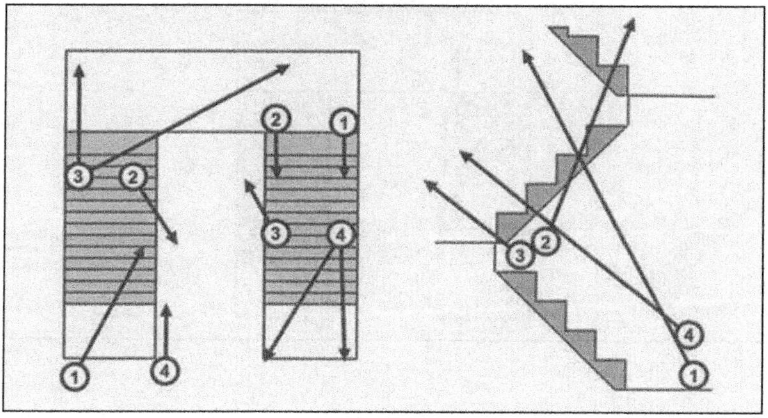

c. **Closed Stairwell**. This is any stairwell with walls separating the flights of stairs. Figure 6-17 shows a closed stairwell.

STEP 1: The # 1 man checks high to ensure there is no opening on the landing or between the stairs.

STEP 2: The # 2 man pulls long security to the next bend or landing.

STEP 3: The # 1 or # 3 man moves up the steps with the # 2 man. As they approach the corner, the # 2 man signals his presence by tapping the # 1 man on the shoulder.

STEP 4: Keying off the # 1 man's movement, the other men simultaneously break around the corner.

STEP 5: If no fire is received, the # 2 man moves to the opposite wall. Both men continue to move up until they reach their objective.

STEP 6: The # 3 and # 4 men will continue to move 3 to 4 steps behind.

Note: Avoid getting locked into a security position such as inside a stairwell. Also, avoid spreading yourselves thin or getting separated by more than one floor of stairs.

Figure 6-17. CLOSED STAIRWELL

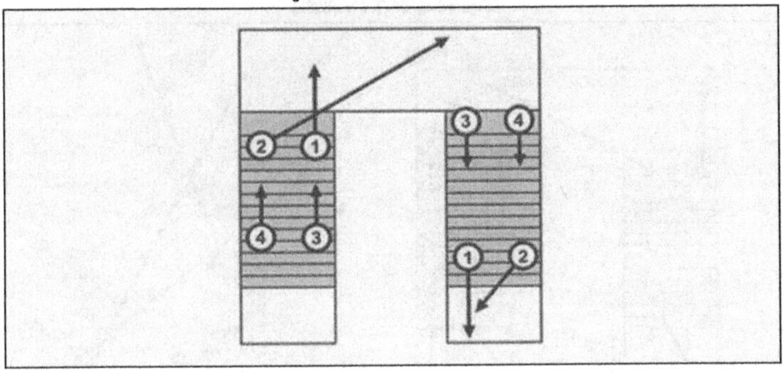

d. **Hallway Movement**. Clearing team(s) move down the hallway using the frontal security (cross cover) technique (Figure 6-18).

Figure 6-18. HALLWAY MOVEMENT

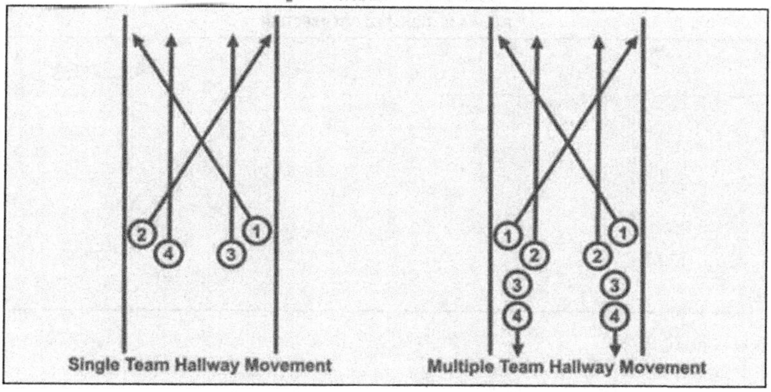

e. **T-Shaped Intersection**. See Figure 6-19. This technique can incorporate the dynamic corner clear (Figure 6-20).

STEP 1: Each # 1 man goes to a knee covering his sector.

STEP 2: On a predetermined signal, each two-man team will break the corner, picking up their sectors of fire.

Figure 6-19. T-SHAPED INTERSECTION

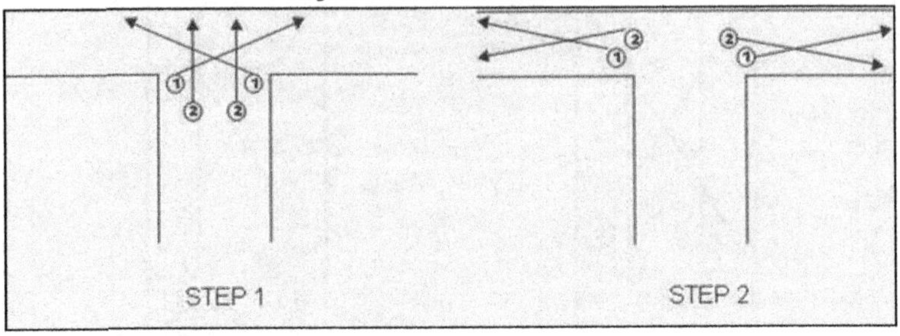

f. **Dynamic Corner**. (See Figure 6-20.)

STEP 1: The #1 and #2 man, as they approach the corner, have to clear. Do not slow down.

STEP 2: The #2 man will tap the #1 man on the shoulder about 2-3 meters away from the corner letting the #1 man know the #2 man is with him.

STEP 3: Keying off of the #1 man's movement, they both break the corner simultaneously.

STEP 4: The #1 man goes low to a knee, the #2 man stays high.

STEP 5: If the Rangers are not receiving fire the #2 man rabbits, or moves, to the far side.

STEP 6: The #1 and #2 man take up sectors of fire.

STEP 7: The #3 and #4 man take long security in the direction of movement.

Figure 6-20. DYNAMIC CORNER

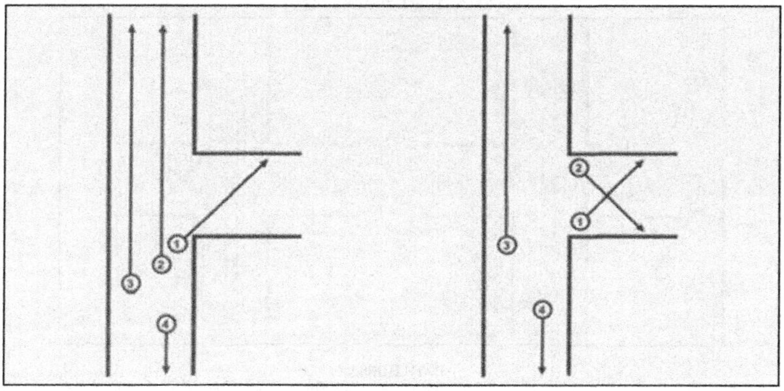

g. **Three-Way Intersection**. (See Figure 6-21.)

STEP 1: The #1 through #4 man will use one of the corner clearing techniques to clear the intersection.

STEP 2: When the intersection is secure, the #5 and #6 man move through the intersection and enter either more hallway or a room. If they are entering more hallway, the #1 through #4 man provide cross coverage as the #5 and #6 man move through the intersection into more hallway. If they are moving into a room, the #5 and #6 man move to one side of the door and signal the #1 and #3 or #2 and #4 man to enter the stack depending on which side of the door they stack on. The remaining members of the team, not in the stack, will continue to provide security down the hallway.

STEP 3: If the #5 and #6 man move into more hallway the #1 through #4 man will enter the stack and proceed down the hallway with the #3 and #4 man providing rear security.

Figure 6-21. THREE WAY INTERSECTION

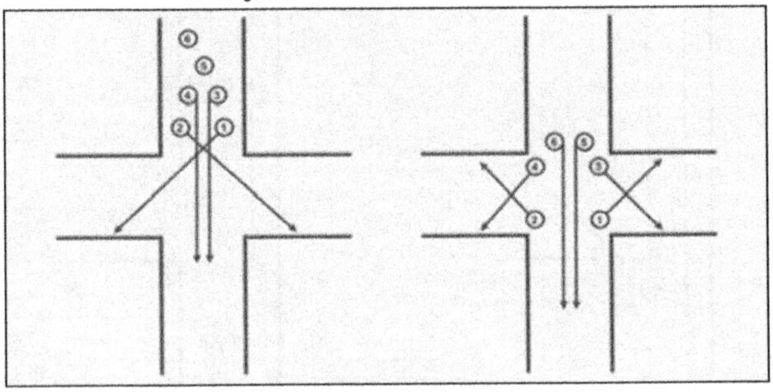

BATTLE DRILL 6
ENTER/CLEAR A TRENCH

SITUATION: The platoon is attacking as part of a larger force and identifies the enemy in a trench line. The platoon deploys and establishes a base of fire. The platoon leader determines that he has sufficient combat power to maneuver and assault the trench line.

REQUIRED ACTIONS: See Figure 6-22 and Figure 6-23.

1. The platoon leader directs one squad to enter the trench and secure a foothold.
2. The platoon leader designates the entry point of the trench line and the direction of movement once the platoon begins clearing.
3. The PSG positions Soldiers and machine guns to suppress the trench and isolate the entry point.
4. Platoon leader directs platoon FO to initiate fire mission, if necessary, in support of assault. They maintain accurate battle tracking of all friendly elements to facilitate quick clearance of fires. The platoon FO ceases or shifts fires to isolate the OBJ as assault team advances.
5. The assaulting squad executes actions to enter the trench and establish a foothold. The squad leader signals to the platoon leader that the foothold is secure, and the follow on elements can move into the trench. The squad leader remains at the entry point and marks it. The platoon follows the success of the seizure of the foothold with the remainder of the platoon as part of the platoon actions to clear a trench line.
6. The platoon leader moves into the trench with the assaulting squad.

7 The platoon leader directs one of the base-of-fire squads to move into the trench and begin clearing it in the direction of movement from the foothold.

8. The base-of-fire element repositions as necessary to continue suppressive fires.

9. The assaulting squad passes the squad that has secured the foothold and executes actions to take the lead and clear the trench.

a. The squad leader designates a lead fire team and a trail fire team.

b. The lead fire team and the squad leader move to the forward most secure corner or intersection. The squad leader tells the team securing that corner or intersection that his squad is ready to continue clearing the trench. The trail fire team follows, maintaining visual contact with the last Soldier of the lead team.

Notes: 1. The fire support element must be able to identify the location of the lead fire team in the trench at all times.

2. Throughout this battle drill, the team leader positions himself at the rear of the fire team to have direct control (physically, if necessary) of his Soldiers. Other Soldiers in the fire team rotate the lead to change magazines and prepare grenades. Rotating the lead provides constant suppressive fires down the trench and maintains the momentum of the attack as the squad clears the trench.

c. The lead fire team passes the element securing the foothold. The following then occurs.

(1) The lead Soldier of the fire team moves abreast of the Soldier securing the corner or intersection, taps him, and announces, "*Taking the lead.*"

(2) The Soldier securing the corner or intersection acknowledges that he is handing over the lead by shouting, "*Okay!*" He allows the fire team to pass him.

d. The lead fire team starts clearing in the direction of movement. They arrive at a corner or intersection. The following then occurs.

(1) Allowing for cook-off (2 seconds maximum) and shouting, "*Frag out,*" the second Soldier prepares and throws a grenade around the corner.

(2) Upon detonation of the grenade, the lead Soldier moves around the corner firing three-round bursts and advancing as he fires. The entire fire team follows him to the next corner or intersection.

e. The squad leader—

- Follows behind the team.
- Ensures that the trailing fire team moves up and is ready to pass the lead at his direction.
- Rotates fire teams as necessary to keep his Soldiers fresh and to maintain the momentum of the attack.
- Requests indirect fires, if required, through the platoon leader. The squad leader also directs the employment of the M203 to provide immediate suppression against positions along the trench line.
- Ensures fire teams maintain sufficient interval to prevent them from being engaged by the same enemy fires.

f. At each corner or intersection, the lead fire team performs the same actions previously described.

g. If the lead Soldier finds that he is nearly out of ammunition before reaching a corner or intersection, he announces, "*Ammo,*" and then--

(1) The lead Soldier stops and moves against one side of the trench, ready to let the rest of the team pass. He continues to aim his weapon down the trench in the direction of movement.

(2) The next Soldier ensures that he has a full magazine, moves abreast of the lead Soldier, taps him, and announces, "*Taking the lead.*"

(3) The lead Soldier acknowledges that he is handing over the lead by shouting, "*Okay.*" Positions rotate and the squad continues forward.

h. The trailing fire team secures intersections and marks the route within the trench as the squad moves forward. The trailing fire team leader ensures that follow-on squads relieve his buddy teams to maintain security.

i. The squad leader reports the progress of the clearing operation. The base-of-fire element must be able to identify the location of the lead fire team in the trench at all times.

10. The platoon leader rotates squads to keep the Soldiers fresh and to maintain the momentum of the assault.

11. The PSG calls forward ammunition resupply and organizes teams to move it forward into the trench.

12. The base-of-fire element ensures that all friendly forces move into the trench only through the designated entry point to avoid fratricide.

13. The platoon leader reports to the company commander that the trench line is secured, or he is no longer able to continue clearing. If trench line is secured, then platoon leader directs platoon FO to develop a fire plan to support the defense of the platoon position.

DANGER

FIRE TEAMS: MAINTAIN SUFFICIENT INTERVALS TO AVOID ENGAGEMENT BY THE SAME ENEMY FIRES.

Figure 6-22. ENTER A TRENCH (SQUAD)

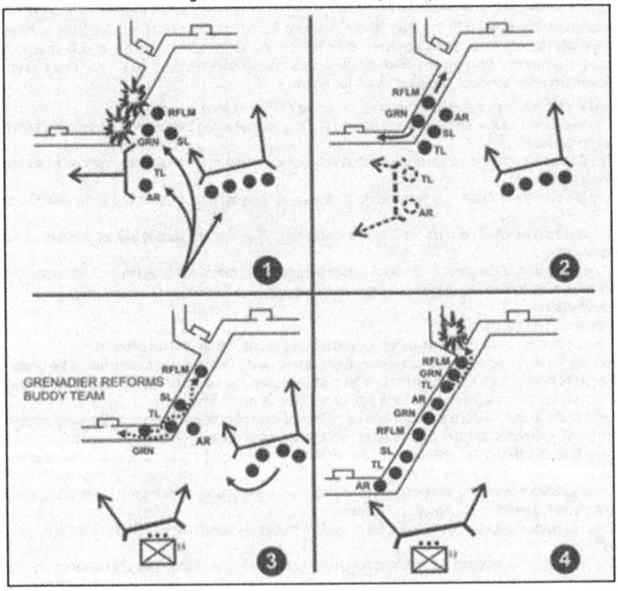

Figure 6-23. CLEAR A TRENCH LINE (PLATOON)

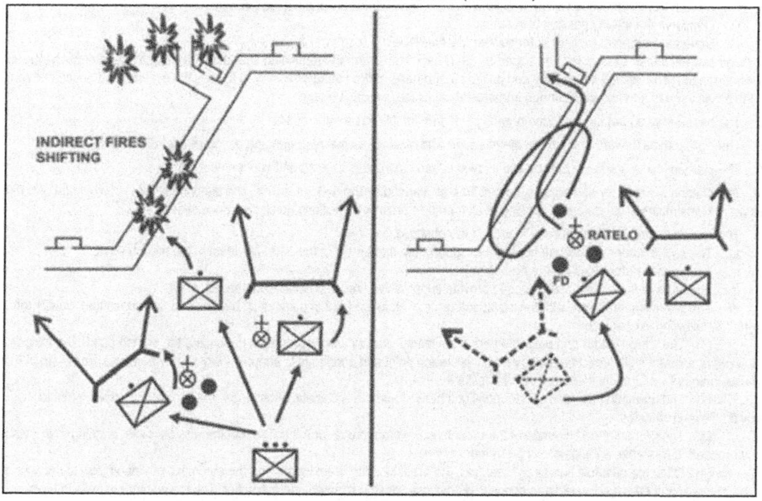

BATTLE DRILL 7
CONDUCT INITIAL BREACH OF A MINED WIRE OBSTACLE (PLATOON)

SITUATION: The platoon is operating as part of a larger force. The lead squad identifies a wire obstacle reinforced with mines that cannot be bypassed and enemy positions on the far side of the obstacle.

REQUIRED ACTIONS: See Figure 6-24 and Figure 6-25.

1. The squad in contact reacts to contact.
2. The platoon gains suppressive fires. The following then occurs.
   a. The squad in contact establishes a base-of-fire position.
   b. The platoon leader, his RTO, platoon FO, and the squad leader of the next squad with one machine gun team move forward to link up with the squad leader of the squad in contact.
3. The platoon leader determines that he can maneuver by identifying—
   - The obstacle and enemy positions.
   - The size of the enemy force engaging the squad, for example, the number of enemy automatic weapons, the presence of any vehicles, and the employment of indirect fires are indicators of enemy strength.
   - A breach point.
   - A covered and concealed route to the breach point.
4. The platoon leader directs the squad in contact to support the movement of another squad to the breach point. He indicates the base-of-fire position and the route to it, the enemy position to be suppressed, and the breach point and route the rest of the platoon will take to it. The platoon leader also clears the location, task, purpose, and method of conducting the fire mission with the platoon FO.

5. On the platoon leader's signal, the base-of-fire squad—

- Destroys or suppresses enemy weapons that are firing effectively against the platoon.
- Obscures the enemy position with smoke.
- Continues suppressive fires at the lowest possible level.

6. The platoon leader designates one squad as the breach squad and the remaining squad as the assault squad once the breach has been made. The assault squad may add its fires to the base-of-fire squad. Normally, it follows the covered and concealed route of the breach squad and assaults through immediately after the breach is made.

7. The base-of-fire squad moves to the breach point and establishes a base of fire.

8. The PSG moves forward to the base-of-fire squad with the second machine gun and assumes control of the squad.

9. The platoon leader leads the breach and assault squads along the covered and concealed route.

10. The platoon FO calls for and adjusts indirect fires as directed by the platoon leader. The platoon leader in conjunction with the platoon FO maintains accurate battle tracking of all friendly elements to facilitate quick clearance of fires.

11. The breach squad executes actions to breach the obstacle.

a. The squad leader directs one fire team to support the movement of the other fire team to the breach point.

b. The squad leader identifies the breach point.

c. The base-of-fire element continues to provide suppressive fires and isolates the breach point.

d. The breaching fire team, with the squad leader, moves to the breach point using the covered and concealed route. From there, the following takes place:

(1) The squad leader and breaching fire team leader employ smoke grenades to obscure the breach point. The platoon base-of-fire element shifts direct fires away from the breach point and continues to suppress key enemy positions. The platoon FO ceases indirect fires or shifts them beyond the obstacle.

(2) The breaching fire team leader positions himself and the automatic rifleman on one flank of the breach point to provide close-in security.

(3) The grenadier and rifleman of the breaching fire team probe for mines and cut the wire obstacle, marking their path as they proceed. If available, a bangalore is preferred.

(4) Once the obstacle has been breached, the breaching fire team leader and the automatic rifleman move to the far side of the obstacle and take up covered and concealed positions with the rifleman and grenadier. The team leader signals to the squad leader when they are in position and ready to support.

(5) The squad leader signals the base-of-fire team leader to move his fire team up and through the breach. He then moves through the obstacle and joins the breaching fire team, leaving the grenadier (or anti-armor specialist) and rifleman of the supporting fire team on the near side of the breach to guide the rest of the platoon through.

(6) Using the same covered and concealed route as the breaching fire team, the base-of-fire team moves through the breach and takes up covered and concealed positions on the far side.

12. The squad leader reports the situation to the platoon leader and posts guides at the breach point.

13. The platoon leader leads the assault squad through the breach in the obstacle and positions them beyond the breach to support the movement of the remainder of the platoon or assaults the enemy position covering the obstacle. Platoon leader directs the platoon FO to shift indirect fires off the enemy position.

14. The breaching squad continues to widen the breach to allow vehicles to pass through.

15. The platoon leader reports the situation to the company commander, and then he directs his base-of-fire squad to move up and through the obstacle. The platoon leader appoints Soldiers to guide the company through the breach point.

16. The PSG brings the remaining elements forward and through the breach on the platoon leader's command.

17. The company follows up the success of the platoon as it conducts the breach and continues the assault against the enemy positions.

Figure 6-24. CONDUCT INITIAL BREACH OF A MINED WIRE OBSTACLE (SQUAD)

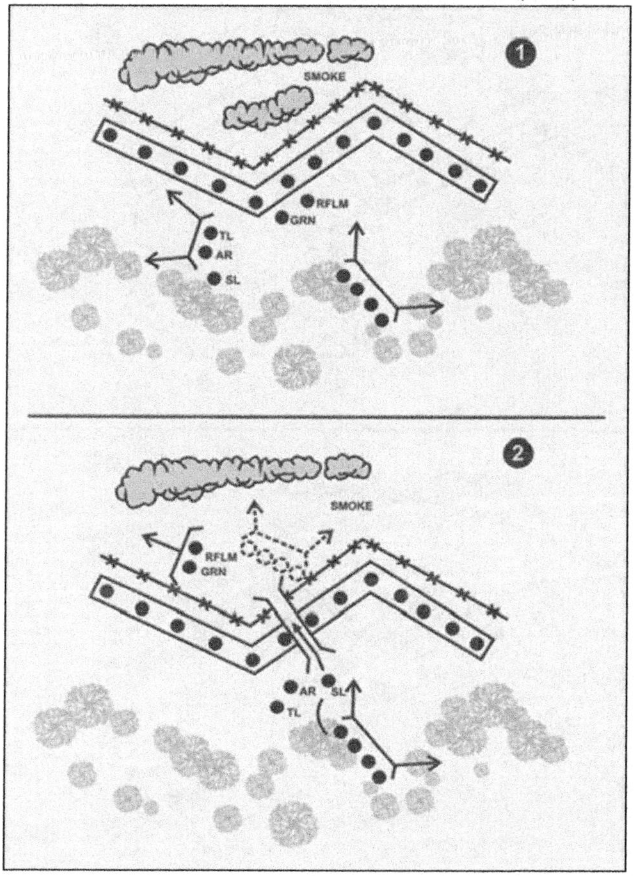

Figure 6-25. CONDUCT INITIAL BREACH OF A MINED WIRE OBSTACLE (PLATOON)

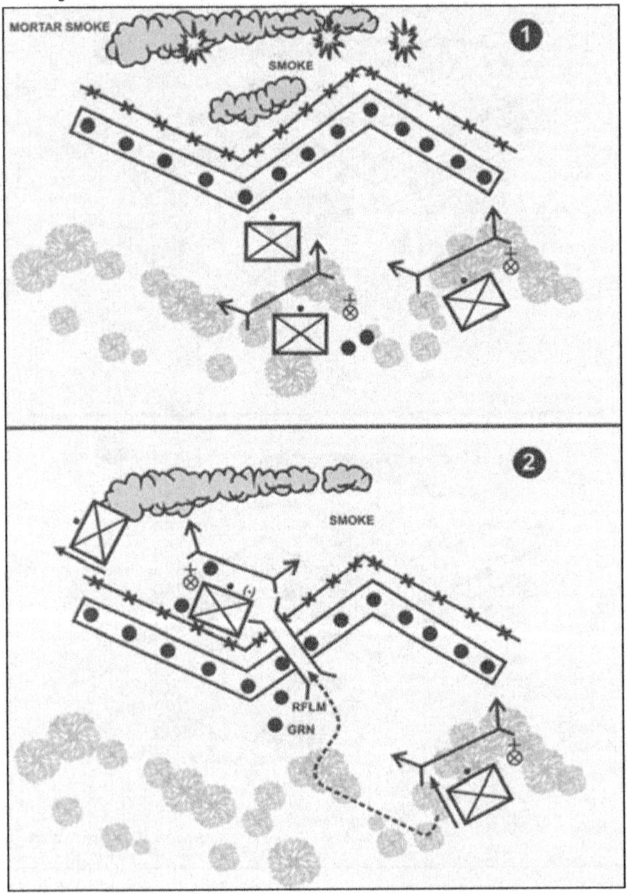

BATTLE DRILL 8
REACT TO INDIRECT FIRE

SITUATION: A platoon or squad while moving or at the halt (not dug in) receives indirect fire.

REQUIRED ACTIONS: NA

Any Soldier announces, "*Incoming.*" Then, if the platoon/squad is moving, the following takes place:

1. The platoon/squad leader gives direction and distance for the platoon/squad to move to a rally point by ordering direction and distance, for example, "*Four o'clock, three hundred meters.*"

2. Platoon/squad members move rapidly along the direction and distance to the rally point.

3. At the rally point, the leader immediately accounts for personnel and equipment, and forms the platoon/squad for a move to an alternate position.

4. The senior leader submits a SITREP to higher headquarters.

5. If the platoon/squad is halted (not dug in) or is preparing to move because they hear incoming artillery, the following takes place:

   a. The platoon/squad leader gives direction and distance for the platoon/squad to move to a rally point by ordering direction and distance, for example, "*Four o'clock, three hundred meters.*"

   b. Platoon/squad members secure all mission-essential equipment and ammunition and move rapidly along the direction and distance to the rally point.

   c. At the rally point, the leader immediately accounts for personnel and equipment, and forms the platoon/squad for a move to an alternate position.

   d. The senior leader submits a SITREP to higher headquarters.

Note: If platoon/squad members are in defensive (dug in) positions, then members will remain in those positions if appropriate. Senior leader submits SITREP.

NOTES

Chapter 7
COMMUNICATIONS

The basic requirement of combat communications is to provide rapid, reliable, and secure interchange of information. Communications are a vital aspect in successful mission accomplishment. The information in this chapter helps the Ranger squad/platoon maintain effective communications and correct any radio antenna problems.

Section I. INTRODUCTION

This section discusses military frequency modulated radios and automated net control devices (ANCDs).

7-1. **MILITARY FREQUENCY MODULATED RADIOS.** Military frequency modulated (FM) radios are composed of a receiver and a transmitter. Together, the receiver-transmitter has many capabilities and features that enable you to perform your mission more effectively. The radio can operate in Single Channel mode or frequency hopping mode. The radio has about 2,320 SC Channels and includes voice and digital communication. The operating voltage for the manpack radio is 13.5 volts from the primary battery. The range of the manpack radio is 5-10 KM on Hi power. This range is based upon line of sight and is derived from averages achieved under normal atmospheric and weather conditions. Ranges depend upon location, sighting, weather, and surrounding noise level, among other factors. the automated net control device (ANCD) is an essential tool. It replaces the signal operating instructions booklet. The ANCD has many capabilities. The main purpose of the ANCD for your mission is to obtain signal operating instructions electronically. The following discusses assembly, operation, and troubleshooting for the AN/PRC-119; provides step-by-step instructions to obtain SOI from an ANCD; and explains how to repair broken antennas:

a. **Manpack Radio Assembly**. To assemble a manpack radio you must first check and install a battery.
   (1) Inspect the battery box for dirt or damage.
   (2) Stand RT on front panel guards.
   (3) Check battery life condition (you will be using the rechargeable BB-390 batteries).
   (4) Place battery in box.
   (5) Close battery cover, and secure using latches.
   (6) Return radio to upright position.
   (7) If used battery was installed, enter the battery life condition into radio by performing the following:
      (a) Set **FCTN** to **LD**.
      (b) Press **BAT**, then **CLR**.
      (c) Enter number recorded on side of battery.
      (d) Press **STO**.
      (e) Set **FCTN** to **SQ ON**.

b. **Antenna**.
   (1) Inspect whip antenna connector on antenna and on radio for damage.
   (2) Screw whip antenna into base.
   (3) Hand tighten.
   (4) Carefully mate antenna base with RT ANT connector.
   (5) Hand tighten.
   (6) Position antenna as needed by bending goose neck.

Note: Keep antenna straight, if possible. If the antenna is bent to a horizontal position, it may be necessary to turn the radio in order to receive and transmit messages.

c. **Handset**.
   (1) Inspect the handset for damage.
   (2) Push handset on **AUD/DATA** and twist clockwise to lock in place.

d. **Field Pack**.
   (1) Place radio transmitter (RT) in field pack with antenna on the left shoulder.
   (2) Fold top flap of field over RT and secure flap to field pack using straps and buckles.

e. **Presets**.
   (1) Set **CHAN** to **1**.
   (2) Set **MODE** to **SC**.
   (3) Set **RF PWR** to **HI**.

(4) Set **VOL** to mid range.
(5) Set **DIM** full clockwise.
(6) Set **FCTN** to **LD**.
(7) Set **DATA RATE** to **OFF**.

f. **Single Channel Loading Frequencies**.
(1) Obtain Ranger SOI.
(2) Set **FCTN** to **LD**.
(3) Set **MODE** to **SC**.
(4) Set **CHAN** to **MAN, Cue**, or desired channel (**1** to **6**) where frequency is to be stored.
(5) Press **FREQ** (display will show "**00000**," or the frequency the RT is currently set to).
(6) Press **CLR** (display will show five lines).
(7) Enter the number of the new frequency.
(8) If you make a mistake with a number, press **CLR**.
(9) Press **STO** (display will blink).
(10) Set **FCTN** to **SQ ON**.

g. **Clearing of Frequencies**.
(1) Set **MODE** to **SC**.
(2) Set **CHAN** to **MAN, Cue** or the channel whose frequency is to be cleared.
(3) Press **FREQ**.
(4) Press **CLR**.
(5) Press **Load**; then press **STO**.
(6) Set **FCTN** to **SQ ON**.

h. **Scanning of Multiple Frequencies**.
(1) Load all desired frequencies using "Single Channel Loading Frequencies" instructions (see 7-2h).
(2) Set **CHAN** to **CUE**.
(3) Set **SC** to **FH**.
(4) Set **FCTN** to **SQ ON**.
(5) Press **STO** (display will say **SCAN**).
(6) Press **8**. You will now be able to scan more than one frequency.

7-2. **AUTOMATED NET CONTROL DEVICE**. The RTO retrieves all necessary COMSEC and information from the ANCD. To retrieve the SOI from an ANCD:

a. Press the **ON** button on the ANCD keypad.
b. Press the letter lock button to unlock the keys on the ANCD.
c. Press **Main Menu** key.
d. Once you press the main menu key, this is what should show on the screen.

Appl Date Time Setup
Util Bit.

e. Using the arrow keys, scroll over to **Appl** (as shown above) and press the enter button on the keypad.

f. Once you have entered into the **Appl** (applications) menu using the arrow keys scroll over to **SOI** and press the enter button on the key pad. For example--

RDS SOI Radio.
(Appl) <--*(Indicates that you are in the applications menu).*

SOI menu will look like this:
Qref Group Net Sufx Pyro
Tmpd Set C/S Find Memo

g. Using the arrow keys scroll over to **Set** and press the enter button on the keypad.
h. Press enter on choose once in the **Set** menu.

> *EXAMPLE:*
> SELECT:
> **Choose Send Receive.**

i. Once in the **Choose** menu press the enter key on set **1-5** or **6-10**.

j. After you choose your selection out of the **Set** menu, the ANCD will automatically return back to the **SOI** menu.

k. Using the keypad scroll over to **Tmpd** and enter the number out of the set you need for that day. After this operation, the ANCD will automatically return back to the **SOI** menu.

l. Using the arrow keys on the keypad scroll over to **Net** and press the enter button on the keypad. The Net menu should look like the following example:

> *EXAMPLE:*
> T05 <--(time period) 1 <--(platoon) / A <--(company) /A3Q <--(prefix).
> C 60.000.

Note: When in **Net** menu, this is where you will obtain your frequency and the first part of your call sign.

m. After you obtain the information you need out of **Net**, press the **Abort** button on the keypad. This will return you to the **SOI** menu.

n. Using the keypad, scroll over to **Sufx** and press the **Enter** button.

o. Once you are in the **Sufx** menu, use the up and down arrow keys on the ANCD to locate your two-digit designator. These two digits will go at the end of your prefix; your prefix plus the two digits equal your call sign.

p. Once you are done, press the **OFF** button on the ANCD to end operation.

Section II. ANTENNAS

This section discusses repair techniques, construction and adjustment, field-expedient antennas, antenna length and orientation, and improvement of marginal communications.

7-3. **REPAIR**. Antennas are sometimes broken or damaged, causing either a communications failure or poor communications. If a spare antenna is available, the damaged antenna is replaced. When there is no spare, the Ranger squad/platoon may have to construct an emergency antenna. The following paragraphs contain suggestions for repairing antennas and antenna supports and the construction and adjustment of emergency antennas.

> **DANGER**
>
> **SERIOUS INJURY OR DEATH CAN RESULT FROM CONTACT WITH THE RADIATING ANTENNA OF A MEDIUM-POWER OR HIGH-POWER TRANSMITTER. TURN THE TRANSMITTER OFF WHILE ADJUSTING THE ANTENNA.**

a. **Whip Antennas**. When a whip antenna is broken into two sections, the part of the antenna that is broken off can be connected to the part attached to the base by joining the sections.

(1) Use the method shown in A, Figure 7-1 when both parts of the broken whip are available and usable.

(2) Use the method in B, Figure 7-1, when the part of the whip that was broken off is lost, or when the whip is so badly damaged that it cannot be used.

(3) To restore the antenna to its original length, a piece of wire is added that is nearly the same length as the missing part of the whip. The pole support is then lashed securely to both sections of the antenna. The two antenna sections are cleaned thoroughly to ensure good contact before connecting them to the pole support. If possible, the connections are soldered.

b. **Wire Antennas**. Emergency repair of a wire antenna may involve the repair or replacement of the wire used as the antenna or transmission line; or the repair or replacement of the assembly used to support the antenna.

(1) When one or more wires of an antenna are broken, the antenna can be repaired by reconnecting the broken wires. To do this, lower the antenna to the ground, clean the ends of the wires, and twist the wires together. Whenever possible, solder the connection.

(2) If the antenna is damaged beyond repair, construct a new one. Make sure that the lengths of the wires on the substitute antenna match the length of the original.

(3) Antenna supports may also require repair or replacement. A substitute item may be used in place of a damaged support and, if properly insulated, can be of any material of adequate strength. If the radiating element is not properly insulated, field antennas may be shorted to ground and be ineffective. Many commonly found items can be used as field-expedient insulators.

(4) The best of these items are plastic or glass to include plastic spoons, buttons, bottle necks, and plastic bags. Though less effective than plastic or glass but still better than no insulator at all are wood and rope. The radiating element--the actual antenna wire-should touch only the antenna terminal and should be physically separated from all other objects, other than the supporting insulator. (Figure 7-2 shows how to make emergency insulators.)

Figure 7-1. USE OF BROKEN ANTENNA

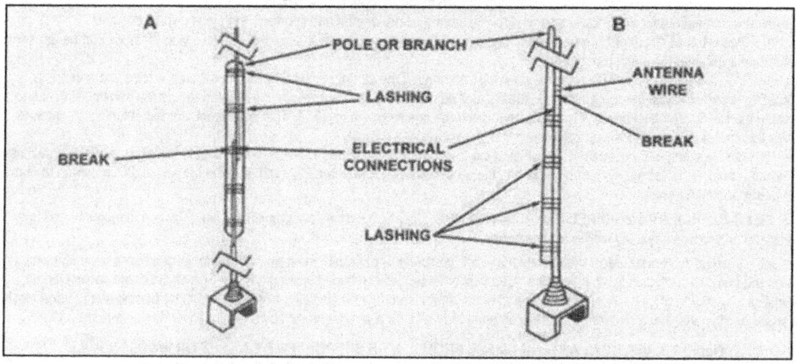

Figure 7-2. METHODS OF CONSTRUCTING EMERGENCY INSULATORS

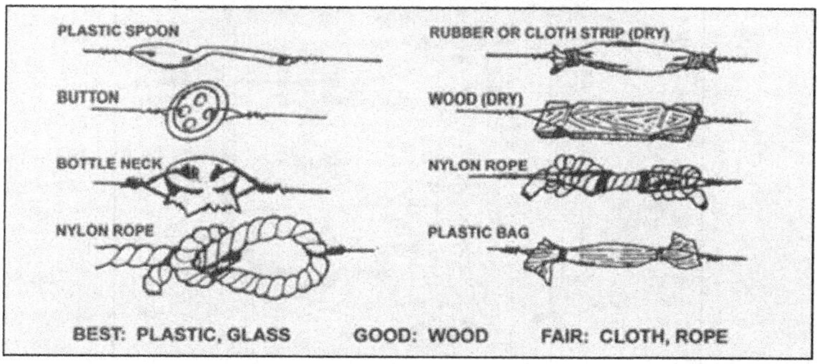

7-4. **CONSTRUCTION AND ADJUSTMENT**. Ranger squad/platoons may use the following methods to construct and adjust antennas:

a. **Construction**. The best kinds of wire for antennas are copper and aluminum. In an emergency, however, Rangers use any type of wire that is available.

(1) The exact length of most antennas is critical. The emergency antenna should be the same length as the antenna it replaces.

(2) Antennas supported by trees can usually survive heavy wind storms if the trunk of a tree or a strong branch is used as a support. To keep the antenna taut and to prevent it from breaking or stretching as the trees sway, the Ranger attaches a spring or old inner tube to one end of the antenna. Another technique is to pass a rope through a pulley or eyehook. The rope is attached to the end of the antenna and loaded with a heavyweight to keep the antenna tightly drawn.

(3) Guidelines used to hold antenna supports are made of rope or wire. To ensure the guidelines will not affect the operation of the antenna, the Ranger cuts the wire into several short lengths and connects the pieces with insulators.

b. **Adjustment**. An improvised antenna may change the performance of a radio set. The following methods can be used to determine if the antenna is operating properly:

(1) A distant station may be used to test the antenna. If the signal received from this station is strong, the antenna is operating satisfactorily. If the signal is weak, the Ranger adjusts the height and length of the antenna and the transmission line to receive the strongest signal at a given setting on the volume control of the receiver. This is the best method of tuning an antenna when transmission is dangerous or forbidden.

(2) In some radio sets, the Ranger uses the transmitter to adjust the antenna. First, he sets the controls of the transmitter to normal; then, he tunes the system by adjusting the antenna height, the antenna length, and the transmission line length to obtain the best transmission output.

7-5. **FIELD-EXPEDIENT OMNIDIRECTIONAL ANTENNAS**. Vertical antennas are omnidirectional. The omnidirectional antenna transmits and receives equally well in all directions.

a. **Vertical Antennas**. Most tactical antennas are vertical. For example, the man-pack portable radio uses a vertical whip and so do the vehicular radios in tactical vehicles. A vertical antenna can be made by using a metal pipe or rod of the correct length, held erect by means of guidelines. The lower end of the antenna should be insulated from the ground by placing it on a large block of wood or other insulating material. A vertical antenna may also be a wire supported by a tree or a wooden pole (Figure 7-3).

Figure 7-3. VERTICAL ANTENNA MADE FROM A WIRE SUPPORTED BY A TREE OR WOODEN POLE

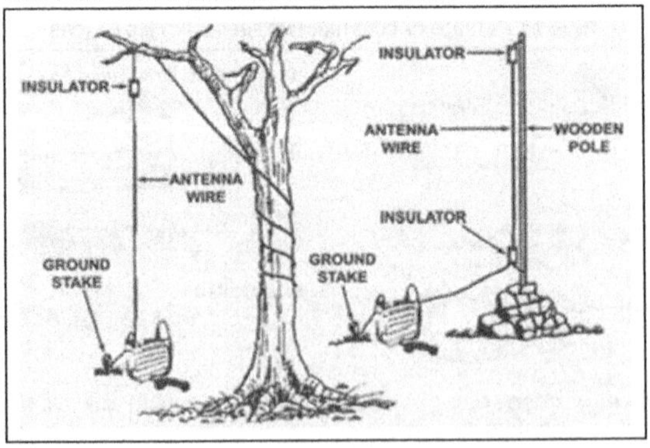

b. **Modification of Connection on Shorter Vertical Antennas**. For short vertical antennas, the pole may be used without guidelines (if properly supported at the base). If the length of the vertical mast is not long enough to support the wire upright, it may be necessary to modify the connection at the top of the antenna (Figure 7-4; see FM 24-18).

Figure 7-4. MODIFIED CONNECTION AT TOP OF ANTENNA

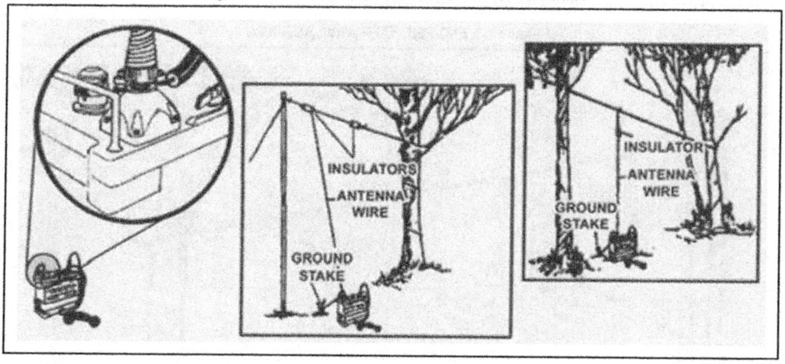

c. **End-Fed, Half-Wave Antenna**. An emergency, end-fed half-wave antenna (Figure 7-5) can be constructed from available materials such as field wire, rope, and wooden insulators. The electrical length of this antenna is measured from the antenna terminal on the radio set to the far end of the antenna. The best performance can be obtained by constructing the antenna longer than necessary and then shortening it, as required, until the best results are obtained. The ground terminal of the radio set should be connected to a good earth ground for this antenna to function efficiently.

Figure 7-5. END-FED, HALF-WAVE ANTENNA

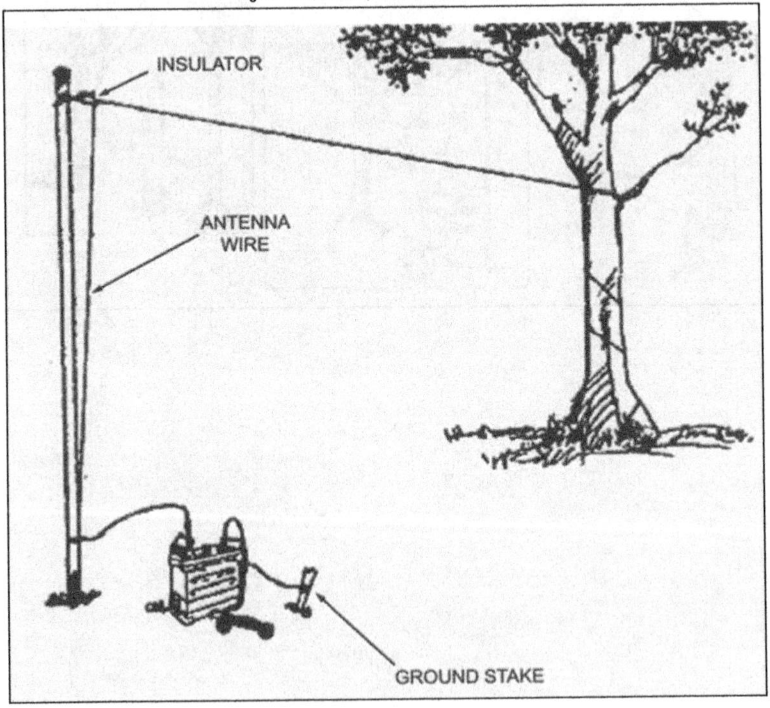

d. **Center-Fed, Doublet Antenna**. The center-fed doublet is a half-wave antenna consisting of two quarter wavelength sections on each side of the center (Figure 7-6). Doublet antennas are directional broadside to their length, which makes the vertical doublet antenna omnidirectional. This is because the radiation pattern is doughnut-shaped and bidirectional.

Figure 7-6. CENTER-FED, DOUBLET ANTENNA

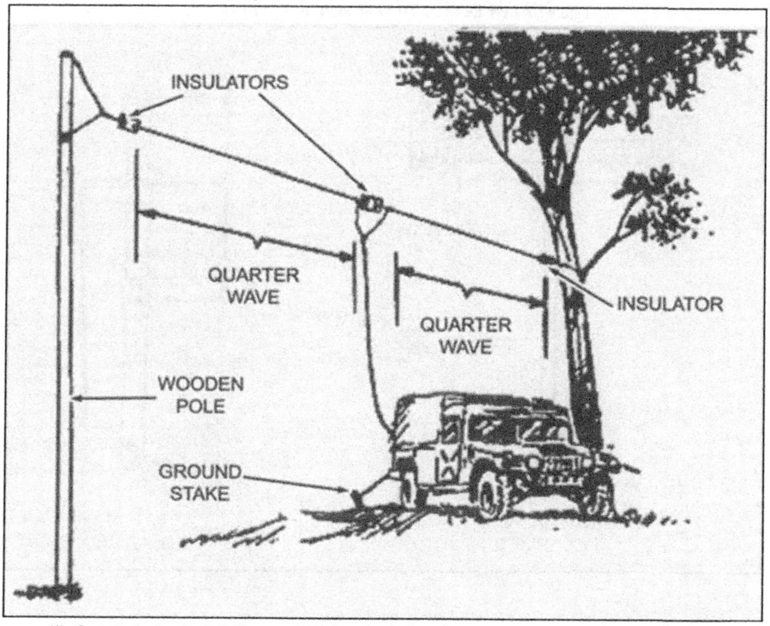

(1) Compute the length of a half-wave antenna by using the formula previously discussed. Cut the wires as close as possible to the correct length; this is very important.

(2) Uses transmission line for conducting electrical energy from one point to another and for transferring the output of a transmitter to an antenna. Although it is possible to connect an antenna directly to a transmitter, the antenna is usually located some distance away.

(3) Support center-fed half-wave FM antennas entirely with pieces of wood. (A horizontal antenna of this type is shown in A, Figure 7-7 and a vertical antenna in B, Figure 7-7.) Rotate these antennas to any position to obtain the best performance.

(a) If the antenna is erected vertically, bring out the transmission line horizontally from the antenna for a distance equal to at least one-half of the antenna's length before it is dropped down to the radio set.

Figure 7-7. CENTER-FED ANTENNAS SUPPORTED WITH WOOD

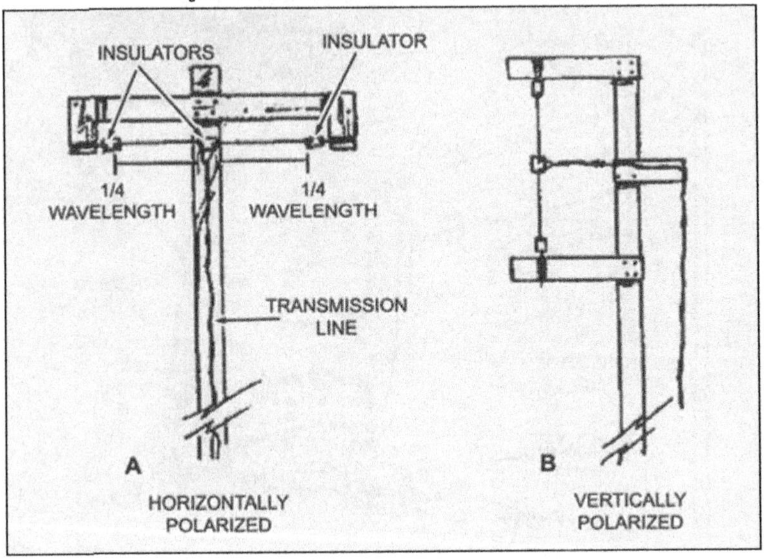

(b) The half-wave antenna is used with FM radios (Figure 7-8). It is effective in heavily wooded areas to increase the range of portable radios. Connect the top guidelines to a limb or pass it over the limb and connect it to the tree trunk or a stake.

Figure 7-8. HALF-WAVE ANTENNA USED WITH FM RADIO

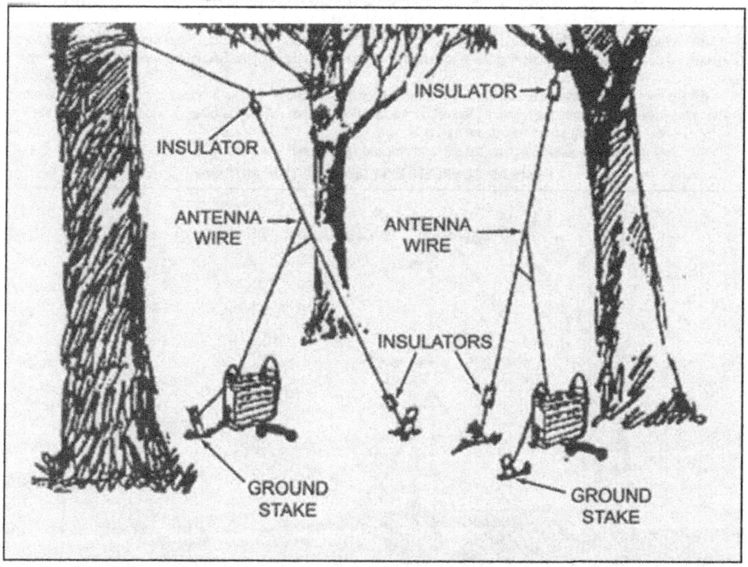

c. **Expedient 292-Type Antenna**. The expedient 292-type antenna was developed for use in the jungle and, if used properly, can increase the team's ability to communicate. In its entirety, the antenna is bulky, heavy, and not acceptable for sniper team operations. The team can, however, carry only the mast head and antenna sections, mounting these on wood poles or hanging them from trees; or, the team can make a complete expedient 292-type antenna (Figure 7-9), using WD-1, wire, and other readily available material. The team can also use almost any plastic, glass, or rubber objects for insulators. Dry wood is acceptable when nothing else is available. (See Figure 7-2 for types of insulators that may be used.) The following describes how to make this antenna:

(1) Use the quick-reference table (Table 7-2) to determine the length of the elements (one radiating and three ground planes) for the frequency that will be used. Cut these elements (A, Figure 7-9) from WD-1 field wire (or similar wire). Cut spacing sticks (B, Figure 7-9) the same length. Place the ends of the sticks together to form a triangle and tie the ends with wire, tape, or rope. Attach an insulator to each corner. Attach a ground-plane wire to each insulator. Bring the other ends of the ground-plane wires together, attach them to an insulator (C, Figure 7-9), and tie securely. Strip about 3 inches of insulation from each wire and twist them together (D, Figure 7-9).

(2) Tie one end of the radiating element wire to the other side of insulator C and the other end to another insulator (D, Figure 7-9). Strip about 3 inches of insulation from the radiating element at insulator C.

(3) Cut enough WD-1 field wire to reach from the proposed location of the antenna to the radio set. Keep this line as short as possible, because excess length reduces the efficiency of the system. Tie a knot at each end to identify it as the *hot* lead. Remove insulation from the *hot* wire and tie it to the radiating element wire at insulator C. Remove insulation from the other wire and attach it to the bare ground-plane element wires at insulator C. Tape all connections and do not allow the radiating element wire to touch the ground-plane wires.

(4) Attach a rope to the insulator on the free end of the radiating element and toss the rope over the branches of a tree. Pull the antenna as high as possible, keeping the lead-in routed down through the triangle. Secure the rope to hold the antenna in place.

(5) At the radio set, remove about 1 inch of insulation from the *hot* lead and about 3 inches of insulation from the other wire. Attach the *hot* line to the antenna terminal (doublet connector, if so labeled). Attach the other wire to the metal case-the handle, for example. Be sure both connections are tight or secure.

(6) Set up correct frequency, turn on the set, and proceed with communications.

Figure 7-9. COMPLETE EXPEDIENT 292-TYPE ANTENNA

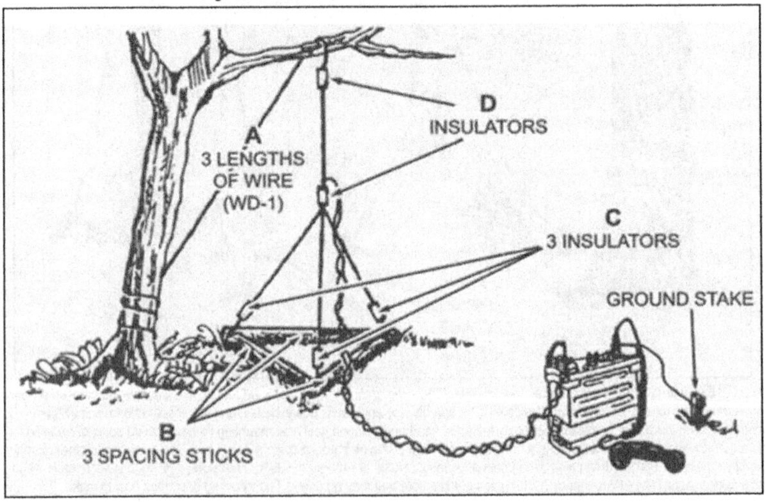

Table 7-2. QUICK REFERENCE TABLE

OPERATING FREQUENCY IN MHz	ELEMENT LENGTH (RADIATING ELEMENTS AND GROUND-PLANE ELEMENTS)	
30	2.38 m	(7 ft 10 in)
32	2.23 m	(7 ft 4 in)
34	2.1 m	(6 ft 11 in)
36	1.98 m	(6 ft 6 in)
38	1.87 m	(6 ft 2 in)
40	1.78 m	(5 ft 10 in)
43	1.66 m	(5 ft 5 in)
46	1.55 m	(5 ft 1 in)
49	1.46 m	(4 ft 9 in)
52	1.37 m	(4 ft 6 in)
55	1.3 m	(4 ft 3 in)
58	1.23 m	(4 ft 0 in)
61	1.17 m	(3 ft 10 in)
64	1.12 m	(3 ft 8 in)
68	1.05 m	(3 ft 5 in)
72	0.99 m	(3 ft 3 in)
76	0.94 m	(3 ft 1 in)

7-6. **FIELD-EXPEDIENT DIRECTIONAL ANTENNAS**. The vertical half-rhombic antenna (Figure 7-10) and the long-wire antenna (Figure 7-11) are two field-expedient directional antennas. These antennas consist of a single wire, preferably two or more wavelengths long, supported on poles at a height of 3 to 7 meters (10 to 20 feet) above the ground. The antennas will, however, operate satisfactorily as low as 1 meter (about 3 feet) above the ground–the radiation pattern is directional. The antennas are used mainly for either transmitting or receiving high-frequency signals.

Figure 7-10. VERTICAL HALF-RHOMBIC ANTENNA

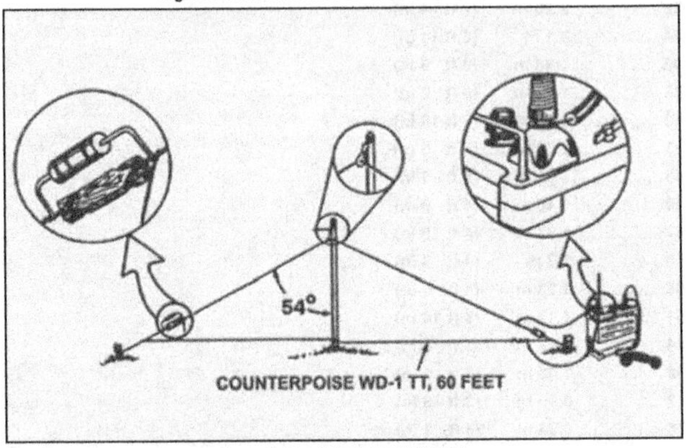

Figure 7-11. LONG-WIRE ANTENNA

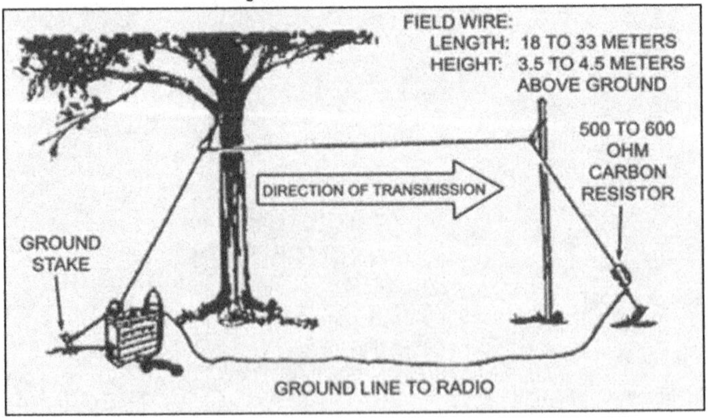

a. The vee antenna (Figure 7-12) is another field-expedient directional antenna. It consists of two wires forming a V with the open area of the "V" pointing toward the desired direction of transmission or reception. To make construction easier, the legs should slope downward from the apex of the "V." This is called a sloping vee antenna (Figure 7-13). The angle between the legs varies with the length of the legs to achieve maximum performance. (Use Table 7-1 to determine the angle and the length of the legs.)

Figure 7-12. VEE ANTENNA

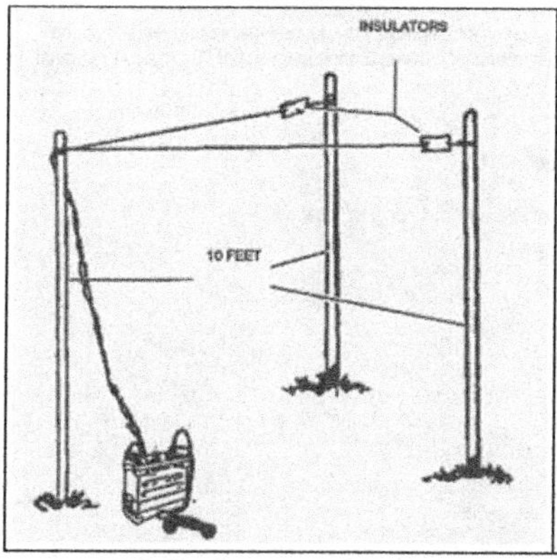

Figure 7-13. SLOPING VEE ANTENNA

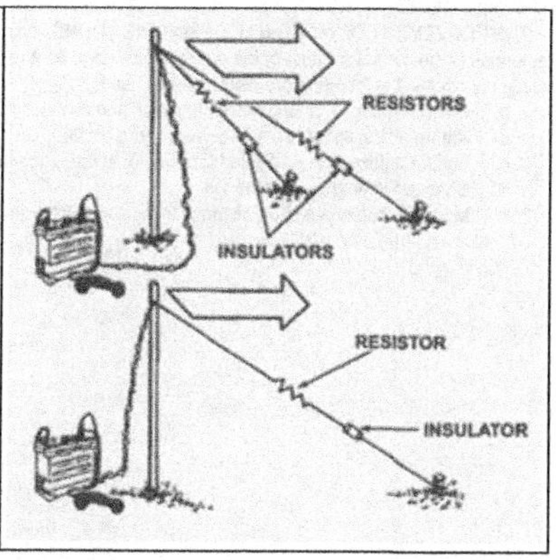

b. When the antenna is used with more than one frequency or wavelength, use an apex angle that is midway between the extreme angles determined by the chart. To make the antenna radiate in only one direction, add noninductive terminating resistors from the end of each leg (not at the apex) to ground. (See TM 11-666.)

7-7. **ANTENNA LENGTH**. The length of an antenna must be considered in two ways: both a physical and an electrical length. These two lengths are never the same. The reduced velocity of the wave on the antenna and a capacitive effect (known as end *effect*) make the antenna seem longer electrically than it is physically. The contributing factors are the ratio of the diameter of the antenna to its length and the capacitive effect of terminal equipment, such as insulators and clamps, used to support the antenna.

a. To calculate the physical length of an antenna, use a correction of 0.95 for frequencies between 3.0 and 50.0 MHz The figures given below are for a half-wave antenna.

b. Use the following formula to calculate the length of a long-wire antenna (one wavelength or longer) for harmonic operation:

EXAMPLE:

If N = the number of half-wavelengths in the total length of the antenna = 3,
and if the frequency in MHz = 7, then--

$$\text{Length (meters)} = \frac{150(N-0.05)}{\text{Frequency in MHz}} = \frac{150(3-.05)}{7} =$$

$$\frac{150 \times 2.95}{7} = \frac{442.50}{7} = 63.2 \text{ meters}$$

7-8. **ANTENNA ORIENTATION**. If the azimuth of the radio path is not provided, the azimuth should be determined by the best available means. The accuracy required in determining the azimuth of the path depends on the radiation pattern of the directional antenna. In transportable operation, the rhombic and V antennas may have such a narrow beam as to require great accuracy in azimuth determination. The antenna should be erected for the correct azimuth. Great accuracy is not required in erecting broad-beam antennas. Unless a line of known azimuth is available at the site, the direction of the path is best determined by a magnetic compass.

7-9. **IMPROVEMENT OF MARGINAL COMMUNICATIONS**. Under certain situations, it may not be feasible to orient directional antennas to the correct azimuth of the desired radio path. As a result, marginal communications may suffer. To improve marginal communications, the following procedure can be used:

a. Check, tighten, and tape cable couplings and connections.
b. Return all transmitters and receivers in the circuit.
c. Ensure antennas are adjusted for the proper operating frequency.
d. Change the heights of antennas.
e. Move the antenna a short distance away and in different locations from its original location.

NOTES

Chapter 8
ARMY AVIATION

Army aviation and infantry units can be fully integrated with other members of the combined arms team to form powerful and flexible air assault task forces that can project combat power throughout the entire depth, width, and breadth of the modern battlefield with little regard for terrain barriers. These combat operations are deliberate, precisely planned, and vigorously executed. They are assigned to strike the enemy when and where he is most vulnerable (See page 2-28, *Air Movement Annex* and page 2-37, *Army Aviation Coordination Checklist*).

8-1. **REVERSE PLANNING SEQUENCE**. Successful air assault execution is based on a careful analysis of METT-TC and detailed, precise reverse planning. Five basic plans that comprise the reverse planning sequence are developed for each air assault operation. The battalion is the lowest level that has sufficient personnel to plan, coordinate, and control air assault operations. When company size or lower operations are conducted, the bulk of the planning takes place at battalion or higher headquarters. They are:

a. **Ground Tactical Plan**. The foundation of a successful air assault operation is the commander's ground tactical plan. All additional plans must support this plan. The plan specifies actions in the objective area to ultimately accomplish the mission and address subsequent operations.

b. **Landing Plan**. The landing plan must support the ground tactical plan. This plan outlines a sequence of events that allows elements to move into the area of operations, and ensures that units arrive at designated locations at prescribed times prepared to execute the ground tactical plan.

c. **Air Movement Plan**. The air movement plan is based on the ground tactical and landing plans. It specifies the schedule and provides instructions for air movement of troops, equipment, and supplies from PZs to LZs.

d. **Loading Plan**. The loading plan is based on the air movement plan. It ensures that troops, equipment, and supplies are loaded on the correct aircraft. Unit integrity is maintained when aircraft loads are planned. Cross-loading may be necessary to ensure survivability of command and control assets, and that the mix of weapons arriving at the LZ is ready to fight.

e. **Staging Plan**. The staging plan is based on the loading plan and prescribes the arrival time of ground units (troops, equipment and supplies) at the PZ in the order of movement

8-2. **SELECTION AND MARKING OF PICKUP AND LANDING ZONES**

a. Small unit leaders should consider the following when selecting a PZ/LZ:

(1) ***Size***. Minimal circular landing point separation from other aircraft and obstacles is needed:

- OH-58D – 25 meters.
- UH-1, AH-1 – 35 meters.
- UH-60, AH-64 – 50 meters.
- Cargo helicopters – 80 meters.

(2) ***Surface Conditions***. Avoid potential hazards such as sand, blowing dust, snow, tree stumps, or large rocks.

(3) ***Ground Slope***.

- 0% - 6 % -- land upslope.
- 7% - 15% -- land sideslope.
- over 15% -- no touchdown (aircraft may hover).

(4) ***Obstacles***. An obstacle clearance ratio of 10 to 1 is used in planning approach and departure of the PZ and LZ, for example, a 10-foot tall tree requires 100 feet of horizontal distance for approach or departure. Obstacles will be marked with a red chemlight at night, or red panels in daytime. Avoid using markings if the enemy would see them.

(5) ***Approach/Departure***. Approach and departure are made into the wind and along the long axis of the PZ/LZ.

(6) ***Loads***. The greater the load, the larger the PZ/LZ must be to accommodate the insertion or extraction.

b. **Marking of PZs and LZs**.

(1) ***Day***. A ground guide will mark the PZ or LZ for the lead aircraft by holding an M16/M4 rifle over his head, by displaying a folded VS-17 panel chest high, or by other coordinated and identifiable means.

(2) ***Night***. The code letter "Y" (inverted "Y") is used to mark the landing point of the lead aircraft at night (Figure 8-1). Chemical lights or "beanbag" lights are used to maintain light discipline. A swinging chemlight may also be used to mark the landing point.

Figure 8-1. INVERTED "Y"

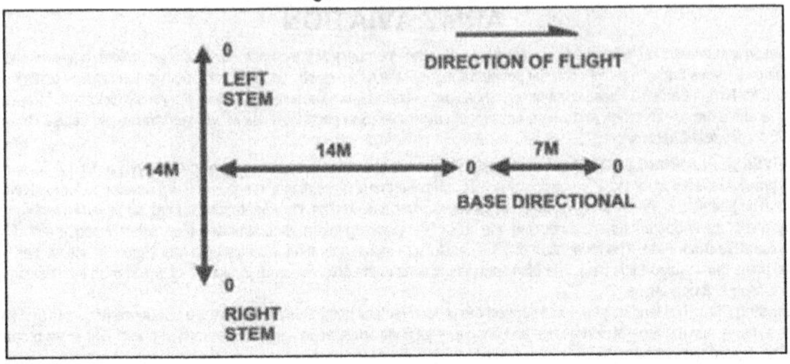

8-3. **AIR ASSAULT FORMATIONS**. Aircraft supporting an operation may use any of the following PZ/LZ configurations which are prescribed by the air assault task force (AATF) commander working with the air mission commander (AMC):

a. ***Heavy Left or Right***. Requires a relatively long, wide landing area; presents difficulty in pre-positioning loads; restricts suppressive fire by inboard gunners; provides firepower to front and flank (Figure 8-2).

Figure 8-2. HEAVY LEFT/HEAVY RIGHT

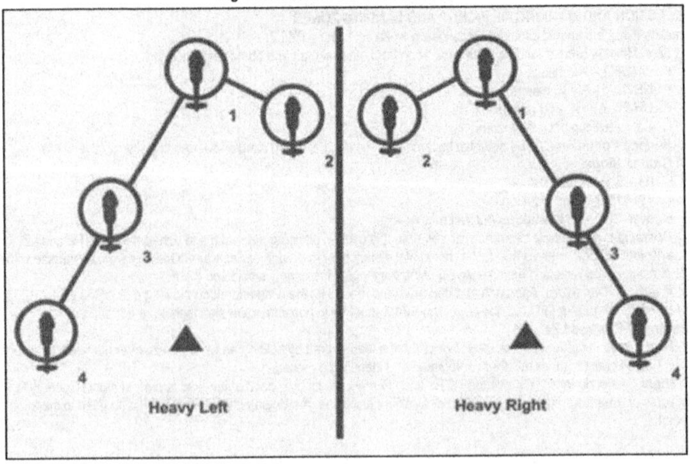

b. **Diamond**. Allows rapid deployment for all-round security; requires small landing area; presents some difficulty in pre-positioning loads; restricts suppressive fire of inboard gunners (Figure 8-3).

Figure 8-3. DIAMOND

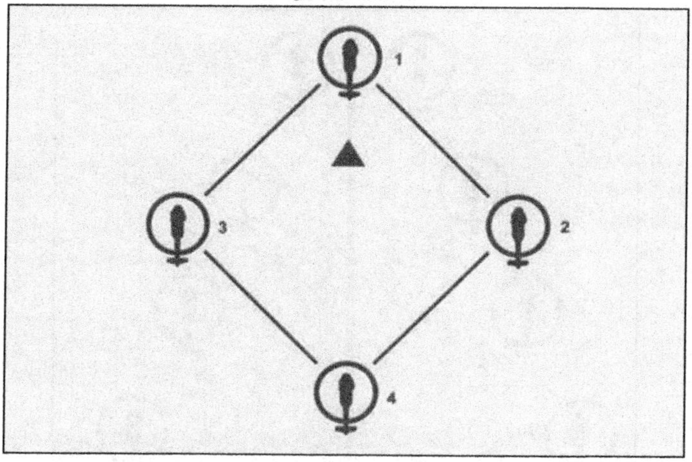

c. **Vee**. Requires a relatively small landing area; allows rapid deployment of forces to the front; restricts suppressive fire of inboard gunners; presents some difficulty in pre-positioning loads (Figure 8-4).

Figure 8-4. VEE

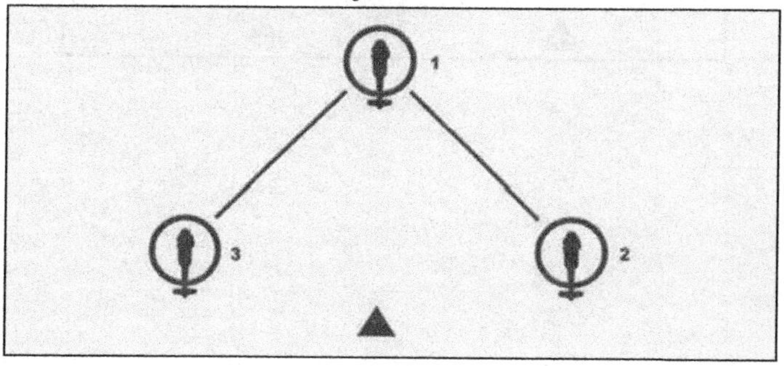

d. **Echelon Left or Right**. Requires a relatively long, wide landing area; presents some difficulty in pre-positioning loads; allows rapid deployment of forces to the flank; allows unrestricted suppressive fire by gunners (Figure 8-5).

Figure 8-5. ECHELON LEFT/ECHELON RIGHT

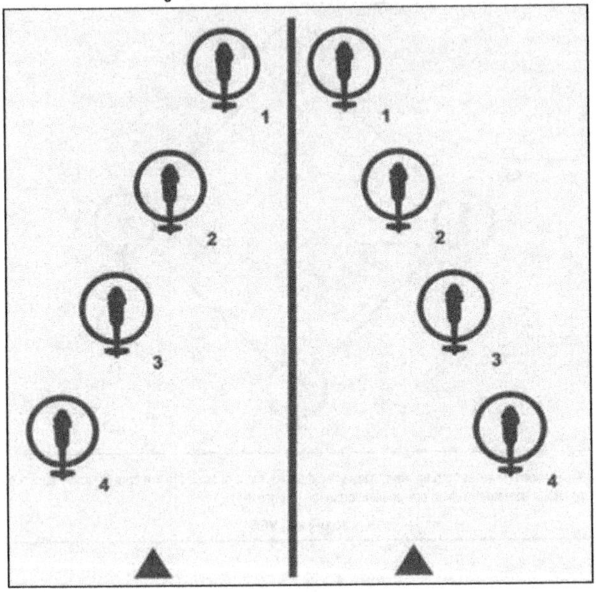

e. **Trail**. Requires a relatively small landing area; allows rapid deployment of forces to the flank; simplifies pre-positioning loads; allows unrestricted suppressive fire by gunners (Figure 8-6).

Figure 8-6. TRAIL

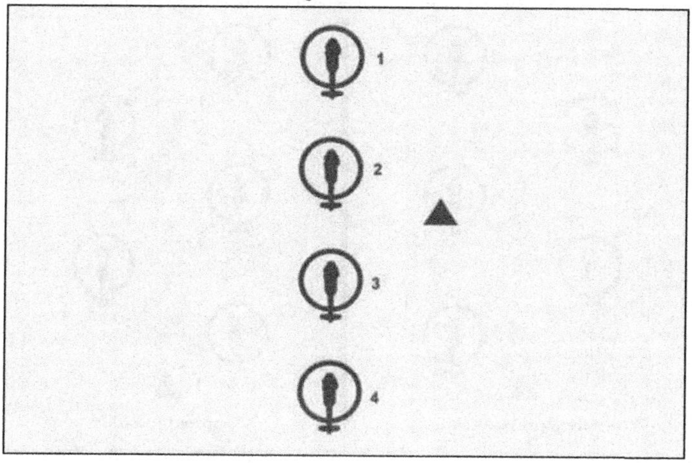

f. **Staggered Trail Left or Right**. Requires a relatively long, wide landing area; simplifies pre-positioning loads; allows rapid deployment for all round security; gunners' suppressive fire restricted somewhat (Figure 8-7).

Figure 8-7. STAGGERED TRAIL LEFT/STAGGERED TRAIL RIGHT

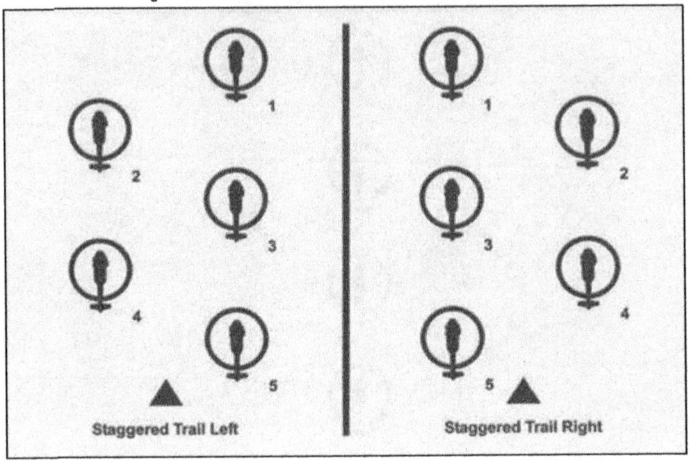

8-4. **PICKUP ZONE OPERATIONS.** Prior to arrival of aircraft, the PZ is secured, PZ control party is positioned, and the troops and equipment are positioned in platoon/squad assembly areas. In occupying a patrol/squad assembly area, the patrol/squad leader does the following. Figure 8-8 shows an example of a large, one-sided PZ. Figures 8-8 through 8-12 show loading/unloading procedures and techniques:

- Maintain all-around security of the assembly area.
- Maintain communications.
- Organize personnel and equipment into chalks and loads.
- Conduct safety briefing and equipment check of troops.

Figure 8-8. LARGE, ONE SIDED PZ

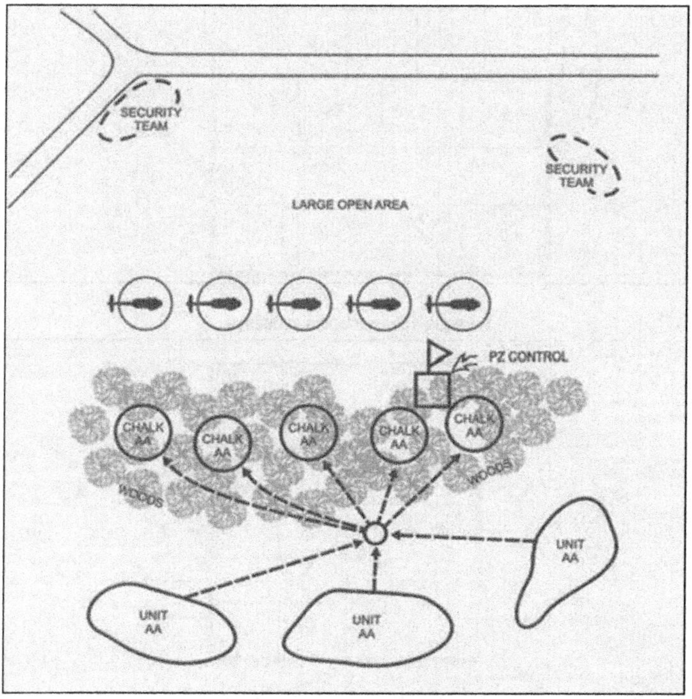

Figure 8-9. UH-60 LOADING SEQUENCE

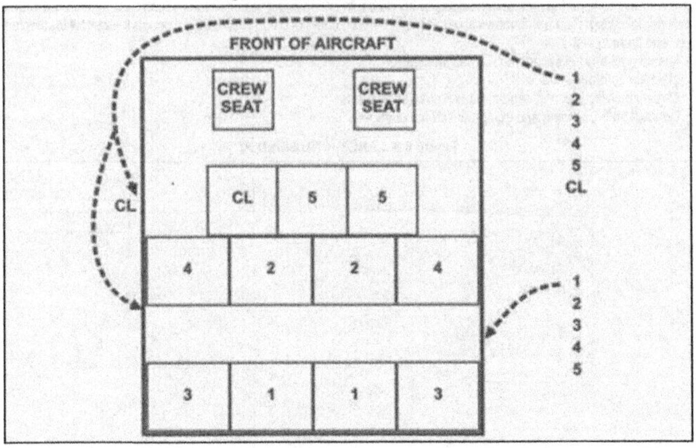

Figure 8-10. UNLOADING SEQUENCE

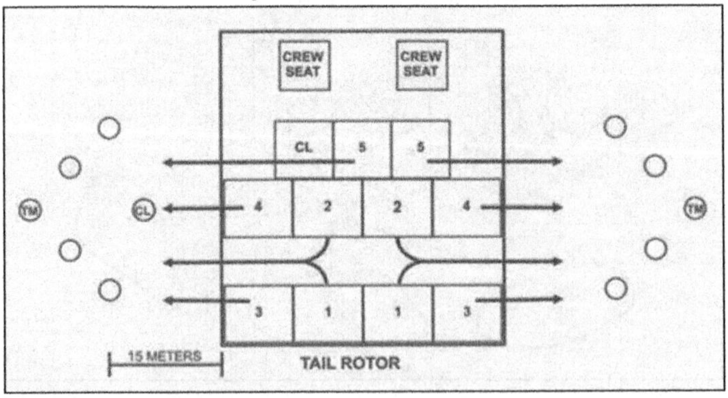

Figure 8-11. TACTICAL LOADING SEQUENCE

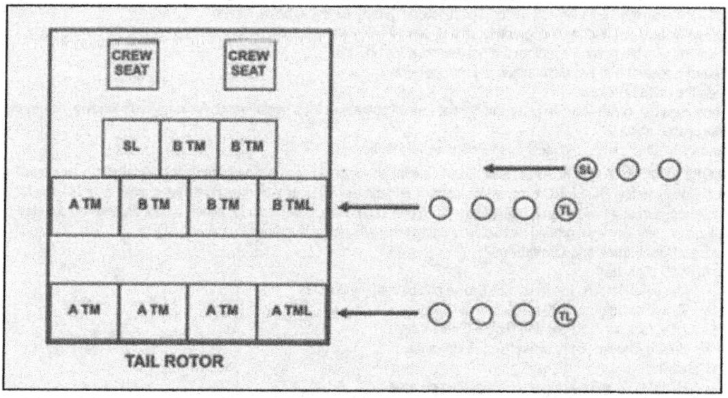

Figure 8-12. TACTICAL UNLOADING USING DOOR NEAREST COVER, CONCEALMENT

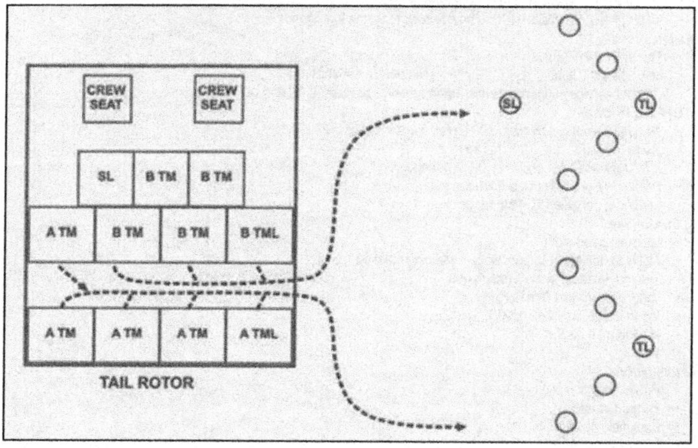

8-5. **SAFETY**. Safety is the primary concern of all leaders when operating in/around aircraft. The inclusion of aircraft into Ranger operations carries an inherent high-risk factor. The following guidelines are to be considered.

a. Approach the aircraft from 90 degrees to 45 degrees off the nose.

b. Weapons with blank firing adapters are pointed muzzles up.

c. Weapons loaded with live ammunition are muzzles down.

d. Wear the ballistic helmet.

e. When possible, conduct an air crew safety brief with all personnel. At a minimum, cover loading/off-loading, emergencies, and egress procedures.

f. Leaders need to carry a manifest and turn a copy into higher.

8-6. **CLOSE COMBAT ATTACK AVIATION**. Close combat attack is defined as a hasty or deliberate attack in support of units engaged in close combat. During CCA, armed helicopters engage enemy units with direct fire that impacts near friendly forces. Targets may range from a few hundred meters to a few thousand meters. Close combat attack is coordinated and directed by a team, platoon or company level ground unit soldiers using standardized CCA procedures in unit SOP's.

a. **Aircraft Capabilities and Limitations**.

(1) AH-64D (Apache):

- Air speed (knots): 164 max, 120 Cruise combat radius: 200 KM
- 30 mm chain gun: 1200 rds, 3,500 m range
- 2.75" rockets: 19 per pod (76), 3-5 km range
- Hellfire Missile: 8 per side (16), 5-8 km range

(2) Capabilities.

- MOBILITY - AO can cover corps or division area
- SPEED - 100-120 knots day/80-100 Knots night
- VERSATILITY - scout-weapons teams vs. pure attack
- LETHALITY - Attack bn can engage 288 targets
- VIDEO RECONNAISSANCE - Provide near real time intelligence

(3) ***Limitations***.

- Threat ID with FLIR
- Low ceilings (Clouds) <500' AGL degrades hellfire capability
- Combat service support consume large amounts of class III, V, and IX

(4) **OH-58D (Kiowa)**.

- Air speed (knots): 125 max, 100 cruise combat radius: 120 KM
- .50 cal MG: 500 rds, 2,000 m range
- 2.75" rockets: 7 per pod (14), 3-5 km range
- Hellfire: 2 per side (4), 5-8 km range
- Stinger: 2 per side (4), 4 km range

(5) ***Capabilities***.

- Mast-mounted sight
- Ability to designate targets while remaining masked
- Thermal imaging system (day/night)
- Laser designator / Aim1 laser
- Video image crosslink (VIXL)
- Moving map display
- Video

(6) ***Limitations***.

- Power limited
- Infrared crossover
- Battlefield obscurants
- Low ceilings - Hellfire
- Remote designation constraints
- Instrument meteorological conditions

c. **CCA Call-for-Fire Format**.

(1) IP/BP/ABF or friendly location__________

- Grid/Lat Long

(2) Known point

- Terrain feature

(3) HDG to TGT _________ (magnetic)

- Specify from IP/BP/ABF or friendly location

(4) DST to TGT __________ (m) (Specify from IP/BP/ABF or friendly location.)

(5) TGT elevation_________ (feet, mean sea level)

(6) TGT Description ________________

(7) TGT location ____________

- Grid
- Known point
- Terrain Feature

(8) Type TGT Mark __________ (day/night)

(9) location of friendly __________

- Omit if previously given
- Grid/Lat Long
- Known point
- Terrain feature

(10) Egress direction___________

- Cardinal direction not to over fly threats

d. **Marking Method Chart**. Table 8-1 shows this chart.

Table 8-1. MARKING METHOD CHART

Method	D	N	NVG	NVS	Friendly	Target
Smoke	X				X	X
Smoke IR	X		X		X	X
Signal Mirror	X				X	
IR Laser			X		X	X
Tracer	X	X	X			X
Glint Tape				X	X	
CID	X				X	
Strobe		X	X		X	
IR Strobe			X		X	
IR Panel				X	X	
Chem Light		X	X		X	
AN/PAQ-4			X			X
VS-17 Panel	X				X	
Spot Light		X	X		X	
MRE Heater				X	X	
AN/PEQ-2			X			X
Hydra 70 Illum	X	X	X	X		X
Ref FM 3-04.111 Appendix Q for more detail						

e. **Risk-Estimate Distances**. REDs for CCA munitions are shown in Table 8-2.

Table 8-2. RISK ESTIMATE DISTANCES FOR CCA MUNITIONS

WEAPON	*DANGER CLOSE*	*MAX EFF RANGE)*
50 cal	150 meters	2,000 meters
30 mm	150 meters	3,500 meters
2.75 rocket	175 meters	8,000 meters
Hellfire Missile	500 meters	8,000 meters
Danger Close = Risk-Estimate Distances for 0.1% (1/10th of a percent) probability of incapacitation (PI) per FM 3-09.32.		

Chapter 9
WATERBORNE OPERATIONS

This chapter discusses rope bridges, poncho rafts, and watercraft.

9-1. **ROPE BRIDGE TEAM.** The Ranger patrol seldom has ready-made bridges, so they must know how to employ field-expedient stream crossing techniques.

a. **Personnel.**

(1) # 1 man................Lead safety swimmer and far side lifeguard.
(2) # 2 man................Rope puller, swims water obstacle pulling 150-foot rope, ties off rope on far side anchor point.
(3) # 3 man................Near side lifeguard is the last man to cross water obstacle.
(4) # 4 man................Bridge team commander (BTC; most knowledgeable person on team).
(5) # 5/6 men............Rope tighteners.

b. **One-Rope Bridge (Wet Crossing).** Special equipment follows:

- Two snaplinks per piece of heavy equipment.
- Two snaplinks for every 120 feet of rope.
- One 14-foot utility rope per person (Swimmer safety line).
- Two snaplinks per person.
- One waterproof bag per RTO.
- Two B-7 life preservers.
- Three flotation work vests.
- Two 150-foot nylon ropes.

c. **Planning.** A stream crossing annex is prepared in conjunction with the unit's operation order. Special organization is accomplished at this time. For a platoon size patrol, a squad is normally given the task of providing the bridge team. The squad leader designates the most technically proficient soldier in the squad as the bridge team commander (BTC).

d. **Rehearsals and Inspections.**

(1) Rehearse the entire stream crossing, emphasizing--

- Security and actions on enemy contact.
- Actual construction of the rope bridge within the 8-minute time standard on dry land.
- Individual preparation.
- Order of crossing.
- All signals and control measures.
- Reorganization.

(2) Conduct rehearsals as realistically as possible.
(3) Ensure personnel are proficient in the mechanics of a stream crossing operation.
(4) Inspect for equipment completeness, correct rigging and preparation, personnel knowledge and understanding of the operation.
(5) Actions of the # 4 man (BTC) during the preparation phase.

(a) Rehearse the bridge team.
(b) Account for all equipment in the bridge kit.
(c) Ensure 150-foot rope is coiled.

e. **Execution Phase.**

(1) Steps for the establishment and conduct of bridge stream crossing:

(a) Leader halts short of the river, local security is established, and a recon is conducted of the area for the presence of the enemy and for crossing site suitability/ necessity. He directs the BTC to construct the bridge.

(b) Security is established up and downstream while unit leader briefs BTC on anchor points. The unit leader counts individuals across.

(c) The bridge team begins to establish the rope bridge while unit members begin individual preparation.

(d) The BTC is responsible for construction of one-rope bridge and selection of the near side anchor point as well as the far side anchor point if visibility permits. He will tie a swimmer's safety line to anchor himself into the bridge. Tying an end of the line bowline around the waist secured with an overhand knot, and on the free running end, an end of line bowline with an overhand knot. A snaplink will be placed on the loop portion of the end of line bowline, which will extend no further than arms length away from the soldier (standard waterborne uniform). This will ensure that the soldier is never more than one arms length away from the rope bridge should he lose his grip.

(e) Noise and light discipline are enforced and security is maintained.

(f) Individual soldiers put a snaplink in their end of the line bowline and the front sight assembly of every M-4/M16 or M203. M240 gunners put a snaplink through the front sight assembly and rear swivel of their M240 MG. RTOs (and others with heavy rucksacks) place an additional snaplink on their rucksack frame, top center.

(2) The responsibilities of the bridge team while constructing the rope bridge are as follows:

(a) # 1 man (Lead safety swimmer and far side lifeguard). He grounds his rucksack (with snaplink through top of frame) to the rear of the near side anchor point. He wears equipment in the following order (body out). He carries a safety line to assume duties of far side lifeguard:

-- Waterborne uniform (top button buttoned, pants unbloused)
-- B7 life preserver or engineer work vest
-- LBV
-- Weapon (across the back)

(b) # 1 man enters the water upstream from the # 2 man and stays an arms length away from the # 2 man on the upstream side. He identifies the far side anchor point upon exiting the water and once the # 2 man has exited the water moves to his far side lifeguard position downstream of the rope bridge with knotted safety line on wrist, LBV LBV/weapon grounded, and work vest held in throwing hand. He continues to wear the B7.

(c) # 2 man (rope puller) in waterborne uniform (same as # 1 man) wears his equipment in the following order. He grounds his rucksack (with snaplink through top of frame) to the rear of the near side anchor point. His duties are to swim across the water obstacle pulling the rope. He ties off the rope on the anchor point identified by the # 1 man with a round turn and two half hitches with a quick release. The direction of the round turn is the same direction as the flow of water (current) to facilitate exit off the rope bridge:

- Work vest.
- LBV.
- Weapon (across the back).
- Swimmer safety line.

(d) # 3 man (near side lifeguard) in the same waterborne uniform as the far side lifeguard. The # 3 man positions himself on the downstream side of the bridge before the # 1/2 men enter the water. He (# 1 man) grounds his rucksack (with snaplink through top of frame) on rear of near-side anchor point. After the PSG crosses and verifies the headcount, the # 1 man unties the quick release at the near side anchor point. The # 3 man reties his safety line into an Australian rappel seat, hooks the end of line bowline into his snaplink, and connects his snaplink to the one on the end-of-line bowline . # 3 man is the last pulled across the water obstacle. Before crossing the water obstacle, he dons his equipment in the following order:

- B7.
- Work vest.
- LBV.
- Weapon.

(e) # 4 man (BTC). He is in the standard waterborne uniform with LBV and sling rope tied in safety line (round the waist bowline with end of line bowline no more than one arms length). He is responsible for construction of rope bridge and organization of bridge team. He is also responsible for back feeding the rope and tying end of line bowlines. He designates the near side anchor point, ties the Figure Eight slip of the transport tightening system, and hooking all personnel to the rope bridge. He ensures that the transport tightening knot is on the upstream side of the rope bridge. He ensures that all individuals are in the waterborne uniform, hooked into the rope facing the current with the safety line routed through the trailing shoulder of the individual's LBV and rucksack. He ensures that the weapon is hooked onto the rope. He controls the flow of traffic on the bridge. He is responsible for crossing with the # 1 man's rucksack. He is generally the next to the last man to cross (follows PSG who is keeping head count).

(f) # 5 and # 6 men (rope tighteners) wear the waterborne uniform with LBV and safety line. They tighten the transport tightening knot. They also take the rucksacks of the # 2 and # 3 men across. Once they reach the far side, # 5 and # 6 men pull the last man (# 3 man) across.

(g) The rucksacks of # 1/2/3 men are transported across by # 4/5/6 men. The rucksacks of the # 1/2/3 men are hooked into the rope by the snaplink through the top of the frames and the # 4/5/6 men pull them across. The weapons of # 4/5/6 men are attached between the # 4/5/6 men and the rucksack that they are pulling across the bridge.

(h) BTC rehearses the bridge team during the planning sessions and directs construction and emplacement. The unit leader selects the crossing site which complements the tactical plan.

- # 3 man positions himself downstream of crossing site.
- # 1 man enters water upstream of # 2. He stays one arm's length from the # 2 man and is prepared to render any assistance to the # 2 man. Both swim in conjunction upstream to compensate for the current. BTC feeds rope out of rucksack positioned on the downstream side of near side anchor point.

(i) The # 1 man exits and identifies the far side anchor point (if BTC cannot identify it for the # 2 man). # 2 man exits on the upstream side of the far side anchor point. The rope is now routed to facilitate movement onto and off the bridge.

(j) Radios and heavy equipment are waterproofed and rigged. All individuals don waterborne uniform and tie safety lines. PSG moves to anchor point and maintains accountability through headcount.

(k) The # 2 man signals the BTC that the rope is temporarily attached to the far side anchor point, and the BTC pulls out excess slack and ties the transport tightening system using a Figure Eight slip. The BTC signals the # 2 man to pull the knot twelve to fifteen feet from the near side anchor point. After this is done the # 2 two man ties round turns 18 to 24 inches off the water with the remaining rope and secures the rope, to itself, with a snaplink. The # 2 man signals the BTC and the pulling team (# 4/5/6) tightens the bridge, pulling the transport tightening system as close as possible to the near side anchor point.

(l) The # 1 man moves downstream and assumes his duties as the far side lifeguard. The bridge team commander will tie off the rope with a round turn and two half hitches around the near side anchor point. The BTC will place himself on the upstream side of the bridge (facing downstream). He begins to hook individuals into the rope inspecting them for safety.

Note: Any Ranger identified as a weak swimmer will cross with no other personnel on the rope bridge. The weak swimmer crosses individually to allow the near and far side lifeguards to focus their attention exclusively on the weak swimmer and not be distracted by other personnel crossing the bridge.

(m) The # 2 man moves to the upstream side of the rope bridge, assists personnel off the rope on the far side, and maintains head count. The # 5 and # 6 men cross with the # 1 and # 2 men's rucksacks..

(n) The BTC maintains the flow of traffic ensuring that no more than three men are on the bridge at any one time (one hooking up, one near the center, and one being unhooked). Once the PSG has accounted for everyone on the near side, he withdraws left and right (L/R) security and sends them across. PSG follows security across. The # 3 man hooks the BTC (with # 3's ruck) onto the rope. Once the BTC has crossed, the # 3 man unhooks near side anchor point and the BTC unties far side anchor point. The # 3 man ties an Australian rappel seat with snaplink to the front, hooks onto the snaplink that is in the end of the line bowline on the 120-foot rope, and then signals the # 4/5/6 men to take in slack. The # 3 man extends arms in front of his head, slightly upstream to fend off debris and is pulled across by the # 4/5/6. All individuals (except # 1/2/3 and RTOs) wear rucksacks across. The # 4/5/6 men hook the rucksacks of the # 1/2/3 men onto the bridge by the snaplink. All the men cross facing upstream.

(o) Once the far side headcount, weapons and equipment are verified (between PSG and # 5 man), personnel reorganize, and then they continue the mission.

(p) Personnel with heavy equipment:

M240 All major groups are tied together with 1/4-inch cord. An anchor line bowline runs through the rear swivel, down the left side of gun. Tie a round turn through the trigger guard. Route the cord down the right side and tie off two half hitches around the forearm assembly with a round turn and two half hitches through the front sight posts. The remainder of the working end is tied off with an end of the rope bowline approximately one foot from the front sight post large enough to place leading hand through. The M240 is secured to the bridge by snaplinks on the front sight post and rear swivel. The M240 is pulled across by the trailing arm of the M240 gunner.

AN-PRC/II9s These are waterproofed before crossing a one-rope bridge. Once far side FM commo is established, the near side RTO breaks down and waterproofs his radio, and prepares to cross the bridge. He puts a snaplink in the top center of the rucksack frame (same as for # 1/2/3 men). The BTC will hook the rucksack to the rope.

Note: Using two snaplinks binds the load on the rope. Adjust arm straps all the way out. RTO pulls radio across the rope bridge.

9-2. **PONCHO RAFT.** Normally a poncho raft is constructed to cross rivers and streams when the current is not swift. A poncho raft is especially useful when the unit is still dry and the platoon leader desires to keep the individuals equipment dry.

a. **Equipment Requirements**.

(1) Two serviceable ponchos.
(2) Two weapons (poles can be used in lieu of weapons).
(3) Two rucksacks per team.
(4) 10 feet of utility cord per team.
(5) One sling rope per team.

b. **Conditions**. Poncho rafts are used to cross water obstacles when any or all of the following conditions are found:

(1) The water obstacle is too wide for 150-foot rope.
(2) No sufficient near or far shore anchor points are available to allow rope bridge construction.
(3) Under no circumstances will poncho rafts be used as a means to cross a water obstacle if an unusually swift current is present.

c. **Choosing a Crossing Site**. Before a crossing site is used, a thorough reconnaissance of the immediate area is made. Analyzing the situation using METT-TC, the patrol leader chooses a crossing site that offers as much cover and concealment as possible and has entrance and exit points that are as shallow as possible. For speed of movement, it is best to choose a crossing site that has near and far shore banks that are easily traversed by an individual Ranger.

d. **Execution Phase**. Steps for the construction of a poncho raft:

(1) Pair off the unit/patrol in order to have the necessary equipment.
(2) Tie off the hood of one poncho and lay out on the ground with the hood up.
(3) Weapons are then placed in the center of the poncho, approximately 18 inches apart, muzzle to butt.
(4) Next, rucksacks and LBV are placed between the weapons with the two individuals placing their rucksacks as far apart as possible.
(5) The two will then start to undress (bottom to top), first with their boots, taking the laces completely out for subsequent use as tie downs if necessary).
(6) The boots are then placed over muzzle/butt of weapon toe in.
(7) Members continue to undress, folding each item neatly and placing it on top of their boots.
(8) Once all of the equipment is placed between the two weapons or poles, the poncho is snapped together. The snapped portion of the poncho is then lifted into the air and tightly rolled down to the equipment. Start at the center and work out to the end of the raft creating pigtails at the end. This is accomplished much easier if done by both soldiers together. The pigtailed ends are then folded in toward the center top of the raft and tied off with a single boot lace.
(9) The other poncho is then laid out on the ground with the hood up and the first poncho with equipment is placed in its center. The second poncho is then snapped, rolled and tied in the same manner as the first poncho. The third and fourth boot laces (or utility cord) are then tied around the raft approximately one foot from each end for added security. The poncho raft is now complete.

Note: The patrol leader must analyze the situation using METT-TC and make a decision on the uniform for crossing the water obstacle such as weapons inside the poncho raft or slung across the back, remaining dressed or stripped down with clothes inside raft.

9-3. **WATERCRAFT.** Use of inland and coastal waterways may add flexibility, surprise, and speed to tactical operations. Use of these waterways will also increase the load carrying capacity of normal dismounted units. Watercraft are employed in reconnaissance and assault operations.

a. **Inflation Method**. Inflatable with foot pumps using four separate valves located on the inside of the buoyancy tubes. Each of the four valves are used to section off the assault boat into eight separate airtight compartments. To pump air into the boat, turn all valves into the "orange" or "inflate" section of the valve. Once the assault boat is filled with air, turn all valves onto the "green" or "navigation" section. This will section the assault boat into eight separate compartments.

b. **Characteristics**.

- Overall length: 15 feet, 5 inches.
- Overall width: 6 feet, 3 inches.
- Weight: 265 pounds.
- Maximum payload: 2,756 pounds.
- Crew : 1 coxswain + either 10 paddlers or a 65 HP short-shaft outboard motor.

c. **Preparation**.

(1) ***Rubber Boat***.

- All rubber boats will have a twelve-foot bowline secured to the front starboard D-Ring. This rope will be tied with an anchor-line bowline, and the knot will be covered with 100-MPH tape.
- All rubber boats will have a fifteen foot center line tied to the rear floor D-Ring. The same procedure for securing the bowline will be used for the centerline.
- All rubber boats will be filled to 240 millibars of air, and checked to ensure that all valve caps are tight, and set in the NAVIGATE position.
- All rubber boats will be equipped with one foot pumps. These will be placed in the front pouches of the rubber boat. If no pouches are present, the foot pumps will be placed on the floor.
- All rubber boats will be inspected with the maintenance chart.

(2) ***Personnel and Equipment***.

- All personnel will wear work vest or kapoks (or another suitable positive flotation device).
- LBV is worn over the work vest, unbuckled at the waist.
- Individual weapon is slung across the back, muzzle pointed down and facing toward the inside of the boat.
- Crew served weapons, radios, ammunition and other bulky equipment is lashed securely to the boat to prevent loss if the boat should overturn. Machine guns with hot barrels are cooled prior to being lashed inside the boats.
- Radios and batteries are waterproofed.
- Pointed objects are padded to prevent puncturing the boat.

NOTES

d. **Positions.** Assign each individual a specific boat position (Figure 9-1).

e. **Duties.**

(1) Designate a commander for each boat, (normally coxswain)

(2) Designate a navigator (normally a leader within the platoon) - observer team as necessary.

(3) Crew is positioned as shown in Figure 9-2.

(4) Duties of the Coxswain.

(a) Responsible for control of the boat and actions of the crew.

(b) Supervises the loading, lashing and distribution of equipment.

(c) Maintains the course and speed of the boat.

(d) Gives all commands.

(5) # 2 paddler (long count) is responsible for setting the pace.

(6) # 1 paddler is the observer, stowing and using the bowline unless another observer is assigned.

f. **Embarkation and Debarkation Procedures.**

(1) When launching, the crew will maintain a firm grip on the boat until they are inside it: similarly, when beaching or debarking, they hold on to the boat until it is completely out of the water. Loading and unloading is done using the bow as the entrance and exit point.

(2) Keep a low center of mass when entering and existing the boat to avoid capsizing. Maintain 3 points of contact at all times.

(3) The long count is a method of loading and unloading by which the boat crew embarks or debarks individually over the bow of the boat. It is used at river banks, on loading ramps, and when deep water prohibits the use of the short count method.

(4) The short count is a method of loading or unloading by which the boat crew embarks or debarks in pairs over the sides of boat while the boat is in the water. It is used in shallow water allowing the boat to be quickly carried out of the water. The short count method of organization is primarily used during surf operations.

(5) Beaching the boat is a method of debarking the entire crew at once into shallow water and quickly carrying the boat out of the water.

g. **Commands**. Commands are issued by the coxswain to ensure the boat is transported over land and controlled in the water. All crew members learn and react immediately to all commands issued by the coxswain. T he various commands are as follows:

(1) "Short Count...count off," Crew counts off their position by pairs, such as 1,2,3,4,5 (passenger # 1, # 2, if applicable) coxswain.

(2) "Long Count...count off," Crew counts off the position by individual, such as 1,2,3,4,5,6,7,8,9,10, (Passenger # 1, # 2, if applicable), coxswain.

(3) "Boat Stations," Crew takes positions along side the boat.

(4) "High Carry...Move," (used for long distance move overland).

(a) On the preparatory command of "High Carry," the crew faces the rear of the boat and squats down grasping carrying handles with the inboard hand.

(b) On the command "Move," the crew swivels around, lifting the boat to the shoulders so that the crew is standing and facing to the front with the boat on their inboard shoulders.

(c) Coxswain guides the crew during movement.

(5) "Low Carry...Move," (used for short distance moves overland).

(a) On preparatory command of "Low Carry," the crew faces the front of the boat, bent at the waist, and grasps the carrying handles with their inboard hands.

(b) On the command of "Move," the crew stands up straight raising the boat approximately six to eight inches off the ground.

(c) Coxswain guides the crew during movement.

(6) "Lower the Boat----Move," Crew lowers the boat gently to the ground using the carrying handles.

(7) "Give Way Together," Crew paddles to front with # 2 setting the pace.

(8) "Hold," Entire crew keeps paddles straight downward motionless in the water stopping the boat.

(9) "Left side hold (Right)," Left crew holds, right continues with previous command.

(10) "Back paddle," Entire crew paddles backward propelling the boat to the rear.

Figure 9-1. BOAT POSITIONS

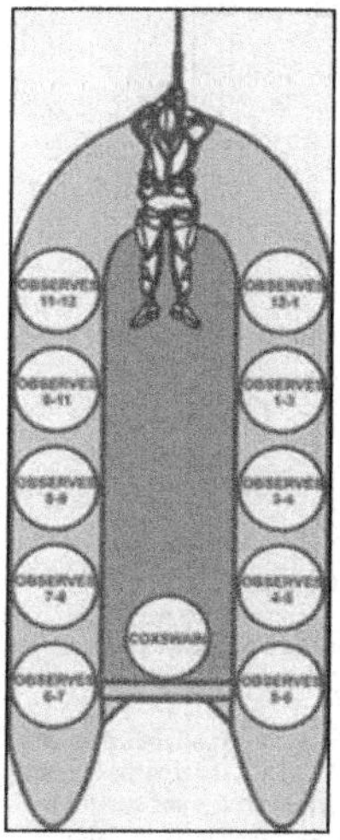

Figure 9-2. CREW POSITIONS, LONG COUNT AND SHORT COUNT

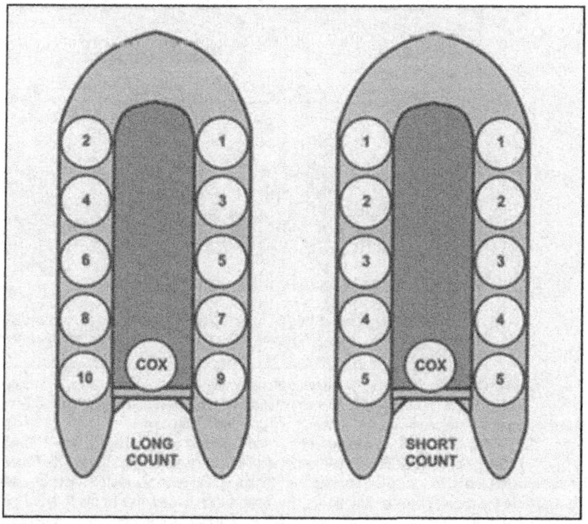

(11) "Back Paddle Left" (Right), Left crew back paddles causing the boat to turn left, right crew continues with previous command.

(12) "Rest Paddles," Crew members place paddles on their laps with blades outboard. This command may be given in pairs such as "# 1s, rest paddles").

h. **Securing of Landing Site.**

(1) If the landing site cannot be secured prior to the waterborne force landing, some form of early warning, such as scout swimmers, is considered. These personnel swim to shore from the assault boats and signal the boats to land. All signals and actions are rehearsed prior to the actual operation.

(2) If the patrol is going into an unsecured landing site, it can provide security by having a security boat land, reconnoiter the landing site and then signal to the remaining boats to land. This is the preferred technique.

(3) The landing site can be secured by force with all the assault boats landing simultaneously in a line formation. While this is the least desirable method of securing a landing site, it is rehearsed in the event the tactical situation requires its use.

(4) Arrival at the debarkation point.

(a) Unit members disembark according to leaders order (Figure 9-3).

(b) Local security is established.

(c) Leaders account for personnel and equipment.

(d) Unit continues movement.

- Soldiers pull security initially with work vest on.
- Coxswains and two men unlash and de-rig rucksacks.
- Soldiers return in buddy teams to secure rucksack and drop off work vest.
- Boats are camouflaged/cached if necessary prior to movement.

Figure 9-3. DEBARKATION

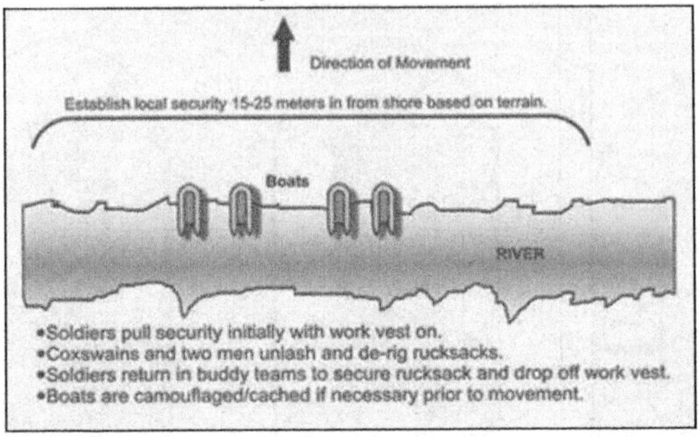

i. **Capsize Drill.** The following commands and procedures are used for capsize drills or to right an over-turned boat,

(1) ***PREPARE TO CAPSIZE***. This command alerts the crew and they raise paddles above their heads, with the blade pointed outward. Before capsizing, the coxswain will conduct a long count.

(2) ***PASS PADDLES***. All paddles are passed back and collected by the # 9/10 men.

(3) ***CAPSIZE THE BOAT***. All personnel slide into the water except the # 3/5/7 men. The # 1 man secures the bowline. The three men in the water grasp the capsize lines (ensuring the lines are routed under the safety lines) and stand on the buoyancy tubes opposite the capsize lines anchor points. The boat is then turned over by the # 3/5/7 men, who lean back and straighten their legs while pulling back on the capsize lines. As the boat lifts off the water, the # 4 man grasps the center carrying handle and rides the boat over. Once the boat is over, the # 4 man helps the # 3/7 men back onto the boat. The # 5 man holds onto the center carrying handle and turns the boat over the same way. The # 5 man rides the boat back over and helps the rest of the crew into the boat. As soon as the boat is capsized, the coxswain commands a long count to ensure that no one sank or was trapped under the boat. Every time the boat is turned over, he conducts another long count.

j. **River Movement.**

(1) ***Characteristics of River***.

(a) Know local conditions prior to embarking on river movement.

(b) A bend is a turn in the river course.

(c) A reach is a straight portion of river between two curves.

(d) A slough is a dead end branch from a river. They are normally quite deep and can be distinguished from the true river by their lack of current

(e) Dead water is a part of the river, due to erosion and changes in the river course that has no current. Dead water is characterized by excessive snags and debris.

(f) An island is usually a pear-shaped mass of land in the main current of the river. Upstream portions of islands usually catch debris and are avoided.

(g) The current in a narrow part of a reach is normally greater than in the wide portion.

(h) The current is greatest on the outside of a curve; sandbars and shallow water are found on the inside of the curve.

(i) Sandbars are located at those points where a tributary feeds into the main body of a river or stream.

(j) Because the # 1 and # 2 men are sitting on the front left and right sides of the boat, they observe for obstacles as the boat moves downriver. If either notices an obstacle on either side of the boat, he notifies the coxswain. The coxswain then adjusts steering to avoid the obstacle.

(2) ***Navigation***. The patrol leader is responsible for navigation. The three acceptable methods of river navigation are--

(a) Checkpoint and general route. These methods are used when the drop site is marked by a well-defined checkpoint and the waterway is not confused by a lot of branches and tributaries. They are best used during daylight hours and for short distances.

(b) Navigator-observer method. This method is the most accurate means of river navigation and is used effectively in all light conditions.

(c) *Equipment Needed*.

- Compass
- GPS
- Photo map (1st choice)
- Topographical map (2nd choice)
- Poncho (for night use)
- Pencil/Grease pencil
- Flashlight (for night use)

(d) *Procedure*. Navigator is positioned in center of boat and does not paddle. During hours of darkness, he uses his flashlight under the poncho to check his map. The observer (or # 1 man) is at the front of the boat.

(a) The navigator keeps his map and compass oriented at all time.

(b) The navigator keeps the observer informed of the configuration of the river by announcing bends, sloughs, reaches, and stream junctions as shown on his map.

(c) The observer compares this information with the bends, sloughs, reaches, and stream junctions he actually sees. When these are confirmed, the navigator confirms the boat's location on his map.

(d) The navigator also keeps the observer informed of the general azimuths of reaches as shown on his map and the observer confirms these with actual compass readings of the river.

(e) The navigator announces only one configuration at a time to the observer and does not announce another until it is confirmed and completed.

(f) A strip-map drawn on clear acetate backed by luminous tape may be used. The drawing is to scale or a schematic. It should show all curves and the azimuth and distance of all reaches. It may also show terrain features, stream junctions, and sloughs.

k. **Formations.** Various boat formations are used (day and night) for control, speed, and security (Figure 9-4). The choice of which is used depends on the tactical situation and the discretion of the patrol leader. He should use hand and arm signals to control his assault boats. The formations are:

- Wedge
- Line
- File
- Echelon
- Vee

Figure 9-4. FORMATIONS

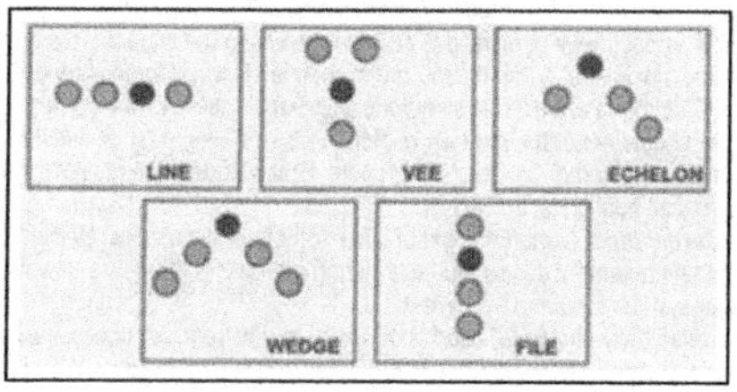

Chapter 10
MILITARY MOUNTAINEERING

In the mountains, commanders face the challenge of maintaining their units' combat effectiveness and efficiency. To meet this challenge, commanders conduct training that provides Rangers with the mountaineering skills necessary to apply combat power in a rugged mountain environment, and they develop leaders capable of applying doctrine to the distinct characteristics of mountain warfare.

10-1. **TRAINING**. Military mountaineering training provides tactical and fundamental mobility skills needed to move units safely and efficiently in mountainous terrain. Rangers should receive extensive training prior to executing combat operations in mountainous environments. Some of the areas to train are as follows:

- Characteristics of the mountain environment.
- Care and use of basic mountaineering equipment.
- Mountain bivouac techniques.
- Mountain communications.
- Mountain travel and walking techniques.
- Mountain navigation, hazard recognition and route selection.
- Rope management and knots.
- Natural and artificial anchors.
- Belay and rappel techniques.
- Installation construction and use such as rope bridges.
- Rock climbing fundamentals.
- Rope bridges and lowering systems.
- Individual movement on snow and ice.
- Mountain stream crossings (to include water survival techniques).

10-2. **DISMOUNTED MOBILITY**. Movement in more technical terrain demands specialized skills and equipment. Before Rangers can move in such terrain, a technical mountaineering team might have to secure the high ground.

10-3. **MOUNTAINEERING EQUIPMENT**. Mountaineering equipment is everything that allows the trained Ranger to accomplish many tasks in the mountains. The importance of this gear to the mountaineer is no less than that of the rifle to the Infantry Soldier.

a. **Ropes and Cords**. Ropes and cords are the most important pieces of mountaineering equipment. They secure climbers and equipment during steep ascents and descents. They are also used to install other ropes and to haul equipment. From WWII until the 1980's, US armed forces mainly used 7/16 *nylon laid rope*, often referred to as *green line* for all mountaineering operations. Since kernmantle ropes were introduced, other, more specialized ropes have in many cases replaced the all-purpose green line. Kernmantle ropes are made much like parachute cord, with a smooth sheath around a braided or woven core. Laid ropes are still used, but should be avoided whenever rope failure could cause injury or loss of equipment.

(1) ***Types***. Kernmantle rope is classified as *dynamic* or *static*:

(a) *Dynamic Ropes*. Ropes used for climbing are classified as *dynamic ropes*. These rope stretch or elongate 8 to 12 percent once subjected to weight or impact. This stretching is critical in reducing the impact force on the climber, anchors, and belayer during a fall by softening the catch. An 11-mm by 150-meter rope is standard for military use. Other sizes are available.

(b) *Static Ropes*. Static ropes are used in situations where rope stretch is undesired, and when the rope is subjected to heavy static weight. Static ropes should never be used while climbing, since even a fall of a few feet could generate enough impact force to injure climber and belayer, and/or cause anchor failure. Static Ropes are usually used when constructing rope bridges, fixed rope installations, vertical haul lines, and so on.

(2) ***Sling Ropes and Cordelettes***. A short section of static rope or static cord is referred to as a sling rope or cordelette. These are a critical piece of personal equipment during mountaineering operations. They are usually 7 to 8mm diameter and up to 21 feet long. The minimum Ranger standard is 8mm by 15 feet.

(3) ***Rope Care***. Ropes used daily should be used no longer that one year. An occasionally used rope can be used generally up to 4 to 5 years if properly cared for.

- Inspect ropes thoroughly before, during, and after use for cuts, frays, abrasion, mildew, and soft or worn spots.
- Never step on a rope or drag it on the ground unnecessarily.
- Avoid running rope over sharp or rough edges (pad if necessary).

- Keep ropes away from oil, acids, and other corrosive substances.
- Avoid running ropes across one another under tension (nylon to nylon contact will damage ropes).
- Do not leave ropes knotted or under tension longer than necessary.
- Clean in cool water, loosely coil, and hang to dry out of direct sunlight. Ultraviolet light rays harm synthetic fibers. When wet, hang rope to drip-dry on a rounded wooden peg, at room temperature (do not apply heat).

(4) ***Webbing and Slings***. Loops of tubular webbing or cord, called slings or runners, are the simplest pieces of equipment and some of the most useful. The uses for these simple pieces are endless, and they are a critical link between the climber, the rope, carabiners, and anchors. Runners are usually made from either 9/16- or 1-inch tubular webbing and are either tied or sewn by a manufacturer.

b. **Carabiners**. The carabiner is one of the most versatile pieces of equipment available in the mountains. This simple piece of gear is the critical connection between the climber, his rope, and the protection attaching him to the mountain. New metal alloys are stronger to hold hard falls; and, they are lighter than steel, which allows the Ranger to carry many at once. Consequently, steel carabiners are used less and less often. Basic carabiners come in several different shapes.

c. **Protection**. This is the generic term used to describe a piece of equipment, natural or artificial, that is used to construct an anchor. Protection, along with a climber, belayer, and climbing rope, forms the lifeline of the climbing team. The rope connects two climbers, and the protection connects them to the rock or ice. The two types of artificial protection are traditional (removable) and fixed (usually permanent) (Figures 10-1 and 10-2).

Figure 10-1. EXAMPLES OF TRADITIONAL (REMOVABLE) PROTECTION USED ON ROCK

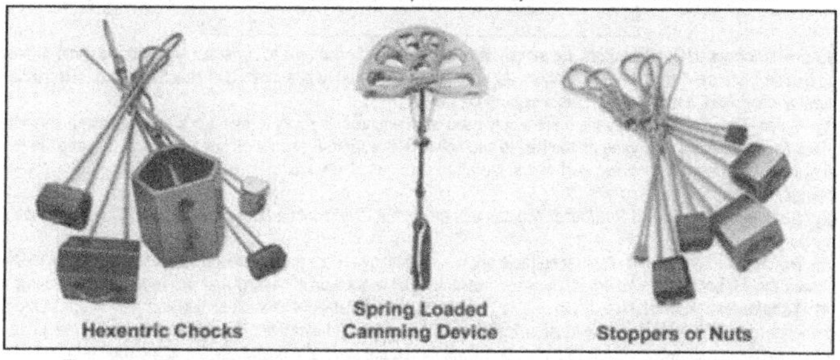

Figure 10-2. EXAMPLES OF FIXED (PERMANENT OR SEMI) PROTECTION USED ON ROCK

10-4. **ANCHORS**. Anchors form the base for all installations and roped mountaineering techniques. Anchors must be strong enough to support the entire weight of the load or of the impact placed upon them. Several pieces of artificial or natural protection may be incorporated together to make one multipoint anchor. Anchors are classified as artificial or natural:

a. **Artificial Anchors**. Artificial anchors are made from man-made materials. The most common ones use traditional or fixed protection (Figure 10-3).

Figure 10-3. CONSTRUCTION OF THREE-POINT, PRE-EQUALIZED ANCHOR WITH FIXED ARTIFICIAL PROTECTION

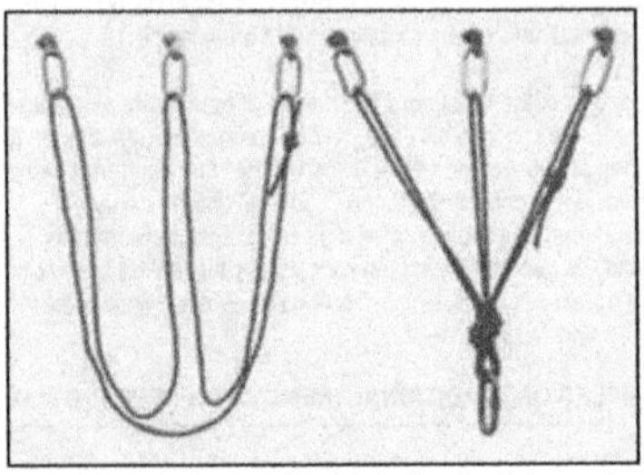

b. **Natural Anchors**. Natural anchors are usually very strong and often simple to construct with little equipment. Trees, shrubs, and boulders are the most common anchors. All natural anchors require is a method of attaching a rope. Regardless of the type of natural anchor used, it must support the weight of the load.

(1) ***Trees***. These are probably the most widely used of all anchors. In rocky terrain, trees usually have a very shallow root system. Check this by pushing or tugging on the tree to see how well it is rooted. Anchor as low as possible to keep the pull on the base of the tree, where it is nearer its own anchor and therefore stronger. Use padding on soft, sap-producing trees to keep sap off ropes and slings.

(2) ***Rock Projections and Boulders***. You can use these, but they must be heavy enough, and they must have a stable enough base to support the load.

(3) ***Bushes and Shrubs***. If no other suitable anchor is available, route a rope around the bases of several bushes. As with trees, place the anchoring rope as low as possible. Make sure all vegetation is healthy and well rooted to the ground.

(4) ***Tensionless Anchor***. Use this to anchor rope on high-load installations such as bridging. The wraps of the rope around the anchor (Figure 10-4) absorb the tension, keeping it off the knot and carabiner. Tie this anchor with at least four wraps around the anchor. A smooth anchor, such as a pipe or rail, might require several more wraps. Wrap the rope from top to bottom. Place a fixed loop into the end of the rope, and attach it loosely back onto the rope with a carabiner.

Figure 10-4. TENSIONLESS NATURAL ANCHOR

10-5. **KNOTS**.

a. **Square Knot**. This joins two ropes of equal diameter (Figure 10-5). It has two interlocking bights; the running ends exit on the same side of the standing portion of the rope. Each tail is secured with an overhand knot on the standing end. When you dress the knot, leave at least a 4-inch tail on the working end.

Figure 10-5. SQUARE KNOT

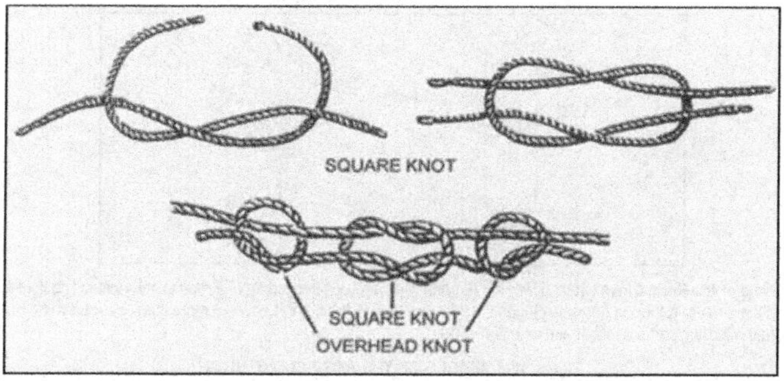

b. **Round-Turn with Two Half Hitches**. This is a constant tension anchor knot (Figure 10-6). The rope forms a complete turn around the anchor point (thus the name round turn), with both ropes parallel and touching, but not crossing. Both half hitches are tightly dressed against the round turn, with the locking bar on top. When you dress the knot, leave at least a 4-inch tail on the working end.

Figure 10-6. ROUND-TURN WITH TWO HALF-HITCHES

c. **End-of-the-Rope Clove Hitch**. The end-of-the-rope clove hitch (Figure 10-7) is an intermediate anchor knot that requires constant tension. Make two turns around the anchor. A locking bar runs diagonally from one side to the other. Leave no more than one rope width between turns of rope. Locking bar is opposite direction of pull. When you dress the knot, leave at least a 4-inch tail on the working end.

Figure 10-7. END-OF-THE-ROPE CLOVE HITCH

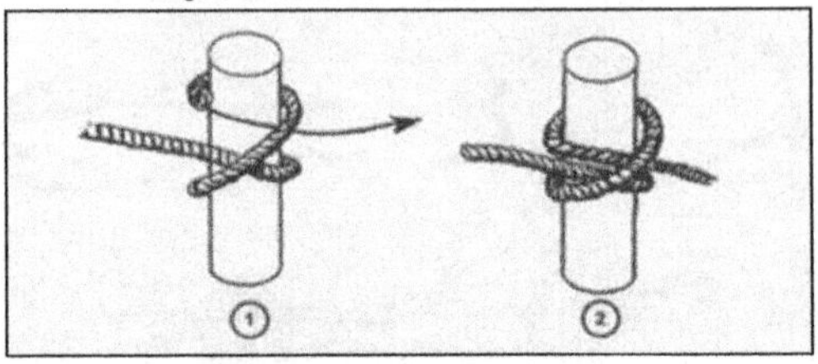

d. **Middle of the Rope Clove Hitch**. This knot (Figure 10-8) secures the middle of a rope to an anchor. The knot forms two turns around the anchor. A locking bar runs diagonally from one side to the other. Leave no more than one rope width between turns. Ensure the locking bar is opposite the direction of pull.

Figure 10-8. MIDDLE-OF-THE-ROPE CLOVE HITCH

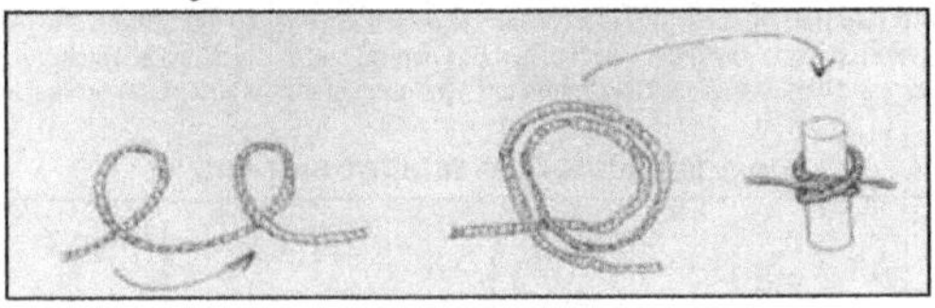

e. **Rappel Seat**. The rappel seat (Figure 10-9) is a rope harness used in rappelling and climbing. It can be tied for use with the left or right hand. Leg straps do not cross, and are centered on buttocks and tight. Leg straps form locking half hitches on rope around waist. A square knot is tied on the right hip and finished with two overhand knots. Tails must be even, no longer than 6 inches. A carabiner is inserted around all ropes, with the gate opening up and away. The carabiner must not contact the square or overhand knot. The rappel seat must be tight enough to prevent a fist from fitting between the rappeller's body and the harness.

Figure 10-9. RAPPEL SEAT

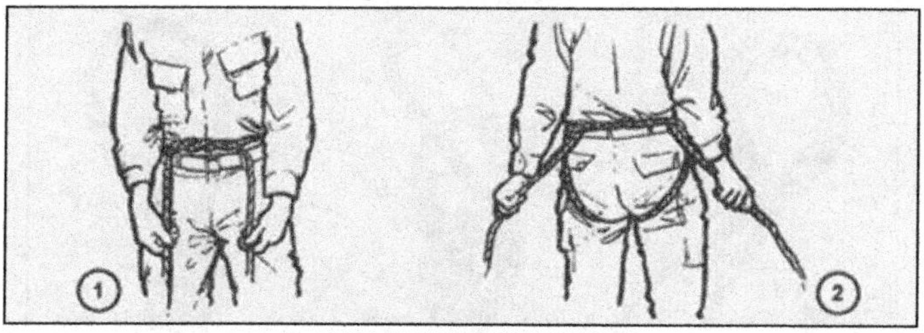

Figure 10-9. RAPPEL SEAT (CONTINUED)

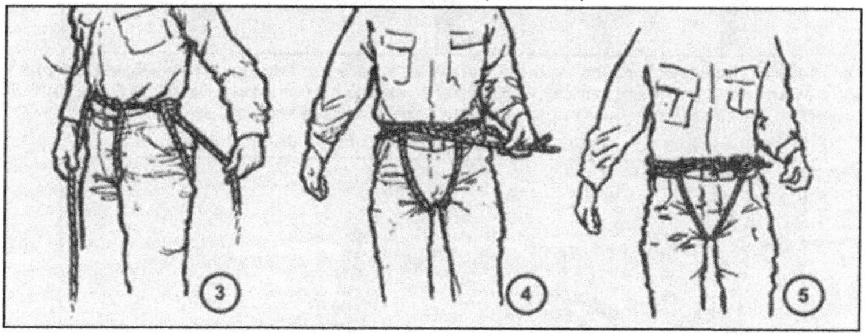

f. **Double Figure Eight**. Use a Figure Eight loop knot (Figure 10-10) to form a fixed loop in the end of the rope. It can be tied at the end of the rope or anywhere along the length of the rope. Figure Eight loop knots are formed by two ropes parallel to each other in the shape of a Figure Eight, no twists are in the Figure Eight. Fixed loops are large enough to insert a carabiner. When you dress the knot, leave at least a 4-inch tail on the working end.

Figure 10-10. DOUBLE FIGURE EIGHT-LOOP KNOT

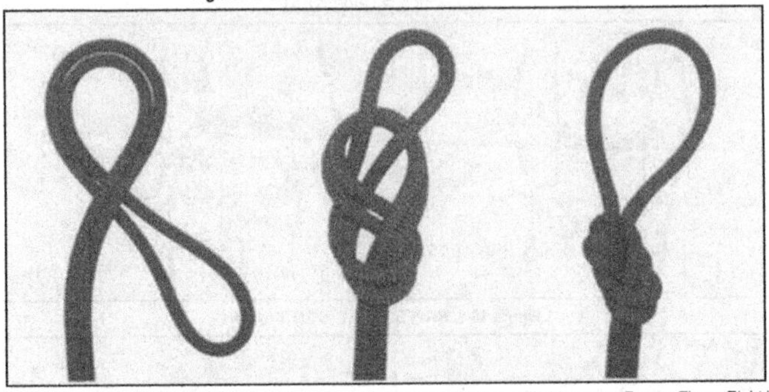

g. **Rerouted Figure Eight Knot**. This anchor knot also attaches a climber to a climbing rope. Form a Figure Eight in the rope, and pass the working end around an anchor. Reroute the end back through to form a double Figure Eight (Figure 10-11). Tie the knot with no twists. When you dress the knot, leave at least a 4-inch tail on the working end.

Figure 10-11. REROUTED FIGURE EIGHT KNOT

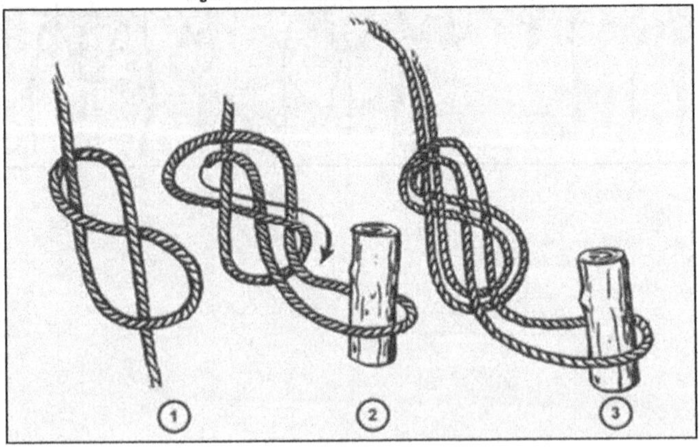

h. **Figure Eight Slip Knot**. The Figure Eight slip knot is used to form an adjustable bight in the middle of a rope. Knot is in the shape of a Figure Eight. Both ropes of the bight pass through the same loop of the Figure Eight. The bight is adjustable by means of a sliding section (Figure 10-12).

Figure 10-12. FIGURE EIGHT SLIP KNOT

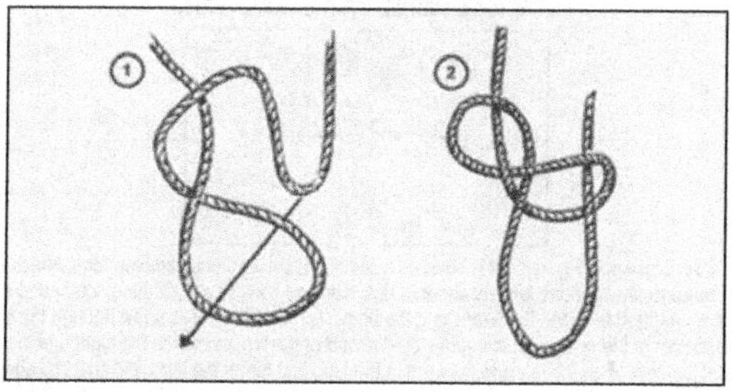

i. **End-of-the-Rope Prusik**. The End-of-the-Rope Prusik (Figure 10-13) attaches a movable rope to a fixed rope. The knot has two round turns, with a locking bar perpendicular to the standing end of the rope. Tie a bowline within 6 inches of the locking bar. When you dress the knot, leave at least a 4-inch tail on the working end.

Figure 10-13. END-OF-THE-ROPE PRUSIK

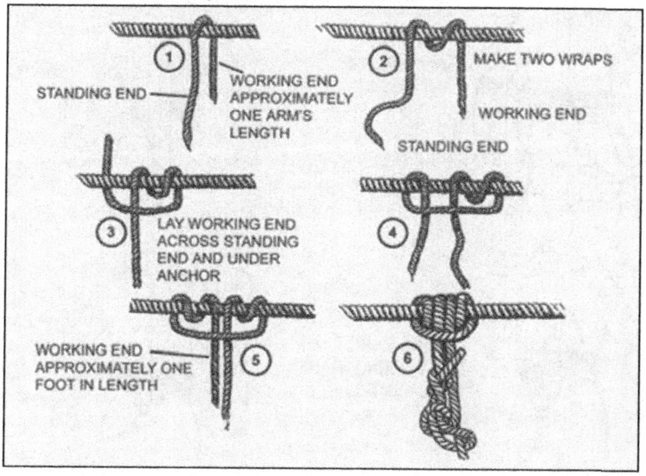

j. **Middle-of-the-Rope Prusik**. The middle-of-the-rope Prusik (Figure 10-14) attaches a movable rope to a fixed rope, anywhere along the length of the fixed rope. Make two round turns with a locking bar perpendicular to the standing end. Ensure the wraps do not cross and that the overhand knot is within 6 inches from the horizontal locking bar. Ensure the knot does not move freely on fixed rope.

Figure 10-14. MIDDLE-OF-THE-ROPE PRUSIK

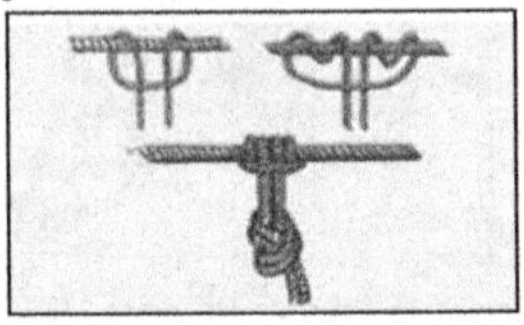

k. **Bowline on a Coil**. This knot (Figure 10-15) secures a climber to the end of a climbing rope. Make at least four parallel wraps around the body, between the hipbone and lower set of ribs. Keep all coils free of clothing, touch and are tight enough to ensure that a fist cannot be inserted between the wraps and the body. Three distinct coils show through the bight of the bowline. The rope coming off the bottom of the coils is on the right side, forward of the hip and forms the bight and the overhand knot. The rope coming off the top of the coils is on the left side, forward of the hips and forms the third and final coil showing through the bight of the bowline. Bowline is centered on the gig line and secured with an overhand knot with a minimum of 4 inch tail remaining after the knot is dressed.

Figure 10-15. BOWLINE ON A COIL

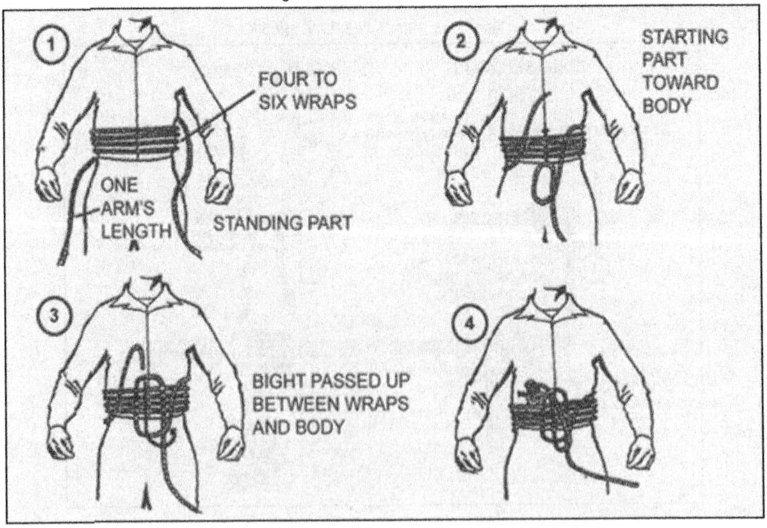

10-6. **BELAYS**. Belaying is any action taken to stop a climber's fall or to control the rate a load descends. The belayer also helps manage the rope for a climber by controlling the amount of rope that is taken out or in, or by controlling a rappeller's rate of descent. The belayer must be stably anchored to prevent him from being pulled out of his position, and losing control of the rope. The two types of belays are body and mechanical belays.

a. **Body Belay**. This belay (Figure 10-16) uses the belayer's body to apply friction by routing the rope around the belayer's body. Caution must be used when belaying from the body since the entire weight of the load may be placed on the belayer's body.

Figure 10-16. BODY BELAY

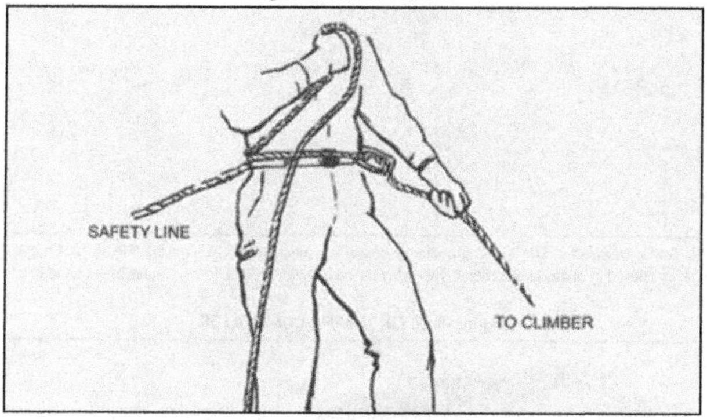

b. **Mechanical Belay**. This belay (Figure 10-17) uses mechanical devices to help the belayer control the rope, as in rappelling. A variety of devices are used to construct a mechanical belay for mountaineering.

Figure 10-17. MECHANICAL BELAY

(1) ***Munter Hitch***. One of the most often used belays, the munter hitch (Figure 10-18), requires very little equipment. The rope is routed through a locking, pear-shaped carabiner, then back on itself. The belayer controls the rate of descent by manipulating the working end back on itself with the brake hand.

Figure 10-18. MUNTER HITCH

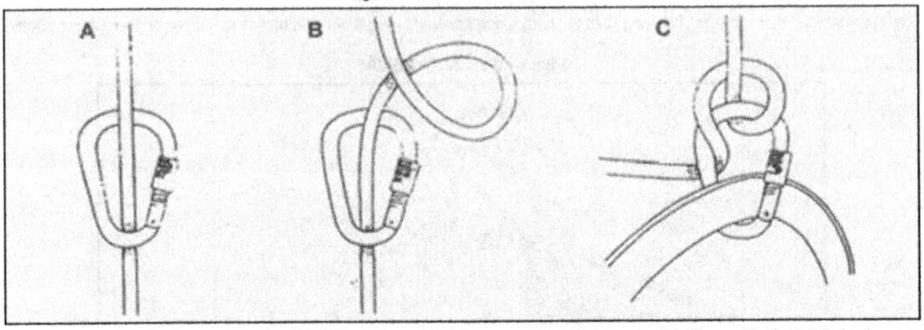

(2) ***Air Traffic Controller***. The ATC (air traffic controller) is a mechanical belay device (Figure 10-19) that locks down on itself when tension is applied in opposite directions. The belayer need apply very little force with his brake hand to control descent or arrest a fall.

Figure 10-19. AIR TRAFFIC CONTROLLER

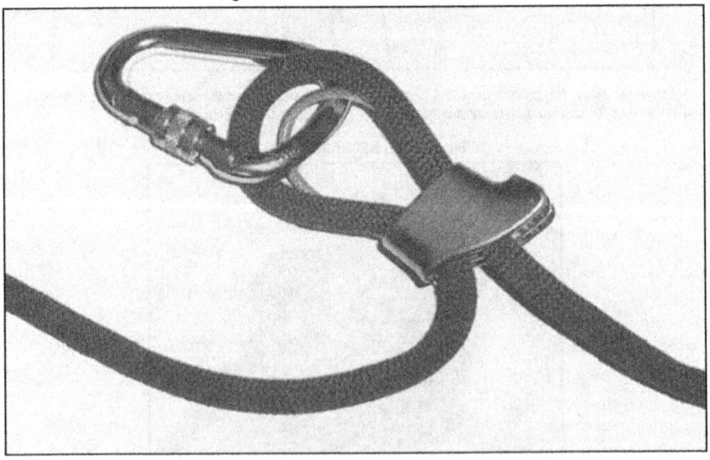

10-7. **CLIMBING COMMANDS**. Table 10-1 shows the sequence of commands used by climber and belayer.

Table 10-1. SEQUENCE OF CLIMBING COMMANDS

SEQUENCE	*COMMAND*	*GIVEN BY*	*MEANING*
1.	BELAY ON, CLIMB	Belayer	Alerts climber that belay is on and climber may climb.
2.	CLIMBING	Climber	Alerts belayer that climber is climbing.
3.	UP-ROPE	Climber	Belayer, remove excess slack in the rope.
4.	BRAKE	Climber	Alerts belayer to immediately apply brake.
5.	FALLING	Climber	Alerts belayer that climber is falling and that the belayer should immediately apply the brake and prepare to arrest the fall.
6.	TENSION	Climber	Alerts belayer to remove all slack from climbing rope until rope is tight, then apply brake and hold position.
7.	SLACK	Climber	Alerts belayer to pull slack into the climbing rope (Belayer may have to assist)
8.	ROCK	Anyone	Alerts everyone that an object is about to fall near them. Belayer immediately applies brake.
9.	POINT	Climber	Alerts belayer that the direction of pull on the climbing rope has changed in the event of a fall.
10.	STAND-BY	Climber or Belayer	Alerts the other to hold position, stand by, I am not ready.
11.	DO YOU HAVE ME?	Climber	Alerts belayer to prepare for a fall or to prepare to lower the climber.
12.	I HAVE YOU	Belay	Alerts climber that the brake is on and the belayer is prepared for the climber to fall, or to lower him.
13.	OFF-BELAY	Climber	Alerts belayer that climber is safetied in or it is safe to come off belay.
14.	3-METERS	Belayer	Alerts climber to the amount of rope between climber and belayer (May be given in feet or meters)
15.	BELAY-OFF	Belayer	Alerts climber that belayer is off belay

10-8. **ROPE INSTALLATIONS**. Rope installations may be constructed by teams to assist units in negotiating natural and manmade obstacles. installation teams consist of a squad sized element, with 2 to 4 trained mountaineers. Installation teams deploy early and prepare the AO for safe, rapid movement by constructing various types of mountaineering installations. Following construction of an installation, the squad, or part of it, remains on site to secure and monitor the system, assist with the control of forces across it, and make adjustments or repairs during its use. After passage of the unit, the installation team may then disassemble the system and deploy to another area as needed.

a. **Fixed Rope Installations**. A fixed rope is anchored in place to help Rangers move over difficult terrain. Its simplest form is a rope tied off on the top of steep terrain. As terrain becomes steeper or more difficult, fixed rope systems may require intermediate anchors along the route. Planning considerations are as follows:

- Does the installation allow you to bypass the obstacle?
- (Tactical considerations) Can obstacle be secured from construction through negotiation, to disassembly?
- Is it in a safe and suitable location? Is it easy to negotiate? Does it avoid obstacles?
- Are natural and artificial anchors available?
- Is the area safe from falling rock and ice?

b. **Vertical Hauling Line**. This installation (Figure 10-20) is used to haul men and equipment up vertical or near vertical slopes. It is often used in conjunction with the fixed rope.

Figure 10-20. VERTICAL HAULING LINE

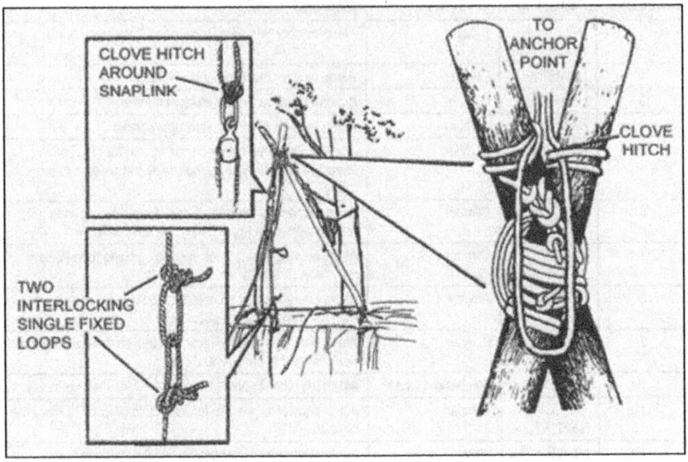

(1) ***Planning Considerations***.

- Does the installation allow you to bypass the obstacle?
- (Tactical considerations) Can you secure the installation from construction through negotiation, to disassembly?
- Does it have good loading and off-loading platforms? Are the platforms natural and easily accessible? Do they provide a safe working area?
- Does it allow sufficient clearance for load? Is there enough space between the slope and the apex of the A-frame to allow easy on-loading and off-loading of troops and equipment?
- Does it have an A-Frame for artificial height?
- Does it allow you to haul line in order to move personnel and equipment up and down slope?
- Does the A frame have a pulley or locking carabiner to ease friction on hauling line?
- Does it have a knotted hand line to help Rangers up the installation?
- Does it allow for Rangers top and bottom to monitor safe operation?

(2) ***Equipment***.

- Three 120-foot (50m) static ropes.
- Three 15-foot sling ropes for constructing A-frame.
- Two A-frame poles, 7 to 9 feet long, 4 to 6 inches in diameter (load dependent).
- Nine carabiners.
- One pulley with steel locking carabiner.

c. **Bridges**. Rope bridges are employed in mountainous terrain to bridge linear obstacles such as streams or rivers where the force of flowing water may be too great or temperatures are too cold to conduct a wet crossing.

(1) ***Construction***. The rope bridge is constructed using a static ropes. The max span that can be bridged is 50 percent of the ropes entire length for a dry crossing, 75 percent for a wet crossing. The ropes are anchored with an anchor knot on the far side

of the obstacle, and are tied off at the near end with a transport tightening system. Rope Bridge Planning Considerations are as follows:

- Does the installation allow you to bypass the obstacle?
- (Tactical considerations) Can you secure the installation from construction through negotiation, to disassembly?
- Is it in the most suitable location such as a bend in the river? Is it easily secured?
- Does it have near and far-side anchors?
- Does it have good loading and off-loading platforms?

(2) ***Equipment (1-Rope Bridge)***.

- One sling rope per-man.
- One steel locking carabineer.
- Two steel ovals.
- Two 120-foot static ropes.

(3) ***Construction Steps***. The first Ranger swims the rope to the far side and ties a tensionless anchor (Fig 10-4) between knee and chest level, with at least 6 to 8 wraps. The BTC ties a transport tightening system (Fig 10-21) to the near side anchor point. Then, he ties a Figure Eight slipknot and incorporates a locking half hitch around the adjustable bight. Insert 2 steel oval carabineers into the bight so the gates are opposite and opposed. The rope is then routed around the near side anchor point at waist level and dropped into the steel oval carabineers.

- A three-Ranger pulling team moves forward from the PLT. No more than three men should tighten the rope. Using more can overtighten the rope, bringing it near failure.
- Once the rope bridge is tight enough, the bridge team secures the transport-tightening system using two half hitches, without losing more than 4 inches of tension.
- Personnel cross using the commando crawl, rappel seat, or monkey crawl method (Figures 10-22 thru 24).

Figure 10-21. TRANSPORT TIGHTENING SYSTEM

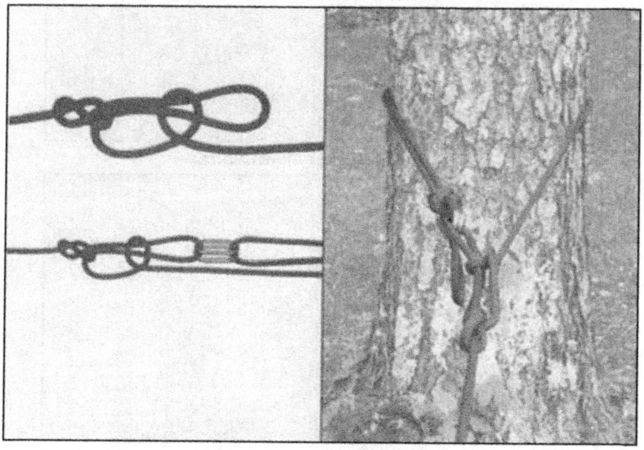

Figure 10-22. COMMANDO CRAWL METHOD

Figure 10-23. RAPPEL SEAT (TYROLEAN TRAVERSE) METHOD

Figure 10-24. MONKEY CRAWL METHOD

(4) ***Bridge Recovery***. Once all but two Rangers have crossed the rope bridge, the bridge team commander (BTC) chooses either the wet or dry method to dismantle it. If he chooses the dry method, he should he should first anchor his tightening system with the transport knot.

- The BTC back-stacks all of the slack coming out of the transport knot, then ties a fixed loop and places a carabiner into the fixed loop.
- The next to last Ranger to cross attaches the carabiner to his rappel seat or harness, and then moves across the bridge using the Tyrolean traverse method.
- The BTC then removes all knots from the system. The far side remains anchored. The rope should now only pass around the near side anchor.
- A three-man pull team, assembled on the far side, takes the end brought across by the next to last Ranger and pulls the rope tight again and holds it.
- The BTC then attaches himself to the rope bridge and moves across.
- Once across, the BTC breaks down the far side anchor, removes the knots, and then pulls the rope across. If the BTC chooses a wet crossing, any method can be used to anchor the tightening system.
- All personnel cross except the BTC or the strongest swimmer.
- The BTC then removes all knots from the system.
- The BTC ties a fixed loop, inserts a carabiner, and attaches it to his rappel seat or harness. He then manages the rope as the slack is pulled to the far side.
- The BTC then moves across the obstacle while being belayed from the far side.

d. **Suspension Traverse**. The suspension traverse is used to move personnel and equipment over rivers, ravines, chasms, and up or down a vertical obstacle. By combining the transport tightening system used during the Rope bridge, an A-Frame used for the Vertical haul Line (Figure 10-25), and belay techniques device, Units can construct a suspension traverse (Figures 10-26 and 10-27). Installation of a suspension traverse can be time-consuming and equipment-intensive. All personnel must be well trained and rehearsed in the procedures.

(1) ***Construction***. The suspension traverse must be constructed using a static ropes. The max span that can be bridged is generally 75 percent length of the shortest rope. Planning considerations are the same as rope bridge and vertical haul line combined.

(2) ***Equipment***.

- Three static installation ropes.
- Seven sling Ropes.
- Nine carabiners.
- One heavy duty double pulley.
- One locking carabiner.
- One canvas pad.

Figure 10-25. ANCHORING THE TRAVERSE ROPE TO THE A-FRAME

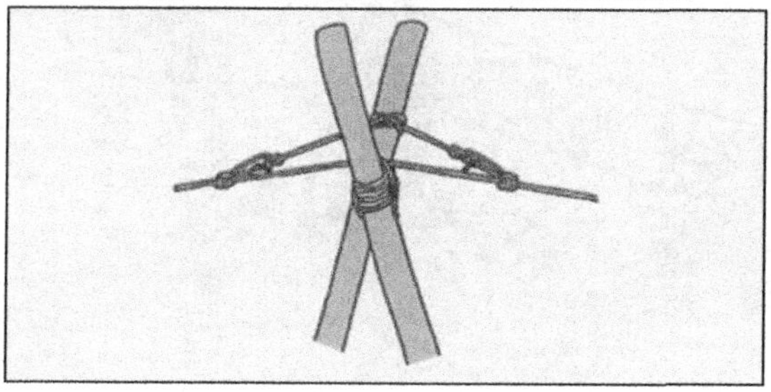

Figure 10-26. CARRYING ROPE FOR USE ON A SUSPENSION TRAVERSE

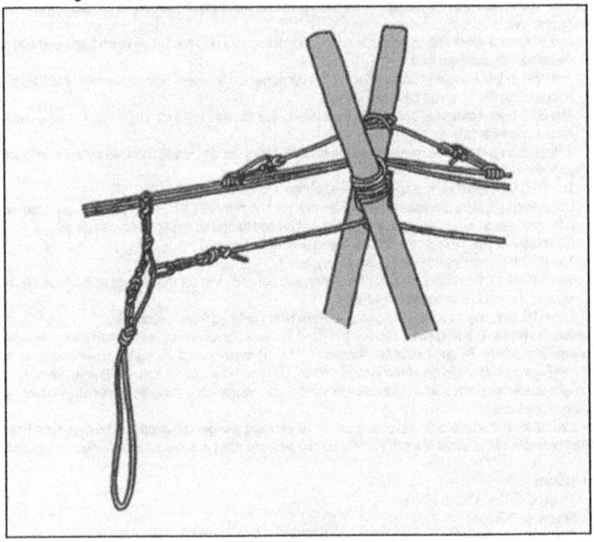

Figure 10-27. SUSPENSION TRAVERSE

10-9. **RAPPELLING**. Rappelling is a quick method of descent but can be extremely dangerous. These dangers include anchor failure, equipment failure, and individual error. Anchors in a mountainous environment should be selected carefully. Great care must be taken to load the anchor slowly and to ensure that no excessive stress is placed on the anchor. The best way to ensure this is to prohibit bounding rappels, and to use only walk-down rappels.

a. **Hasty and Body Rappels**. Hasty (Figure 10-28) and body rappels (Figure 10-29) are easier and faster than other methods, but should only be used on moderate pitches--never on vertical or overhanging terrain. Wear gloves to prevent rope burns.

Figure 10-28. Hasty Rappel

Figure 10-29. Body Rappel

b. **Seat-Hip Rappel**. The seat-hip rappel uses a mechanical rappel device that is inserted in a sling rope seat and is fastened to the rappeller to provide the necessary friction to remain in control. This method provides a faster and more controlled descent than any other method (Figure 10-30). There are several mechanical rappel devices (Figure 10-31 and 10-32) that can be used for the seat-hip rappel.

Figure 10-30. FIGURE EIGHT DESCENDER

Figure 10-30. FIGURE EIGHT DESCENDER (CONTINUED)

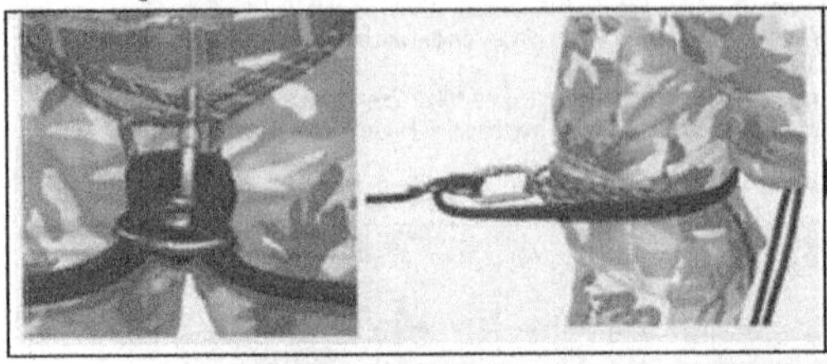

Figure 10-31. CARABINER WRAP

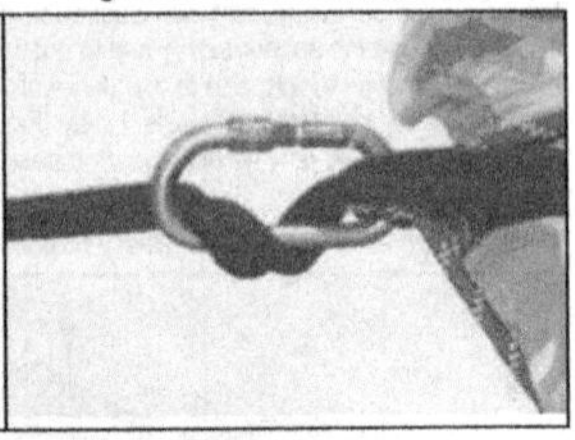

Figure 10-32. SEAT HIP RAPPEL WITH CARABINER WRAP DESCENDER

c. **Site Selection**. The selection of the rappel point depends on factors such as mission, cover, route, anchor points, and edge composition (loose or jagged rocks). There must be good anchors (primary and secondary). The anchor point should be above the rappeller's departure point. Suitable loading and off-loading platforms should be available.

- Each rappel point has primary and secondary anchors.
- Rappel point has equal tension between all anchor points.
- Double rope is used when possible.
- Ropes must reach the off loading platform.
- Site has suitable on and off loading platforms.
- Personnel working near the edge are tied in.
- Select a smooth route free of loose rock and debris.

Chapter 11

EVASION AND SURVIVAL

This chapter will cover basic material dealing with evading the enemy and survival techniques. Field expedient methods of acquiring food/ water, land navigation, constructing shelters, and making fires will be presented.

11-1. **EVASION**. When you become isolated or separated in a hostile area, either as an individual or as a group, your evasion and survival skills will determine whether or not you return to friendly lines.

a. When unable to continue the mission or unable to rejoin your unit, leave the immediate area, and move to your last rally point.

b. Observe activity in the area and form a plan.

c. Traveling alone offers the least possibility of detection, but traveling in groups of two to three is more desirable.

d. Plan a primary and alternate route. Consider distance, cover, food, and water. The easiest and shortest route may not be the best.

e. Food and water are daily requirements. You can do without food for several days; water, however, is essential.

f. Move at night. Use the daylight to observe, plan, and rest in a hide position.

g. Linkup only during daylight hours. Place friendly lines under observation.

h. Attempt to identify the unit you approach, note their movements and routine.

i. After carefully considering your approach route, make voice contact with the unit as soon as possible.

11-2. **SURVIVAL**.

With training, equipment, and the ***will to survive***, you can overcome *any* obstacle you may face. ***You will survive.*** Understand the emotional states associated with survival. "Knowing thyself" is extremely important in a survival situation. It bears directly on how well you cope with serious stresses, anxiety, pain, injury, illness; cold, heat, thirst, hunger, fatigue, sleep deprivation, boredom, loneliness and isolation.

a. **Memory Aid**. You can overcome and reduce the shock of being isolated behind enemy lines if you keep the key word S-U-R-V-I-V-A-L foremost in your mind. Its letters can help guide you in your actions.

(1) S - Size up the situation, the surroundings, your physical condition, and your equipment.

(2) U - Undue haste makes waste; don't be too eager to move. Plan your moves.

(3) R - Remember where you are relative to friendly and enemy units and controlled areas; water sources (most important in the desert); and good cover and concealment. This information will help you make intelligent decisions.

(4) V - Vanquish fear and panic.

(5) I – Improvise/Imagine. You can improve your situation. Learn to adapt what is available for different uses. Use your imagination.

(6) V - Value living. Remember your goal - to get out alive. Remain stubborn. Refuse to give in to problems and obstacles. This will give you the mental and physical strength to endure.

(7) A - Act like the natives; watch their daily routines and determine when, where, and how they get their food and water.

(8) L - Live by your wits. Learn basic skills.

b. **Navigation**. In a survival situation, you might find yourself without a compass. The ability to determine direction can enable you to navigate back to your unit or to a friendly sanctuary. In sunlight, there are two simple ways to determine direction: the shadow-tip and the watch.

(1) ***Shadow Tip***. Use the sun to find approximate true North. Use this in light bright enough to cast shadows. Find a fairly straight stick about 3 feet long, and follow the diagram below (Figure 11-l).

Figure 11-1. SHADOW-TIP METHOD

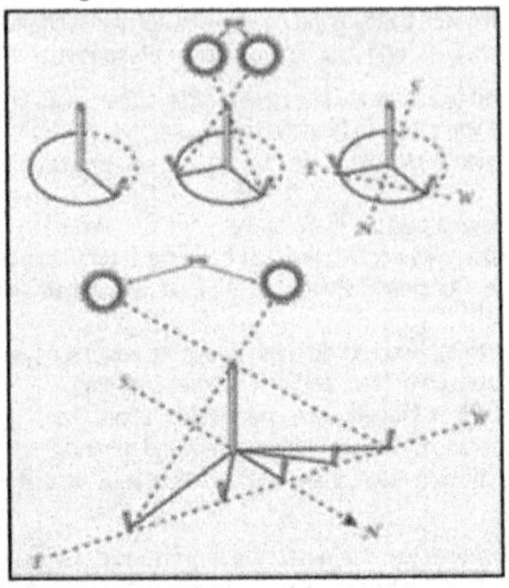

(2) ***Watch Method***. You can also determine direction using a watch (Figure 11-2). The steps you take will depend on whether you are in the northern Temperate Zone or in the southern Temperate Zone. The northern temperate zone is located between 23.4 north and 26.6 north. The southern Temperate Zone is located between 23.4 south and 66.6 south.

Figure 11-2. WATCH METHOD

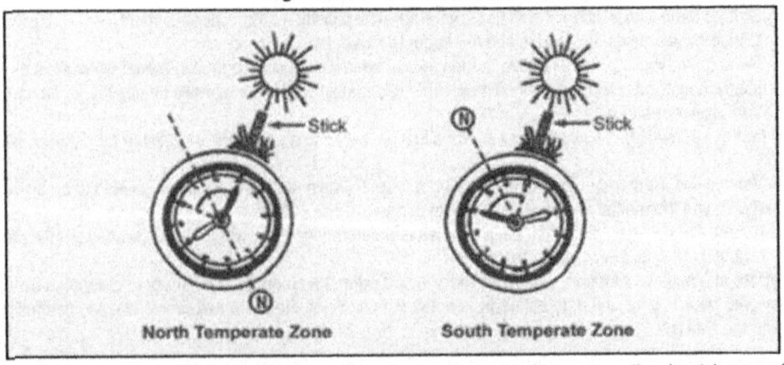

c. **Northern Temperate Zone**. Procedures in the northern temperate zone using a conventional watch are as follows:

(1) Place a small stick in the ground so that it casts a definite shadow.

(2) Place your watch on the ground so that the hour hand points toward and along the shadow of the stick.

(3) Find the point on the watch midway between the hour hand and 12 o'clock and draw an imaginary line from that point through and beyond the center of the watch. This imaginary line is a north-south line. You can then tell the other directions.

Note: If your watch is set on daylight savings time, then use the midway point between the hour hand and 1 o'clock to draw your imaginary line.

d. **Southern Temperate Zone**. Procedures in the southern temperate zone using a conventional watch are as follows:

(1) Place a small stick in the ground so that it casts a definite shadow.

(2) Place your watch on the ground so that 2 o'clock points to and along the shadow.

(3) Find the midway point between the hour and 12 o'clock and draw an imaginary line from the point through and beyond the center of the watch. This is a north-south line.

e. A hasty shortcut using a conventional watch is simply to point the hour hand at the sun in the northern temperate zone (or point the 12 at the sun in the southern temperate zone) and then follow the last step of the watch method above to find your directions. This shortcut, of course, is not as accurate as the regular method but quicker. Your situation will dictate which method to use.

11-3. **WATER**. Water is one of your most urgently needed resources in a survival situation. You can't live long without it, especially in hot areas where you lose so much through sweating. Even in cold areas, you need a minimum of 2 quarts of water a day to maintain efficiency. More than three-fourths of your body is composed of fluids. Your body loses fluid as a result of heat, cold, stress, and exertion. The fluid your body loses must be replaced for you to function effectively. So, one of your first objectives is to obtain an adequate supply of water.

a. **Purification**. Purify all water before drinking. Either--

(1) Boil it for at least one minute (plus 1 more minute for each additional 1,000 feet above sea level) or for a maximum of 10 minutes anywhere.

(2) Use water purification tablets.

(3) Add eight drops of 2-1/2 percent iodine solution to a quart (canteen full) of water. Let it stand for 10 minutes before drinking.

(4) Collect rain water directly in clean containers or on plants. This is generally safe to drink without purifying. Never drink urine or sea water; the salt content is too high. Use old, bluish sea ice, but newer, grayer ice may be salty. Glacier ice is safe to melt and drink.

b. **Desert Environment**. In a desert environment, water has a huge effect on Rangers. If a unit fails to plan properly for water, and resupply is unavailable, then they can run out of water. In the desert, look for four signs of water: animal trails, vegetation, birds, and civilization. Adequate water is critical in a hot environment if a unit is to survive and maintain the physical condition necessary to accomplish the mission. Unit leaders must enforce water discipline and plan for water resupply. The leader can use the following planning considerations for water resupply:

(1) Units average water consumption.

(2) Drop sites.

(3) Aviation support.

(4) DZ and LZ parties.

(5) Caches.

(6) Targets of opportunity (enemy).

c. **Survival Water Still**. Dig a belowground still (Figure 11-3). Select a site where you believe the soil will contain moisture (such as a dry stream bed or a spot where rain water has collected). The soil should be easy to dig, and be in sunlight most of the day:

(1) Dig a bowl-shaped hole about 3 feet across and 2 feet deep.

(2) Dig a sump in center of the hole. The depth and the perimeter of the sump will depend on the size of the container that you have to set inside of it. The bottom of the sump should allow the container to stand upright.

(3) Anchor the tubing to the bottom of the container by forming a loose overhand knot in the tubing.

(4) Place the container upright in the sump.

(5) Extend the unanchored end of the tubing up, over, and beyond the lip of the hole.

(6) Place plastic sheeting over the hole covering the edge with soil to hold it in place.

(7) Place a rock in the center of the plastic.

(8) Allow the plastic to lower into the hole until it is about 15 inches below ground level. The plastic now forms an inverted cone with the rock at its apex. Make sure that the apex of the cone is directly over your container. Also, make sure the plastic cone does not touch the sides of the hole, because the earth will absorb the condensed water.

(9) Put more soil on the edges of the plastic to hold it securely in place and to prevent loss of moisture.

(10) Plug the tube when not being used so that moisture will not evaporate.

d. You can drink water without disturbing the still by using the tube as a straw. You may want to use plants in the hole as a moisture source. If so, when you dig the hole you should dig out additional soil from the sides of the hole to form a slope on which to place the plants. Then proceed as above.

Figure 11-3. SURVIVAL WATER STILL

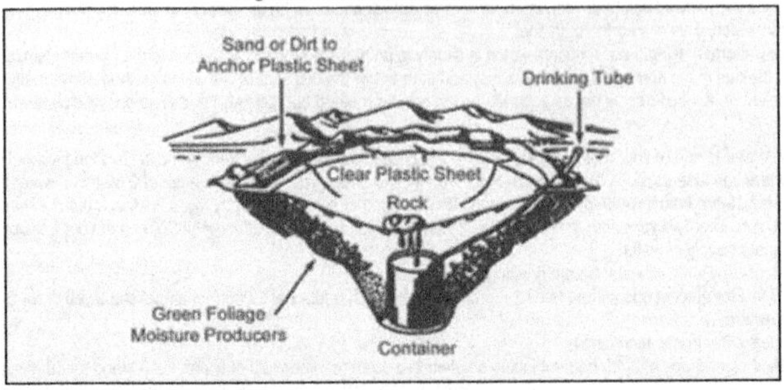

11-4. **PLANT FOOD**. There are many plants throughout the world. Tasting or swallowing even a small portion of some can cause severe discomfort, extreme internal disorders, or death. Therefore, if you have the slightest doubt as to the edibility of a plant, apply the universal edibility test described below before eating any part of it.

a. **Universal Edibility Test**. Before testing a plant for edibility, make sure there are a sufficient number of plants to make testing worth your time and effort. You need more than 24 hours to apply the edibility test outlined below:

(1) Test only one part of a potential food plant at a time.

(2) Break the plant into its basic components, leaves, stems, roots, buds, and flowers.

(3) Smell the food for strong or acid odors. Keep in mind that smell alone does not indicate a plant is edible.

(4) Do not eat for 8 hours before starting the test.

(5) During the 8 hours you are abstaining from eating, test for contact poisoning by placing a piece of the plant you are testing on the inside of your elbow or wrist. Usually 15 minutes is enough time to allow for reaction.

(6) During the test period, take nothing by mouth except purified water and the plant part being tested.

(7) Select a small portion and prepare it the way you plan to eat it.

(8) Before putting the prepared plant part in your mouth, touch a small portion (a pinch) to the outer surface of the lip to test for burning or itching.

(9) If after 3 minutes there is no reaction on your lip, place the plant part on your tongue, holding there for 15 minutes.

(10) If there is no reaction, thoroughly chew a pinch and hold it in your mouth for 15 minutes. *Do not swallow.*

(11) If no burning, itching, numbing, stinging, or other irritation occurs during the 15 minutes, swallow the food.

(12) Wait 8 hours. If any ill effects occur during this period, induce vomiting and drink a lot of water.

(13) If no ill effects occur, eat 1/2 cup of the same plant part prepared the same way. Wait another 8 hours. If no ill effects occur, the plant part as prepared is safe for eating.

b. **Poisonous Plants**. ***Do not*** eat unknown plants that have the below characteristics:

(1) Have a milky sap or a sap that turns black when exposed to air.

(2) Are mushroom like.

(3) Resemble onion or garlic.

(4) Resemble parsley, parsnip, or dill.

(5) Have carrot-like leaves, roots, or tubers.

11-5. **ANIMAL FOOD**. Animal food contains the most food value per pound. Anything that creeps, crawls, swims, or flies is a possible source of food, however you must first catch, kill and butcher it before this is possible. There are numerous methods for catching fish and animals in a survival situation. You can catch fish by using a net across a small stream, (Figure 11-4) or by making fish traps and baskets. Improvise fish hooks and spears as indicated in Figure 11-5, and use them for conventional fishing, spearing and digging.

Figure 11-4. SETTING A GILL NET IN THE STREAM

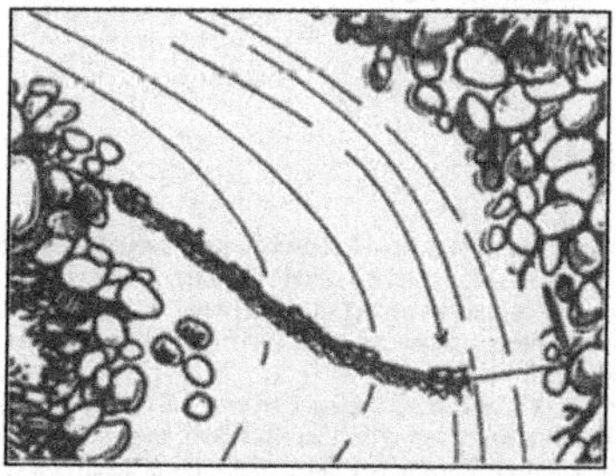

Figure 11-5. SPEARS AND FISH HOOKS

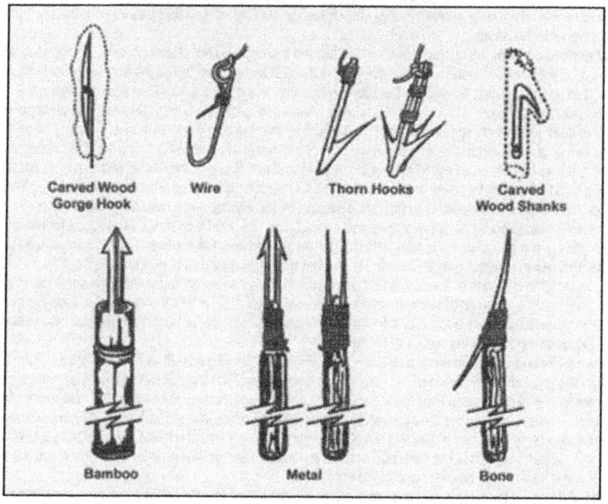

11-6. **TRAPS AND SNARES**. For an unarmed survivor or evader, or when the sound of a rifle shot could be a problem, trapping or snaring wild game is a good alternative. Several well-placed traps have the potential to catch much more game than a Ranger with a rifle is likely to shoot.

a. To be effective with any type of trap or snare, you must---

(1) Know the species of animal you intend to catch.

(2) Know how to construct a proper trap.

(3) Avoid alarming the prey with signs of your presence.

b. There are no catchall traps you can set for all animals. You must determine what species are in a given area and set your traps specifically with those animals in mind. Look for the following:

(1) Runs and trails.
(2) Tracks.
(3) Droppings.
(4) Chewed or rubbed vegetation.
(5) Nesting or roosting sites.
(6) Feeding and watering areas.

c. Position your traps and snares where there is proof that animals pass through. You must determine if it is a "run" or a "trail." A trail will show signs of use by several species and will be rather distinct. A run is usually smaller and less distinct and will only contain signs of one species. You may construct a perfect snare, but it will not catch anything if haphazardly placed in the woods. Animals have bedding areas, waterholes, and feeding areas with trails leading from one to another. You must place snares and traps around these areas to be effective.

d. An evader in a hostile environment must conceal traps and snares. It is equally important, however, to avoid making a disturbance that will alarm the animal and cause it to avoid the trap. Therefore, if you must dig, remove all fresh dirt from the area. Most animals will instinctively avoid a pitfall-type trap. Prepare the various parts of a trap or snare away from the site, carry them in, and set them up. Such actions make it easier to avoid disturbing the local vegetation, thereby alerting the prey. Do not use freshly cut, live vegetation to construct a trap or snare. Freshly cut vegetation will "bleed" sap that has an odor the prey will be able to smell. It is an alarm signal to the animal.

e. You must remove or mask the human scent on and around the trap you set. Although birds do not have a developed sense of smell, nearly all mammals depend on smell even more than on sight. Even the slightest human scent on a trap will alarm the prey and cause it to avoid the area. Removing the scent from a trap is difficult but masking it is relatively easy. Use the fluid from the gall and urine bladders of previous kills. Do not use human urine. Mud, particularly from an area with plenty of rotting vegetation, is also good. Use it to coat your hands when handling the trap and to coat the trap when setting it. In nearly all parts of the world, animals know the smell of burned vegetation and smoke. It is only when a fire is actually burning that they become alarmed. Therefore, smoking the trap parts is an effective means to mask your scent. If one of the above techniques is not practical, and if time permits, allow a trap to weather for a few days and then set it. Do not handle a trap while it is weathering. When you position the trap, camouflage it as naturally as possible to prevent detection by the enemy and to avoid alarming the prey.

f. Traps or snares placed on a trail or run should use canalization. To build a channel, construct a funnel-shaped barrier extending from the sides of the trail toward the trap, with the narrowest part nearest the trap. Canalization should be inconspicuous to avoid alerting the prey. As the animal gets to the trap, it cannot turn left or right and continues into the trap. Few wild animals will back up, preferring to face the direction of travel. Canalization does not have to be an impassable barrier. You only have to make it inconvenient for the animal to go over or through the barrier. For best effect, the canalization should reduce the trail's width to just slightly wider than the targeted animal's body. Maintain this constriction at least as far back from the trap as the animal's body length, then begin the widening toward the mouth of the funnel.

(1) Use a treadle snare against small game on a trail (Figure 11-6). Dig a shallow hole in the trail. Then drive a forked stick (fork down) into the ground on each side of the hole on the same side of the trail. Select two fairly straight sticks that span the two forks. Position these two sticks so that their ends engage the forks. Place several sticks over the hole in the trail by positioning one end over the lower horizontal stick and the other on the ground on the other side of the hole. Cover the hole with enough sticks so that the prey must step on at least one of them to set off the snare. Tie one end of a piece of cordage to a twitch-up or to a weight suspended over a tree limb. Bend the twitch-up or raise the suspended weight to determine where you will tie a 5 centimeter or so long trigger. Form a noose with the other end of the cordage.

(2) Route and spread the noose over the top of the sticks over the hole. Place the trigger stick against the horizontal sticks and route the cordage behind the sticks so that the tension of the power source will hold it in place. Adjust the bottom horizontal stick so that it will barely hold against the trigger. As the animal places its foot on a stick across the hole, the bottom horizontal stick moves down, releasing the trigger and allowing the noose to catch the animal by the foot. Because of the disturbance on the trail, an animal will be wary. You must therefore use canalization.

Figure 11-6. TREADLE SNARE

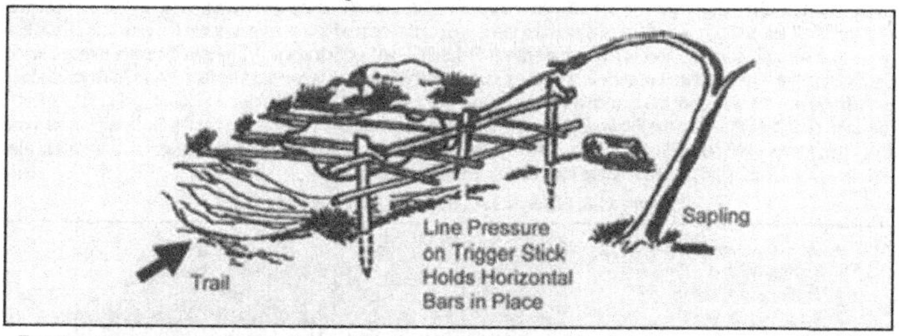

g. Trapping game can be accomplished through the use of snares, traps, or deadfalls. A snare is a noose that will slip and strangle or hold any animal caught in it. You can use inner core strands of parachute suspension lines, wire, bark of small hardwood saplings as well as hide strips from previously caught animals to make snares.

(1) The drag noose snare, Figure 11-7, is usually the most desirable in that it allows you to move away from the site, plus it is one of the easiest to make and fastest to set. It is especially suitable for catching rabbits. To make the drag noose snare, make a loop in the string using a bowline or wireman's knot. (When using wire, secure the loop by intertwining the end of the wire with the wire at the top of the loop). Pull the other end of the string (or wire) through the loop to form a noose that is large enough for the animal's head but too small for its body; tie the string (or attach the wire) to a sturdy branch. The branch should be large enough to span the trail and rest on the bush or support (two short forked sticks) you have selected. A snared animal will dislodge the drag stick, pulling it until it becomes entangled in the brush. Any attempt to escape tightens the noose, strangling or holding the animal.

Figure 11-7. DRAG NOOSE SNARE

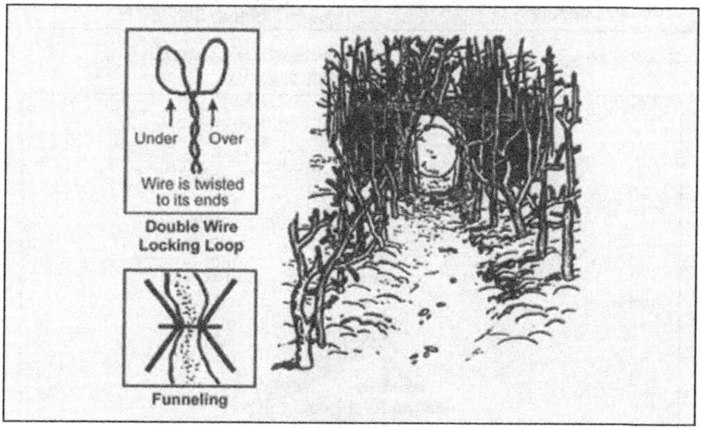

(2) Another type snare is the locking type snare loop (Figure 11-8) that will lock when pulled tight, ensuring the snared animal cannot escape. Use lightweight wire to make this snare, such as, trip wire or the wire from a vehicle or aircraft electrical system. To construct this snare, cut a piece of wire twice the length of the desired snare wire. Double the wire and attach the running ends to a securely placed object, such as the branch of a tree. Place a stick about 1/2 inch in diameter through the loop end of the wire; holding the wire taut, turn the stick in a winding motion so that the wire is twisted together. You should have four to five twists per inch. Detach the wire from the branch and then remove the loop from the stick; make a Figure Eight in the l/2-inch loop by twisting the loop over itself then fold the Figure Eight so the small loops are almost overlapping; run the loose wire ends through these loops. This forms a stiff noose that is strong. Tie the loose end to the stick (for a drag noose square) or branch you are using to complete the snare. This is an excellent snare for catching large animals.

Figure 11-8. FORMING A LOCKING-TYPE SNARE LOOP

(3) Another means of obtaining game is the use of the deadfall trap (Figure 11-9 and Figure 11-10).

Figure 11-9. TRIGGER WITH DEADFALL

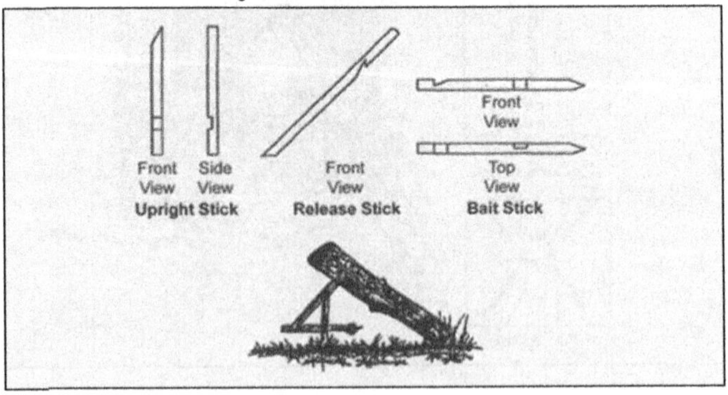

Figure 11-10. TRIP-STRING DEADFALL TRAP

h. Once you have obtained your fish or game, you must clean/butcher and cook/store it. Improper cleaning storing can result in inedible fish and game.

(1) ***Fish***. You must know how to tell if fish are free of bacterial decomposition that makes the fish dangerous to eat. Although cooking may destroy the toxin from bacterial decomposition, do not eat fish that appear spoiled.

(a) Signs of spoilage are:

- A peculiar odor.
- A suspicious color. (Gills should be red or pink. Scales should be a pronounced-not faded shade of gray).
- A dent remaining after pressing the thumb against the flesh.
- A slimy rather than moist or wet body.
- A sharp or peppery taste.

(b) Eating spoiled or poisoned fish may cause diarrhea, nausea, cramps, vomiting, itching; paralysis, or a metallic taste in the mouth. These symptoms appear suddenly 1 to 6 hours after eating. If you are near the sea, drink sea water immediately upon on set of such symptoms and force yourself to vomit.

(c) Fish spoil quickly after death, especially on a hot day, so prepare fish for eating as soon as possible after you catch them.

(d) Cut out the gills and large blood vessels that lie next to the backbone. (You can leave the head if you plan to cook the fish on a spit).

(e) Gut fish that are more than 4 inches long cut along the abdomen and scrape out the intestines.

(f) Scale or skin the fish.

(g) You can impale a whole fish on a stick and cook it over an "open fire." However, boiling the fish with the skin on is the best way to get the most food value. The fats and oil are under the skin, and by boiling the fish, you can save the juices for broth. Any of the methods used for cooking plant food can be used for cooking fish. Fish is done when the meat flakes off.

(h) To dry fish in the sun, hang them from branches or spread them on hot rocks. When the meat has dried, splash it with sea water, if available, to salt the outside. Keep seafood only if it is well dried or salted.

(2) ***Snakes***. All poisonous and nonpoisonous fresh water and land snakes are edible.
To prepare snakes for eating use the following steps (Figure 11-11):

(a) Grip the snake firmly behind the head and cut off the head with a knife.

(b) Slit the belly and remove the innards. (You can use the innards for baiting traps and snares).

(c) Skin the snake. (You can use the skin for improvising, belts, straps, or similar items).

CAUTION

Take extreme care in securing snakes--the bite of some poisonous snakes can be fatal. Even after a snake's head is cut off, its reflex action can cause it to bite, injecting poison. The best time to capture snakes is in the early morning or late evening when temperatures are low and they move slow. Kill or use a long stick to pin down its head and capture it. To pick up a snake, place the index finger on the top rear of its head with your thumb and middle finger on either side of the head behind the jaws. Keep your index finger on top of snake's head to prevent it from turning inside its skin and biting you.

Figure 11-11. CLEANING A SNAKE

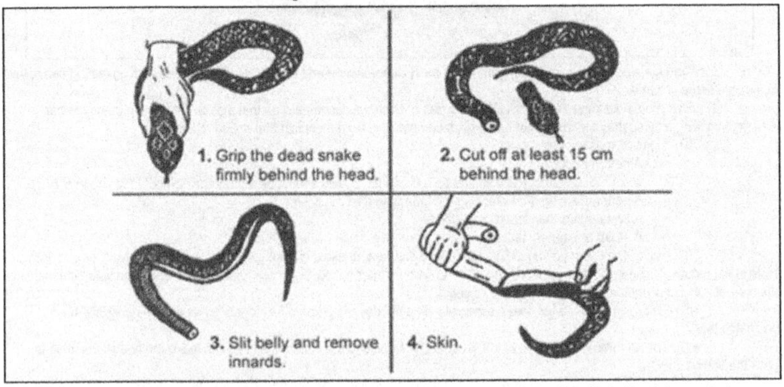

(3) ***Fowl***. Your first step after killing a fowl for eating or preserving is to pluck its feathers. If plucking is impractical, you can skin the fowl. Keep in mind, however, that a fowl cooked with the skin on retains more food value. Waterfowl are easier to pluck while dry, but other fowl are easier to pluck after scalding. After you pluck the fowl--

(a) Cut off its neck close to the body.

(b) Cut an incision in the abdominal cavity and clean out the insides. Save the neck, liver, and heart for stew. Thoroughly clean and dry the entrails to use for cordage.

(c) Wash out the abdominal cavity with fresh clean water. You can boil fowl or cook it on a spit over a fire. You should boil scavenger birds such as vultures and buzzards for at least 20 minutes to kill any parasites. Use the feathers from fowl for insulating your shoes clothing, or bedding. You can also use feathers for fish lures.

(4) **Medium-Sized Mammals**. The game you trap or snare will generally be alive when you find it and therefore dangerous. Be careful when you approach a trapped animal. Use a spear or club to kill it so you can keep a safe distance from it. After you kill an animal, immediately bleed it by cutting its throat. If you must drag the carcass any distance, do so before you cut off the hide so that the carcass is protected from dirt and debris that might contaminate it. Clean the animal near a stream if possible so that you can wash and cool the carcass and edible parts. Fleas and parasites will leave a cooled body so if the situation allows, wait until the animal cools before cleaning and dressing the carcass. To skin and dress the animal (Figure 11-12 and 11-13).

Figure 11-12. SKINNING AND BUTCHERING LARGE GAME

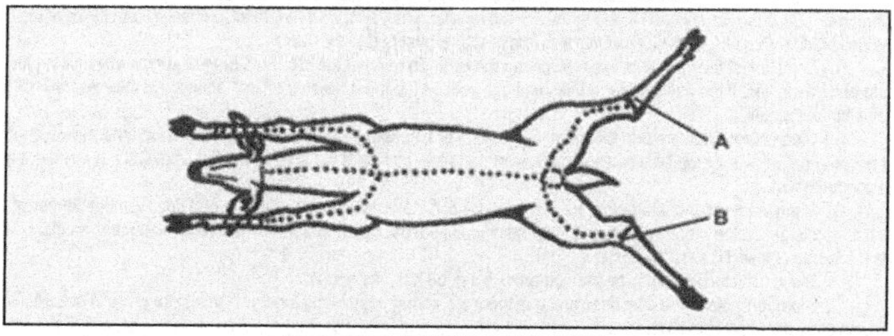

Figure 11-13. SKINNING SMALL GAME

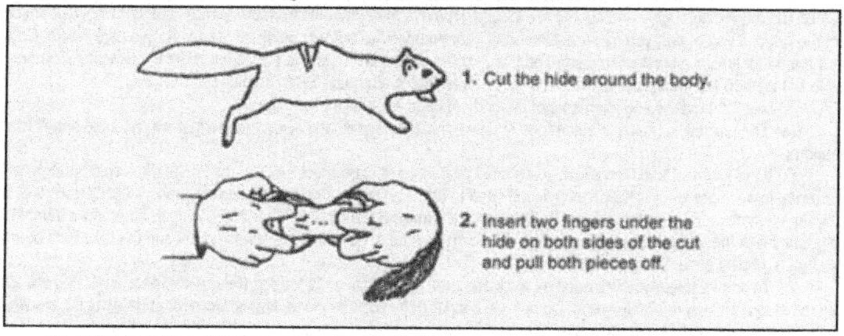

(a) Place carcass, belly up, on a slope if available. You can use rocks or brush to support it.

(b) Remove genitals or udder.

(c) Remove musk glands to avoid tainting meat.

(d) Split hide from tail to throat. Make the cut shallow so that you do not pierce the stomach.

(e) Insert your knife under the skin, taking care not to cut into the body cavity. Peel the hide back several inches on each side to keep hair out of the meat.

(f) Open the chest cavity by splitting the sternum. You can do this by cutting to one side of the sternum where the ribs join.

(g) Reach inside and cut the windpipe and gullet as close to the base of the skull as possible.

(h) With the forward end of the intestinal tract free, work your way to the rear lifting out internal organs and intestines. Cut only where necessary to free them.

(i) Carefully cut the bladder away from the carcass so that you do not puncture the bladder (urine can contaminate meat). Pinch the urethra tightly and cut it beyond the point you are pinching.

(j) Remove the bladder.

(k) From the outside of the carcass, cut a circle around the anus.

(l) Pull the anus into the body cavity and out of the carcass.

(m) Lift or roll the carcass to drain all blood.

Note: Try to save as much blood as you can as it is a vital source of food and salt. Boil the blood.

(n) Remove the hide, and make cuts along the inside of the legs to just above the hoof or paw. Peel the skin back, using your knife in a slicing motion to cut the membrane between the skin and meat. Continue this until the entire skin is removed.

(o) Most of the entrails are usable. The heart, liver, and kidneys are edible. Cut open the heart and remove the blood from its chambers. Slice the kidneys and if enough water is available, soak or rinse them. In all animals except those of the deer family, the gall bladder (a small, dark-colored, clear-textured sac) is attached to the liver.

(p) Sometimes, the sac looks like a blister on the liver. To remove the sac, hold the top portion of it and cut the liver around and behind the sac. If the gall bladder breaks and gall gets on the meat, wash it off immediately so the meat will not become tainted. Dispose of the gall.

(q) Clean blood splattered on the meat will glaze over and help preserve the meat for a short time. However, if an animal is not bled properly, the blood will settle in the lowest part of its body and will spoil in a short time. Cut out any meat that becomes contaminated.

(r) When temperatures are below 40 degrees, you can leave meat hanging for several days without danger of spoilage. If maggots get on the meat, remove the maggots and cut out the discolored meat. The remaining meat is edible. Maggots, which are the larvae of insects, are also edible.

(s) Blood, which contains salts and nutrients, is a good base for soups.

(t) Thoroughly clean the intestines and use them for storing or smoking food or lashings for general use. Make sure they are completely dry to preclude rotting.

(u) The head of most animals contains a lot of meat, which is relatively easy to get. Skin the head, saving the skin for leather. Clean the mouth thoroughly and cut out the tongue. Remove the outer skin from the tongue after cooking. Cut or scrape the meat from the head. If you prefer, you can roast the head over an open fire before cutting off the meat. Eyes are edible. Cook them but discard the retina (this is a plastic like disc). The brain is also edible; in fact, some people consider it a delicacy. The brain is also used to tan leather, the theory being that the brain of an animal is adequate to tan its hide.

(v) Use the tendons and ligaments of the body of large animals for lashings.

(w) The marrow in bones is a rich-food source. Crack the bones and scrap out the marrow, and use bones to make weapons.

(x) If the situation and time allow, you should preserve the extra meat for later use. If the air is cold enough, you can freeze the meat. In warmer climates however, you will need to use a drying or smoking process to preserve it. One night of heavy smoking will make meat edible for about 1 week. Two nights will make it remain edible for 2 to 4 weeks. To prepare meat for drying or smoking, cut it with the grain in quarter inch strips. To air dry the meat, hang it in the wind and hot sun out the reach of animals; cover it so that blow flies cannot land on it.

(y) To smoke meat, you will need an enclosed area – for instance, a teepee (Figure 11-14) or a pit. You will also need wood from deciduous trees, preferably green. Do not use conifer trees such as pines, firs, spruces, or cedars, as the smoke from these trees give the meat a disagreeable taste.

(z) When using the para-teepee or other enclosed area with a vent at the top, set the fire in the center and let it burn down to coals, then stoke it with green wood. Place the strips of meat on a grate or hang them from the top of the enclosure so that they are about 2 feet above the smoking coals. To use the pit method of smoking meat dig, a hole about 3 feet (1 meter) deep and 1 1/2 feet (1/2 meter) in diameter. Make a fire at the bottom of the hole. After it starts burning well, add chipped green wood or small branches of green wood to make it smoke. Place a wooden grate about 1 1/2 feet (1/2 meter) above the fire and lay the strips of meat on the grate. Cover the pit with poles, boughs, leaves, or other material.

11-7. **SHELTERS**. A shelter can protect you from the sun, insects, wind, rain, snow, hot or cold temperatures, and enemy observation. In some areas your need for shelter may take precedence over your need for food, possibly even your need for water. After determining your shelter site, you should keep in mind the type of shelter (protection) you need. You need to know how to make different types of shelters. Only two are described in this handbook. Additional information is available in FM 21-76.

a. **Planning Considerations**.

- How much time and effort are needed to build the shelter?
- Will the shelter adequately protect you from the elements (rain, snow, wind, sun, and so on)?
- Do you have tools to build it? If not, can you improvise tools from materials in the area?
- Do you have the type and amount of manmade materials needed to build it? If not, are there sufficient natural materials in the area?

Figure 11-14. SMOKING MEAT

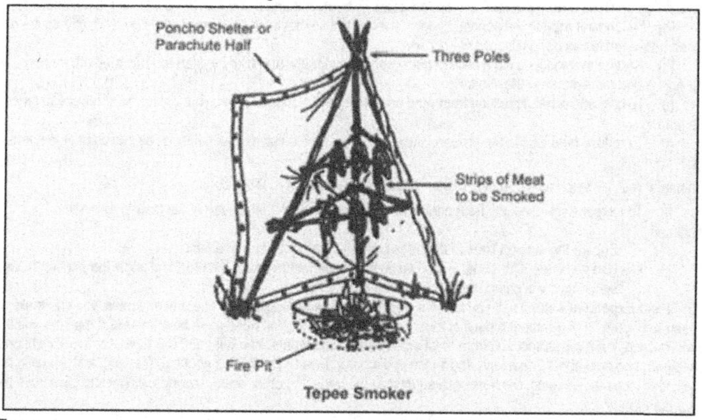

b. **Types**.

(1) ***Poncho Lean-To***. It takes only a short time and minimal equipment to build this lean-to (Figure 11-15). You need a poncho, 6 to 10 feet of rope, three stakes about 6 inches long, and two trees (or two poles) 7 to 9 feet apart. Before you select the trees you will use (or decide where to place the poles), check the wind direction. Make sure the back of your lean-to will be into the wind. To make the lean-to:

(a) Tie off the hood of the poncho. To do this, pull the draw cord tight; roll the hood long ways, fold it into thirds, and tie it with the draw cord.

(b) Cut the rope in half. On one long side of the poncho, tie half of the rope to one corner grommet, and the other half to the other corner grommet.

(c) Attach a drip stick (about a 4-inch stick) to each rope 1/4 to 3/4 inches away from the grommet. These drip sticks will keep rainwater from running down the ropes into the lean-to. Using drip lines is another way to prevent dripping inside the shelter. Tie lines or string about 4 inches long to each grommet along the top edge of the shelter. This allows water to run to and down the line without dripping into the shelter.

(d) Tie the ropes about waist high on the trees (uprights). Use a round turn and two half hitches with quick-release knot.

(e) Spread the poncho into the wind and anchor to the ground. To do this, put three sharpened sticks through the grommets and into the ground.

Figure 11-15. PONCHO LEAN-TO

(f) If you plan to use the lean-to for more than one night, or if you expect rain, make a center support to the lean-to. You can do this by stretching a rope between two upright poles or trees that are in line with the center of the poncho.

(g) Tie another rope to the poncho hood; pull it upward so that it lifts the center of the poncho, and tie it firmly to the rope stretched between the two uprights.

(h) Another method is to cut a stick to place upright under the center of the lean-to. This method, however, will restrict your space and movements in the shelter.

(i) To give additional protection from wind and rain, place boughs, brush, your rucksack, or other equipment at the sides of the lean-to.

(j) To reduce heat loss to the ground, place some type of insulating material, such as leaves or pine needles, inside your lean-to.

Note: When at rest, as much as 80 percent of your body heat can be lost to the ground.

(l) To increase your security from enemy observation, lower the silhouette of the lean-to by making two modifications.

- Secure the support lines to the trees knee-high rather than waist-high.
- Use two knee-high sticks in the two center grommets (sides of lean-to), and angle the poncho to the ground, securing it with sharpened sticks as above.

(2) ***Field Expedient Lean-To.*** If you are in a wooded area and have sufficient natural materials, you can make an expedient lean-to (Figure 11-16) without the aid of tools or with only a knife. You need more time to make it than the shelter previously mentioned, but it will protect you from most environmental elements. You will need two trees, (or two upright poles), about 6 feet apart; one pole about 7 feet long and 1 inch in diameter. Five to eight poles about 10 feet long and 1 inch in diameter for beams, cord or vines for securing, the horizontal support to the trees and other poles, saplings, or vines to crisscross the beams. To make this lean-to:

Figure 11-16. FIELD-EXPEDIENT LEAN-TO

(a) Tie the 7-foot pole to the two trees at point about waist to chest high. This is your horizontal support. If there is a fork in the tree, you can rest the pole in it instead of tying the pole in place. If a standing tree is not available, construct a bipod using an Y-shaped sticks or two tripods.

(b) Place one end of the beams (10-foot poles) one side of the horizontal support. As with all lean-to type shelters, make sure the backside of the lean-to is placed into the wind.

(c) Crisscross sapling or vines on the beams.

(d) Cover the framework with brush, leaves, pine needles, or grass, starting at the bottom and working your way up like shingling.

(e) Place straw, leaves, pine needles, or grass inside the shelter for bedding.

(f) In cold weather, you can add to the comfort of your lean-to by building a fire-reflector wall (Figure 11-16). Drive four stakes about 4 feet long into the ground to support the wall. Stack green logs on top of one another between the support stales. Bind the top of the support stakes so the green logs will stay in place. Fill in the spaces between the logs with twigs or small branches. With just a little more effort, you can have a drying rack. Cut a few 3/4-inch diameter poles. The length depends on distance between the lean-to support and the top of the fire-reflector wall. Lay one end of the poles on the lean-to horizontal support and the other ends on top of the reflector wall. Place and tie into place smaller sticks across these poles. You now have a place to dry clothes, meat, or fish.

11-8. **FIRES**. A fire can full fill several needs. It can keep you warm, it can keep you dry: you can use it to cook food, to purify water, and to signal. It can also cause you problems when you are in enemy territory: it creates smoke, which can be smelled and seen from a long distance: It causes light which can be seen day or night and it leaves signs of your presence. Remember you should always weigh your need for a fire against your need to avoid enemy protection. When operating in remote areas you should always take a supply of matches in a waterproof case and keep them on your person.

a. **Selection**. When selecting a site to build a fire, you should consider the following:

(1) The area (terrain and climate) in which you are operating.
(2) The material and tools available.
(3) How much time you have.
(4) Why you need a fire.
(5) The nearness of the enemy.

b. **Preparation**. If you are in a wooded or brush-covered area, clear brush away, and scrape the surface soil from the spot you selected. The cleared circle should be at least 3 feet (1 meter) in diameter so that there is little chance of the fire spreading. To prepare the site for a fire, ensure that it is dry and that it--look for a dry spot that--

(1) Offers protection from the wind.
(2) Is suitably placed in relation to your shelter (if any).
(3) Concentrates the heat in the direction you desire.
(4) Has a supply of wood or other fire burning material.

c. **Dakota Fire Hole**. In some situations, an underground fireplace will best meet your needs. It conceals the fire to some extent and serves well for cooking food. To make an underground fireplace or Dakota fire hole (Figure 11-17):

Figure 11-17. DAKOTA FIRE HOLE

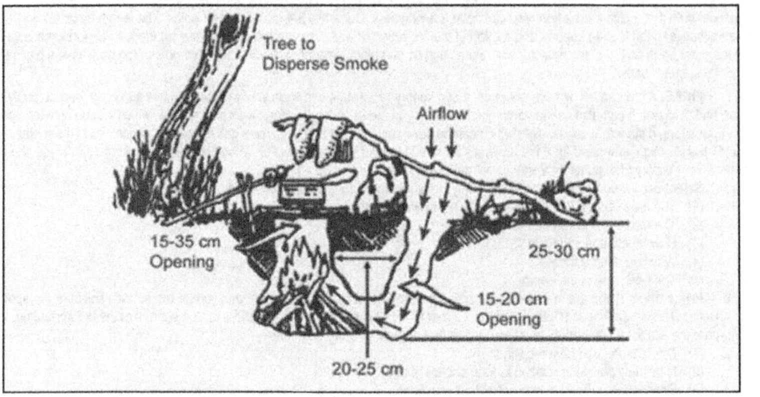

(1) Dig a hole in the ground.

(2) On the upwind side of this hole, poke one large connecting hole for ventilation.

d. **Aboveground fire**. If you are in a snow covered or wet area, you can use green logs to make a dry base for your fire (Figure 11-18). Trees with wrist-size trunks are easily broken in extreme cold. Cut or break several green logs and lay them side by side on top of the snow. Add one or two more layers, laying the top layer logs in a direction opposite those of the layer below it.

Figure 11-18. BASE FOR FIRE IN SNOW COVERED AREA

e. **Methods**. There are several methods for laying a fire for quick fire making. Three easy methods are Tepee, lean-to, and cross-ditch. Tepee (Figure 11-19). Arrange tinder and a few sticks of kindling in the shape of a cone. Fire the center. As the cone burns away, the outside logs will fall inward, feeding the heart of the fire. This type of fire burns well even with wet wood.

(1) ***Lean-To*** (Figure 11-19). Push a green stick into the ground at a 30 degree angle. Point the end of the stick in the direction of the wind. Place some tinder (at least a handful) deep inside this lean-to stick. Light the tinder. As the kindling catches fire from the tinder, add more kindling.

Figure 11-19. METHODS FOR LAYING A FIRE

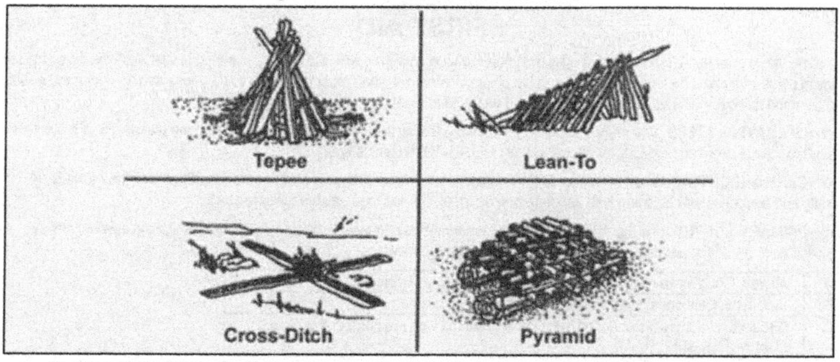

(2) ***Cross-Ditch*** (Figure 11-19). Scratch a cross about 1 foot in size in the ground. Dig the cross 3 inches deep. Put a large wad of tinder in the middle of the cross. Build a kindling pyramid above the tinder. The shallow ditch allows air to sweep under the fire to provide a draft.

NOTES

Chapter 12
FIRST AID

Patrolling, more than some other types of missions, puts Rangers in harms way. CASEVAC planning is vital. Also, because trained medical personnel might be unavailable at the initial point of injury, everyone must know how to diagnose and treat injuries, wounds, and common illnesses. The unit should also have a plan for handling KIAs.

12-1. **LIFESAVING STEPS**. Whatever the injury, (1) open the airway and restore breathing; (2) stop the bleeding and protect the wound; (3) check, treat and monitor for shock; and (4) MEDEVAC the casualty.

12-2. **CARE UNDER FIRE**. When still under fire, (1) maintain situational awareness; (2) return fire; (3) protect the casualty; (4) Identify and control severe bleeding with bandage or tourniquet; (5) and move the casualty to cover.

12-3. **PRIMARY SURVEY**. Use the ABC's to help you remember how to identify and deal with life-threatening injuries, airway complications, breathing problems, or uncontrolled hemorrhaging (bleeding).

A	Airway. Open airway by patient position or with airway adjuncts.
B	Breathing. Seal open chest wounds with occlusive dressing.
C	Circulation. Identify uncontrolled bleeding and control with pressure or tourniquet. Start IV if needed.
D	Disability. Determine Level of consciousness.
E	Exposure. Fully expose patient. (Environment dependent)

12-4. **AIRWAY MANAGEMENT**. The *airway* is most often obstructed at the base of the tongue.

a. If this happens, open the airway using the chin lift (nontrauma, Figure 12-1) or jaw thrust (trauma, Figure 12-2).

Figure 12-1. CHIN LIFT

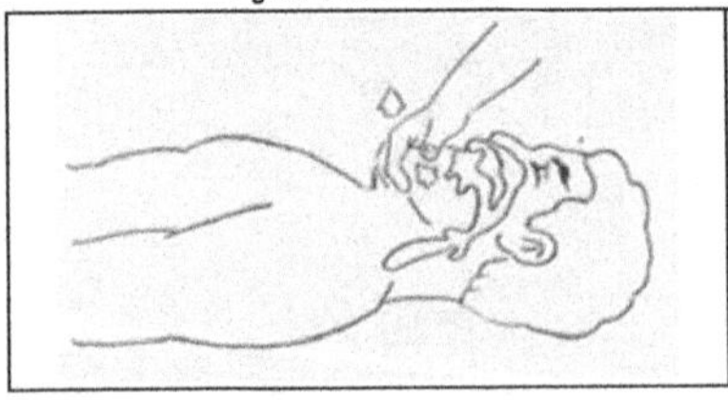

Figure 12-2. JAW THRUST

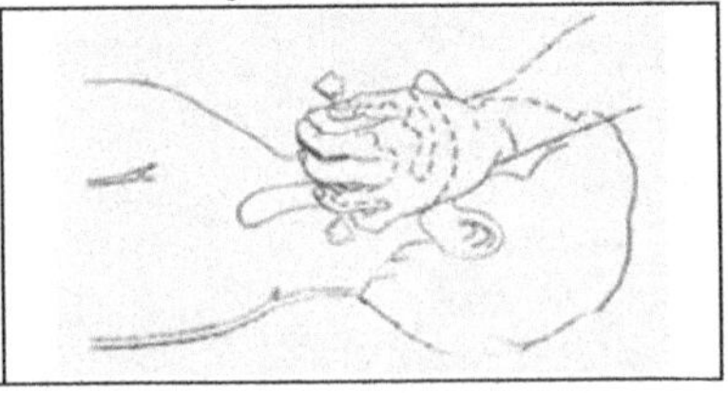

b. **Remove** debris (teeth, blood clots, bone) from the oral cavity; use suction if you have it; and place airway adjuncts to allow the victim to breathe through their nose (Figure 12-3) or mouth (Figure 12-4).

Figure 12-3. NASAL AIRWAY

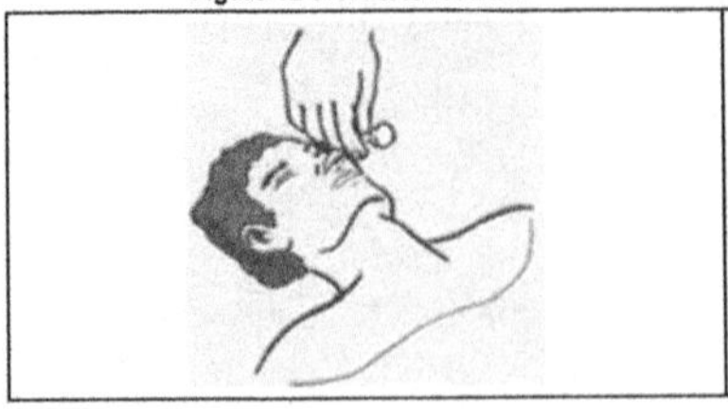

Figure 12-4. MOUTH AIRWAY

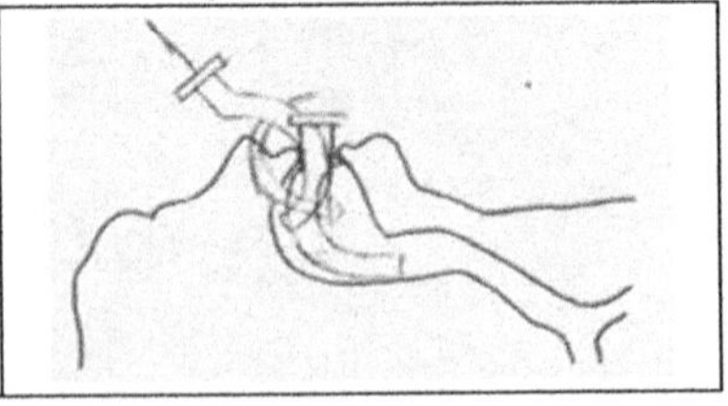

12-5. **BREATHING**. If the patient is having difficulty breathing, (1) expose the chest and identify open chest injuries; (2) place occlusive dressing (plastic and tape, or a COTS chest dressing) over open entry and exit chest wounds; and (3) place the patient on his injured side, or position him where he can breathe most comfortably.

12-6. **BLEEDING**. Quickly identify and control bleeding. (1) Apply a standard or pressure dressing. (2) If this does not control the bleeding, immediately apply a tourniquet. (3) Reassess dressings frequently to ensure bleeding has been controlled.

12-7. **SHOCK**. Shock causes an inadequate flow of oxygen to the tissues. Shock can cause bleeding (hemorrhagic) or not (nonhemorrhagic). Signs and symptoms of shock can include increased pulse, increased respiration; and decreased level of consciousness. (1) Open the victim's airway; (2) restore his breathing; (3) control the bleeding; (4) initiate an IV or a saline lock; and (5) monitor his condition.

12-8. **EXTREMITY INJURIES**. Identify and control bleeding. If you suspect a fracture, splint it as it lies. Do not reposition injured extremity.

12-9. **ABDOMINAL INJURIES**. Identify and control bleeding. Treat for shock. (1) If internal organs are exposed, cover them with dry, sterile dressing. Do not place organs back in wound area. Do not handle internal organs. (2) Place patient in position of comfort. Flex knees to relax abdomen. (3) Do not give anything by mouth to the patient.

12-10. **BURNS**. Remove patient from burn source. Then (1) remove all clothing and jewelry from the area; (2) cover with dry, sterile dressings, and ensure fingers and toes have dressings between them before covering entire area; and (3) Evacuate immediately any casualties with burns of the face, neck, hands, genitalia, or over 20 percent of his body surface.

12-11. **WEATHER (HEAT AND COLD) INJURIES**. Tables 12-1, 12-2, and 12-3 show first aid for (respectively) heat injuries, cold injuries, and environmental injuries.

Table 12-1. HEAT INJURIES

INJURY	*SIGNS/SYMPTOMS*	*FIRST AID*
Heat Cramps	Casualty experiences muscle cramps in arms, legs and/or stomach, may also have wet skin and extreme thirst.	1. Move the casualty to a shaded area and loosen clothing. 2. Allow casualty to drink 1 quart of cool water slowly per hour. 3. Monitor casualty and provide water as needed. 4. Seek medical attention if cramps persist.
Heat Exhaustion	Casualty experiences loss of appetite, headache, excessive sweating, weakness or faintness, dizziness, nausea, muscle cramps. The skin is moist, pale, and clammy.	1. Move the casualty to a cool, shaded area and loosen clothing. 2. Pour water on casualty and fan to increase cooling effect of evaporation. 3. Provide at least one quart of water to replace lost fluids. 4. Elevate legs. 5. Seek medical aid if symptoms continue.
Heatstroke (Sunstroke)	Casualty stops sweating (hot, dry skin), may experience headache, dizziness, nausea, vomiting, rapid pulse and respiration, seizures, mental confusion. Casualty may suddenly collapse and lose consciousness. THIS IS A MEDICAL EMERGENCY!	1. Move casualty to a cool, shaded area, loosen clothing, and remove outer clothing if the situation permits. 2. Immerse in cool water. If cool bath is not available, massage arms and legs with cool water. Fan casualty to increase the cooling effect of evaporation. 3. If conscious, slowly consume one quart of water. 4. SEEK MEDICAL AID AND EVACUATE AS SOON AS POSSIBLE. Perform any lifesaving measures.

Table 12-2. COLD INJURIES

INJURY	*SIGNS/SYMPTOMS*	*FIRST AID*
Chilblain	Red, swollen, hot, tender, itching skin. Continued exposure may lead to infected (ulcerated bleeding) skin lesions.	1. Area usually responds to locally applied warming (body heat). 2. Do Not rub or massage area. 3. Seek medical treatment.
Immersion (trench) foot	Affected parts are cold, numb, and painless. As parts warm, they may be hot, with burning and shooting pains. Advanced stage: skin pale with bluish cast: pulse decreases, blistering, swelling, heat hemorrhages, and gangrene may follow.	1. Gradual warming by exposure to warm air. 2. DO NOT massage or moisten skin. 3. Protect affected parts from trauma. 4. Dry feet thoroughly: avoid walking. 5. Seek medical treatment.
Frostbite	***SUPERFICIAL*** Redness, blisters in 24-36 hours and sloughing of the skin. ***DEEP*** Preceded by superficial frostbite; skin is painless, pale- yellowish, waxy, "wooden" or solid to touch, blisters form in 12-36 hours	***SUPERFICIAL*** 1. Keep casualty warm; gently warm affected parts. 2. Decrease constricting clothing, increase exercise and insulation. ***DEEP*** 1. Protect the part from additional injury. 2. Seek medical treatment as fast as possible.
Snow Blindness	Red, scratchy, or watery eyes; headache; increased pain in eyes with exposure to light.	1. Cover the eyes with a dark cloth. 2. Seek medical treatment.
Dehydration	Similar to heat exhaustion.	1. Keep warm, loosen clothes. 2. Replace lost fluids, rest, and additional medical treatment.
Hypothermia	Casualty is cold, shivers uncontrollably until shivering stops. Rectal (core) temp less 95° F can affect consciousness. Uncoordinated movements, shock, and coma may occur as body temperature drops.	***MILD HYPOTHERMIA*** 1. Warm body evenly and without delay. (Heat source must be provided.) 2. Keep dry, protect from elements. 3. Warm liquids may be given to conscious casualty only. 4. Be prepared to start CPR. 5. Seek medical treatment immediately. ***SEVERE HYPOTHERMIA*** 1. Quickly stabilize body temperature. 2. Attempt to prevent further heat loss. 3. Handle the casualty gently. 4. Evacuate to nearest medical treatment facility as soon as possible.

Table 12-3. ENVIRONMENTAL INJURIES

TYPE	*FIRST AID*
Snake bite	1. Get the casualty away from the snake. 2. Remove all rings and bracelets from the affected extremity. 3. Reassure the casualty and keep him quiet. 4. Apply constricting band(s) 1 to 2 finger widths close to the bite. You should be able to slip 1 finger between the band and skin. ***ARM OR LEG BITE*** - Place one band above and one band below the bite site. ***HAND OR FOOT BITE*** - Place one band above the wrist or ankle. 5. Immobilize the affected limb below the level of the heart. 6. Kill the snake, if possible, (without damaging its head or endangering yourself) and send it with the casualty. 7. Seek medical treatment immediately.
Brown recluse or black widow, spider bite	1. Keep the casualty calm. 2. Wash the area. 3. Apply ice or a freeze pack, if available. 4. Seek medical treatment.
Tarantula bite, scorpion sting, ant bite	1. Wash the area. 2. Apply ice or a freeze pack, if available. 3. Apply baking soda, calamine lotion, or meat tenderizer to the bite site to relieve pain and itching. 4. If site of bite(s) or sting(s) is on the face, neck (possible airway blockage), or genital area, or if reaction is severe, or if the sting is by the dangerous Southwestern scorpion, keep the casualty as quiet as possible, administer the epinephrine auto-injector (EAI) if needed, and then seek immediate medical aid.
Wasp or bee sting	1. If the stinger is present, remove by scraping with a knife or finger nail. DO NOT squeeze venom sack on stinger, more venom may be injected. 2. Wash the area. 3. Apply ice or freeze pack, if available. 4. If allergic signs or symptoms appear, be prepared to administer the EAI, and then seek medical assistance.
Human and other animal bites	1. Cleanse the wound thoroughly with soap or detergent solution. 2. Flush bite well with water. 3. Cover bite with a sterile dressing. 4. Immobilize injured extremity. 5. Transport casualty to a medical treatment facility. 6. For animal bites, without endangering yourself or damaging the animal's head, kill the animal and send its head with the casualty. 7. For human bites, try to extract some of the attacker's saliva from the wound and send that in a sealed, identified container with the casualty.

Table 12-3. ENVIRONMENTAL INJURIES (CONTINUED)

Poison Ivy, Oak, Sumac	1. Gently clean affected area. Clean areas 2-3 times daily. Wash clothing. 2. Apply topical anti-itch lotion or ointment as needed, and cover. 3. Avoid scratching or itching the area. 4. Observe for signs of infection (increasing redness, tenderness, warm to touch). 5. Seek medical attention if rash persists or signs of infection develop.

12-12. POISONOUS PLANT IDENTIFICATION

Poison plants include, among others, poison ivy, oak, and sumac, as well as a few more such as stinging nettles which are not discussed here.

a. **Poison Ivy**. Poison ivy is grows in two forms, a vine or shrub. The compound leaves of poison ivy consist of three pointed leaflets; the middle leaflet has a much longer stalk than the two side ones. The leaflet edges can be smooth or toothed but are rarely lobed. The leaves vary greatly in size, from 8 to 55 mm (0.31" to 2.16") in length. They are reddish when they emerge in the spring, turn green during the summer, and become various shades of yellow, orange, or red in the autumn. Small greenish flowers grow in bunches attached to the main stem close to where each leaf joins it. Later in the season, clusters of poisonous, berrylike drupes form. They are whitish, with a waxy look.

b. **Poison Oak**. Poison oak is a widespread deciduous shrub throughout mountains and valleys of North America, generally below 5,000 feet elevation. It commonly grows as a climbing vine with aerial (adventitious) roots that adhere to the trunks of oaks and sycamores. Poison oak also forms dense thickets in chaparral and coastal sage scrub. Poison oak leave typically have three leaflets (sometimes five), the terminal one on a slender rachis (also called a stalk or petiolule). Eastern poison ivy often has a longer rachis and the leaflet margins tend to be less lobed and serrated (less "oak-like"). Like many members of the sumac family (Anacardiaceae) new foliage and autumn leaves often turn brilliant shades of pink and red.

c. **Poison Sumac**. Poison Sumac is a woody perennial shrub or small tree growing from 5 - 25 feet tall that grows in peat bogs and swamps. To identify Poison Sumac, look for the fruit that grows between the leaf and the branch. Leaves are arranged in an alternate pattern on the vine. There are about 7-13 leaflets forming a feather-like appearance. The foliage has brilliant orange or scarlet coloring in the fall.

Figure 12-5. POISONOUS PLANTS

12-13. **FOOT CARE.** Use moleskin to prevent blisters prior to movement or foot march. Drain large blisters. Clean area, puncture with needle, drain blister. Place moleskin over area. Observe for signs of infection. Keep feet as clean and dry as possible. Use foot powder and change socks. Let feet air dry as mission permits.

12-14. **CASUALTY RESCUE AND TRANSPORT SYSTEM.** Remove the hoistable CRTS litter from its pack, and place the litter on the ground. Unfasten the retainer strap, step on the foot end of the litter, and unroll the litter completely to the opposite end. Then--

a. Bend the CRTS litter in half and backroll it. Repeat with the opposite end of the litter. The litter will now lay flat. Lift the sides of the litter, and fasten the four cross-straps to the buckles directly opposite the straps.

b. Place the CRTS litter next to the casualty. Insure the head end of the stretcher is adjacent to the head of casualty. Place the cross-straps under the litter. Logroll the casualty and slide the litter as far under the casualty as possible. Gently roll the casualty down on the litter. Slide the casualty to the center of the litter. Be sure to keep his spinal column as straight as possible.

c. Pull the straps out from under the CRTS litter and fasten it to the buckles. Position the foot end of the litter at the head of the casualty. Have one rescuer straddle the litter and support the casualty's head, neck, and shoulders.

d. Grasp the foot straps of the litter, and slide the litter under the casualty. Center the casualty on the litter, and feed foot straps thru the unused grommets at the foot end of the litter. Fasten the straps to the buckles.

12-15. **HYDRATION AND ACCLIMATIZATION**

a. Table 12-4 shows strategies for minimizing dehydration and increasing acclimatization and good hydration practices.

Table 12-4. HYDRATION MANAGEMENT AND ACCLIMATIZATION

STRATEGY	*SUGGESTIONS FOR IMPLEMENTATION*
Start early	1. Start at least 1 month prior to School 2. Be flexible and patient: performance benefits take longer than the physiological benefits
Mimic the training environment climate	1. In warm climates, acclimatize in the heat of day. 2. In temperate climates workout in a warm room wearing sweats.
Ensure adequate heat stress	1. Induce sweating. 2. Work up to 100 minutes of continuous physical exercise in the heat. Be patient. The first few days, you may not be able to go 100 minutes without resting. 3. Once you can comfortably exercise for 100 minutes in the heat, then continue for at least 7-14 days with added exercise intensity (loads or training runs).
Teach yourself to drink and eat	1. Your thirst mechanism will improve as you become heat acclimatized, but you will still under-drink if relying on thirst sensation. 2. Heat acclimatization will increase your water requirements. 3. Dehydration will negate most benefits of physical fitness and heat acclimatization. 4. You will sweat out more electrolytes when not acclimatized, so add salt to your food, or drink electrolyte solutions during the first week of heat acclimatization. 5. A convenient way to learn how much water your body needs to replace is to weigh yourself before and after the 100 minutes of exercise in the heat. For each pound lost, you should drink about one-half quart of fluid. 6. Do not skip meals, as this is when your body replaces most of its water and salt losses.

12-16. **WORK, REST, AND WATER CONSUMPTION.** Table 12-5 shows a work, rest, and water consumption table. The guidance applies to the average size, heat acclimated Soldier wearing ACU, not hot weather gear, except as specified otherwise. The work and rest times and fluid replacement volumes shown will help the Soldier sustain his performance and hydration for at least 4 hours of work in the specified heat category. Fluid needs can vary based on individual differences (±1 quart per hour).

a. "NL" means that there is no limit to work time per hour. "Rest" means minimal physical activity such as sitting or standing, preferably in the shade.

b. Consume no more than 1.5 quarts of fluid per hour, and no more than 12 quarts per day.

c. If you are wearing body armor in a humid climate, then add 5° F to the WBGT. if wearing MOPP 4 clothing, add 10° F to the WBGT.

d. Work categories include easy, moderate, and hard.

1) ***Easy Work***. This includes, for example, maintaining weapons; walking on hard surfaces at 2.5 mph with a load weighing no more than 30 pounds; participating in marksmanship training; and participating in drills or ceremonies.

2) ***Moderate Work***. This includes, for example, walking in loose sand at 2.5 mph (maximum) or under with no load; walking on a hard surface at 3.5 mph (maximum) with a load weighing no more than 40 pounds; performing calisthenics; patrolling; or conducting individual movement techniques such as the low or high crawl.

3) ***Hard Work***. This includes, for example, walking on a hard surface at 3.5 mph with a load weighing 40 or more pounds; walking in loose sand at 2.5 mph while carrying a load; and conducting field assaults.

Table 12-5. WORK, REST, AND WATER CONSUMPTION TABLE

HEAT CATEGORY	WBGT INDEX, F°	*EASY WORK*		*MODERATE WORK*		*HARD WORK*	
		WORK/ REST	WATER INTAKE (QT/H)	WORK/ REST	WATER INTAKE (QT/H)	WORK/ REST	WATER INTAKE (QT/H)
1	78° to 81.9°	NL	0.50	NL	0.75	40/20	0.75
2 (Green)	82° to 84.9°	NL	0.50	50/10	0.75	30/30	1.00
3 (Yellow)	95° to 87.9°	NL	0.75	40/20	0.75	30/30	1.00
4 (Red)	88° to 89.9°	NL	0.75	30/30	0.75	20/40	1.00
5 (Black)	90° or more	50/10 min	1.00	20/40	1.00	10/50	1.00

NOTES

Chapter 13
DEMOLITIONS

This chapter introduces Rangers to the characteristics of explosives, to initiation systems, MDI components, detonation systems, safety considerations, expedient explosives, breaching charges, and timber cutting charges (FM 5-250). The two categories of explosives are low and high (Table 13-1).

- *Low explosives* have a detonating velocity up to 1,300 feet per second, which produces a pushing or shoving effect.
- *High explosives* have a detonating velocity of 3,280 to 27,888 feet per second, which produces a shattering effect.

Table 13-1. CHARACTERISTICS OF US DEMOLITIONS EXPLOSIVES

NAME	APPLICATIONS	DETONATION VELOCITY		RE FACTOR*	FUME TOXICITY	WATER RESISTANCE
		MIN/SEC	FT/SEC			
AMMONIUM NITRATE	CRATERING CHG	2,700	8.800	0.42	DANGEROUS	POOR
PETN	DET CORD BLASTING CAPS DEMOLITION CHGS	8,300	27,200	1.66	SLIGHT	EXCELLENT
RDX	BLASTING CAPS COMPOSITION EXPLOSIVE	8,350	27,400	1.60	DANGEROUS	EXCELLENT
TRINITROTOLUENE (TNT)	DEMOLITION CHG COMPOSITION EXPLOSIVE	6,900	22,600	1.00	DANGEROUS	EXCELLENT
TETRYL	BOOSTER CHG COMPOSITION EXPLOSIVE	7,100	23,300	1.25	DANGEROUS	EXCELLENT
NITROGLYCERIN	COMMERCIAL DYNAMITE	7,700	25,200	1.50	DANGEROUS	GOOD
BLACK POWDER	TIME FUSE	400	1,300	0.55	DANGEROUS	POOR
AMATOL 80/20	BURSTING CHG	4,900	16,000	1.17	DANGEROUS	POOR
COMPOSITION A3	BOOSTER CHG BURSTING CHG	8,100	26,500	---	DANGEROUS	GOOD
COMPOSITION B	BURSTING CHG	7,800	25,600	1.35	DANGEROUS	EXCELLENT
COMPOSITION C4 (M112)	CUTTING CHG BREACHING CHG	8.040	26,400	1.34	SLIGHT	EXCELLENT
COMPOSITION H6	CRATERING CHG	7,190	23,600	1.33	DANGEROUS	EXCELLENT
TETRYTOL 75/25	DEMOLITION CHG	7,000	23,000	1.20	DANGEROUS	EXCELLENT
PENTOLITE 50/50	BOOSTER CHG BURSTING CHG	7,450	24,400	---	DANGEROUS	EXCELLENT

NAME	APPLICATIONS	DETONATION VELOCITY		RE FACTOR*	FUME TOXICITY	WATER RESISTANCE
		MIN/SEC	FT/SEC			
M1 DYNAMITE	DEMOLITION CHG	6,100	20,000	0.92	DANGEROUS	FAIR
DET CORD	PRIMING DEMOLITION CHG	6,100 TO 7,300	20,000 TO 24,000	---	SLIGHT	EXCELLENT
SHEET EXPLOSIVE M118 AND M186	CUTTING CHG	7,300	24,000	1.14	DANGEROUS	EXCELLENT
BANGALORE TORPEDO M1A2	DEMOLITION CHG	7,800	25,600	1.17	DANGEROUS	EXCELLENT
SHAPED CHARGES M2A3, M2A4, AND M3A1	CUTTING CHG	7,800	25,600	1.17	DANGEROUS	EXCELLENT
* TNT = 1.00 RELATIVE EFFECTIVENESS						

13-1. **INITIATING (PRIMING) SYSTEMS**. The best way to prime demolition systems is with modernized demolition initiators (MDI). These consist of blasting caps attached to various lengths of time fuse or shock tube. MDI can be used with a fuse igniter and detonating cord to create many firing systems. In the absence of MDI, field expedient methods may be used.

a. **Shock Tube**.

(1) Thin, plastic tube of extruded polymer with a layer of special explosive material on the interior surface.

(2) Explosive material propagates a detonation wave which moves along the shock tube to a factory crimped and sealed blasting cap.

(3) Detonation is normally contained within the plastic tubing. However, burns may occur if the shock tube is held.

(4) Advantages of shock tube.

(a) The shock tube offers the instantaneous action of electric initiation without the risk of accidental initiation caused by radio transmitters, static electricity, and so on.

(b) Extremely reliable.

(c) May be extended using left over pieces from previous operations.

b. **Blasting Caps**. Five MDI blasting caps are available which replace the M6 electric and M7 Non-electric blasting cap. Three of these are high-strength caps and two are low-strength. High-strength blasting caps can be used to prime all standard military explosives (including detonating cord) or to initiate the shock tube of other MDI blasting caps.

(1) ***M11***.

- Factory crimped to 30 feet of shock tube.
- A movable "J" hook is attached for quick and easy attachment to detonating cord.
- A red flag is attached 1 meter from the blasting cap and a yellow flag 2 meters from the blasting cap.

(2) ***M14***.

- Factory crimped to 7.5 feet of time fuse.
- May be initiated using a fuse igniter or match.
- Burn-time for total length is about five minutes.
- Yellow bands indicate calibrated one-minute time intervals.

(3) ***M15***.

- Two blasting caps factory crimped to 70 feet of shock tube.
- Each blasting cap has delay elements to allow for staged detonations.
- Low-strength blasting caps. Used as a relay device to transmit a shock tube detonation impulse from an initiator to a high strength-blasting cap.

(4) ***M12***. Factory crimped to 500 feet of shock tube on a cardboard spool.

(5) ***M13***. Factory crimped to 1,000 feet of shock tube.

c. **Matches**. If fuse igniter is unavailable, light the time blasting fuse with a match. Split the fuse at the end (Figure 13-1), and place the head of an unlit match in the powder train. Light the inserted match head with a flaming match, or rub the abrasive on the match box against it. It may be necessary to use two match heads during windy conditions. Burn time will increase with altitude and colder temperatures.

d. **M81 Fuse Igniter**. This is used to ignite time blasting fuse or to initiate the shock tube of MDI blasting caps.

Figure 13-1. LIGHTING TIME FUSE WITH A MATCH

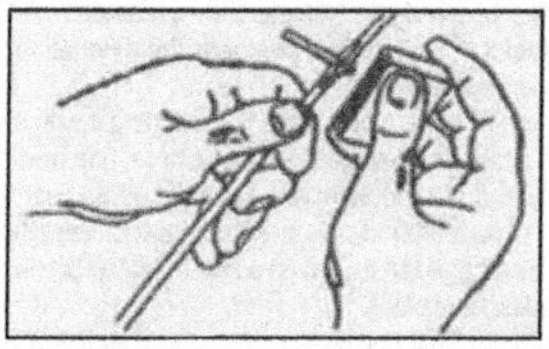

Notes: Low strength blasting caps cannot reliably set off explosives. They should only be used to set off additional shock tubes. The M60 fuse igniter may still be used to ignite time blasting fuse. However, it will not reliably initiate the shock tube.

13-2. **DETONATION (FIRING) SYSTEMS**. The two types of firing systems are MDI alone, or MDI plus detonating cord.

a. **MDI Alone**. An MDI firing system is one in which the initiation set, transmission and branch lines are constructed using MDI components and the explosive charges are primed with MDI blasting caps. Construct the charge in the following manner.

(1) Emplace and secure explosive charge (C4, TNT, cratering charge, and so on.) on target.

(2) Place a sandbag or other easily identifiable marker over the M11, M14, or M15 blasting cap to be used.

(3) Connect to an M12 or M13 transmission line if desired.

(4) Connect blasting cap with shock tube to an M14 cap with time fuse. Cut the time blasting fuse to desired delay time.

(5) Prime the explosive charge by inserting the blasting cap into the charge.

(6) Visually inspect firing system for possible misfire indicators such as cracks, bulges, or corrosion.

(7) Return to the firing point and secure a fuse igniter to the cut end of the time fuse.

(8) Remove the safety cotter pin from the igniter's body.

(9) Actuate the charge by grasping the igniter body with one hand while sharply pulling the pull ring.

b. **MDI and Detonating Cord**. Construct the charge using the above steps for MDI stand-alone system. Incorporate detonating cord branch lines into the system using the "J" hooks of the M11 shock tube. Taping the ends of the Detonation Cord reduces the effect of moisture on the system.

13-3. **SAFETY**. MDI is not recommended for below ground use, except in quarry operations with water-gel or slurry explosives. Use detonating cord when it is necessary to bury primed charges.

a. Do not handle misfires downrange until the required 30 minute waiting period for both primary and secondary initiation systems has elapsed and other safety precautions have been accomplished.

b. Never yank or pull hard on the shock tube. This may actuate the blasting cap.

c. Do not dispose of used shock tubes by burning because of potentially toxic fumes given off from the burning plastic.

d. Do not use M1 dynamite with the M15 blasting cap. The M15 delay-blasting cap should be used only with water-gel or slurry explosives.

e. Always use protective equipment when handling demolitions. Minimum protection consists of leather gloves, ballistic eye protection, and helmet.

13-4. **EXPEDIENT EXPLOSIVES--IMPROVISED SHAPED CHARGE**. An improvised shaped charge (Figure 13-2) concentrates the energy of the explosion released on a small area, making a tubular or linear fracture in the target.

a. The versatility and simplicity of these charges make them effective against targets, especially those made of concrete or those with armor plating.

(1) Bowls, funnels, cone shaped glasses, (champagne glasses with stem removed) used as cones. Champagne or cognac bottles are excellent.

(2) Charge characteristics.

(a) *Cavity Liners*. These are made of copper, tin, or zinc. If none is available, cut a cavity out of the plastic explosive.

(b) *Cavity Angle*. This will work with 30 to 60 degree angles. The cavity angle in most high-explosive antitank (HEAT) ammunition is 42 to 45 degrees.

(c) *Explosive Height (In Container)*. 2 x the height of the cone measured from the base of the cone to the top of the explosive.

(d) *Standoff*. Normal standoff is one and one half the cone's diameter.

(e) *Detonation Point*. The exact top center of the charge is the detonation point. Cover the blasting cap with a small amount of C4 if any part of the blasting cap is exposed.

b. Remove the narrow neck of a bottle or the stem of a glass by wrapping it with a piece of soft, absorbent twine or by soaking the string in gasoline and lighting it. Place two bands of adhesive tape, one on each side of the twine, to hold the twine firmly in place. The bottle or stem must be turned continuously with the neck up, to heat the glass uniformly.

c. A narrow band of plastic explosive placed around the neck and burned gives the same result. After the twine or plastic has burned, submerge the neck of the bottle in water and tap it against some object to break it off. Tape the sharp edge of the bottle to prevent cutting hands while tamping the explosive in place.

d. Do not immerse the bottle in water before the plastic has been completely burned, or it could detonate.

Figure 13-2. IMPROVISED SHAPED CHARGE

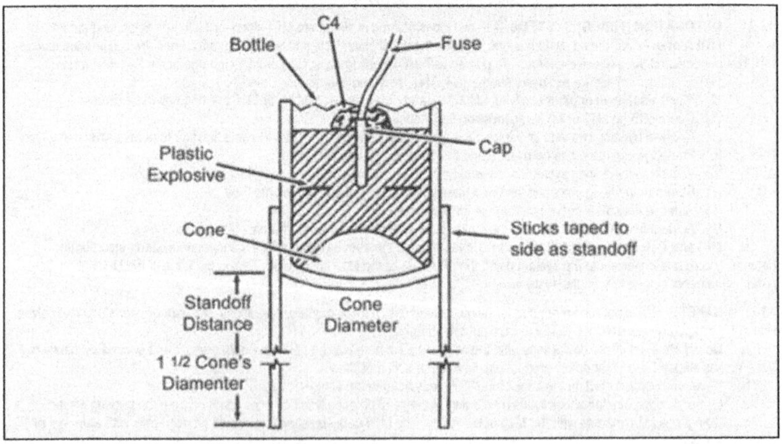

13-5. **EXPEDIENT EXPLOSIVES--PLATTER CHARGE**. This device turns a metal plate into a powerful blunt-nosed projectile (Figure 13-3). The plate should be steel, preferably round but square will work, and should weigh from 2 to 6 pounds.

a. The weight of the explosive should equal the weight of the platter.

b. Uniformly pack the explosive behind the platter. You will only need a container if the explosives fail to remain firmly against the platter. You can use tape to anchor the explosives, if needed.

c. Prime the charge at the exact, rear center of the charge. If any part of the blasting cap is exposed, cover the blasting cap with a small quantity of C4.

d. Aim charge at the direct center of the target, and ensure that the charge is on the opposite side of the platter from the target. Effective range is 35 yards for a small target. With practice, you might hit a 55-gallon drum at 25 yards 90 percent of the time. A gutted fuse igniter can serve as an expedient aiming device.

Figure 13-3. PLATTER CHARGE

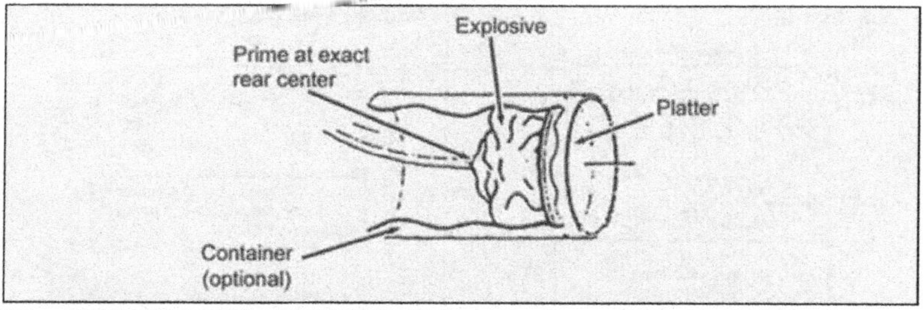

13-6. **EXPEDIENT EXPLOSIVES--GRAPESHOT CHARGE**. To use this antipersonnel fragmentation mine (Figure 13-4)--

a. **Hole**. Create a hole in the center, bottom of the container, for the blasting cap.

b. **Explosives**. Place explosives evenly on the bottom of the container. Remove all voids and air pockets by pressing the C4 into place using a non-sparking instrument.

c. **Buffer**. Place buffer material directly over top of the explosives.

d. **Projectiles**. Place projectiles over top of the buffer materials, then cover to prevent spilling from movement.

e. **Aim**. Aim at target from about 100 feet. Use a small amount of C4 on any exposed portion of the blasting cap.

Figure 13-4. GRAPESHOT CHARGE

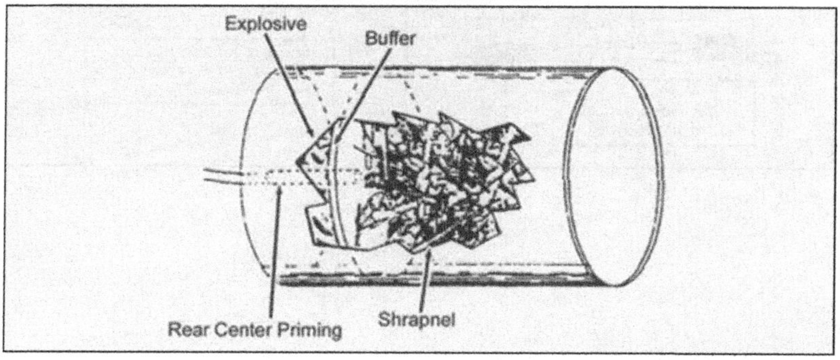

13-7. **DEMOLITION KNOTS.** Several knots are used in demolitions. Figures 13-5 and 13-6 show a few simple knots that can join demolitions to detonation cord.

Figure 13-5. VARIOUS JOINING KNOTS USED IN DEMOLITIONS

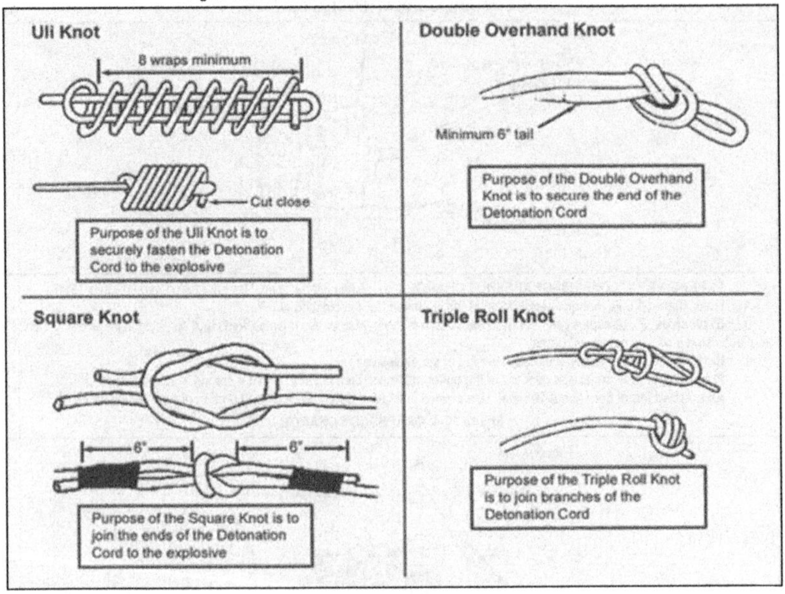

Figure 13-6. BRITISH JUNCTION

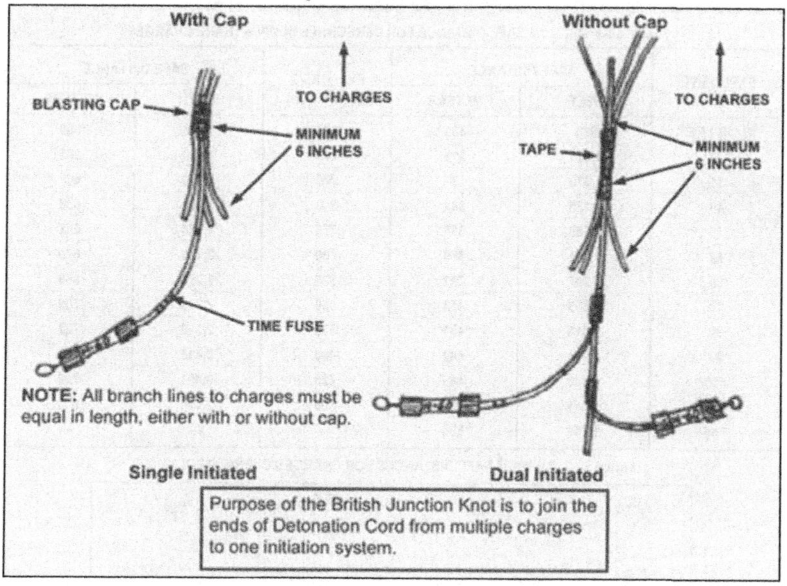

13-8. **MINIMUM SAFE DISTANCES**. Rangers must remain especially aware of their situation when using demolitions. (Table 13-2 shows minimum safe distances for employing demolitions.) For charges over 500 pounds, use the formulas shown in Figure 13-7.

Table 13-2. MINIMUM SAFE DISTANCE FOR PERSONNEL IN OPEN (BARE CHARGE)

EXPLOSIVE WEIGHT (LB)	SAFE DISTANCE		EXPLOSIVE WEIGHT (LB)	SAFE DISTANCE	
	FEET	METERS		FEET	METERS
27 OR LESS	985	300	175	1,838	560
30	1,021	311	200	1,920	585
35	1,073	327	225	1,999	609
40	1,123	342	250	2,067	630
45	1,168	356	275	2,136	651
50	1,211	369	300	2,199	670
60	1,287	392	325	2,258	688
70	1,355	413	350	2,313	705
80	1,415	431	375	2,369	722
90	1,474	449	400	2,418	737
100	1,526	465	425	2,461	750
125	1,641	500	500	2,625	800
150	1,752	534	---	---	---

Figure 13-7. MINIMUM SAFE DISTANCES FOR CHARGES OVER 500 POUNDS

$$\text{safe distance (meters)} = 100\sqrt[3]{\text{pounds of explosive}}$$

$$\text{safe distance (feet)} = 300\sqrt[3]{\text{pounds of explosive}}$$

Notes: For charges on targets, the minimum radius of danger is 1,000 meters. Minimum safe distance when in a missile-proof shelter from the point of detonation is 100 meters.

13-9. **BREACHING CHARGES**. For Table 13-3, the left column represents the thickness of reinforced concrete wall. The remaining 7 columns show the number of packages of C4 required to breach the wall using the charge placements shown in the drawings above the columns.

a. Use Table 13-3, 13-4 and 13-5 for breaching charges.

b. Use the formula in Figure 13-8 to calculate the charges.

c. Multiply number of packages of C4 from Table 13-3 by conversion factor from Table 13-4 for materials other than reinforced concrete.

Table 13-3. BREACHING CHARGES FOR REINFORCED CONCRETE

Rein-forced-Concrete Thick-ness (ft)	Placement Methods						
	Placed in center of mass	Tamped or stemmed (Fill)	Deep water	Elevated untamped	Shallow water	Earth tamping	Ground placed, untamped

	C = 1.0	C = 1.0	C = 1.0	C = 1.8	C = 2.0	C = 2.0	C = 3.6
	Packages of M112 (C4)						
2.0	1	5	5	9	10	10	17
2.5	2	9	9	17	18	18	33
3.0	2	13	13	24	26	26	47
3.5	4	21	21	37	41	41	74
4.0	5	31	31	56	62	62	111
4.5	7	44	44	79	88	88	157
5.0	9	48	48	85	95	95	170
5.5	12	63	63	113	126	126	226
6.0	13	82	82	147	163	163	293
6.5	17	104	104	186	207	207	372
7.0	21	111	111	200	222	222	399
7.5	26	137	137	245	273	273	490
8.0	31	166	166	298	331	331	595

NOTE: The results of all calculations for this table have been rounded UP to the next whole package.

Table 13-4. CONVERSION FACTORS FOR MATERIALS OTHER THAN REINFORCED CONCRETE

MATERIAL	*CONVERSION FACTOR*
• EARTH	0.1
• ORDINARY MASONRY • HARDPAN • SHALE • ORDINARY CONCRETE • ROCK • GOOD TIMBER • EARTH CONSTRUCTION	0.5
• DENSE CONCRETE • FIRST-CLASS MASONRY	0.7

Table 13-5. MATERIAL FACTOR (K) FOR BREACHING CHARGES

Material	R	K
Earth	All values	0.07
Poor masonry, shale, hardpan, good timber, and earth construction	Less than 1.5 m (5 ft)	0.32
	1.5 m (5 ft) or more	0.29
Good masonry, concrete block, and rock	0.3 m (1 ft) or less	0.88
	Over 0.3 m (1 ft) to less than 0.9 m (3 ft)	0.48
	0.9 m (3 ft) to less than 1.5 m (5 ft)	0.40
	1.5 m (5 ft) to less than 2.1 m (7 ft)	0.32
	2.1 m (7 ft) or more	0.27
Dense concrete and first-class masonry	0.3 m (1 ft) or less	1.14
	Over 0.3 m (1 ft) to less than 0.9 m (3 ft)	0.62
	0.9 m (3 ft) to less than 1.5 m (5 ft)	0.52
	1.5 m (5 ft) to less than 2.1 m (7 ft)	0.41
	2.1 m (7 ft) or more	0.35
Reinforced concrete (factor does not consider cutting steel)	0.3 m (1 ft) or less	1.76
	Over 0.3 m (1 ft) to less than 0.9 m (3 ft)	0.96
	0.9 m (3 ft) to less than 1.5 m (5 ft)	0.80
	1.5 m (5 ft) to less than 2.1 m (7 ft)	0.63
	2.1 m (7 ft) or more	0.54

Figure 13-8. FORMULA FOR COMPUTING SIZES OF CHARGES TO BREACH CONCRETE, MASONRY, AND ROCK

$$P = R^3KC$$

where—

P = *TNT required, in pounds*

R = *breaching radius, in feet*

K = *material factor, which reflects the strength, hardness, and mass of the material to be demolished (Table 3-5)*

C = *tamping factor, which depends on the location and tamping of the charge (Figure 3-15)*

13-10. **TIMBER CUTTING CHARGES**. Table 13-6 shows timber-cutting charge sizes. Figures 13-9 through 13-15 show the types of charges and the formulas to use with each.

Table 13-6. TIMBER-CUTTING CHARGE SIZE

Charge Type	Packages of C4 Required (1.25-lb Packages) by Timber Diameter (in)											
	6	8	10	12	15	18	21	24	27	30	33	36
Internal	1	1	1	1	1	1	2	2	2	3	3	4
External	1	1	2	3	4	5	7	9	11	14	17	20
Abatis	---	---	---	---	---	---	---	7	9	11	14	16
NOTE: Packages required are rounded UP to the next whole package.												

Figure 13-9. ABATIS

Figure 13-10. FORMULA FOR FALLEN TREE OBSTACLES OR TEST SHOT

$$P = \frac{D^2}{50} = P = 0.02D^2$$

where—

P = *TNT required per tree, in pounds*

D = *diameter or least dimension of dimensioned timber, in inches*

Figure 13-11. TIMBER-CUTTING RING CHARGE

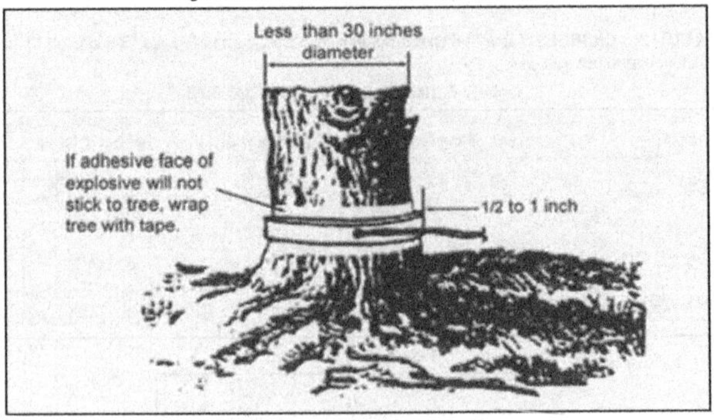

Figure 13-12. TIMBER-CUTTING CHARGE (EXTERNAL)

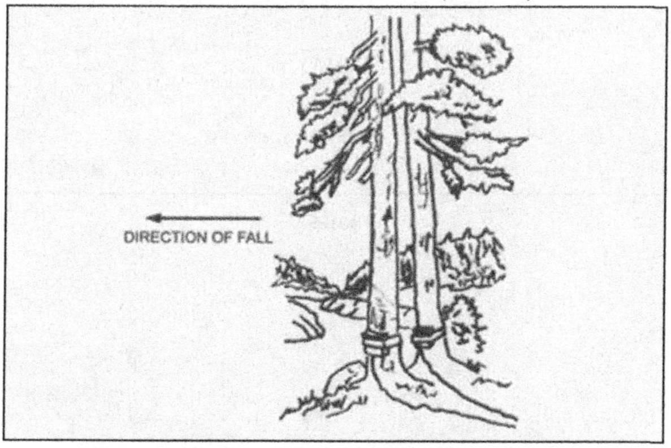

Figure 13-13. FORMULA FOR EXTERNAL TIMBER-CUTTING CHARGE

$$P = \frac{D^2}{40} \text{ or } P = 0.025D^2$$

where—

P = *TNT required per target, in pounds*

D = *diameter or least dimension of dimensioned timber, in inches*

Figure 13-14. TIMBER-CUTTING CHARGE (INTERNAL)

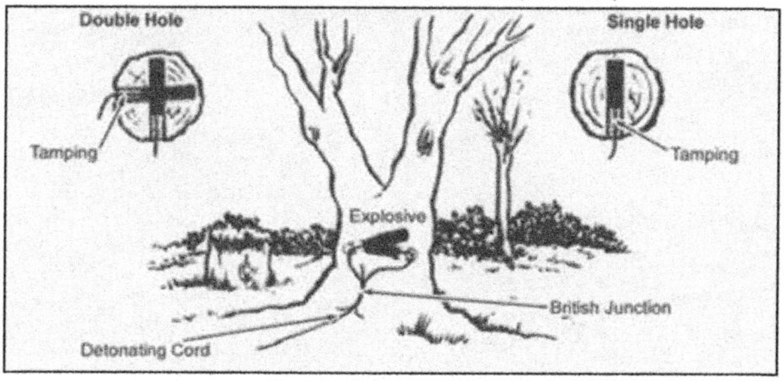

Figure 13-15. FORMULA FOR INTERNAL TIMBER-CUTTING CHARGE

$$P = \frac{D^2}{250} \text{ or } P = 0.004D^2$$

where—

P = *TNT required per tree, in pounds*

D = *diameter or least dimension of dressed timber, in inches*

NOTES

Chapter 14
RANGER URBAN OPERATIONS

Urban operations are planned and conducted in an area of operations (AO) that includes one or more urban areas. An urban area is a topographical complex where man-made construction or high population density is the dominant feature (FM 30-06.11). Urban terrain is likely to be one of the most significant future areas of operations for American forces throughout the world. Expanding urban development affects military operations as the terrain is altered. The increasing focus on stability and support operations, urban terrorism, and civil disorder emphasizes that combat in urbanized areas is unavoidable. Urban areas are the power centers, the centers of gravity, and thus the future battlefield. For further study, see FM 30-06.11, FM 90-10, FM 90-1, FM 7-8, 75th Ranger Regiment Advanced UO SOP, and Ranger Training Circular 350-1-2.

14-1. **TYPES.** Infantry units must be trained to conduct urban combat under high-intensity conditions. High-intensity urban combat requires the employment of combat power of the joint combined arms team. An Infantry unit's mission is normally to recon, isolate, penetrate, systematically clear, defend the urban area, and engage and defeat the enemy with decisive combat power. Although the changing world situation may have made urban combat under high-intensity conditions less likely for US forces, it represents the high end of the combat spectrum, and units must be trained for it. High-intensity urban operations can be casualty-intensive for both sides. With the integrated firepower of the joint, combined arms team, leaders must make every attempt to limit unnecessary destruction of critical infrastructure and casualties among noncombatants.

a. **Precision Conditions**. Infantry units train to defeat an enemy that is mixed with non-combatants in precision urban combat. Leaders plan to limit civilian casualties and collateral damage through the establishment of strict rules of engagement (ROE) and the employment of precision weapons and munitions. The ROE provides the focus for the use and restraint of combat power. The ROE may be significantly more restrictive than under high-intensity conditions.

b. **Surgical Conditions**. Operations conducted under surgical conditions include special-purpose raids, small precision strikes, or small-scale personnel seizure or recovery operations in an urban environment (for example, hostage rescue). Joint special operation forces usually conduct these operations. They may closely resemble US police operations performed by Special Weapons and Tactics (SWAT) teams. They may even involve cooperation between US forces and host nation police. Though regular units may not usually be involved in the actual surgical operation, they may support it by isolating the area, by providing security or crowd control, or providing search and rescue teams.

c. **Transitions**. Leaders must always be prepared to transition rapidly from one type of urban combat to another, and back. Real-world combat shows us that urban operations can deteriorate rapidly and without warning. A force involved in stability and support operations can suddenly find itself in a high-intensity combat situation.

14-2. **PRINCIPLES**

a. **Surprise**. Strike the enemy at a time or place or in a manner for which he is unprepared. Key to success: gives the assaulting element the advantage.

b. **Security**. Never permit the enemy to acquire unexpected advantage.

(1) Maintain during all phases of the operation.

(2) Four-dimensional battlefield (height, depth, width, subterranean).

(3) Always maintain 360 degree security (include elevated and subterranean areas).

(4) Mission is never complete as long as you remain in the urban environment. The status of actors in the urban environment does not afford the sense of security offered by "open" terrain. The key to survivability is a constant state of situational awareness.

c. **Simplicity**. Prepare clear, uncomplicated plans, and provide subordinates with concise orders to ensure thorough understanding.

(1) Always keep plans simple.

(2) Ensure everyone understands the mission and the commander's intent.

(3) Plan and prepare for the worst.

d. **Speed**. Rate of military action.

(1) Act as security.

(2) Move in a careful hurry.

(3) Smooth is fast and fast is smooth.

(4) Never move faster than you can accurately engage targets.
(5) Exercise tactical patience.

e. **Violence of Action**. Eliminate the enemy with sudden, explosive force.
(1) Combined with speed gives surprise.
(2) Prevents enemy reaction.
(3) Both physical and mental.

14-3. **METT-TC**. To effectively plan combat operations in urban environments, leaders must use Troop Leading Procedures and conduct a thorough analysis utilizing METT-TC factors. The following lists specific planning guidance that must be incorporated when planning for urban operations. For more specifics on mission planning, refer to Chapter 2, Ranger Handbook.

a. **Mission**. Know correct Task Organization to accomplish the mission (Offense, Defense, or Stability and Support Operations).

b. **Enemy**.

(1) ***Disposition***. Analyze the arrayal of enemy forces in and around your objective, known and suspected. Example: Known or suspected locations of minefields, obstacles, and strong points.

(2) ***Composition and Strength***. Analyze the enemy's task organization, troops available, suspected strength, and amount of support from local civilian populace based on intelligence estimates. Is the enemy a conventional or unconventional force?

(3) ***Morale***. Analyze the enemy's current operational status based on friendly intelligence estimates. For example, is the enemy well supplied; have they had recent success against friendly forces, taken many casualties; and what is the current weather?

(4) ***Capabilities***. Determine what the enemy can employ against your forces. Example: Enemy's weapons, artillery assets, engineer assets, air defense assets, NBC threats, thermal/NVG capabilities, close air support, armor threat, and so on.

(5) ***Probable Course(s) of Action***. Based on friendly intelligence estimates, determine how the enemy will fight within his area of operation (in and around your area of operation).

c. **Terrain**. Leaders conduct a detailed terrain analysis of each urban setting, considering the types of built-up areas and composition of existing structures. They use OCOKA when analyzing terrain, in and around the area of operation.

(1) ***Observation and Fields of Fire***. Always be prepared to conduct urban operations under limited visibility conditions. This includes the effects of reduced illumination, as well as natural and manmade obscuration. Leaders should ensure that Rangers are equipped with adequate resources, which allow them to successfully operate in the urban environment under these types of conditions. (2) ***Cover and Concealment***. Leaders must perform a thorough analysis of peripheral as well as intra-urban areas. Leaders should identify routes to objectives, which afford assault forces with the best possible cover and concealment. Additionally, leaders should take advantage of limited visibility conditions, which would allow forces to move undetected to their final assault / breaching positions. When in the final assault position, forces should move as rapidly as tactically possible to access structures, which afford additional cover and concealment. Leaders must learn to properly employ obscurants and exercise "tactical patience" to fully take advantage of these effects. Finally, all members of the urban force must practice noise and light discipline. Soldiers must avoid unnecessary voice communications, learn the proper use of white light, and limit contact with surfaces that may alert the enemy of their presence.

(3) ***Obstacles***. There are many manmade and natural obstacles on the periphery, as well as within the urban environment. Leaders should conduct a detailed reconnaissance of routes and objectives (this must include subterranean complexes), taking into consideration route adjustments and special equipment needs.

(4) ***Key Terrain***. Analyze which buildings, intersections, bridges, LZ/PZ, airports, and elevated areas that provide a tactical advantage to you or the enemy. Additionally, the leader must identify critical infrastructure within his area of operations, which would provide the enemy with a tactical advantage on the battlefield. These may include, but are not limited to, communication centers, medical facilities, governmental facilities, and facilities that are of psychological significance.

(5) ***Avenues of Approach***. Consider roads, intersections, inland waterways, and subterranean constructions (subways, sewers, and basements). Leaders should classify areas as go, slow go, or no-go based on the navigability of the approach.

Note: Military maps may not provide enough detail for urban terrain analysis or reflect the underground sewer systems, subways, water systems, or mass transit routes.

(6) ***Troops***. Analyze your forces utilizing their disposition, composition, strength, morale, capabilities, and so on. Leaders must also consider the type and size of the objective to plan effective use of troops available.

(7) ***Time***. Operations in an urban environment have a slower pace and tempo. Leaders must consider the amount of time required to secure, clear, or seize the urban objective and stress and fatigue Rangers will encounter. Additional time must also be allowed for area analysis efforts; these may include, but are not limited to--

- Maps and urban plans Recon and analysis.
- Hydrological data analysis.
- Line-of sight surveys.
- Long Range Surveillance and Scout reconnaissance.
- Artillery.
- Armor.
- Aviation.
- Armor.
- Engineers.

(8) ***Civilians***. Authorities such as the National Command will establish the Rules of Engagement. Commanders at all levels may provide further guidance regarding civilians occupying the area of operations (AO). Leaders must daily reiterate the ROE to subordinates, and immediately inform them of any changes to the ROE. Rangers must have the discipline to identify the enemy from noncombatants and ensure civilians understand and follow all directed commands.

Note: Civilians may not speak English, may be hiding (especially small children), or dazed from a breach. Civilians must not be given the means to resist. Rehearse how clearing/search teams will react to these variables. Never compromise the safety of your Rangers.

14-4. **CLOSE QUARTERS COMBAT.** Due to the very nature of a CQC encounter, engagements will be very close (within 10 meters) and very fast (targets exposed for only a few seconds). Most close quarter's engagements are won by who hits first and puts the enemy down. It is more important to knock a Ranger down as soon as possible than it is to kill him. In order to win a close quarters engagement, Rangers must make quick, accurate shots by mere reflex. This is accomplished by reflexive fire training. Remember, no matter how proficient you are, always fire until the enemy goes down. All reflexive fire training is conducted with the eyes open.

Note: Research has determined that, on average, only three out of ten people actually fire their weapons when confronted by an enemy during room clearing operations. Close quarters combat success for the Ranger begins with the Ranger being psychologically prepared for the close quarter's battle. The foundation for this preparedness begins with the Ranger's proficiency in basic rifle marksmanship. Survival in the urban environment does not depend on advanced skills and technologies. Rangers must be proficient in the basics.

14-5. **REHEARSALS**. Similar to the conduct of other military operations, leaders need to designate time for rehearsals. Urban operations require a variety of individual, collective, and special tasks, which are not associated with operations on less complex terrain. These task require additional rehearsal time for clearing, breaching, obstacle reduction, casualty evacuation, and support teams. Additionally, rehearsal time must be identified for rehearsals with combined arms elements. These may include, but are not limited to--

a. **Stance**. Feet are shoulder width apart, toes pointed straight to the front (direction of movement). The firing side foot is slightly staggered to the rear of the non-firing foot. Knees are slightly bent and the upper body leans slightly forward. Shoulders are not rolled or slouched. Weapon is held with the butt stock in the pocket of the shoulder. The firing side elbow is kept in against the body. The stance should be modified to ensure the Ranger maintains a comfortable boxer stance.

(1) ***Low-Carry Technique***. The butt stock of the weapon is placed in the pocket of the shoulder. The barrel is pointed down so the front sight post and day optic is just out of the field of vision. The head is always up identifying targets. This technique is safest and is recommended for use by the clearing team once inside the room.

(2) ***High-Carry Technique***. The butt stock of the weapon is held in the armpit. The barrel is pointed slightly up with the front sight post in the peripheral vision of the individual. Push out on the pistol grip and thrust the weapon forward and pull straight back into the pocket of the shoulder to assume the proper firing position. This technique is best suited for the line-up outside the door. Exercise caution with this technique always maintaining situational awareness, particularly in a multi-floored building.

Note: Muzzle awareness is critical to the successful execution of close quarter's operations. Rangers must never point their weapons or cross the bodies of their fellow Rangers at any time. Additionally Rangers should always avoid exposing the muzzle of their weapons around corners; this is referred to as "flagging."

b. **Malfunction**. If a Ranger has a malfunction with his weapon during any CQC training, he will take a knee to conduct immediate action. Once the malfunction is cleared, there is no need to immediately stand up to engage targets. Rangers can save precious seconds by continuing to engage from one knee. Whenever other members of the team see a Ranger down, they must automatically clear his sector of fire. Before rising to his feet, the Ranger warns his team members of his movement and only rises after they acknowledge him. If a malfunction occurs once committed to a doorway, the Ranger must enter the room far enough to allow those following him to enter and move away from the door. This drill must be continually practiced until it is second nature.

c. **Approach to a Building or Breach Point**. One of the trademarks of ranger operations is the use of limited visibility conditions. Whenever possible, breaching and entry operations should be executed during hours and conditions of limited visibility. Rangers should always take advantage of all available cover and concealment when approaching breach and entry points. When natural or manmade cover and concealment is not available, Rangers should employ obscurants to conceal their approach. There are times when Rangers will want to employ obscurants to enhance existing cover and concealment. Members of the breach / entry team should be numbered for identification, communication, and control purposes.

(1) The # 1 Ranger should always be the most experienced and mature member of the team, usually the team leader. The # 1 Ranger is responsible for frontal and entry and breach point security. The team leader is responsible for initiating all voice and physical commands. The team leader must exercise situational awareness at all time with respect to the task, friendly force, and enemy activity.

(2) The # 2 Ranger is directly behind the # 1 Ranger in the order of movement, and he moves through the breach point in the opposite direction from the #1 Ranger.

(3) The # 3 Ranger will simply go opposite the # 2 Ranger inside the room at least 1 meter from the door.

(4) The # 4 Ranger moves the opposite the # 3 Ranger and is responsible for rear security (and is normally the last Ranger into the room). An additional duty of the # 4 Ranger is breaching.

Note: Consider how much firepower each Ranger delivers. Where do you put the SAW gunner in the order? You must weigh the benefits of firepower against those of quick, accurate shots. If the # 4 Ranger has breaching responsibilities, it should not be the SAW gunner, because this would reduce your firepower.

d. **Actions Outside the Point of Entry**. Entry point position and individual weapon positions are important. The clearing team members should stand as close to the entry point as possible, ready to enter. Weapons are oriented in such a manner that the team provides itself with 360 degree security at all times. Team members must signal to one another that they are ready at the point of entry. This is best accomplished by sending up a "squeeze." If a tap method is used, an inadvertent bump may be misunderstood as a tap.

e. **Enter Building/Clear Room**. See Battle Drill 5, Chapter 6.

f. **Locking Down the Room**.

(1) Control the situation within the room.

(2) Use clear, concise arm and had signals. Voice commands should be kept to a minimum to reduce the amount of confusion and to prevent the enemy--who might be in the next room--from discerning what is going on. This enhances the opportunity for surprise and allows the assault force the opportunity to detect any approaching force.

(3) Physically and psychologically dominate.

(4) Establish security and report status.

(5) Cursory search of the room to include the ceiling (3 Dimensional Fight).

(6) Identify the dead using reflexive response techniques (Eye thump method).

(7) Search the room for PIR, precious cargo as per the mission and time available.

(8) Evacuate personnel.

(9) Mark room clear using chemical lights, engineer tape, chalk, paint, VS-17 panels, and so on.

14-6. **TTPS FOR MARKING BUILDINGS AND ROOMS**. Units have long identified a need to mark specific buildings and rooms during UO. Sometimes rooms need to be marked as having been cleared, or buildings need to be marked as containing friendly forces. The US Army Infantry School is currently testing a remote marking device that can be used to mark doors from as far away as across a wide street. In the past, units have tried several different field-expedient marking devices; some with more success than others. Chalk has been the most common. It is light and easily obtained but not as visible as other markings. Some of the other techniques have been to use spray paint, and paintball guns.

a. **NATO Standard Marking SOP**. The North Atlantic Treaty Organization (NATO) has developed a standard marking SOP for use during urban combat. It uses a combination of colors, shapes, and symbols. These markings can be fabricated from any material available. (Figure 14-1 shows examples.)

Figure 14-1. EXAMPLE NATO STANDARD MARKINGS

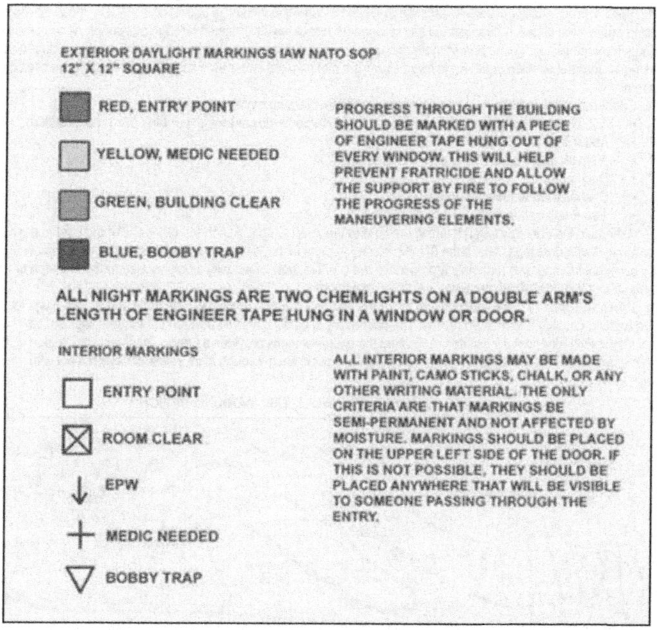

b. **Spray Paint**. Canned spray paint is easily obtained and comes in a wide assortment of colors including florescent shades that are highly visible in daylight. It cannot be removed once used. Cans of spray paint are bulky and hard to carry with other combat equipment. Paint is not visible during darkness nor does it show up well through thermal sights.

c. **Paintball Guns**. Commercial paintball guns have been purchased by some units and issued to small unit leaders. Some models can be carried in standard military holsters. They can mark a building or door from about 30 meters. The ammunition and propellant gas is not easily obtainable. The ammunition is fragile and often jams the gun if it gets wet. The available colors are not very bright, and just like spray paint, cannot be seen at night or through thermal sights.

d. **Wolf Tail**. A simple, effective, easy-to-make, lightweight device called a "wolf tail" can be fabricated to mark buildings, doorways, and windows (Figure 14-1). A unit has changed its tactical SOP to require that each Infantryman carry one of these devices in his ACU cargo pocket. Wolf tails, when used IAW a simple signaling plan understood by all members of the unit, can aid in command and control, reduce the chances of fratricide, and speed up casualty collection during urban combat.

(1) The wolf tail marking device is simple to make and versatile. It can be used together with the NATO marking scheme. Rolled up, it makes a small, easily accessible package that can be carried in the cargo pocket of the ACUs. It can be recovered easily and used again if the situation changes. All its components can be easily obtained through unit supply. It combines a variety of visual signals (colored strapping and one or more chemical lights of varying colors) with a distinctive heat signature that is easily identified through a thermal weapon sight. An infrared chemlight can be used either as a substitute for the colored chemlight(s) or in addition to them.

(2) Constructing the wolf-tail marking device requires the following material:

- A 2-foot length of nylon strap (the type used for cargo tie-downs) (engineer tape can be substituted).
- About 5 feet of 550 cord.
- A small weight such as a bolt or similar object.
- Duct tape.
- Chemical lights (colored or IR).
- Two 9-volt batteries.

(3) Assemble the items by tying or taping the cord to the small weight. Attach the other end of the cord to the nylon strapping, securing it with duct tape. Attach the 9-volt batteries in pairs to the lower end of the strapping with several wraps of duct tape, making sure that the negative terminals are opposite the positive, but not actually touching. Use more duct tape to attach the chemical lights, about 2 inches above the batteries, to the strapping.

(4) When you want to mark your position, push the batteries together firmly until the male and female plugs lock. This shorts out the battery, causing it to heat up rapidly. The hot battery is easily identified through the thermal sights of tanks or BFVs. The batteries stay visible for about 45 minutes. Activating the chemical lights provides an easily identified light source visible to the naked eye. You can use infrared chemical lights if you want them to be seen through night vision devices but not with the naked eye.

Figure 14-2. EXAMPLE OF A WOLF-TAIL MARKING DEVICE

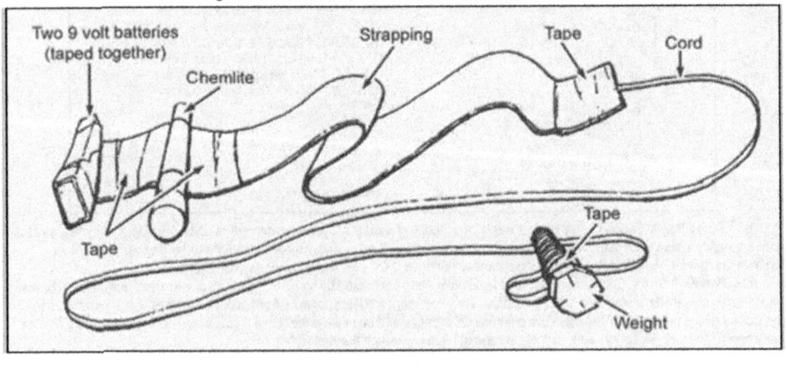

Note: One option is to place chemical lights and batteries at both ends of the wolf tail to mark the inside and outside of a building or room.

(5) Use the cord and the small weight to hold the wolf tail in position by tying or draping it out a window or hanging it on a door, wherever it is best seen by other friendly troops. Squads or platoons can vary the numbers and colors of chemical lights, or use multiple battery sets to identify precisely what unit is in which building.

(6) Medics and combat lifesavers can carry a standardized variation that can be used to clearly identify a building as containing wounded personnel needing evacuation. This could be a white strap with multiple red chemical lights, or any other easily identified combination.

14-7. **URBAN ASSAULT BREACHES.** Understanding how to employ and incorporate breaching as part of a leader's planning process is an important part of urban operations. It is imperative that elements of your patrol be skilled in the art of breaching. Whether infiltrating or exfiltrating from an objective, leaders must plan for either option. One constant disadvantage with the employment of explosives is that placement requires Rangers to expose themselves to possible enemy fires. Breach teams need to be supported by fires or obscurants and breaching operations should be performed during hours of limited visibility whenever possible. Breaching classes follow:

a. **Mechanical**. Mechanical breaching should be an important part of a leader's breaching training program because it is usually an option. Mechanical breaching is best described as gaining access by the use of tools or saws. Although most tools and saws used are recognizable and self explanatory to the individual Ranger, one must practice on various techniques to increase speed an effectiveness. This reduces fatigue and expedites the actual assault.

(1) **Tools**.

(a) Hooligan tools..........................Doors/windows of all types.

(b) Sledge hammer.......................Heavy duty doors, locks, and window frames.

(c) Picket pounder.........................Doors of all types, light walls.

(d) Bolt Cutters..............................Chain link fence, locks, and wire obstacles.

(e) Pick Ax.....................................Lightweight doors and locks.

(f) Saws...Fences, light doors, locks.

b. **Ballistic**. Ballistic breaching is defined as a forced entry or exit by the use of weapons. Whether using shotguns, M16A2/M4, M249 SAW, specific considerations must be addressed:

(a) Type of round and ricochet factor.

(b) Composition of the breaching point.

(c) Composition of the floor beyond the door.

(d) Personnel behind the door (friendly/enemy).

(e) Always shoot at a 45-degree angle.

c. **Explosive**. Explosive breaching is the most viable because it is the most effective. When employing explosives during breaching operations, leaders must consider three major factors.

(1) ***Overpressure***. The amount of PSI released from the concussion of the blast.

(2) ***Missile Hazard***. Fragmentation or projectiles sent at tremendous speed from the explosion area. This occurs from either the charge or target being breached.

(3) ***Minimum Safe Distance Requirements***. Use of explosives in the urban environment must consider the presence of noncombatants and friendly forces. Additionally, the are many hazardous materials located in the urban environment, these may include chemicals as well as construction materials. There is always a risk of secondary explosions and fires, when employing explosive breaching techniques.

(4) ***Charges***. Various charges can be used for explosive breaching. Leaders must conduct extensive training on the use of the charges to get proper target feedback. Figure 14-3 shows examples of charges used for explosive breaching.

Figure 14-3. EXAMPLE CHARGES FOR EXPLOSIVE BREACHING

Charge	Target
Water Impulse................	Steel/wood doors
Flexible linear.................	Wood doors
Ranger wall breach........	Masonry/brick walls
Chain link ladder............	Chain link fence
E-Type silhouette...........	Wood doors
Brashier breacher..........	Concertina wire

Chapter 15
VEHICLE CONVOY OPERATIONS

This chapter outlines a technique for conducting vehicle convoy operations. Convoy operations present a challenge to the Ranger leader. Trucks and other combat vehicles produce a large signature on the battlefield and increase your unit's value as a target. Vehicle movement is restricted to roads and terrain that they can traverse; therefore, a sound plan must be implemented to minimize the possibility of compromise.

15-1. **PLANNING CONSIDERATIONS**. This paragraph explains how to plan a vehicle convoy IAW the eight steps of the TLP.

a. **Receive the Mission**. (PL will extract the following information from Company OPORD).
(1) Truck Support (number and type of Truck, ACL).
(2) Weather. Road conditions.
(3) Truck pickup and drop-off location / marking.
(4) Truck movement Timeline (Pick-up time, roll time, H-hour).
(5) Truck Routes (primary and alternate, Check points).
(6) Drop-off locations (primary and alternate).
(7) IED/Contact compromise and contingencies.

b. **Issue the Warning Order**. During the WARNO the PL gives a basic overview of the Truck Movement and gives specific Task to Maneuver Units.
(1) Manifest Personnel (PSG).
(2) Pick up point security and set-up (PSG).
(3) Primary and alternate routes (Security SL).

c. **Make a tentative Plan**. PL considers METT-TC.
(1) ***Mission***. Develop your Ground Tactical Plan and a then develop a truck plan that best supports it.
(2) ***Enemy***. Consider enemy patrols and activity in AO.
(3) ***Terrain***. Evaluate routes and dropoff and pickup points using OCOKA.
(4) ***Troops***.
(a) Truck ACL.
(b) Chalks and chalk leaders identified.
(c) Chalks planed with tactical integrity & self sufficiency.
(d) Key leaders and weapons are cross-loaded.
(e) Pickup point marking team identified.
(f) Pickup point security teams identified.
(5) ***Time***. PL considers the following when evaluating time available.
(a) Movement to Pickup point.
(b) Recon of Pickup point.
(c) Emplacement of security.
(d) Marking the Pickup point.
(e) Pickup point Posture time.

d. **Initiate Necessary Movement**. PL decides whether to move the platoon to the pickup location or to conduct a map reconnaissance and complete the plan.

e. **Conduct Reconnaissance**. (Is not necessary if pickup point is secured and established by Higher HQ's).
(1) Leader's Recon consists of PL, RTO, Chalk Leaders and Pickup point Security element.
(2) PL determines suitability of Pickup point.
(3) PL determines security plan (either overwatch or strong point).

f. **Complete the Plan**. Complete Truck Movement Annex on Page ??? the Ranger Handbook.

g. **Issue the OPORD**.

h. **Supervise**.
(1) Back brief from chalk leaders.
(2) Rock drill of truck movement and contingency plans.

15-2. **FIVE PHASES OF TRUCK MOVEMENT**. Each phase must support the ground tactical plan, which specifies actions in the objective area to accomplish the commander's intent for the assigned mission, be it a raid, ambush, recon or other follow-on missions.

a. **Staging Plan**.
   (1) Secure the pickup point (strongpoint or overwatch).
   (2) Mark pickup point (day/night).

b. **Loading Plan**. Each Ranger is assigned a truck.

c. **Movement Plan**.
   (1) Troops awake and alert during movement.
   (2) Platoon leader and chalk leaders tracking route progress.
   (3) Compromise and contingency plan.
   (a) React to IED.
   (b) React to ambush.
   (c) Truck breakdown.

d. **Dropoff Plan**.
   (1) Establish security of dropoff point (overwatch or strongpoint).
   (2) Dismount vehicles.
   (3) Recon and secure assembly area.
   (4) Adjust perimeter as chalks arrive.
   (5) PSG clears the trucks when last chalk departs.

e. **Ground Tactical Plan**. Execute after platoon is reconsolidated or minimum force required is assembled.

15-3. **CONVOY TECHNIQUES**. The following convoy techniques have been included for your convenience:

- Convoy warning order
- Convoy brief
- React to Ambush Near/Far
- Convoy forced to stop (Methods 1 and 2)
- Break contact
- Recovery/CASEVAC Operations

Convoy Warning Order

1. SITUATION: A brief statement of the enemy and friendly situation. Who, What, Where
2. MISSION: Who, What, When, Where, Why
3. TASK ORGANIZATION: Based on tasking from higher WARNO.
4. INITIAL TIME SCHEDULE:

When	Who	What

5. SPECIAL INSTRUCTIONS: PCC/PCI Guidance, Rehearsals, additional tasks to be accomplished.
6. SERVICE SUPPORT:

Class I: (Rations & Water)

Class III: (POL)

Class V: (Ammunition /Pyrotechnics)

Weapon System	Rounds	Type

Pyrotechnic Device	Number	Location

Class VIII: (Medical)

7. UNIFORM AND EQUIPMENT COMMON TO ALL.

CONVOY BRIEF **(Modified from FM 55-30)**

Movement Order No. ____

References: A. ______________________________ *(Maps, tables and relevant documents)*

B. ______________________________

TASK ORGANIZATION: (Internal organization for convoy – Manifest)

1. SITUATION:

a. **Enemy Forces:**

(1) Weather. General forecast.

(2) Light Data (EENT, % Illumination, MR, MS, BMNT)

(3) Discuss Enemy.

Identification of enemy (If known).

Composition / capabilities / strength / equipment

Location

(Hot Spots highlighted on map)

b. **Friendly Forces:**

(1) Operational support provided by higher headquarters.

Helicopter / Gunships

Quick Reaction Forces (QRFs)

MP Escorts / Rat Patrols

Fire Support elements

c. **Attachments:** (From outside the organization)

2. MISSION: (WHO, WHAT, WHEN, WHERE, & WHY)

3. Employment Tactics, Techniques and Procedures

a. Readiness Level
- i. Ride / Scanning (Observation)
- ii. At the Ready

b. Scanning / Sector of Fire
- i. Driver
  1. Sector of Scan is 9 – 1 clock position
  2. Observation with Mirrors
  3. Sector of Fire is 9 – 11 o'clock
- ii. TC
  4. Sector of Scan is 11 – 3 clock position
  5. Sector of Fire is 1 – 3 clock position
- iii. Other Systems are based off vehicle type and load

b. Target ID
- i. Communicate / Signal

c. Body Positioning
- i. Engage as you Train (Right or Left handed Firing)
- ii. Firing-side Shoulder Down
- iii. Maintain your Body Position
- iv. Weapon to Head

d. Rules of Engagement Concerns – Employ the appropriate systems based off the threat.

e. Point of Aim
- i. Moving Platform – Stationary Target requires aim to the rear (trail) and low dependant upon speed.

CONVOY BRIEF (Modified from FM 55-30) **(Continued)**

3. Employment Tactics, Techniques and Procedures

e. Point of Aim
   - ii. Moving Platform – Moving Target requires aim directly on and low.
   - iii. Stationary Platform – Moving Target requires aim to the front (lead) and low dependent upon speed.

f. Rate of Fire
   - i. Burst Mode
   - ii. Steady Suppression (ROE)

g. Magazine Awareness
   - i. Serviceability / maintenance
   - ii. Tracer Mix
   - iii. Magazine Storage / Placement
   - iv. Mounted, Reload when time is available

Dismounted, Seek cover prior to the need to Reload

REACT TO AMBUSH (NEAR)

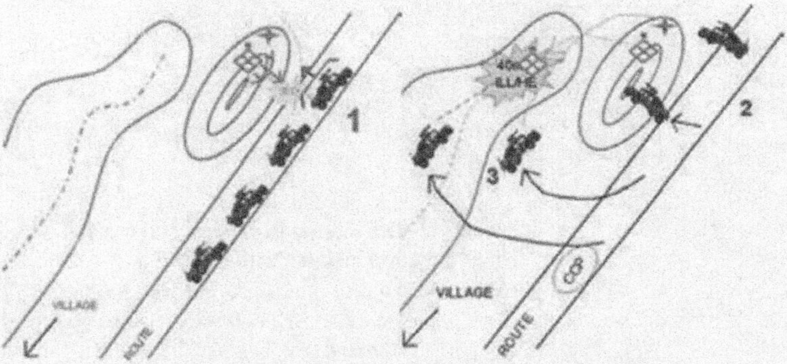

(Enemy Initiates Weapons Fire / RPG / IED)

1. Lead element of convoy reacts to contact and immediately returns fire. Lead fire team fires 40mm HE (at night also white star if NVGs drown out by vehicle headlights – leader calls it. FYI, most travel at night will be with headlights on).
2. Lead elements assualt through ambush – fire team dismounts while gun trucks use M249, MK-19, and .50 cal in close support.
3. Trail fire team and trail gun truck move to block most likely enemy escape route. Fire team uses 40mm HE/Illumination if needed. (Use of 40mm becomes less likely as units get closer to urban areas due to collateral damage risk.) Establish CCP upon action completion or enemy neutralized.

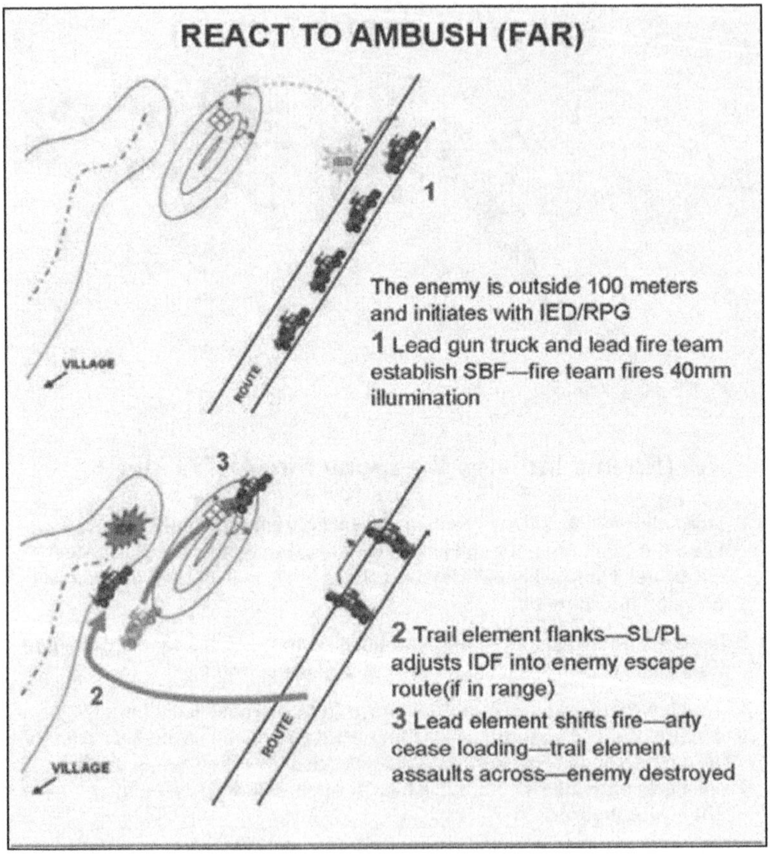
REACT TO AMBUSH (FAR)
1
VILLAGE
ROUTE
The enemy is outside 100 meters and initiates with IED/RPG
1 Lead gun truck and lead fire team establish SBF—fire team fires 40mm illumination
3
2
VILLAGE
ROUTE
2 Trail element flanks—SL/PL adjusts IDF into enemy escape route(if in range)
3 Lead element shifts fire—arty cease loading—trail element assaults across—enemy destroyed

CONVOY FORCED TO STOP

(Weapons Fire / RPG / IED / Indirect)

1. Vehicles forced to stop. Activate vehicle turn signal to indicate direction of contact.
2. Vehicle(s) / Personnel not in direct contact, report on internal communication, identifying truck number, type of contact, clock direction and eight digit grid if available.

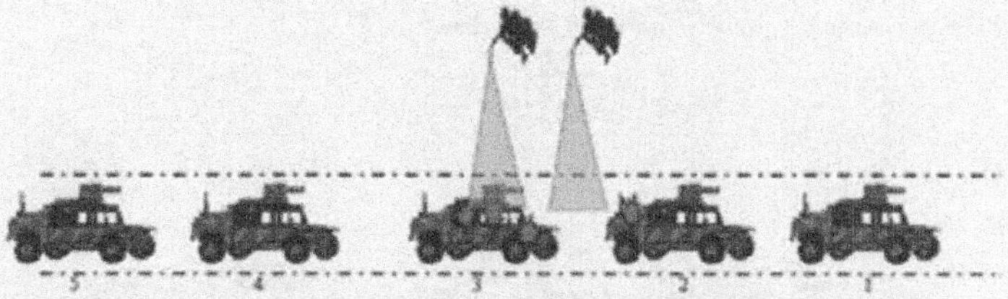

3. Personnel on vehicle(s) forced to stop dismount on the non-contact side, assume covered position(s) and provide initial base of fire.
4. The <u>entire</u> convoy halts; personnel will dismount vehicle(s) on the non-contact side and provide additional base of fire on the enemy if in range. Vehicle(s) not in contact will reposition to attack the attackers. M16's should now be on <u>Semi-Automatic</u> in order to conserve ammunition.
5. PL/CC/SC will move to better assess the situation and position the Gun Truck(s) in order to best suppress the enemy while maintaining <u>standoff</u>. Gun trucks will close with and engage the enemy.
6. Once the PL/CC/SC determines the convoy has either gained fire superiority or defeated the enemy contact, Recovery / CASEVAC operations will begin.
7. If the PL/CC/SC determines the convoy cannot gain fire superiority, leader will then conduct Break Contact procedures

CONVOY FORCED TO STOP - METHOD 2

(Weapons Fire /RPG / IED / Indirect)

1. Vehicle(s) forced to stop. Activate vehicle turn signal to indicate direction of contact.

*** ALL PERSONNEL STAY IN VEHICLES ***

2. All vehicles immediately drive forward out of the kill zone.

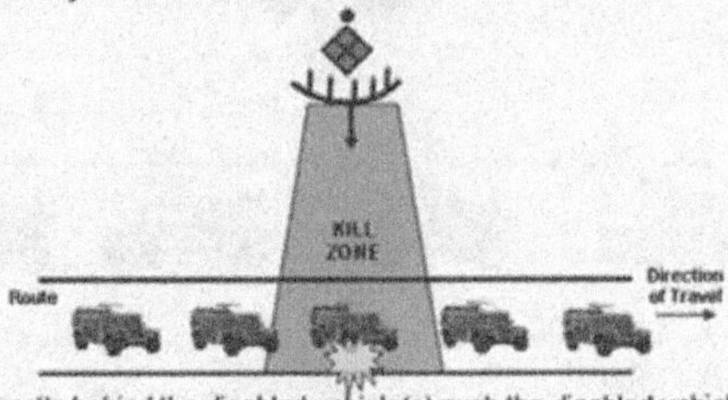

3. The vehicle(s) directly behind the disabled vehicle(s) push the disabled vehicle(s) out of the kill zone.
4. The vehicle(s) that are not disabled will establish a base of fire towards the suspected/known contact.
5. If fire superiority can be gained the PL will use the minimum force necessary to destroy the enemy.
6. If the PL determines the convoy cannot gain fire superiority, the leader will conduct break contact.

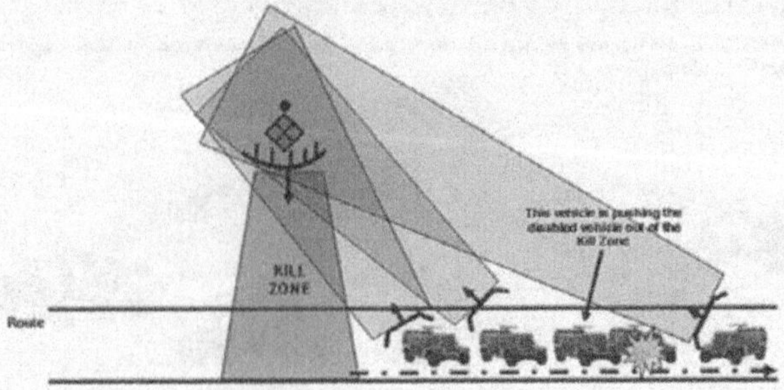

BREAK CONTACT

Always try to close with and destroy the enemy first. This way he cannot come back later on to attack you or another convoy again. However – if you must...

1. The PL/CC/SC determines the convoy cannot gain fire superiority and the decision has been made to Break Contact.
2. The PL/CC/SC designates that either Rally Point "Rear" or "Forward" will be used. If necessary, both Rally Points may be used. Communication systems and appropriate pyrotechnic signals will be used to communicate Break Contact and Rally Point.
3. Personnel will deploy obscuration measures if available. Utilizing cover and concealment, Aid & Litter team(s) will evacuate all casualty(s) under support of Gun Truck and other protective fire(s).
4. Personnel will maintain position and suppression in contact zone and assist Aid & Litter team(s) as necessary.
5. Disabled vehicle(s) will be hauled back or destroyed as directed by leaders. (Thermite or explosives)

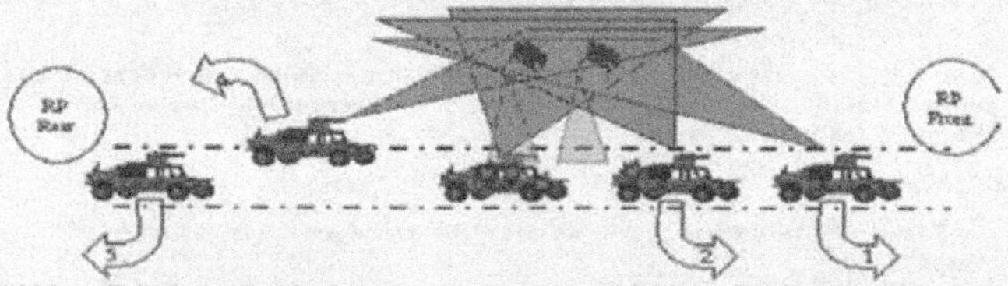

6. Vehicles will displace either backwards or forwards through the convoy lines under control of leaders. The most forward vehicle in the contact zone moves first, then the next most forward vehicle moves second. Vehicles will continue to displace. As vehicles displace, Gun Truck repositions as necessary until contact is broken.

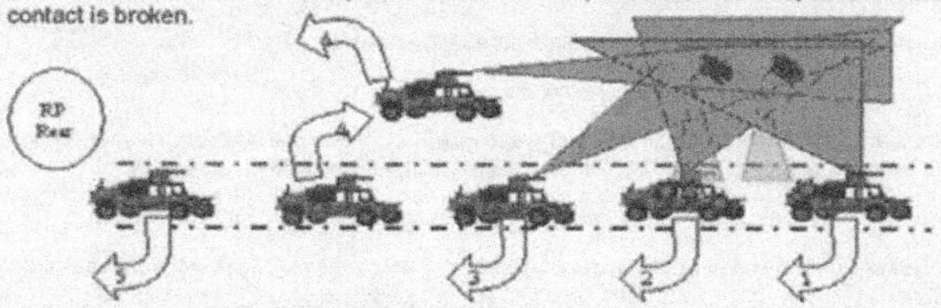

7. If Break Contact occurs with vehicles on both sides of the contact zone, displacement of vehicles will occur using an alternating displacement technique.
8. Upon occupation of the Rally Point, leaders will immediately position vehicles, setting 360 degree security, and conduct Consolidation and Reorganization.
9. If the convoy vehicles get separated when not in contact with the enemy, personnel and vehicles stay together and move to the closest Rally Point or Check Point.

RECOVERY / CASEVAC OPERATIONS

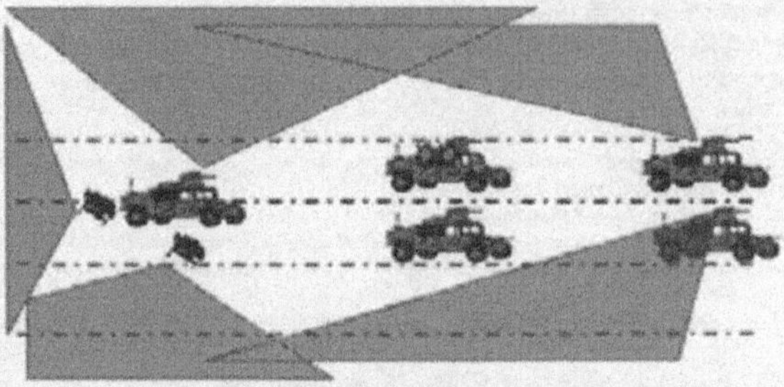

1. Once the leader assesses the enemy threat to be destroyed, neutralized, defeated and the area secured – Recovery / CASEVAC operations will begin. This keeps soldiers focused first on defeating and destroying the threat.
2. CASEVAC:
   a. Aid & Litter team will position on the safe side of the vehicle and extract casualty(s) and personnel.
   b. Treatment of casualty(s) will occur once they are safely removed from the contact area.
3. Vehicle Recovery Procedures:
   a. Recovery team will position on the safe side of the disabled vehicle.
   b. TC will dismount and assess the disabled vehicle.
   c. If determined the vehicle can be safely recovered, TC guide the recovery vehicle into position and conduct a hasty hook-up. TC will operate the disabled vehicle is necessary.
   d. Upon exiting the contact area, complete and correct hook-up procedures will occur.
   e. If assessment results in outside support necessary for recovery, leader will contact higher for guidance.
4. Once recovery operations are complete, the team will displace and conduct link-up with the convoy at the Rally Point.
5. Disabled vehicle(s) will be abandoned or destroyed as directed by leaders. (Thermite or explosives)

GLOSSARY

1SG	first sergeant
5 W's	Who, What, Where, When, and Why?
5-S Rule	Search, Silence, Segregate, Safeguard, and Speed to rear
AA	avenue(s) of approach
AAR	after-action review
AATF	air assault task force
ABCDE	method of identification and response to life-threatening conditions: Airway, Breathing, Circulation, Disability, Exposure
ACE	ammunition, casualties, and equipment
ACL	allowable combat load
ACP	aerial checkpoint
ACU	Army combat uniform
ADA	air defense artillery
AG	assistant gunner
ATC	air traffic controller; a mechanical belay device that locks down on itself when tension is applied in opposite directions
ALT	alternate
AMC	air mission commander
ammo	ammunition
ANCD	automated net-control device
AO	area of operations
AOO	actions on the objective
AR	automatic rifleman
ABF	attack by fire
ATL	Alpha team leader
ATM	Alpha Team
bangalore torpedo	A manually emplaced, 1.5-meter long explosive-filled tube used to breach wire and detonate simple, pressure-activated antipersonnel mines. Ten tubes clear a 1- by 15-meter lane.
belay	any action taken to stop a climber's fall or to control the rate a load descends
binos	binoculars
BMNT	begin morning nautical twilight
BN	battalion
body belay	belay that uses the belayer's body to apply friction by routing the rope around the his body
bowline on a coil	knot used to secure a climber to the end of a climbing rope
BP	battle position
British junction knot	knot used to join the ends of detonation cords from multiple charges to one initiation system
BTC	bridge team commander
BTL	Bravo Team leader
BTM	Bravo Team

C2	command and control
CAS	close air support
CASEVAC	casualty evacuation
CCA	close combat attack
CCIR	commander's critical information requirements
CCP	casualty collection point
CDR	commander
CDS	Camp Darby Special (map)
CLS	combat lifesaver
CO	company
COA	course of action
commo	communications
COMSEC	communications security
cordelette	short section of static rope or static cord. Also called "sling rope"
COTS	commercial off the shelf
CP	command post
CPR	cardiopulmonary resuscitation
CQC	close quarters combat
CS	combat support
CSS	*obsolete: now referred to as* sustainment
CTT	common task test
DAR	designated area of recovery
DOL	direction of landing
double figure eight knot	knot used to form a fixed loop in the end of the rope; loops are large enough to insert a carabiner
double overhand knot	knot used to secure the end of detonation cord
DP	duty position; decision point (depending on context)
DST	distance
DTG	date-time group
dynamic ropes	one of two classifications of kernmantle rope; used for climbing; *see also* static rope
DZ	drop zone
EA	engagement area
EEFI	essential elements of friendly information
EENT	end evening nautical twilight
end-of-the-rope clove hitch	intermediate anchor knot that requires constant tension
end-of-the-rope Prusik	knot used to attach a movable rope to a fixed rope; *see also* middle-of-the-rope Prusik
ENY	enemy
EPW	enemy prisoner of war
FA	field artillery
FDC	fire direction center
FFIR	friendly force information requirements

figure 8 slip knot	knot used to form an adjustable bight in the middle of a rope
FLIR	forward looking infrared
FLOT	forward line of own troops
FO	forward observer
FPF	final protective fires
FRAGO	fragmentary order
FSC	fire support coordinator
FSO	fire support officer
GOTWA	G - Where leader is GOING O - Others he is taking with him T - Time he plans to go W - WHAT to do if the leader does not return in time A - The unit's and the leader's ACTIONS on chance contact while the leader is gone
GPS	global positioning system
GRN	grenadier
HDG	heading
HE	high explosive
H-Hour	hit hour (the time the unit plans to accomplish the mission)
HI	high temperature (weather)
HPT	high-payoff target
HQ	headquarters
IAW	in accordance with
ICM	improved conventional munitions
ID	identification
IP	initial point
IR	intelligence requirements
ERRP	en route to release point
JAAT	joint air attack team
JD	Julian date
KIA	killed in action
LAW	light antiarmor weapon
LBV	load-bearing vest
LD	line of departure
LDA	linear danger area
LO	low temperature (weather)
LOGSTAT	logistical status
LP	listening post
LZ	landing zone
material factor	the strength, hardness, and mass of the material to be demolished
mb	millibar (a metric unit used to measure air pressure)

MDI	modernized demolition initiator
ME	main effort
mechanical belay	a belay that uses mechanical devices to help the belayer control the rope, as in rappelling
MEDEVAC	medical evacuation
METL	mission-essential task list
METT-TC	mission, enemy, terrain (and weather), troops (and support) available, time available, and civil considerations
middle of the rope clove hitch	knot that secures the middle of a rope to an anchor
middle-of-the-rope Prusik	knot that attaches a movable rope to a fixed rope, anywhere along the length of the fixed rope; *see also* end-of-the-rope Prusik
MG	machine gun/ner
MOPP	mission-oriented protective posture
MR	moonrise
MRE	meal, ready to eat
MS	moonset
MSD	minimum safe distance
MSL	mean sea level
munter hitch	commonly used belay that requires little equipment
NATO	North Atlantic Treaty Organization
NAV	navigation
NFA	no-fire area
NLT	no later than
NVD	night-vision device
NVG	night vision goggles
NVS	night vision system
OAKOC	observation and fields of fire, avenues of approach, key terrain, observation, and cover and concealment
OBJ	objective
OD	olive drab
OOM	order of movement
OP	observation post
OPORD	operations order
OPSKED	operational schedule
ORP	objective rally point
OT	observer-target
PB	patrol base
PCC	precombat checks
PCI	precombat inspection
PDF	principal direction of fire

PI	probability of incapacitation
PIR	priority intelligence requirements
PL	platoon leader
PLD	probable line of deployment
PLOT-CR	purpose, location, observer, trigger, communication method, resources (a format for planning fire support)
PLT	platoon
POL	petroleum, oils, and lubricants
PRI	primary
protection	a piece of equipment, natural or artificial, that is used to construct an anchor
PSG	platoon sergeant
PSI	pounds per square inch
PZ	pickup zone
R	rifleman
R&S	reconnaissance and surveillance
RACO	rear area combat operations
RAP	rocket-assisted projectile
rappel seat	a rope harness used in rappelling and climbing
RED	risk-estimate distance
REQ	required
rerouted figure 8 knot	anchor knot that also attaches a climber to a climbing rope
RFA	restrictive fire area
RFL	restrictive fire line
RFLM	rifleman
RHB	Ranger Handbook
ROE	rules of engagement
round-turn with two half hitches	a constant tension anchor knot
RP	release point
RTO	radio operator
S-2	intelligence staff officer
S-3	operations staff (and training) officer
SALUTE	Size, Activity, Location, Unit/Uniform, Time, and Equipment
SAW	squad automatic weapon
SBF	support by fire (position)
SDT	self-development test
SE	supporting effort
SEAD	suppression of enemy air defenses
SITREP	Situation Report
SITTEMP	situational template

SL	squad leader
sling rope	short section of static rope or static cord. Also called "cordelette"
SLLS	Stop, Look, Listen, Smell
SOC	succession of command
SOI	signal operating instructions
SOP	standing operating procedures
SP	start point
square knot	knot used to join two ropes of equal diameter; used to join the ends of the detonation cord to the explosive
SR	sunrise
SS	sunset
STANO	surveillance, target acquisition, and night observation
static ropes	one of two classifications of kernmantle rope; used where rope stretch is undesired, and when the rope is subjected to heavy static weight. *See also* dynamic rope
SURVIVAL	S Size up the situation, your surroundings, your physical condition, and your equipment. U Undue haste makes waste; don't be too eager to move. Plan your moves. R Remember where you are in relation to important friendly and locations and critical resources V Vanquish fear and panic. I Improvise. You can improve your situation. Use what you have. Use your Imagination. V Value your life. Remember your goal: to get out alive. Remain stubborn. Refuse to give in to problems and obstacles that face you. This will give you the mental and physical strength to endure. A Act like the natives; watch their daily routines. When, where, and how do they get food? Where they get water? L Live by your wits. Learn basic skills.
suspension traverse	used to move personnel and equipment over rivers, ravines, chasms, and up or down a vertical obstacle
TAC	tactical air controller
tamping factor	depends on the location and tamping of the charge
technical climbing	using safe and proper equipment and techniques to climb on a rock formation in parties of two or more
tensionless anchor	used to anchor rope on high-load installations such as bridging
TL	team leader
TLP	troop-leading procedures
TL	team leader
TOC	tactical operations center
triple roll knot	knot used to join branches of detonation cord
TTP	tactics, techniques, and procedures
uli knot	knot used to securely fasten detonation cord to explosive
VIXL	video image crosslink
WARNO	warning order
WBGT	wet bulb globe temperature

WFFs	warfighting functions (fire support, movement and maneuver, protection, command and control, and sustainment)
WIA	wounded in action
XO	executive officer
WSL	weapons squad leader

NOTES

INDEX

Appendix A
RESOURCES

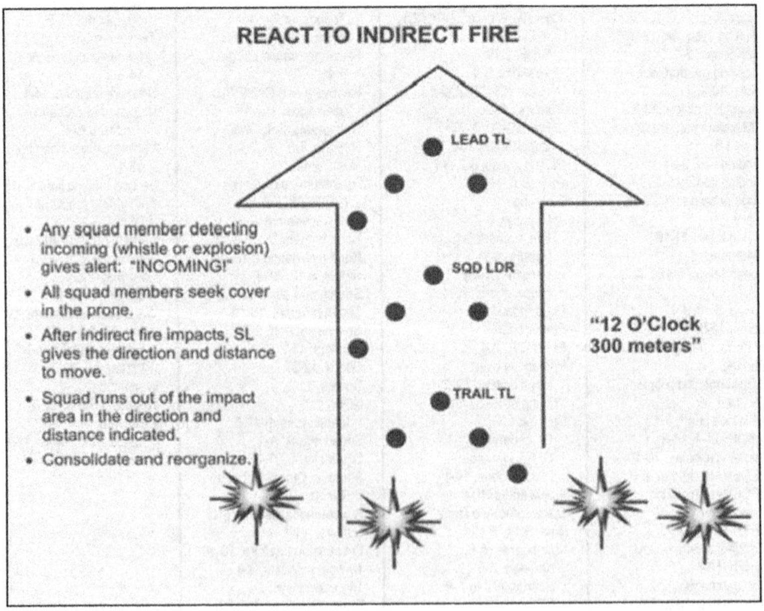

REACT TO CONTACT

- Seek nearest cover.
- Return fire (known or suspected enemy position).
- TM LDRs control fire by using fire commands.
- Report enemy situation (number/position).
- Maintain contact (visual/oral) with team members.
- SQD LDR moves to team in contact and makes an assessment of the situation.
- Factors of his assessment:
  - Can squad move out to engagement area.
  - Can squad gain and maintain suppressive fire.
  - Location of enemy.
  - Size of enemy.
  - Vulnerable flanks.
  - Covered/concealed flanking routes.
- SQD LDR determines COA (break contact, squad attack, etc.)
- Report situation to PL

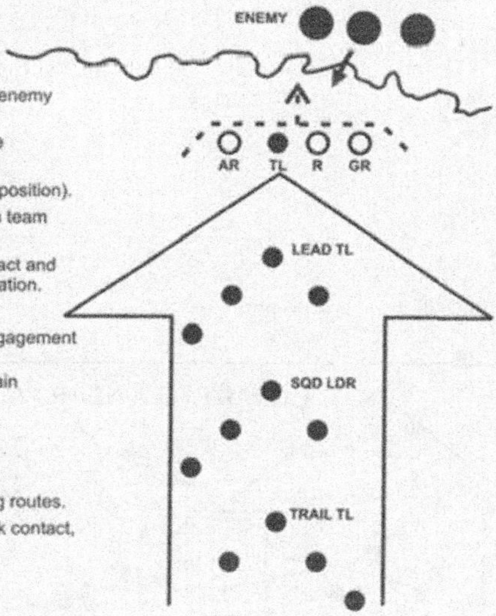

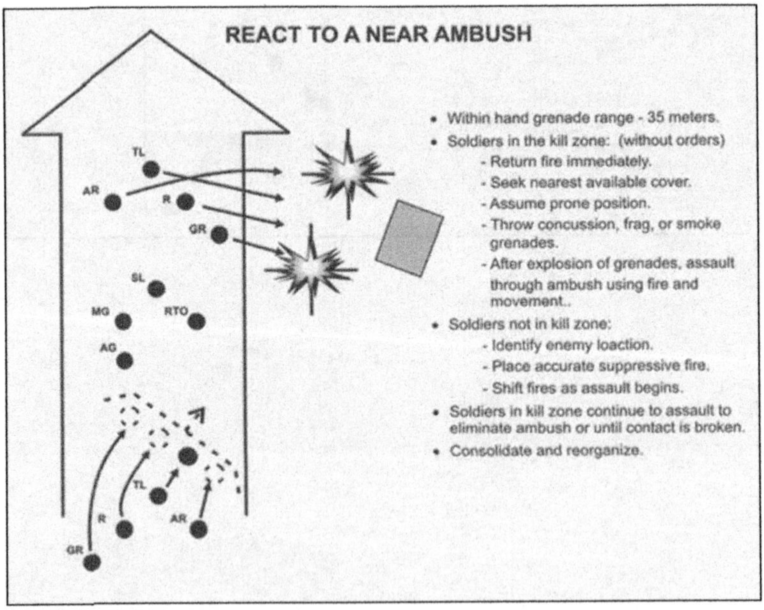
REACT TO A NEAR AMBUSH
TL
AR
R
GR
SL
MG
RTO
AG
• Within hand grenade range - 35 meters.
• Soldiers in the kill zone: (without orders)
- Return fire immediately.
- Seek nearest available cover.
- Assume prone position.
- Throw concussion, frag, or smoke grenades.
- After explosion of grenades, assault through ambush using fire and movement..
• Soldiers not in kill zone:
- Identify enemy loaction.
- Place accurate suppressive fire.
- Shift fires as assault begins.
• Soldiers in kill zone continue to assault to eliminate ambush or until contact is broken.
• Consolidate and reorganize.

REACT TO A FAR AMBUSH

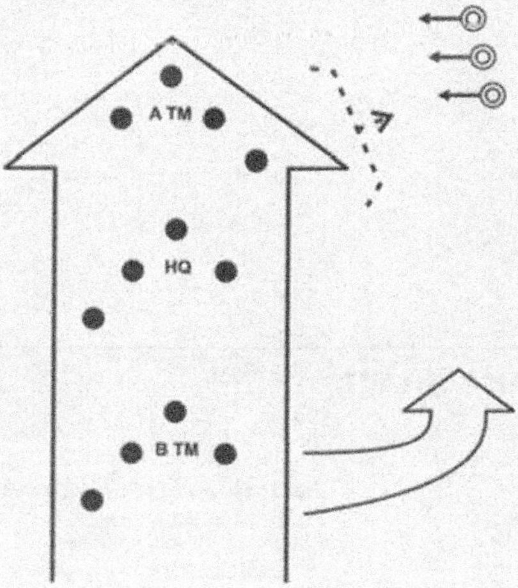

- More than 35 meters.
- TM in the kill zone: (without orders)
  - Return fire immediately.
  - Seek cover and concealment.
  - Suppress enemy (overwatch).
- SL assess situation.
  - Determines COA (flank).
- TM not in contact:
  - Moves along covered and concealed route.
  - Assaults enemy on weak flank.
  - Suppress enemy (overwatch).
- Overwatch TM continues to suppress shifts/cease fire as bounding team enters sector.
- Bounding team continues to assault through enemy.
- SL may request indirect fire.
- Consolidate and reorganize.

BREAK CONTACT

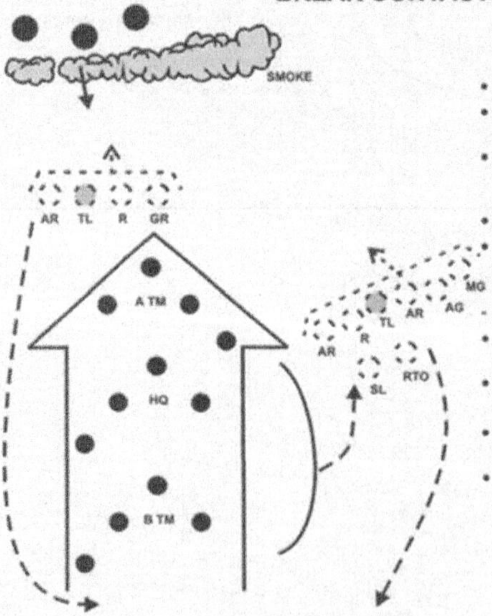

- Squad Leader orders: "Break Contact".
- Squad Leader designates SPT element and maneuver element.
- SL issues distance and direction or a terrain feature for the maneuver element.
- SBF suppresses enemy position.
- Maneuver uses smoke to mask movement.
  - Takes up overwatch position.
  - Begins to suppress enemy
- Squad Leader directs SBF to break contact.
- SBF uses smoke to screen movement.
  - Takes up overwatch position.
- Squad continues to bound away until contact is broken.
- Consolidate / reorganize.

FORMATIONS AND ORDERS OF MOVEMENT

I. Movement Formation: Fire Team Wedge; MG Team attached.

II. Three Movement Techniques utilized:

A. Traveling technique used behind FFL when contact is not likely.

B. Traveling Overwatch forward of the FFL when enemy contact is possible.

C. Bounding Overwatch used forward of the FFL when enemy contact is expected.

III. Distances are based on but not dictated by visibility, terrain, and vegetation.

IV. Actions at Night: Modified Wedge

V. Actions at the Halt: Short and Long Halt (GV/LV)

VI. Leader Location: Fixed/Unfixed

FIRE TEAM WEDGE

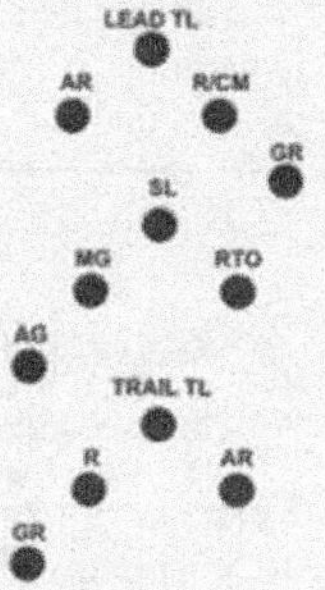

MODIFIED WEDGE

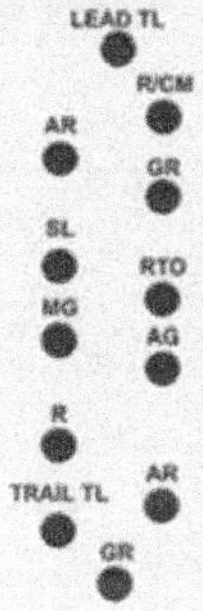

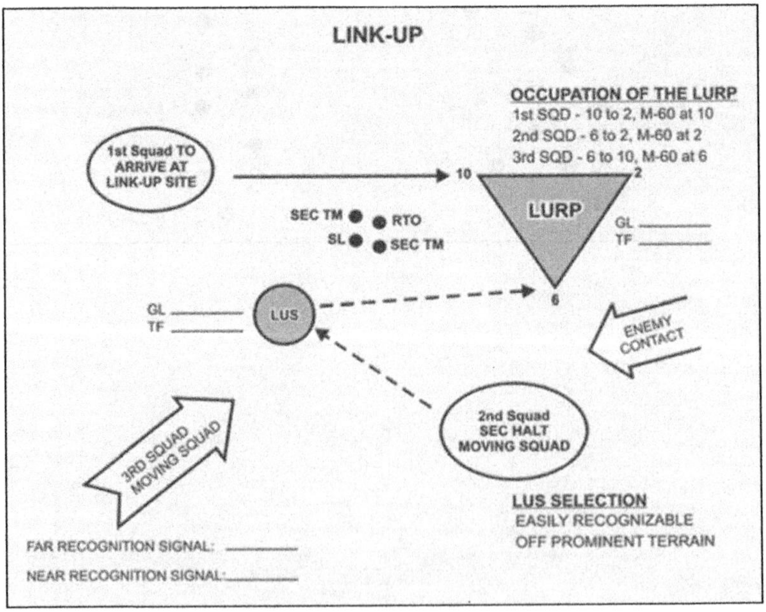

LINK-UP
OCCUPATION OF THE LURP
1st SQD - 10 to 2, M-60 at 10
2nd SQD - 6 to 2, M-60 at 2
3rd SQD - 6 to 10, M-60 at 6
1st Squad TO ARRIVE AT LINK-UP SITE
10
2
LURP
GL
TF
SEC TM
RTO
SL
SEC TM
6
GL
TF
LUS
ENEMY CONTACT
2nd Squad SEC HALT MOVING SQUAD
3RD SQUAD MOVING SQUAD
LUS SELECTION
EASILY RECOGNIZABLE
OFF PROMINENT TERRAIN
FAR RECOGNITION SIGNAL:
NEAR RECOGNITION SIGNAL:

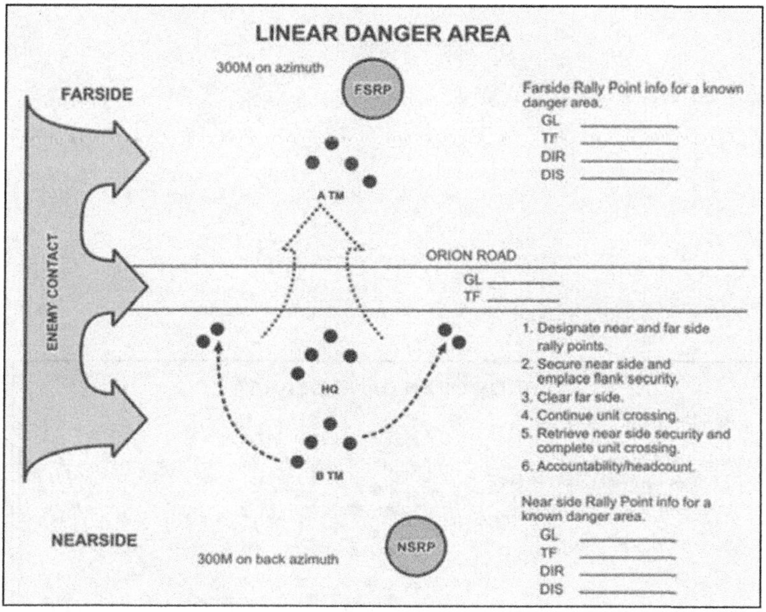

LINEAR DANGER AREA
300M on azimuth
FSRP
FARSIDE
Farside Rally Point info for a known danger area.
GL
TF
DIR
DIS
A TM
ENEMY CONTACT
ORION ROAD
GL
TF
1. Designate near and far side rally points.
2. Secure near side and emplace flank security.
3. Clear far side.
4. Continue unit crossing.
5. Retrieve near side security and complete unit crossing.
6. Accountability/headcount.
HQ
B TM
Near side Rally Point info for a known danger area.
GL
TF
DIR
DIS
NEARSIDE
300M on back azimuth
NSRP

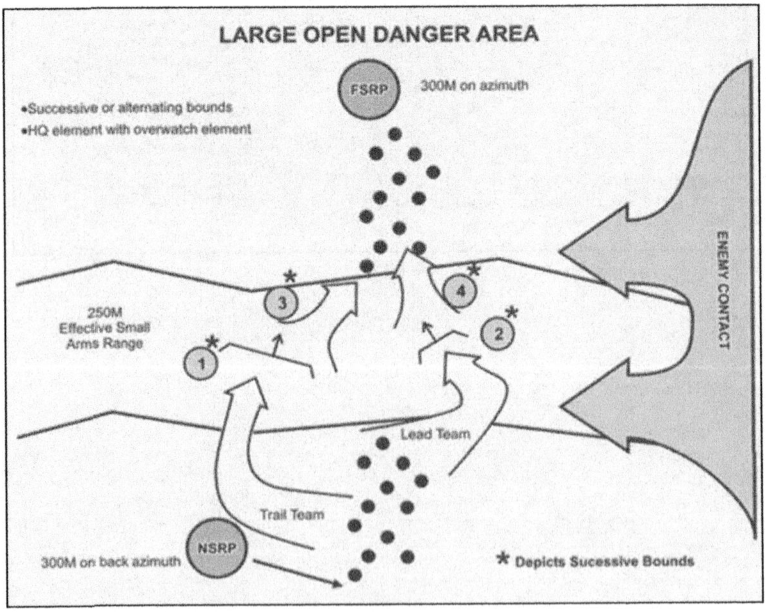

LARGE OPEN DANGER AREA
FSRP
300M on azimuth
•Successive or alternating bounds
•HQ element with overwatch element
ENEMY CONTACT
250M
Effective Small
Arms Range
1
2
3
4
Lead Team
Trail Team
NSRP
300M on back azimuth
* Depicts Sucessive Bounds

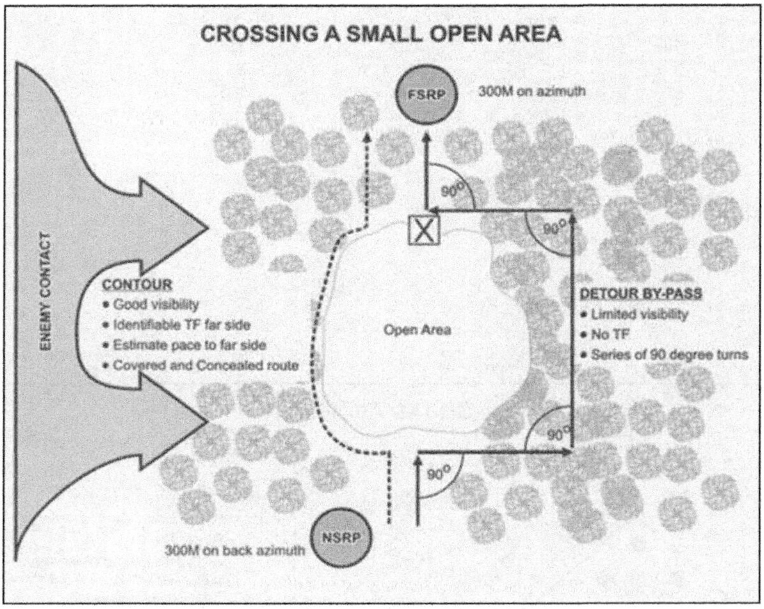
CROSSING A SMALL OPEN AREA
FSRP
300M on azimuth
ENEMY CONTACT
CONTOUR
• Good visibility
• Identifiable TF far side
• Estimate pace to far side
• Covered and Concealed route
Open Area
DETOUR BY-PASS
• Limited visibility
• No TF
• Series of 90 degree turns
90°
90°
90°
90°
NSRP
300M on back azimuth

SQUAD ATTACK

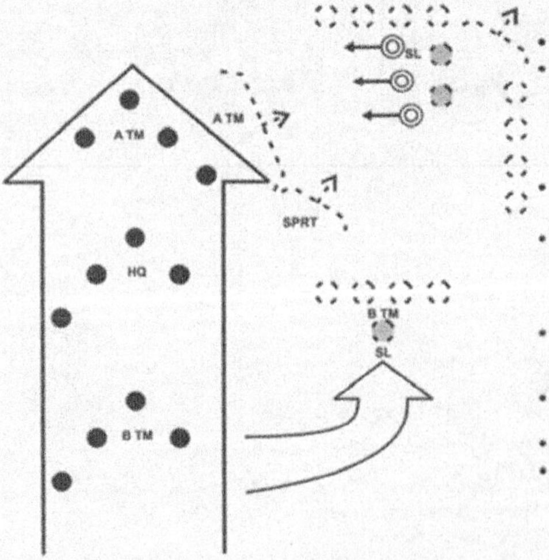

- React to contact.
- TM in the kill zone: (without orders)
  - Return fire.
  - Seek cover and concealment.
  - Suppress enemy calls out 3 D's becomes overwatch.
- SL assess situation.
  - Determines COA (flank/attack).
- TM not in contact: (with SL)
  - Moves along covered and concealed route.
  - Assaults enemy on weak flank.
- Overwatch TM continues to suppress shifts/cease fire as bounding team enters sector.
- Bounding team continues to assault through enemy.
- SL may request indirect fire.
- Consolidate and reorganize.

RAID BOARDS (LEFT)
GENERAL INFO

Raid References:
- RHB Chapter 5
- FM 3-21.8

Purposes:
1. Destroy
2. Liberate
3. Collect Intel

Planning Considerations:
- Mission
- Enemy
- Troops
- Terrain / OAKOC
- Time
- Civilians

STUDENT INPUT WRITTEN HERE

Definition:
A surprise attack on a fixed position or installation ending in a planned withdrawal.

Characteristics:
1. Surprise - Gain
2. Coordinated fires - Maintain
3. Violence of action - Retain
4. Planned withdrawal

INITIATIVE

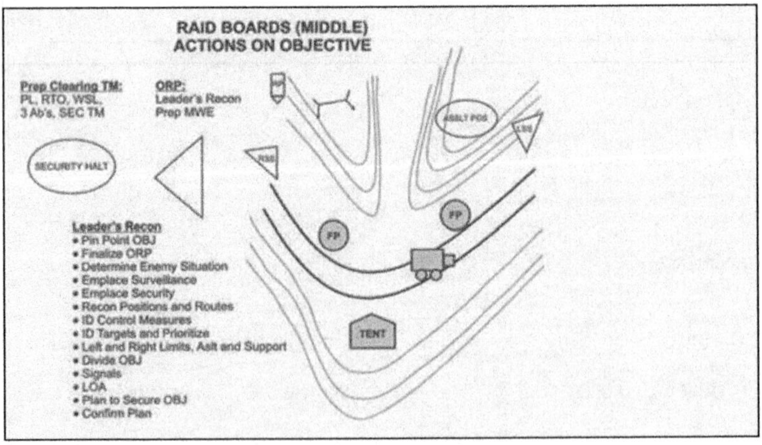
RAID BOARDS (MIDDLE)
ACTIONS ON OBJECTIVE
Prep Clearing TM:
PL, RTO, WSL,
3 Ab's, SEC TM
ORP:
Leader's Recon
Prep MWE
ASSLT POS
LSS
SECURITY HALT
RSS
FP
FP
TENT
Leader's Recon
• Pin Point OBJ
• Finalize ORP
• Determine Enemy Situation
• Emplace Surveillance
• Emplace Security
• Recon Positions and Routes
• ID Control Measures
• ID Targets and Prioritize
• Left and Right Limits, Aslt and Support
• Divide OBJ
• Signals
• LOA
• Plan to Secure OBJ
• Confirm Plan

RAID BOARDS (MIDDLE)
TASK ORGANIZATION

ELEMENT	WHO	WHY	WHAT
HQ	PL, APL RTO, FO, MEDIC	C2/FACILITATE SIGNALS, PEQ-2, STANO	1x119, ICOM, SKEDCO, PLUGGER
ASSLT 1	SQUAD (+)(-)	DESTROY	ICOM, STANO, M18A1, AT4, SPECIAL TEAM KITS
ASSLT 2	SQUAD (+)(-)	DESTROY	ICOM, STANO, M18A1, AT4, SPECIAL TEAM KITS
SUPPORT	WPN S SQD (+)(-)	SUPPRESS	3xM240B COMPLETE, ICOM, PEQ-2, PVS-4, AT, STANO, SIGNALS, MAP
SECURITY	SQUAD (+)(-)	DELAY	ICOMs, M18A1, AT, STANO, MAP SIGNALS

RAID BOARDS (RIGHT)
SOP'S

CONTINGENCIES

1. COMPROMISE
   a. ORP
   b. LEADERS RECON
   c. OCCUPATION

2. MASS CAL

3. COUNTERATTACK

SOP's

1. EPW SEARCH

2. AID LITTER

3. MED EVAC

4. CCP

5. WITHDRAW PLAN

AMBUSH SOPS (LEFT)

References:
- RHB Chapter 5
- FM 3-21.8

Ambush:
A surprise attack from a covered and concealed position on a moving or temporarily halted target.

Types:
- Point
- Area

Fundamentals:
1. Surprise - Gain
2. Fire superiority - Maintain
3. Violence of action - Retain

Purposes:
1. Disrupt/Destroy
2. Collect Intelligence
3. Block or Deny Access
4. Canalize

Categories:
Deliberate: has specific target at a predetermined time and location.

Hasty: Platoon makes visual contact with the enemy and has time to establish an ambush without being detected.

Planning Considerations:
- Mission - Task/Purpose
- Enemy - MPCOA, MDCOA
- Terrain - Map Recon/Ldrs Recon
- Time - 1/3 2/3 Rule
- Troops - Task Org
- Civilians -

AMBUSH BOARDS (MIDDLE)

HQ	PL, PL RTO, PSG, MEDIC, FO	COMMAND AND CONTROL	PRC-119s, ICOM, NVGs, BINO's, SKEDCO, PEQ-2, SIGNALS, MAP
SECURITY	MINIMUM OF 3-2 MAN TEAMS	EARLY WARNING, SEALS OFF OBJ, LFS, RFS, ORP SECURITY	COMMO, NVGs, CLAYMORES, AT WPNS, BINO's, SIGNALS, MAP
SUPPORT	WSL, 3 GUN TEAMS COMPLETE +/-	SUPPRESSIVE FIRE ON KILL ZONE, SEC ON OBJ DURING RECON/REORG	MG's, COMMO, NVGs, AT WPNS, BINO's, SKEDCO, PEQ-2, SIGNALS., MAP
ASSAULT	2 SQUADS +/-	MAIN EFFORT, BLOCK, DESTROY, CANALIZE, CAPTURE PIR	NVGs, BINO's, CLAYMORE, AT, FLEX CUFFS, POLELESS LITTER, MARKINGS, SIGNALS

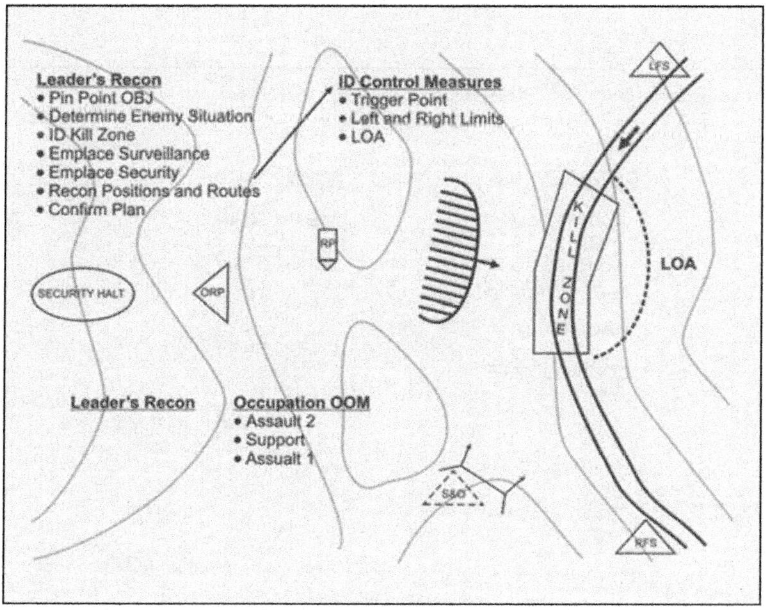

Leader's Recon
• Pin Point OBJ
• Determine Enemy Situation
• ID Kill Zone
• Emplace Surveillance
• Emplace Security
• Recon Positions and Routes
• Confirm Plan
ID Control Measures
• Trigger Point
• Left and Right Limits
• LOA
LFS
RP
KILL ZONE
LOA
SECURITY HALT
ORP
Leader's Recon
Occupation OOM
• Assault 2
• Support
• Assault 1
S&O
RFS

AMBUSH BOARDS (RIGHT)
SOP'S

CONTINGENCIES

1. COMPROMISE
   a. ORP
   b. LEADERS RECON
   c. OCCUPATION
2. MASS CAL
3. COUNTERATTACK

SOP's

1. EPW SEARCH
2. AID LITTER
3. MASS CAL
4. MED EVAC
5. CCP

NOTES

Appendix B

QUICK REFERENCE CARDS

CASUALTY FEEDER CARD

NAME ____________________ ROS # __________

COMPANY ________ ALLEGERIES____________

LOCATION/EVENT ____________________

TIME ____________ TIME ____________

TEMP ____________ TEMP ____________

PULSE ____________ PULSE ____________

RESP ____________ RESP ____________

B/P ____________ B/P ____________

MENTAL STATUS ____________________

A & O x 1 2 3

NPA OPA ET Tube

SAO^2 ____________ GLUC ____________

IV_____ AC FOREARM HAND OTHER

1000-1000-500-500-500- NS RL D_5W

NOTES: ____________________________

____

TROOP-LEADING PROCEDURES
1. Receive mission.
2. Issue warning order.
3. Make a tentative plan.
4. Start movement.
5. Reconnoiter.
6. Complete plan.
7. Issue plan.
8. Supervise.

FIRE REQUEST (FM 6-30)
1. Identification.
2. Warning order.
3. Target location.
4. Target description.
5. Method of engagement.
6. Method of control.

ESTIMATE OF SITUATION (FM 7-10)
1. Conduct a detailed mission analysis.
2. Analyze the situation and develop courses of action.
3. Analyze courses of action (wargame).
4. Compare courses of action.
5. Make a decision.

SPOT REPORT
1. Size.
2. Activity.
3. Location.
4. Unit/uniform.
5. Time.
6. Equipment.

SHELL REPORT (FM 6-121)
1. Observer identification.
2. Location (coded).
3. Azimuth to flash or sound.
4. Time (from and to).
5. Area shelled.
6. Nature of fire.
7. Type rounds received (artillery, mortars, etc.).
8. Damage (coded).

GTA 07-01-038
INFANTRY LEADERS' REFERENCE CARD
January 1995

HEADQUARTERS, DEPARTMENT OF THE ARMY
References: FM 3-3, FM 6-30, FM 6-121, FM 7-8, FM 7-10, FM 7-90
(Supersedes GTA 7-1-31)

OPERATION ORDER (FM 7-10)

Task Organization
1. Situation.
   a. Enemy.
   b. Friendly.
   c. Attachments/detachments.
2. Mission
   - Who, what, where, when, and why.
   - Task and purpose.
3. Execution.
   a. Concept of operation.
      (1) Maneuver.
      (2) Fire support.
      (3) Engineer.
   b. Tasks to maneuver units.
   c. Tasks to combat support units.
   d. Coordinating instructions.
4. Service support.
   a. General.
   b. Materiel and services.
   c. Casualty evacuation.
   d. Miscellaneous.
5. Command and signal.
   a. Command.
   b. Signal.

WEAPONS (FM 7-8)

TYPE	MAX EFF RANGE (m)
M16A2	580 (pt) 800 (area) 200 (mov)
M203	150(pt) 350 (area)
M249	600 (pt) 800 (area)
M136 (AT4)	300
M47 (Dragon)	1,000 (sta) 100 (mov)
MK19	1500 (pt) 2,212 (area)
M3 RAAWS	700 (sta) 60 (mov)
M60 MG	1,100 (600 grazing)
.50 Caliber MG	1,800 (1,000 grazing)
TOW	3,000 (plng purposes)
TOW 2	3,750
105-mm	11,500
105-mm Tank	*2 to 2.5 km
120-mm Tank	*2 to 2.5 km
25-mm BFV	2,200
155-mm M109A3	18,100
M198	24,000
8-in Howitzer	22,900

WEAPONS (MORTAR) HE ONLY (FM 7-90)

	MIN	MAX
60-mm	70m	3,500m
81-mm (M252)	80m	5,600m
81-mm (M29A1)	73m	4,790m
4.2-inch	770m	6,840m
120-mm	200m	7,200m

FPFs (FM 7-90)

GUNS	MORT	WIDTH		DEPTH
2	60-mm	60m	x	30m
4	81 (M252)	150m	x	50m
4	81 (M29A1)	140m	x	40m
6	4.2-in	240m	x	40m
6	120-mm	360m	x	60m
Plt	155-mm	200m	x	50m
Btry	155-mm	400m	x	50m

*Optimum engagement ranges

MEDEVAC REQUEST

1. Requesting unit identification.
2. Location.
3. Number of patients by type (litter or ambulatory).
4. Type of injuries.
5. Special equipment needed.
6. Tactical situation.

AIRCRAFT REQUEST

1. Identification.
2. Precedence/priority.
3. Target description.
4. Target location.
5. Target time/date.
6. Desired ordnance/results.
7. Final control.

DELIBERATE ATTACK CONSIDERATIONS

1. Reconnoiter, pinpoint objective/enemy positions/obstacles.
2. Determine weak points; designate supporting positions.
3. Assign platoon/squad objectives—identify the decisive point.
4. Determine main attack, supporting attack, reserve.
5. Assign breach, support, assault missions.
6. Designate fire control measures.
7. Coordinate indirect/direct fires and CAS to time of attack.
8. Control measures during attack.
9. Secure (ground and air).
10. Consolidate and reorganize.

NBC 1 OBSERVER'S INITIAL *OR FOLLOW-UP REPORT*

Instructions

1. Line items D and H are mandatory for NBC 1 reports.
2. Line items A, E, G, I, K, L, M, S, Y, and ZA are optional for NBC 1 reports.
3. Line items B, C, F, PAR, and PBR are reported if data is available.

Section I. Chemical or Biological Only

A. Strike serial number, if known (assigned by NBCE).
B. Position of observer.
C. Azimuth of attack from observer (state degrees or mils).
D. Date and time attack started (Zulu, local, or letter zone).
E. Time attack ended, if known.
F. Location of attack (UTM or place) (state actual or estimated).
G. Means of delivery, if known.
H. Type of agent and height of burst, if known.
I. Type and number of munitions or aircraft (state which).
K. Description of terrain (bare, scrubby vegetation, wooded, urban, or unknown).
S. Date and time contamination detected (Zulu, local, or letter zone).
Y. Representative downwind direction—4 digits (state degrees or mils), wind speed—3 digits (data kmph or knots).
ZA. Temperature (centigrade)—2 digits, cloud cover—1 digit, significant weather phenomena—1 digit, air stability—1 digit.
ZB. Remarks

Section II. Nuclear Only

A. Strike serial number, if known (assigned by NBCE).
B. Position of observer.
C. Azimuth of attack from observer (state degrees or mils and grid or magnetic).
D. Date and time attack started (Zulu, local, or letter zone).
F. Location of attack (UTM or place) (state actual or estimated).
G. Means of delivery, if known.
H. Type of burst (state air, surface, or unknown).
J. Flash-to-bang time (seconds).
K. Crater diameter (meters), if known.
L. Cloud width at H +5 minutes (degrees or mils).
M. Cloud angle(top or bottom) or cloud height(top or bottom) at H +10 minutes (state degrees, mils, meters, or feet).
PAR. Location of radioactive cloud outline (UTM).
PAB. Downwind direction of radioactive cloud (state degrees or mils).
ZB. Remarks.

DEFENSE PLANNING CONSIDERATIONS

1. Establish security (OP/patrols, PWs, M8),
2. Position key weapons:
   a Coordinate with units on left and right
   b. Establish FPF or PDF for machine gun.
   c. Ensure mutual support between machine guns.
   d. Cover armor approaches with antiarmor systems.
   e. Establish fire control measures.
3. Prepare positions:
   a. Check sectors of fire.
   b. Check overhead cover and view positions from enemy vantage.
   c. Position in depth and achieve mutual support between positions.
   d. Select/prepare alternate and supplementary positions.
4. Integrate indirect fires, CAS, and obstacles with direct and indirect fire.
5. Check communications and establish emergency signals.
6. Designate ammunition, supply, PW, and casualty points.

IED / UXO

Procedures when IEDs are found

Security - Maintain 360 degrees. Scan close in and far out, up high and down low.

Always - Scan you immediate surroundings for more IEDs.

Move - Move away. Vary safe distances, but plan for 300m minimum safe distance and adapt to your METT-TC.

Attempt - To confirm suspected IEDs using optics while staying back as far as possible.

Cordon - Off the area. Direct people out of danger area. Do not allow anyone to enter except for EOD. Question, search, and detain suspects as defined by your Existing ROE.

Report - Your situation using the 9 line IED / UXO spot report format.

This could be your hand if you try to dispose of UXOs or IEDs. The enemy has developed anti-handling to catch you when you try defusing. Leave it to experts!

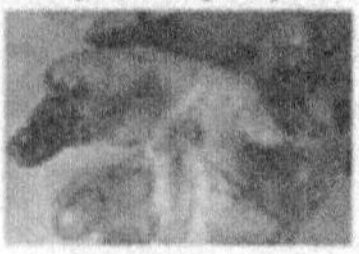

Call EOD
- Don't be a Hero!

IED / UXO Report

LINE 1. DATE-TIME-GROUP: When the item was discovered.
LINE 2. REPORT ACTIVITY AND LOCATION: Unit and grid location of the IED/UXO.
LINE 3. CONTACT METHOD: Radio frequency, call sign, POC, and telephone number.
LINE 4. TYPE OF ORDNANCE: Dropped, projected, placed, or thrown. Give the number of items, if more than one.
LINE 5. NBC CONTAMINATIONS: Be as specific as possible.
LINE 6. RESOURCES THREATENED: Equipment, facilities, or other assets that are threatened.
LINE 7. IMPACT ON MISSION: Short description of current tactical situation and how the IED/UXO affects the status of the mission.
LINE 8. PROTECTIVE MEASURES: Any measures taken to protect personnel and equipment.
LINE 9. RECOMMENDED PRIORITY: Immediate, Indirect, Minor, No Threat.

Priority

Immediate: Stops unit's maneuver and mission capability or threatens critical assets vital to the mission.
Indirect: Stops the unit's maneuver and mission capability or threatens critical assets important to the mission.
Minor: Reduces the unit's maneuver and mission capability or threatens non-critical assets of value.
No Threat: Has little or no effect on the unit's capabilities or assets.

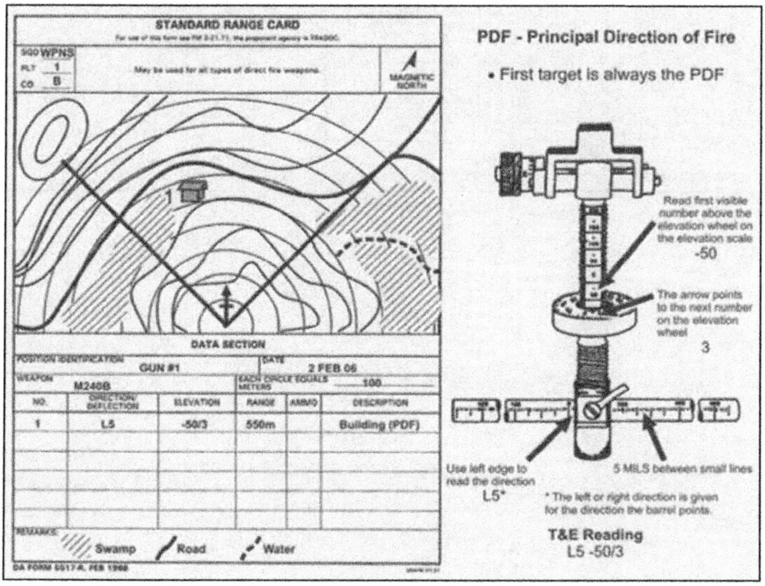
STANDARD RANGE CARD
SQD WPNS
PLT 1
CO B
May be used for all types of direct fire weapons.
MAGNETIC NORTH
1
DATA SECTION
POSITION IDENTIFICATION GUN #1
DATE 2 FEB 06
WEAPON M240B
EACH CIRCLE EQUALS 100 METERS
NO. | DIRECTION/DEFLECTION | ELEVATION | RANGE | AMMO | DESCRIPTION
1 | L5 | -50/3 | 550m | | Building (PDF)
REMARKS: Swamp Road Water
DA FORM 5517-R, FEB 1986
PDF - Principal Direction of Fire
• First target is always the PDF
Read first visible number above the elevation wheel on the elevation scale
-50
The arrow points to the next number on the elevation wheel
3
Use left edge to read the direction
L5*
5 MILS between small lines
* The left or right direction is given for the direction the barrel points.
T&E Reading
L5 -50/3

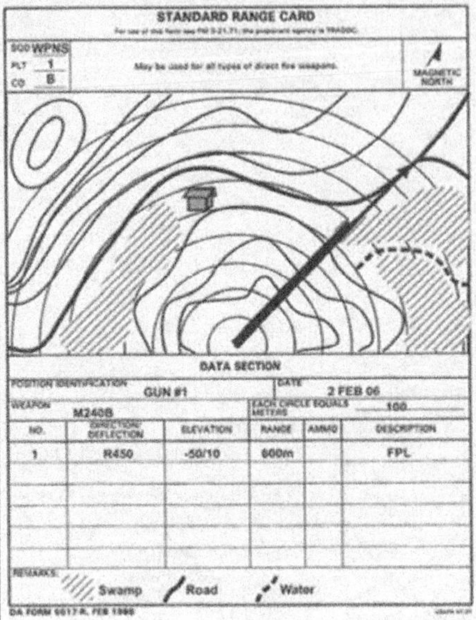

FPL - Final Protective Line

- FPL is always target #1
- FPL will always be metal-to-metal
- Represented by thick line
  - Break in thick line for dead space out to 600 meters
  - The gap is equal to the width of the dead space

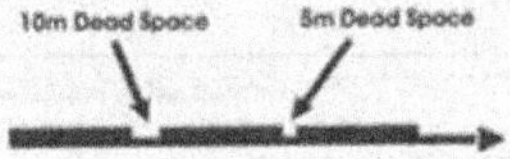

FPL - Final Protective Line

STANDARD RANGE CARD

For use of this form see FM 3-21.71; the proponent agency is TRADOC.

SQD ______
PLT ______
CO ______

May be used for all types of direct fire weapons.

MAGNETIC NORTH

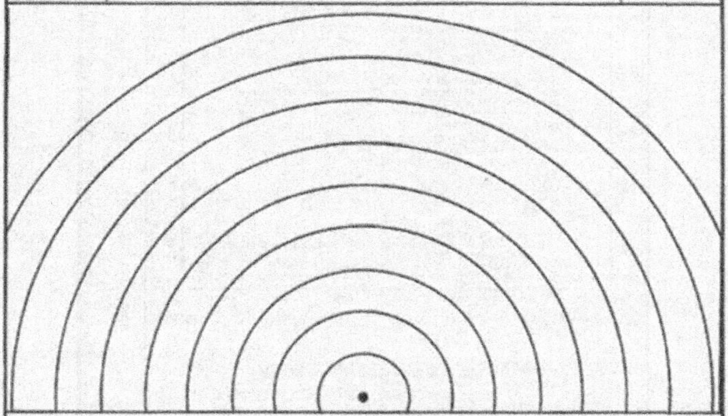

DATA SECTION

POSITION IDENTIFICATION	DATE
WEAPON	EACH CIRCLE EQUALS ______ METERS

NO.	DIRECTION/ DEFLECTION	ELEVATION	RANGE	AMMO	DESCRIPTION

REMARKS:

DA FORM 5517-R, FEB 1986 USAPA V1.01

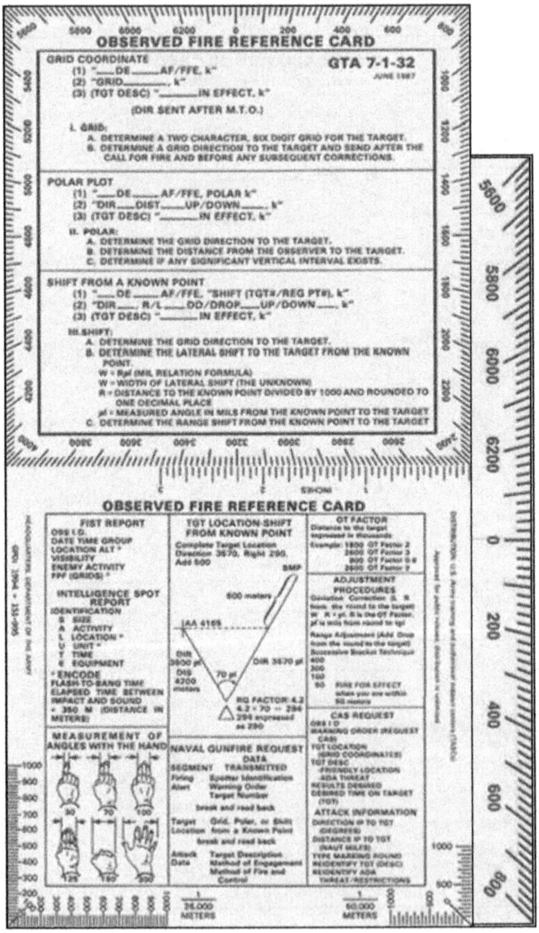

OBSERVED FIRE REFERENCE CARD

GTA 7-1-32
JUNE 1987

GRID COORDINATE
(1) "____DE____AF/FFE, k"
(2) "GRID________, k"
(3) (TGT DESC) "________IN EFFECT, k"

(DIR SENT AFTER M.T.O.)

I. GRID:
A. DETERMINE A TWO CHARACTER, SIX DIGIT GRID FOR THE TARGET.
B. DETERMINE A GRID DIRECTION TO THE TARGET AND SEND AFTER THE CALL FOR FIRE AND BEFORE ANY SUBSEQUENT CORRECTIONS.

POLAR PLOT
(1) "____DE____AF/FFE, POLAR k"
(2) "DIR____DIST____UP/DOWN____, k"
(3) (TGT DESC) "________IN EFFECT, k"

II. POLAR:
A. DETERMINE THE GRID DIRECTION TO THE TARGET.
B. DETERMINE THE DISTANCE FROM THE OBSERVER TO THE TARGET.
C. DETERMINE IF ANY SIGNIFICANT VERTICAL INTERVAL EXISTS.

SHIFT FROM A KNOWN POINT
(1) "____DE____AF/FFE, "SHIFT (TGT#/REG PT#), k"
(2) "DIR____, R/L____DD/DROP____UP/DOWN____, k"
(3) (TGT DESC) "________IN EFFECT, k"

III. SHIFT:
A. DETERMINE THE GRID DIRECTION TO THE TARGET.
B. DETERMINE THE LATERAL SHIFT TO THE TARGET FROM THE KNOWN POINT.
W = Rµl (MIL RELATION FORMULA)
W = WIDTH OF LATERAL SHIFT (THE UNKNOWN)
R = DISTANCE TO THE KNOWN POINT DIVIDED BY 1000 AND ROUNDED TO ONE DECIMAL PLACE
µl = MEASURED ANGLE IN MILS FROM THE KNOWN POINT TO THE TARGET
C. DETERMINE THE RANGE SHIFT FROM THE KNOWN POINT TO THE TARGET

OBSERVED FIRE REFERENCE CARD

FIST REPORT
OBS I.D.
DATE TIME GROUP
LOCATION ALT *
VISIBILITY
ENEMY ACTIVITY
FPF (GRIDS) *

INTELLIGENCE SPOT REPORT
IDENTIFICATION
S SIZE
A ACTIVITY
L LOCATION *
U UNIT *
T TIME
E EQUIPMENT
* ENCODE
FLASH-TO-BANG TIME
ELAPSED TIME BETWEEN IMPACT AND SOUND × 350 M (DISTANCE IN METERS)

MEASUREMENT OF ANGLES WITH THE HAND
30 70 100
125 180 300

TGT LOCATION-SHIFT FROM KNOWN POINT
Complete Target Location Direction 3570, Right 290, Add 500
BMP
500 meters
AA 4165
DIR 3800 µl
DIS 4200 meters
70 µl
DIR 3570 µl
RG FACTOR 4.2
4.2 × 70 = 294
294 expressed as 290

NAVAL GUNFIRE REQUEST
DATA

SEGMENT	TRANSMITTED
Firing Alert	Spotter Identification Warning Order Target Number
	break and read back
Target Location	Grid, Polar, or Shift from a Known Point
	break and read back
Attack Data	Target Description Method of Engagement Method of Fire and Control

OT FACTOR
Distance to the target expressed in thousands
Example: 1800 OT Factor 2
2400 OT Factor 3
900 OT Factor 0.9
2600 OT Factor 3

ADJUSTMENT PROCEDURES
Deviation Correction (L, R from the round to the target) W = R × µl. R is the OT Factor, µl is mils from round to tgt
Range Adjustment (Add, Drop from the round to the target) Successive Bracket Technique
400
200
100
50 FIRE FOR EFFECT when you are within 50 meters

CAS REQUEST
OBS I D
WARNING ORDER (REQUEST CAS)
TGT LOCATION (GRID COORDINATES)
TGT DESC
-FRIENDLY LOCATION
-ADA THREAT
RESULTS DESIRED
DESIRED TIME ON TARGET (TOT)

ATTACK INFORMATION
DIRECTION IP TO TGT (DEGREES)
DISTANCE IP TO TGT (NAUT MILES)
TYPE MARKING ROUND
REIDENTIFY TGT (DESC)
REIDENTIFY ADA THREAT/RESTRICTIONS

1/25,000 METERS
1/50,000 METERS

HEADQUARTERS, DEPARTMENT OF THE ARMY
GPO: 1994 - 151-995

9 Line MEDEVAC

LINE ITEM	EXPLANATION
1. Location of Pickup Site.	Encrypt grid coordinates. When using *DRYAD Numeral Cipher, the same SET line* will be used to encrypt grid zone letters and coordinates. To preclude misunderstanding, a statement is made that grid zone letters are included in the message (unless unit SOP specifies its use at all times).
2. Radio Frequency, Call Sign, Suffix.	Encrypt the frequency of the radio at the pickup site, *not* a relay frequency. The call sign (and suffix if used) of person to be contacted at the pickup site may be transmitted in the clear.
3. No. of Patients by Precedence.	Report only applicable info & encrypt brevity codes. A = Urgent, B = Urgent-Surg, C = Priority, D = Routine, E = Convenience. (If 2 or more categories reported in same request, insert the word "*break*" between each category).
4. Spec Equipment.	Encrypt applicable brevity codes. A = None, B = Hoist, C = Extraction equipment, D = Ventilator.
5. No. of Patients by Type.	Report only applicable information and encrypt brevity code. If requesting MEDEVAC for both types, insert the word "*break*" between the litter entry and ambulatory entry: L + # of Pnt - Litter; A + # of Pnt - Ambul (sitting).
6. Security Pickup Site (Wartime).	N = No enemy troops in area, P = Possibly enemy troops in area (approach with caution), E = Enemy troops in area (approach with caution), X = Enemy troops in area (armed escort required).
6. Number and type of Wound, Injury, Illness (Peacetime).	Specific information regarding patient wounds by type (gunshot or shrapnel). Report serious bleeding, along with patient blood type, if known.
7. Method of Marking Pickup Site.	Encrypt the brevity codes. *A* = Panels, *B* = Pyrotechnic signal, *C* = Smoke Signal, *D* = None, *E* = Other.
8. Patient Nationality and Status.	Number of patients in each category need not be transmitted. Encrypt only applicable brevity codes. A = US military, B = US civilian, C = Non-US mil, D = Non-US civilian, E = EPW.
9. NBC Contamination (Wartime).	Include this line only when applicable. Encrypt the applicable brevity codes. N = nuclear, B = biological, C = chemical.
9. Terrain Description (Peacetime).	Include details of terrain features in and around proposed landing site. If possible, describe the relationship of site to a prominent terrain feature (lake, mountain, tower).

Reference: FM 8-10-6, *Medical Evacuation in a Theater of Operations, pages 7-7 through 7-9.*

Made in United States
Orlando, FL
04 November 2026

90201503R00361